Rice
Improvement in the Genomics Era

Rice
Improvement in the Genomics Era

Edited by
Swapan K. Datta

CRC Press
Taylor & Francis Group
Boca Raton London New York

CRC Press is an imprint of the
Taylor & Francis Group, an **informa** business

CRC Press
Taylor & Francis Group
6000 Broken Sound Parkway NW, Suite 300
Boca Raton, FL 33487-2742

First issued in paperback 2019

ISBN-13: 978-1-56022-952-0 (hbk)
ISBN-13: 978-0-367-38713-6 (pbk)

Library of Congress Cataloging-in-Publication Data

Rice improvement in the genomics era / [edited by] Swapan K. Datta.
    p. cm.
    ISBN: 978-1-56022-952-0 (hard : alk. paper)
    1. Rice—Genetics. 2. Rice—Genetic engineering. I. Datta, S. K.

SB191.R5R459 2007
633.1'8233—dc22   2007047013

Visit the Taylor & Francis Web site at
http://www.taylorandfrancis.com

and the CRC Press Web site at
http://www.crcpress.com

CONTENTS

ABOUT THE EDITOR

Swapan K. Datta, PhD, is Rash Behari Ghosh Professor at the University of Calcutta, Kolkata, India. He was former Senior Scientist and CGIAR HarvestPlus Rice crop leader at the International Rice Research Institute, Manila, Philippines from 1993-2005. He was awarded prestigious Fellowships/assignments such as DAAD Fellow (Germany) FMI, Fellow (Switzerland), Senior Scientist and Group leader at ETH-Zurich (Switzerland), and Visiting Associate Professor at the University of California–Davis, USA. He has mentored 25 PhD students and published over 115 research papers in journals such as *Nature, Nature Biotechnology, Plant Journal, Plant Biotechnology Journal, Theoretical Applied Genetics, Molecular Genetics and Genomics, Crop Science,* and *Plant Molecular Biology.* He is editor/associate editor of two international journals and edited a book, *Pathogenesis Related Proteins in Plants,* published by the CRC Press. He pioneered the genetic engineering approach in *indica* rice in 1990 and reported homozygous fertile transgenic *indica* rice with genes of agronomic importance such as *Bt,* Pr-Protein, *Xa21,* ferritin, genes for beta-carotene biosynthesis, etc. He has extended this work for both biotic and abiotic stress tolerance in rice in collaboration with JIRCAS, Japan, KSU, USA, Syngenta, University of California–Davis, etc. Dr. Datta is a recipient of Fellowships from the National Academy of Sciences, National Academy of Agricultural Sciences and awarded TATA Innovation Fellow as recognition for his work on translational research in transgenic rice from the Department of Biotechnology, Government of India. He has also taken numerous public policy and scholarly activities that address the topics such as need of agriculture-biotechnology, bio- and food-safety of genetically modified crops, and intellectual property rights and crop management in developing countries.

CONTRIBUTORS

M. Agarwal is affiliated with the Department of Plant Molecular Biology, University of Delhi South Campus, Benito Juarez Road, Dhaula Kuan, New Delhi-110021, India.

S. Agarwal is affiliated with the Department of Plant Molecular Biology, University of Delhi South Campus, Benito Juarez Road, Dhaula Kuan, New Delhi-110021, India.

Baltazar A. Antonio is Senior Researcher, Genome Research Center, National Institute of Agrobiological Sciences, 2-1-2, Kannondai, Tsukuba, Ibaraki 305-8602, Japan.

N. Baisakh is Project Scientist, Plant Breeding, Genetics, and Biotechnology Division, International Rice Research Institute, DAPO 7777, Metro Manila, Philippines.

J. S. Bao is Associate Professor, Institute of Nuclear Agricultural Sciences, College of Agriculture and Biotechnology, Zhejiang University, Hua Jiachi Campus, Hangzhou, 310029, China. (e-mail: jsbao@zju.edu.cn).

C. Bergman is Research Chemist, United States Department of Agriculture, Agriculture Research Service, Rice Research Unit, Beaumont, TX 77707 (e-mail: cbergman@ag.tamu.edu).

A. Chandramouli is affiliated with the Department of Plant Molecular Biology, University of Delhi South Campus, Benito Juarez Road, Dhaula Kuan, New Delhi-110021, India.

B. M. Chassy is Assistant Dean for Biotechnology Outreach, Office of Research and also Executive Associate Director of the Biotechnology Center, College of Agricultural, Consumer and Environmental Sciences, University of Illinois at Urbana-Champaign, IL 61801 (e-mail: bchassy@uiuc.edu).

K. Datta was Senior Scientist and Plant Biotechnologist, Plant Breeding, Genetics, and Biotechnology Division, International Rice Research Institute, DAPO Box-7777, Metro Manila, Philippines. (e-mail: krbdatta@ yahoo.com).

M. Dey is Postdoctoral Fellow, Department of Molecular Biology and Genetics, Cornell University, Ithaca, NY 14853 (e-mail: deymoul@ hotmail.com).

D. Endo is affiliated with the National Institute of Agrobiological Sciences, 2-1-2, Kannondai, Tsukuba, Ibaraki 305-8602, Japan (e-mail: skikuchi@ nias.affrc.go.jp).

R. Fjellstrom is Research Geneticist, United States Department of Agriculture, Agriculture Research Service, Rice Research Unit, Beaumont, TX 77707 (e-mail: rfjellstrom@tamu.edu).

G. B. Gregorio is Rice Breeder IRS for Africa in Plant Breeding, Genetics, and Biotechnology Division, International Rice Research Institute (IRRI), DAPO Box 777, Metro Manila, Philippines (e-mail: g.gregorio@cgiar.org).

A. Grover is affiliated with the Department of Plant Molecular Biology, University of Delhi South Campus, Benito Juarez Road, Dhaula Kuan, New Delhi-110021, India (e-mail: grover_anil@hotmail.com).

N. Huang is affiliated with Ventria Bioscience, 4110 North Freeway Boulevard, Sacramento, CA 95834.

M. Ishikawa is affiliated with the National Institute of Agrobiological Sciences, 2-1-2, Kannondai, Tsukuba, Ibaraki 305-8602, Japan (e-mail: skikuchi@nias.affrc.go.jp).

Y. Jia is affiliated with Dale Bumpers National Rice Research Center, P. O. Box 1090, Stuttgart, AR 72160, USA (e-mail: yjia@spa.ars.usda.gov).

S. Katiyar-Agarwal is affiliated with the Department of Plant Molecular Biology, University of Delhi South Campus, Benito Juarez Road, Dhaula Kuan, New Delhi-110021, India.

S. Kikuchi is affiliated with the National Institute of Agrobiological Sciences, 2-1-2, Kannondai, Tsukuba, Ibaraki 305-8602, Japan (e-mail: skikuchi@nias.affrc.go.jp).

N. Kishimoto is affiliated with the National Institute of Agrobiological Sciences, 2-1-2, Kannondai, Tsukuba, Ibaraki 305-8602, Japan.

I. Kumar is International Research Fellow, Plant Breeding, Genetics, and Biotechnology Division, International Rice Research Institute, DAPO Box 7777, Metro Manila, Philippines (e-mail: i.kumar@cgiar.org).

D. J. Mackill is Head, Plant Breeding, Genetics & Biotechnology Division, International Rice Research Institute (IRRI), DAPO Box 7777, Metro Manila, Philippines (e-mail: d.mackill@cgiar.org).

H. P. Moon is Deputy Administrator, Rural Development Administration, 250 Seodundong, Suwon, Kyounggido 441-707, Republic of Korea (e-mail: moonhp@rda.go.kr).

S. Nandi is affiliated with Ventria Bioscience, 4110 North Freeway Boulevard, Sacramento, CA 95834, USA (e-mail: snandi@ventriabio.com).

S. Appa Rao is Associate Coordinator, Nutritious Millets Project, M. S. Swaminathan Research Foundation, Chennai, India (e-mail: s.apparao@cgiar.org).

C. Sahi is affiliated with the Department of Plant Molecular Biology, University of Delhi South Campus, Benito Juarez Road, Dhaula Kuan, New Delhi-110021, India

Takuji Sasaki is Head, Japan's Rice Genome Research Program and director of the National Institute of Agrobiological Sciences, 2-1-2, Kannondai, Tsukuba, Ibaraki 305-8602, Japan (e-mail: tsasaki@nias.affrc.go.jp).

M. S. Swaminathan is UNESCO Chair in Ecotechnology and Chairman, M. S. Swaminathan Research Foundation, Chennai, India (e-mail: msswami@ mssrf.res.in).

L. B. Torrizo is Associate Scientist, Plant Breeding, Genetics, and Biotechnology Division, International Rice Research Institute (IRRI), DAPO Box 777, Metro Manila, Phillippines (e-mail: 1.torrizo@cgiar.org).

B. Valent is affiliated with the Department of Plant Pathology, 4024 Throckmorton Plant Sciences Center, Kansas State University, Manhattan, KS 66506 (e-mail: BVALENT@plantpath.ksu.edu).

S. S. Virmani is Principal Plant Breeder, Plant Breeding, Genetics, and Biotechnology Division, International Rice Research Institute, DAPO Box 7777, Metro Manila, Philippines (e-mail: s.virmani@cgiar.org).

J. Yazaki is affiliated with the National Institute of Agrobiological Sciences, 2-1-2, Kannondai, Tsukuba, Ibaraki 305-8602, Japan.

Foreword

Rice is the most important food crop; it feeds half the world's population. Almost 25% of the calories consumed by the world's population come from this crop. Productivity of the crop has been improved continuously since its domestication about 10,000 years ago. Rice farmers themselves initiated rice improvement when they started selecting for improved traits from variable populations. This conscious and unconscious selection resulted in about 150,000 land races by the time Mendel's laws of heredity were rediscovered at the beginning of this century. Several rice breeding stations were established after the rediscovery of Mendel's laws. The main emphasis in rice improvement was pure line selection and intervarietal hybridization. However, there was little improvement in the yield potential of rice, and per hectare yields remained stagnant.

After the establishment of the International Rice Research Institute (IRRI) in the Philippines in 1960, a major breakthrough in the yield potential of rice occurred. Incorporation of a dwarfing gene led to improvements in harvest index and nitrogen responsiveness, doubling the yield potential. This innovation was widely accepted, and rice breeders at IRRI and national rice improvement programs developed numerous high-yielding varieties of rice. These varieties are now planted on more than 80% of the world's rice land, and rice production has kept ahead of population growth. However, the rate of growth of rice production has now slowed down, but the population of rice consumers continues to grow at the rate of more than 2% a year.

For continued food security, rice varieties with higher yield potential and greater yield stability are needed. Recent breakthroughs in cellular and molecular biology and advances in genomics are poised to help meet this challenge. This book documents some of the applications of biotechnology and genomics to rice improvement. The authors of various chapters are authorities in their respective fields. T. Sasaki, who led the International Rice Genome Sequencing Project, discusses the implications of advances in genomics for rice improvement. Microarrays are the latest tools for investigating the functions of rice genes. Kikuchi and colleagues have reviewed the application of this technology in Chapter 2. Molecular markers have emerged as useful tools for genetic and breeding research. David Mackill discusses this subject in Chapter 3. Regeneration of doubled haploids from anther culture of F_1 plants has been employed to shorten the time for varietal development. Swapan Datta and colleagues have reviewed the applications of this technology in Chapter 4. Hybrid rice technology, another tool to speed up varietal development, is covered by Virmani and Kumar in Chapter 5.

Transgenic approaches to introduce novel genes for disease and insect resistance and abiotic stress tolerance are discussed in Chapters 6, 7, 8, and 9. Four hundred million people in the world suffer from vitamin A deficiency. Introduction of two genes—one from daffodil and another from *Erivinia uredovora,* through genetic engineering, has led to the establishment of a biosynthetic pathway for production of β-carotene in rice endosperm. β-carotene is the precursor of vitamin A. Swapan Datta has discussed the present status of development of this so-called golden *indica* rice in Chapter 11. Pharmaceuticals can be produced more efficiently and economically in plants. Production of human milk proteins in rice grains is an excellent example, reviewed by Nandi and Huang in Chapter 12. Inorganic nitrogen is a major input in rice production. It has been a dream of some to introduce the capability to fix nitrogen from the atmosphere in rice. However, progress in achieving this goal has been slow. The student-teacher team of Dey and Datta has discussed the present status of this dream in Chapter 13. All the technologies discussed in this volume must be applied while keeping in view the sustainability of production systems. The sagacious advice of Swaminathan and Appa Rao in Chapter 14 should be kept in mind. Finally, B. M. Chassy discusses the impact of the perceived hazards of transgenic plant crops on the public acceptance of these new technologies in Chapter 15.

Overall, this book provides an excellent and balanced coverage of the subject. I feel confident that this book will prove useful for students as well as for researchers. Swapan Datta should be congratulated for preparing this volume.

Gurdev S. Khush
Former Principal Plant Breeder at IRRI, Philippines
Adjunct Professor, University of California
at Davis

Preface

The journey of rice started long ago, when a cereal grain was found in Asia. A new Civilization began about ten thousand years ago with the adoption of cereals as foods. In 1876, Mendel introduced the concept of genetics, which eventually led to the development of modern plant breeding and biotechnology. In recent history, Norman Borlaug's exciting "Green Revolution" from 1960 onward led to greatly improved crop yields based on natural resources, and the first such improved rice product was developed at the International Rice Research Institute (IRRI) in 1964. Further improved rice varieties were developed and grown across Asia and later cultivated worldwide. One such rice variety alone, IR36, has been grown extensively in Asia. The more common the use of a single variety, the greater the chance of a breakdown of the plant's resistance against diseases and pests. Therefore, the search for new genes, varieties, and technology led to the development of high-yielding varieties with plant protection. The past few decades have brought several pioneering milestones in crop improvement, including haploidy, chromosome mapping, marker-assisted breeding, and the combination of these different tools in plant breeding. The most striking breakthrough came in 1990, when fertile homozygous transgenic rice was created using protoplasts with foreign genes. The further improvement of rice using different agronomically important genes continues worldwide, and genetically improved rice can now be found in the field in China, India, the Philippines, Spain, the United States, and elsewhere. Genome sequencing in rice is the latest modern tool of science, which will enable the discovery and cloning of many useful genes and will eventually be used in genetic engineering to make rice more competitive and functional for human welfare as the most important staple food. This rice improvement is the subject of this book, written by some of the most well-known scientists in this area.

Initially, this book was intended to be completed in 2004, during the "International Year of Rice" declared by the UNO. However, it took more time than I expected to complete and refine with the help of some of the world's most highly reputed scientists, including Drs. T. Sasaki, S. Kikuchi,

D. Mackill, K. Datta, B. Valent, S. Virmani, H. P. Moon, A. Grover, C. Bergman, N. Huang, M. S. Swaminathan, and B. M. Chassy. I am grateful to all these people, who have made enormous contributions to rice research, for making this book possible.

Dr. Gurdev S. Khush, one of the most qualified and renowned scientists in contemporary rice research, has generously assented to write the Foreword for this book. I am sincerely grateful to him not only for this kind gesture, but also for having the privilege of working with him for more than ten years at IRRI.

Personally, I thank my colleagues at IRRI and elsewhere, including Drs. Ingo Potrykus, Gerhard Wenzel, Gary Toenniessen, Qifa Zhang, Henry Daniell, S. Muthukrishnan, T. Murashige, H. Uchimiya, Mahabub Hossain, Darshan S. Brar, D. Senadhira, David Dawe, Glenn Gregorio, Seiji Yanagihara, Yoshimichi Fukuta, Surapong Sarkarung, Tim Setter, Osamu Ito, Tom Mew, Mike Cohen, Paul Teng, John Bennett, M. Reddy, Hei Leung, and Bill Hardy. I am also grateful to Dr. Mangla Rai, Director General of the Indian Council of Agricultural Research (ICAR), for his contributions in leadership and constructive criticism regarding rice and agricultural research.

I thank all of my research co-workers in the laboratory, including Drs. Nouchine Soltanifer, Mohammad Firoz Alam, Jumin Tu, Niranjan Baisakh, Moul Dey, Narayanan Narayanan, Marta Vasconcelos, Mayank Rai, Vilas Parkhi, M. Khalekuzzaman, Sayeda Rehana, R. Thet, S. Krishnan, S. Balachandran, Aindya Bandyapadhyay and particularly Michelle Viray, for technical assistance. I would also like to thank a few other researchers, including Lina Torrizo, Editha Abrigo, Normal Oliva, M. Alamgir, and many more visiting scientists who contributed their thoughts, time, and dedication to rice research. I also would like to thank my teacher Professor R. P. Purakayastha with whom I started rice work in 1976 in Kolkata. I am grateful to Dr. A. K. Banerjee, Vice Chancellor of Calcutta University, for his personal support in the completion this book. I also take the opportunity to thank the Director General of IRRI, Dr. Robert Zeigler, with whom I had the privilege to work and play tennis at IRRI. Special thanks are also due to Lina Torrizo, who assisted me in editing the book.

My grateful thanks go to the Publisher, The Haworth Press, Taylor & Francis Group, and particularly a few people like Rebecca Browne, Donna Barnes, Amy Rentner, Vidya Jayaprakash, Jessica Smith, and Angela Ralano for their cooperation and help during various stages of development and completion of the book. A special thanks is also due to the Department of Biotechnology, Government of India for providing program support on "Translational Research on Transgenic Rice" at CU, Kolkata.

Last but not least, I thank Dr. Karabi Datta for her tremendous contribution as a scientist and her untiring support as wife with two wonderful children, Mome and Baiku, who are always curious to know more about rice, science and its beneficial aspects, and concerns about the food safety of transgenic rice. They constantly remind me of the importance of this crop and the value of the public perception of any food crop and its safe use.

Chapter 1

The High-Quality Rice Genome Sequence and Its Impact on Cereal Genomics

Takuji Sasaki
Baltazar A. Antonio

In the past decade, a new wave of scientific innovations has emerged that has changed our thinking and strategies in tackling rice research. This wave was primarily based on the sequencing of the blueprint of life also known as DNA. Although the main flow of genetic information from DNA to protein via RNA, which defines the central dogma of molecular biology, has been elucidated using many types of living systems, a large discrepancy still exists between theory and applications, particularly in plants. Two of the most important discoveries in genetics, namely, the fundamental law of inheritance elucidated by Gregor Mendel and the existence of transposable elements promulgated by Barbara McClintock, were established by studying plants as experimental systems. Thus the next challenge in genetics should focus on making new scientific discoveries by taking advantage of the DNA or genetic information to further advance research in plant science.

ROAD TO RICE GENOME SEQUENCING

The shift from traditional genetics to molecular genetics was brought about by the realization that nucleotide variations in DNA were inheritable and could be effectively used for genetic analysis (Botstein et al., 1986). Medical researchers succeeded in identifying the location of a serious, genetically dominant disease known as Huntington's disease based on the principle of restriction fragment length polymorphism (RFLP) in the DNA

(Allitto et al., 1991). This approach was soon introduced in rice genetics and led to the development of a linkage map with molecular markers instead of phenotypic traits (McCouch et al., 1988). Soon enough, other types of inheritable polymorphisms in DNA such as simple sequence repeat (SSR) and randomly amplified polymorphic DNA (RAPD) were subsequently elucidated and used as molecular markers. A genetically reliable map is indispensable for further analysis of the genome structure of the rice plant. In particular, the map is very useful in assigning any DNA fragment unambiguously to the correct original position in the genome. Several rice molecular genetic maps with sequence information of mapped DNA markers are now available. The most saturated genetic map of rice was developed by the Rice Genome Research Program (RGP) in Japan from a single F2 population composed of 186 individuals obtained by crossing a *japonica* variety, *Nipponbare* and an *indica* variety, *Kasalath.* Construction of this map has been greatly facilitated by a chemoluminescence detection technique that allowed repeated Southern hybridization, thus resulting in a high-density linkage map with nearly 3,300 DNA markers (Harushima et al., 1998; Rice Genome Research Program, 2000). In addition, polymerase chain reaction (PCR) based markers such as cleaved amplified polymorphic sequence (CAPS) markers (Konieczny and Ausubel, 1993) have also been developed using nucleotide sequence information of RFLP markers, and they have been useful in saturation mapping. The progress in genome sequence determination greatly helps to generate many new markers based on this information.

In addition to molecular genetic maps, other tools for analyzing the rice genome include a wide collection of expressed sequence tags (EST), several types of genomic libraries, and physical maps using clones of these libraries. Currently, the number of rice ESTs registered with the National Center for Biotechnology Information (NCBI) exceeds 800,000 entries. Among them, more than 60,000 ESTs were generated by RGP from approximately 15 types of cDNA libraries derived from various tissues and organs such as leaves, roots, panicles, and calli, using *Nipponbare* as resource (Yamamoto and Sasaki, 1997). ESTs from *indica* rice varieties have also been generated by various groups that, together with the ESTs from *japonica* rice, provide important resources for mapping and genome analysis of many rice varieties. These EST sequences from the major subspecies of rice will be useful for detecting single nucleotide polymorphisms (SNPs) by comparing corresponding sequences.

Genomic libraries for large DNA fragments ranging from 100 to 300 kb are also indispensable for genome analysis. The most commonly used vectors for library construction are bacterial artificial chromosomes (BAC) (Shizuya et al., 1992), P1-derived artificial chromosome (PAC) (Ioannou et al., 1994) with *Escherichia coli* as host, and yeast artificial chromosome (YAC) (Burke et al., 1987) with *Saccharomyces cerevisiae* as host. Historically, YAC was the first vector to be developed and eventually used for physical mapping of the human genome. However, this vector system has a tendency to generate chimeras of alien DNA fragments and the rate of amplification depends on the growth rate of *S. cerevisiae*. BAC and PAC vectors were also developed for the cloning of human genome fragments, mainly as resources for genome sequencing, because they can stably harbor alien DNAs with much more rapid amplification rates than yeast. In case of rice, RGP constructed a YAC library (Umehara et al., 1995) which has been effectively used as the framework for a saturated physical map of the rice genome (Saji et al., 2001). As major templates for genome sequencing, PAC and BAC libraries of the *Nipponbare* genome were also constructed (Baba et al., 2000).

In addition to these resources, high throughput instruments to systematically sequence DNA sequences as well as computers for storage and high speed processing of huge amounts of data have become indispensable in genome analysis. Fortunately, the capillary type of automated DNA sequencers were introduced at the same time as the start of the extensive rice genome sequencing project in Japan. This type of sequencer greatly facilitated the production of a large amount of data in a short period of time with less labor. The most advanced automated sequencer can generate about 500 kb of reliable sequence data a day. At RGP, a robotic system for sample preparation using a 96-titer plate was also constructed to complement this high-efficiency sequencer. With these combined resources, the sequencing of the rice genome has been accelerated at a much lower cost than initially anticipated.

HISTORY OF IRGSP

The 1990s paved the way for the sequencing of many prokaryotic and eukaryotic organisms, including the model plant *Arabidopsis*. On September 23, 1997, a Rockefeller-sponsored workshop on the feasibility of an international format to sequence the rice genome was held at the 5th International Congress of Plant Molecular Biology in Singapore. More than 100 researchers and scientists participated in a general discussion, which focused on how to organize an international collaboration to sequence the rice genome. Given a relatively small size of the rice genome as compared with other cereals, it was concluded that sequencing the rice genome was a feasible undertaking in view of available technology. However, with a genome size of approximately 400 Mb, everyone agreed that the task was so great that it was unlikely that any one country could devote the resources to sequence the rice genome in the next ten years. Participants therefore agreed to take part in an international collaboration that includes the sharing of clones and the timely release of sequence and mapping information. The *Nipponbare* variety used to produce STS maps, cDNA sequences, and YAC clones was selected as a single DNA source. In addition, PAC and BAC libraries for sequencing would be simultaneously constructed in Japan and the United States.

On February 5, 1998, the rice genome working group with representatives from Japan, United States, United Kingdom, China, and Korea met in Tsukuba, Japan, to finalize specific details on the sequencing collaboration including the methodologies, data release policies, scientific standards, time frames, and chromosomal sharing of the 12 chromosomes. At that time, only five chromosomes could be assigned to the member countries because of uncertainty about securing enough funding to sequence all the chromosomes. In particular, only Japan, China, and Korea had definite public funding for sequencing at that time. Soon enough, funding agencies began to realize the significance of rice genome sequencing to secure a stable food supply. The National Science Foundation and USDA particularly provided several grants to U.S. researchers. Other participating groups also obtained funding from their respective governments.

From then on, each participating country embarked on the arduous task of sequencing the chromosomes or chromosomal regions assigned to them. Working group meetings were organized twice a year. The meeting in February was always held in Tsukuba, whereas the meeting in the third quarter of the year was held in the countries of other IRGSP members. At each meeting, the participating groups reported on the progress of physical map construction, sequencing, and annotation of the sequence. The chromosomal assignments were also constantly reviewed to adjust to the availability of funding obtained by participating groups. In the succeeding years, several other countries also became interested in joining this collaboration. From December 2002, a total of ten countries and regions became actively involved in the sequencing of entire rice chromosomes or regions of specific chromosomes (Figure 1.1).

IRGSP SEQUENCING STRATEGY

Because rice plants are widely cultivated in the world under various conditions and environments, a large number of local cultivars exist. Moreover,

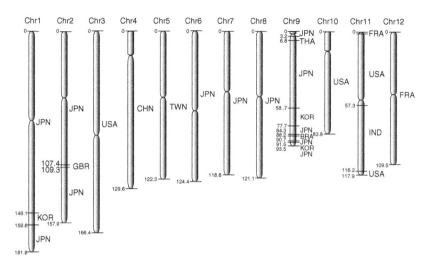

FIGURE 1.1. Chromosomal sharing of the 12 rice chromosomes among the ten participating countries in the International Rice Genome Sequencing Project (IRGSP). Japan was in charge of sequencing almost half of the entire genome. Some countries were involved in sequencing specific regions of one or more chromosomes.

varying preferences of farmers and consumers have also contributed to the diversity among cultivated varieties. Thus, in the course of natural and in-duced selection, it is rational to expect allelic variations in the nucleotide sequence of the different rice varieties. In order to identify the differences in sequences among the different varieties of cultivated rice, *Oryza sativa*, and among wild *Oryza* species, it is necessary to obtain an accurate genome sequence of a single rice cultivar. An accurate genome sequence with ac-companying positional information can be achieved only through the clone-by-clone sequencing strategy.

The IRGSP adopted the hierarchical shotgun method for sequencing the rice genome. This strategy utilized nine genomic libraries from the *O. sativa* ssp. *japonica* cultivar *Nipponbare* to establish the physical map of rice and to aid in gap-filling. These included a PAC and three BAC libraries. During the course of sequencing, Monsanto (Barry, 2001) and Syngenta (Goff et al., 2002) donated their draft sequences of the *Nipponbare* genome to the IRGSP. Monsanto also contributed BACs and collaborated in the con-struction of the BAC-based physical map. Syngenta contigs were used for extending contigs and filling both physical and sequence gaps. Two complementary strategies were used to establish an accurate physical map. The RGP constructed a transcript map (Wu et al., 2002) with 6,591 STS/EST markers derived mostly from 3'-UTR sequences of rice cDNAs. These markers were used to associate PAC/BAC clones with specific regions of the rice genome.

A sequence-ready physical map was constructed by screening the PAC/ BAC libraries with genetic markers or mapped ESTs in order to select clones that would comprise the minimum tiling path. Draft sequences of aligned clones were used to search for minimally overlapping clones from the BAC-end sequence database and FPC contigs with end-sequences that matched the sequences of seed BACs. These were added to the physical maps, thereby facilitating extension of the contigs. A PCR screening method was used to search for clones that filled the remaining gaps. In addition, two 10 kb insert genomic libraries and a 40 kb fosmid library were also constructed and utilized as an additional resource for gap-filling clones.

The PAC/BAC clones that comprised the minimum tiling path were subjected to shotgun sequencing using both universal primers and the dye-terminator or dye-primer methods. The sequences were assembled by PHRED and PHRAP software packages or with the TIGR Assembler (http:// www.tigr.org/software/assembler/). Sequence gaps were resolved by full sequencing of gap-bridge clones, PCR fragments, or direct sequencing of BACs. Sequence ambiguities indicated by low PHRAP scores were resolved by confirming the sequence data using alternative chemistries or different polymerases.

This strategy led to the acceleration of the rice genome sequencing efforts. A draft sequence of the genome was completed in December 2002. From then on, the sequencing collaboration shifted on raising the sequence to "finished" quality. Chromosome 1, the largest among the 12 chromosomes, was one of the first chromosomes to be completely sequenced (Sasaki et al., 2002).

SEQUENCE DATA ANALYSIS

The sequencing of the rice genome was completed in December 2004 and a detailed analysis of the genome was published in *Nature* (International Rice Genome Sequencing Project, 2005). The calculated genome size of rice is 389 Mb and was 260 Mb larger than the fully sequenced dicot plant model, *Arabidopsis thaliana*. A total of 370 Mb finished sequence was generated, representing 95% coverage of the genome and virtually all of the euchromatic regions (Figure 1.2). From this sequence, a total of 37,544 nontransposable element-related protein encoding sequences were detected among which approximately 2,859 genes appear to be unique to rice and the other cereals. Between 0.38% and 0.43% of the nuclear genome contained organellar DNA fragments that represent the repeated and ongoing transfer of organellar DNA to the nuclear genome. The transposon content of rice was calculated to be at least 35% with representatives from all known transposon superfamilies. Approximately 80,127 polymorphic sites that distinguish between two cultivated rice subspecies, *japonica* and *indica,* were also detected from the completed sequence.

Annotation of the sequence data is an integral part of genome analysis in order to characterize the structure and function of the rice genome, and to extract biologically useful information from the sequence. Primarily, it is necessary to predict the position of genes and to determine the existence of transposable elements. It is also important to identify the coding and non-coding regions accurately as well as the junction between them. At RGP, the different prediction programs have been incorporated on an automated annotation system known as RiceGAAS (Rice Genome Automated Annotation

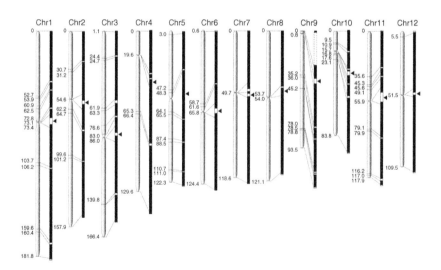

FIGURE 1.2. The completed sequence of the rice genome. For each chromosome, the right and left bars indicate the genetic map and physical map, respectively. The arrow in each chromosome indicates the position of the centromere. The centromeres of chromosomes 4, 5, and 8 have been fully sequenced and characterized.

System, http://RiceGAAS.dna.affrc.go.jp/) to facilitate analysis of any rice genome sequence on a regular basis (Sakata et al., 2002). The final annotations of the completed BAC/PAC sequence were manually curated. All manually curated annotations of genome sequences generated by RGP can be accessed using the graphical viewer INE (INtegrated Rice Genome Explorer, http://rgp.dna.affrc.go.jp/giot/INE.html) (Sakata et al., 2000). INE was implemented on Javascript and shows the sequence data primarily based on map information of rice genome, such as genetic and physical maps (Figure 1.3).

With the completion of the genome sequence, an annotation jamboree was held in Tsukuba, Japan, in December 2004 with the aim of annotating and manually curating all the predicted genes in the rice genome. The Rice Annotation Project Database (RAP-DB) was developed to provide access to the curated annotation data (Ohyanagi et al., 2006). The RAP-DB has two different types of genome browsers (GBrowse and G-integra). Basic local alignment search tool (BLAST) search facilitates links of annotations to other rice genomics data, such as full-length cDNAs and *Tos17* mutant lines. A comprehensive analysis of the RAP annotation data showed that 19,969 (70%) of predicted genes could be assigned with specific functions using cDNA sequences obtained from rice and other representative cereals (Rice Annotation Project, 2007). Manual curation of the annotation is also updated

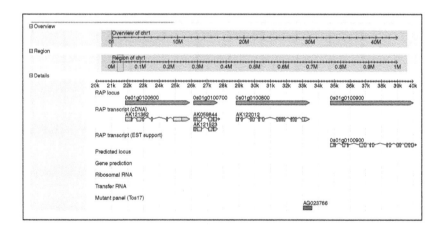

FIGURE 1.3. A graphical view of manually curated annotation of the rice genome. Each RAP locus represents gene models manually curated gene using information from rice full-length cDNA sequences and ESTs from various cereal species. See corresponding Plate 3.1 in the central gallery.

on a regular basis to provide the most accurate information on the structure and function of rice genes.

APPLICATION TO RICE RESEARCH

The high-quality genome sequence can be used in many research fields including rapid and accurate gene identification, establishment of new types of genetic markers, elucidation of the diversity of *Oryza* species, and use as a sequence reference for cereal genome analysis. Both the forward and reverse genetic methods require the application of an accurate sequence for gene identification. In particular, the forward genetic method relies on the plant material that carries the target phenotype. If this phenotype is genetically purified as a single Mendelian factor, the so-called map-based cloning method can be applied. Thus, the *Nipponbare* genome sequence can be used for narrowing the candidate genomic region by finding more recombination points and by identification of a candidate gene (Wang et al., 1999; Ashikari et al., 1999). This strategy has been used even for the identification of genes involved in quantitative traits such as flowering time (Yano et al., 2000).

The most easily and extensively applied strategy for gene identification is by gene-disruption using alien factors carrying the known nucleotide sequence such as the transposable element *Ac/Ds* (Izawa et al., 1997) of maize, T-DNA of *Agrobacterium* (Jeon et al., 2000), or rice endogenous retrotransposon, *Tos17* (Hirochika, 2001). Once these factors are inserted into the rice genome, some change in the plant's appearance or phenotype is expected. The disrupted gene has been identified from the phenotype by tagging or from the nucleotide sequence flanking the inserted sequence. These approaches will be further enhanced using the genome sequence with positional information. In particular, *Tos17* promises to be an indispensable tool to overcome the problem of genetically modified rice plants for cultivation in open paddy fields. So far nearly 50,000 lines have been generated and flanking sequences of *Tos17* for about 20,000 lines have been analyzed (Miyao et al., 2007). The phenotype data for each mutant line together with the flanking sequence data can be accessed through the *Tos17* Mutant Panel Database (http://tos.nias.affrc.go.jp/).

Another important application of the accurate *Nipponbare* genome sequence is in identifying the relationships between variations in nucleotide sequence and phenotype. This type of variation does not correspond to a discontinuous mutation caused by a sequence variation leading to the loss of gene function but rather to variation that causes subtle changes in expression level, and may be found in the promoter region, intron or exon region outside the active site of the gene. To elucidate the involvement of such variation in phenotype, sequence comparison among many rice cultivars for the gene corresponding to the target phenotype is required. This comparison should identify single nucleotide polymorphism (SNP) and insertion/deletion (In/Del). These types of sequence variation normally occur for phenotypes controlled by multigenes, or quantitative traits. Agronomically important traits such as flowering time, culm height, or yield are known to be controlled by multigenes, and the elucidation of SNP and/or In/Del is crucial to understand their contribution to the diversity of cultivars in *japonica* and *indica* ecotypes. The comparison should be extended to other *Oryza* species in the AA genome, and to other genomes within the grass family in order to clarify gene diversity and to understand the origin of cultivated rice.

The introduction of novel genes by crossing cultivated rice and its wild relatives could be also greatly facilitated by the genome sequence information. The accurate genome-wide establishment of SNPs and SSRs will be useful for the easy identification of introgressed alien genes to cultivated rice (McCouch et al., 2002; Nasu et al., 2002). These markers can help to validate the results of crossing and must be used for reliable selection on breeding.

Finally, the accurate rice genome sequence is also useful in searching for similar proteins in *A. thaliana,* a model dicot plant that was sequenced in the year 2000. A total of 26,000 genes have been predicted in *Arabidopsis* and among them, 10% has been actually identified and functionally characterized in detail. Therefore, once an important gene is isolated in *Arabidopsis,* it is reasonable to search for its rice homolog within the rice genome sequence. Using this approach, several rice genes involved in gibberellin biosynthesis have been identified as responsible for dwarf mutation (Itoh et al., 2001; Ashikari et al., 2002). Rice homologs involved in flowering time have also been identified using the information available in *Arabidopsis* (Izawa et al., 2003). Since this method is a shortcut to access the target gene, genetic and physiological analysis are necessary to confirm if the selected homolog correspond to a true ortholog in rice.

APPLICATION TO CEREAL RESEARCH

It is generally known that rice shows colinearity in gene order with other cereal genomes such as wheat, maize, barley, and sorghum (Moore et al., 1995). This syntenic relationship was clarified by cross-genetic mapping of cDNAs with a limited number of probes. An accurate rice genome sequence will provide a reliable clue to prove this relationship in detail and to justify rice as a model plant for cereal genomics. So far, only limited corresponding genomic regions of other cereal crops such as sorghum and maize have been sequenced; therefore, the comparison among these sequences has been of very little use. However, it has been commonly indicated in several comparative analyses that each genome has unique characteristics, even if the gene order is conserved. For instance, the maize genome has an extra gene in its rice ortholog along its corresponding genome sequence (Song et al., 2002), and the corresponding sorghum gene has further divided into extra exons (Ramakrishna et al., 2002). Of course, the expanded genome sequences of these species are rich in transposable elements. Extensive genome sequence information must be collected for sorghum, maize, and wheat to elucidate a micro-level syntenic relationship among the major cereal crops.

The rice genome sequence provides a complete catalog of all rice genes, which could be used as the foundation for organizing the gene organization of other cereals. Identification of orthologous genes will accelerate analysis of agronomically important genes in major cereal crops. Elucidating the functions of rice genes could also lead in understanding the biological processes involved in these traits, which could be used for crop improvement. Furthermore, the rice genome sequence will serve as the most important reference for ongoing sequencing efforts in other cereal genomes.

PERSPECTIVES

The completion of an accurate map-based sequence for rice is one of the most important achievements in basic biology as well as in agriculture. A blueprint of the entire genome showing the distribution and location of all the genes, repetitive sequences, centromeres, telomeres, transposable elements, and other components may be essential in understanding the biology of rice. The pseudomolecules of the 12 rice chromosomes can now be accessed in public databases. This will provide rice researchers with a useful tool to evaluate the genome diversity and accelerate many strategies in rice improvement.

The *Nipponbare* genome sequence must be a standard to understand the rice genome structure and the complete sequence must give us a large amount of basic information that can be used in elucidating the structure of other cereal genomes. The genome sequence is just like an encyclopedia containing a catalog of all the genetic information defining the species. We must decode the information by combining prediction and experimental results. The rice plant is suitable for genetic experiment because it is a self-pollinating plant and produces many seeds. The knowledge obtained by Gregor

Mendel using the common pea, or by Barbara McClintock using maize, paved the way for a new era of genetics. In the same way, the knowledge obtained from rice genome sequencing should be used for improvement of rice and other cereal crops that support life on earth.

REFERENCES

Allitto, B.A., MacDonald, M.E., Bucan, M., Richards, J., Romano, D., Whaley, W.L., Falcone, B., Ianazzi, J., Wexler, N.S., and Wasmuth, J.J. (1991). Increased recombination adjacent to the Huntington disease-linked D4S10 marker. *Genomics 9:* 104-112.

Ashikari, M., Sasaki, A., Ueguchi-Tanaka, M., Itoh, H., Nishimura, A., Datta, S., Ishiyama, K., Saito, T., Kobayashi, M., Khush, G.S., et al. (2002). Loss-of-function of a rice gibberellin biosynthetic gene, *GA20 oxidase (GA20ox-2)*, led to the rice "Green Revolution." *Breeding Science 52:* 143-150.

Ashikari, M., Wu, J., Yano, M., Sasaki, T., and Yoshimura, A. (1999). Rice gibberellin-insensitive dwarf mutant gene *Dwarf1* encodes the alpha-subunit of GTP-binding protein. *Proceedings of the National Academy of Sciences of the USA 96:* 10284-10289.

Baba, T., Katagiri, S., Tanoue, H., Tanaka, R., Chiden, Y., Saji, S., Hamada, M., Nakashima, M., Okamoto, M., Hayashi, M., et al. (2000). Construction and characterization of rice genome libraries: PAC library of *japonica* variety, *Nipponbare* and BAC library of *indica* variety. *Bulletin of NIAR 14:* 41-49.

Barry, G.F. (2001). The use of the Monsanto draft rice genome sequence in research. *Plant Physiology 125:* 1164-1165.

Botstein, D., White, R.L., Skolnick, M., and Davis, R.W. (1986). Construction of a genetic linkage map in man using restriction fragment length polymorphisms. *American Journal of Human Genetics 32:* 314-331.

Burke, D.T., Carle, G.F., and Olson, M.V. (1987). Cloning of large segments of exogenous DNA into yeast by means of artificial chromosome vectors. *Science 236:* 806-812.

Goff, S.A., Ricke, D., Lan, T.H., Presting, G., Wang, R., Dunn, M., Glazebrook, J., Sessions, A., Oeller, P., Varma, H., et al. (2002). A draft sequence of the rice genome (*Oryza sativa* L.ssp. *japonica*). *Science 296:* 92-100.

Harushima, Y., Yano, M., Shomura, A., Sato, M., Shimano, T., Kuboki, Y., Yamamoto, T., Lin, S.Y., Antonio, B.A., Parco, A., et al. (1998). A high-density rice genetic linkage map with 2,275 markers using a single F2 population. *Genetics 148:* 479-494.

Hirochika, H. (2001). Contribution of the *Tos17* retrotransposon to rice functional genomics. *Current Opinion in Plant Biology 4:* 118-122.

International Rice Genome Sequencing Project (2005). The map-based sequence of the rice genome. *Nature 436:* 793-800.

Ioannou, P.A., Amemiya, C.T., Garnes, J., Kroisel, P.M., Shizuya, H., Chen, C., Batzer, M.A., and de Jong, P.J. (1994). A new bacteriophage P1-derived vector for the propagation of large human DNA fragments. *Nature Genetics 6:* 84-89.

Itoh, H., Ueguchi-Tanaka, M., Sentoku, N., Kitano, H., Matsuoka, M., and Kobayashi, M. (2001). Cloning and functional analysis of two gibberellin 3 beta-hydroxylase genes that are differently expressed during the growth of rice. *Proceedings of the National Academy of Sciences of the USA 98:* 8909-8914.

Izawa, T., Ohnishi, T., Nakano, T., Ishida, N., Enoki, H., Hashimoto, H., Itoh, K., Terada, R., Wu, C., Miyazaki, C., et al. (1997). Transposon tagging in rice. *Plant Molecular Biology 35:* 219-229.

Izawa, T., Takahashi, Y., and Yano, M. (2003). Comparative biology comes into bloom: Genomic and genetic comparison of flowering pathways in rice and *Arabidopsis. Current Opinion in Plant Biology 6:* 113-120.

Jeon, J.S., Lee, S., Jung, K.H., Jun, S.H., Jeong, D.H., Lee, J., Kim, C., Jang, S., Yang, K., Nam, J., et al. (2000). *T-DNA* insertional mutagenesis for functional genomics in rice. *The Plant Journal 22:* 561-570.

Konieczny, A. and Ausubel, F.M. (1993). A procedure for mapping *Arabidopsis* mutations using co-dominant ecotype-specific PCR-based markers. *The Plant Journal 4*: 403-410.

McCouch, S.R., Kochert, G., Yu, Z.H., Wang, Z.Y., Khush, G.S., Coffman, W.R., and Tanksley, S.D. (1988). Molecular mapping of rice chromosomes. *Theory of Applied Genetics 76:* 818-820.

McCouch, S.R., Teytelman, L., Xu, Y., Lobos, K.B., Clare, K., Walton, M., Fu, B., Maghirang, R., Li, Z., Xing, Y., et al. (2002). Development and mapping of 2240 new SSR markers for rice (*Oryza sativa* L.). *DNA Research 9:* 199-207.

Miyao, A., Iwasaki, Y., Kitano, H., Itoh, J.I., Maekawa, M., Murata, K., Yatou, O., Nagato, Y., and Hirochika, H. (2007). A large-scale collection of phenotypic data describing an insertional mutant population to facilitate functional analysis of rice genes. *Plant Molecular Biology 63:* 625-635.

Moore, G., Devos, K.M., Wang, Z., and Gale, M.D. (1995). Grassers, line up and form a circle. *Current Biology 5:* 737-739.

Nasu, S., Suzuki, J., Ohta, R., Hasegawa, K., Yui, R., Kitazawa, N., Monna, L., and Minobe, Y. (2002). Search for and analysis of single nucleotide polymorphisms (SNPs) in rice *(Oryza sativa, Oryza rufipogon)* and establishment of SNP markers. *DNA Research 9:* 163-171.

Ohyanagi, H., Tanaka, T., Sakai, H., Shigemoto, Y., Yamaguchi, K., Habara, T., Fujii, Y., Antonio, B.A., Nagamura, Y., Imanishi, T., et al. (2006). The Rice Annotation Project Database (RAP-DB): Hub for *Oryza sativa* ssp. *japonica* genome information. *Nucleic Acids Research 34:* D741-D744.

Ramakrishna, W., Dubcovsky, J., Park, T.J., Busso, C., Emberton, J., San Miguel, P., and Bennetzen, J.L. (2002). Different types and rates of genome evolution detected by comparative sequence analysis of orthologous segments from four cereal genomes. *Genetics 152:* 1389-1400.

Rice Annotation Project (2007). Curated genome annotation of *Oryza sativa* ssp. *japonica* and comparative genome analysis with *Arabidopsis thaliana*. *Genome Research 17:* 175-183.

Rice Genome Research Program (2000). http://rgp.dna.affrc.go.jp/publicdata/ geneticmap2000/index.html.

Saji, S., Umehara, Y., Antonio, B.A., Yamane, H., Tanoue, H., Baba, T., Aoki, H., Ishige, N., Wu, J., Koike, K., et al. (2001). A physical map with yeast artificial chromosome (YAC) clones covering 63% of the 12 rice chromosomes. *Genome 44:* 32-37.

Sakata, K., Antonio, B.A., Mukai, Y., Nagasaki, H., Sakai, Y., Makino, K., and Sasaki, T. (2000). INE: A rice genome database with an integrated map view. *Nucleic Acids Research 28:* 97-101.

Sakata, K., Nagamura, Y., Numa, H., Antonio, B.A., Nagasaki, H., Idonuma, A., Watanabe, W., Shimizu, Y., Horiuchi, I., Matsumoto, T., et al. (2002). Rice GAAS: An automated annotation system and database for rice genome sequence. *Nucleic Acids Research 30:* 98-102.

Sasaki, T., Matsumoto, T., Yamamoto, Y., Sakata, K., Baba, T., Katayose, Y., Wu, J., Niimura, Y., Cheng, Z., Nagamura, Y., et al. (2002). The genome sequence and structure of rice chromosome 1. *Nature 420:* 312-316.

Shizuya, H., Birren, B., Kim, U.J., Mancino, V., Slepak, T., Tachiiri, Y., and Simon, M. (1992). Cloning and stable maintenance of 300-kilobase-pair fragments of human DNA in *Escherichia coli* using an F-factor-based vector. *Proceedings of the National Academy of Sciences of the USA 89:* 8794-8797.

Song, R., Llaca, V., and Messing, J. (2002). Mosaic organization of orthologous sequences in grass genomes. *Genome Research 12:* 1549-1555.

Umehara, Y., Inagaki, A., Tanoue, H., Yasukochi, Y., Nagamura, Y., Saji, S., Otsuki, Y., Fujimura, T., Kurata, N., and Minobe, Y. (1995). Construction and characterization of a rice YAC library for physical mapping. *Molecular Breeding 1:* 79-89.

Wang, Z.X., Yano, M., Yamanouchi, U., Iwamoto, M., Monna, L., Hayasaka, H., Katayose, Y., and Sasaki, T. (1999). The *Pib* gene for rice blast resistance belongs to the nucleotide binding and leucine-rich repeat class of plant disease resistance genes. *The Plant Journal 19:* 55-64.

Wu, J., Maehara, T., Shimokawa, T., Yamamoto, S., Harada, C., Takazaki, Y., Ono, N., Mukai, Y., Koike, K., Yazaki, J., et al. (2002). A comprehensive rice transcript map containing 6591 expressed sequence tag sites. *Plant Cell 14:* 525-535.

Yamamoto, K., and Sasaki, T. (1997). Large-scale EST sequencing in rice. *Plant Molecular Biology 35:* 135-144.

Yano, M., Katayose, Y., Ashikari, M., Yamanouchi, U., Monna, L., Fuse, T., Baba, T., Yamamoto, K., Umehara, Y., Nagamura, Y., et al. (2000). *Hd1,* a major photoperiod sensitivity quantitative trait locus in rice, is closely related to the *Arabidopsis* flowering time gene *CONSTANS*. *Plant Cell 12:* 2473-2484.

Chapter 2

Comprehensive Analysis of Rice Gene Expression Using the Microarray System: What We Have Learned from the Microarray Project

S. Kikuchi
J. Yazaki
N. Kishimoto
M. Ishikawa
D. Endo

INTRODUCTION

The complete genomic sequence of *Arabidopsis*, which is a good model system for studies addressing dicotyledoneous plants, was finished at the end of 2000 (*Arabidopsis* Genome Initiative, 2000), and draft complete genomic sequences obtained through "shotgun sequencing" efforts for the *japonica* (Goff et al., 2002) and *indica* (Yu et al., 2002) subspecies of rice were published in April 2002. Furthermore, the International Rice Genome Sequencing Project (IRGSP) announced that it would finish the majority of its project addressing the *japonica* genome based on its physical map in December 2002 (Feng et al., 2002; Rice Chromosome 10 Sequencing Consortium, 2003; Sasaki et al., 2002). Following the completion of the large-scale expressed sequence tag (EST) collection project during Phase One (1991-1997) of the Rice Genome Project (RGP), the rice full-length cDNA project was launched at the beginning of the year 2000. This project has already collected a set of nearly 30,000 unique, full-length rice cDNA clones, and as of July 2003, the complete nucleotide sequence of 28,469 of these clones has been determined (Rice Full-Length cDNA Consortium, 2003; see also KOME [Knowledge-Based *Oryza* Molecular Biological Encyclopedia]; National Institute of Agrobiological Sciences, 2002).

The consolidation of the aforementioned genomic tools prompted the inception of many other genome-related projects, such as the microarray project, gene isolation project using transposable element insertion lines, proteome project, large-scale DNA transformation project, and bioinformatics project. As collaborations between various national institutes, public institutes, universities, and private companies, these projects address the acceleration of molecular genetic, molecular biological, and biochemical analyses and the application of plant biotechnology to the improvement of crops.

The Rice Genome Microarray Project started in April 1999. This project combines the efforts of the Microarray Center (MAC)—a team of investigators from the department of molecular genetics of the National Institute of Agrobiological Sciences and the Society for Techno-Innovation of Agriculture, Forestry, and Fisheries (STAFF)—with those of about 60 independent research groups. This report introduces our microarray system, the Rice Expression Database (RED; National Institute of Agrobiological Sciences, 2003; Yazaki et al., 2002), and the progress of each research group using microarray technology. Our Web site, the Rice Microarray Opening Site (RMOS; National Institute of Agrobiological Sciences, 2001) provides more information regarding the various projects and results.

THE RICE MICROARRAY SYSTEM

The MAC prepares glass plates, each containing probe DNA spots for the microarray analysis; performs the hybridization experiments after receiving mRNA or total RNA from project members; quantifies the spot data; and then enters them in the RED. Originally, we used about 11,000 nearly nonredundant EST clones (from the Rice Genome Project), of which we used the entire cDNA insert as well as the 3' untranslated region (UTR) as probes. These probes comprised two sets: the first (1,265 sequences) yields higher-intensity signals, whereas the second (8,987 sequences) is better for the discrimination of gene expression among gene families (Yazaki et al., 2000).

With the progress of the rice full-length cDNA project, we are now using as probes the 3'UTR region of full-length cDNA clones instead of EST clones. Furthermore, we are changing the array system to enable us to "customize" our screening efforts. The global array system aims to represent all of the genes in the genome, whereas the purpose of the custom array system is to present a categorized gene set that reflects a researcher's needs. The hardware for microarray analyses was originally provided under contract through the Microarray Technology Access Program (MTAP) with Amersham Pharmacia and Affimetrix. We are using a third-generation microarray spotter, which can spot duplicated sets of 4,608 clones on a single glass plate (25×75 mm^2). In the near future, the fourth-generation microarray spotter will be introduced. This instrument is very important for the reproducible spotting of DNA fragments onto glass plates. The microarray scanner we have introduced is made by Fujifilm (FLA-8000); its leading advantage is its wide dynamic response over the range of signal intensity. A database for the results of microarray experiments has recently been established by the bioinformatics personnel in our research group. The methods of background subtraction, normalization between the different glass plates, and statistical treatment are important aspects of the included data, which can be accessed through our Web site. According to our policy, for the first year after the completion of the experiment, the data can be seen only by the contributing researchers; however, after this time, the data are made available worldwide.

KINDS OF KNOWLEDGE ACCUMULATED

During the past three years, more than 60 research group members have contributed to the microarray project. More than 500 microarray hybridization experiments have been performed, and gene expression data for rice and other plants have been accumulated. These efforts can be grouped into several areas, including (1) studies addressing rice physiology by using the rice microarray system and (2) studies targeting the physiology of plants other than rice by using the microarray hardware, but independently prepared probes. Although some are very close to being applied studies, most of the analyses focus on fundamental plant physiological research. Plants other than rice include barley, wheat, Italian ryegrass, guineagrass, *Lotus japonicus,* apple, peach, citrus, *Ipomea,* and others. Research themes can be classified into groups, including responses to environmental stimuli (e.g., UV irradiation, oxidative stress, cold temperature, nutrient deficiency, anaerobic stress, pest infection, and excess dosage of phytohormones); stage- or organ-specific gene expression (cell cycle dependency, sink-source organ-specific events, differentiation of sex organs, and apomixis); and the analyses of various mutants. These studies focus on those genes showing specific changes in their expression. One of the important and advantageous aspects of the overall research project is that many research groups are using a common microarray system. Therefore, the results of each research group can be compared horizontally, enabling the comparison of changes in gene expression under various physiological conditions. The following sections present the results of the individual research projects.

STUDIES USING THE MICROARRAY SYSTEM ON RICE

UVB and Gamma Irradiation

UVB irradiation is harmful to plants as well as animals. Increased UVB irradiation through the hole in the ozone layer has recently been shown to have decreased the harvest of land crops in tropical areas (Ballare et al., 2001; Rousseaux et al., 1999). In *Arabidopsis,* gamma irradiation induces DNA damage, the scavenging system for reactive oxygen species, and the differentiation of trichomes (Nagata et al., 1999). To understand the physiological response of plants to UVB or gamma irradiation, we used 8,987 rice microarrays to compare changes in transcription levels between irradiated and nonirradiated rice seedlings.

The rice plants (cv. *Nipponbare*) were grown by hydroponics (1000-fold dilution of HYPONeX 10-3-3) in a Sanyo growth cabinet (28°C) on a 16/8 h light/dark cycle after sowing. For studies on UVB irradiation, half of the plants were transferred to a UVB chamber ($1.06 \ \mu J/[m^2 \times sec]^{-1}$) on day 13 after sowing and irradiated for 4.8 min ($0.304 \ J/m^2$). During the irradiation, the root parts were surrounded with tissue paper saturated with culture liquid. The other half of the plants (nonirradiated seedlings) was treated in the same way as were the irradiated seedlings, but they were not irradiated. For the gamma-irradiation studies, the plants were grown in the same way as for the UVB experiments. On day 14 after sowing, half of the plants were transferred to a cobalt-60 (9.2 Gy/min) irradiator and irradiated for 8.5 min (78.5 Gy). During the irradiation, the seedlings were kept on tissue paper saturated with culture liquid. During the irradiation, the remaining plants (nonirradiated seedlings) were kept in the room housing the irradiator. After both types of irradiation, all seedlings were returned to fresh hydroponic solution.

Changes in gene expression (more than threefold difference in signal intensity) were assessed at 8, 24, and 96 h after gamma irradiation. The difference of gene expression was defined as more than a threefold change of spot data after microarray analysis. Of the 54 genes whose expression were upregulated 8 h after gamma irradiation, most were salt-stress-inducible and chitinase genes; most of the 37 genes whose expression was downregulated were phytosynthesis- and respiration/glycolysis-related genes. At 24 and 96 h after irradiation, different and fewer genes showed changes in their expression levels. At 24 h, only one gene was upregulated, and 21 genes were downregulated, whereas at 96 h, 6 genes were upregulated and 9 were downregulated. Quantitative real-time reverse transcription-polymerase chain reaction (RT-PCR) analyses confirmed the changes in the expression of 14 of the 16 genes identified by the microarray analysis.

UVB-mediated changes in gene expression were assessed 24 and 96 hours after irradiation. Of the 58 genes upregulated, most represented the PR (pathogenesis related) proteins and stress-inducible genes. Most of the 15 downregulated genes were related to metabolism. At 96 h after irradiation, different and fewer genes showed changes in expression: four genes were upregulated, and seven were downregulated.

Both gamma and UVB irradiation induce reactive oxygen species and DNA damage. However, our preliminary data suggested that these stimuli are qualitatively different. Using the microarray system to compare oxidative stresses, such as UVB irradiation and ozone treatment, might yield valuable information. At present, we are focusing on the further functional analysis of some of the genes we identified during our screening efforts.[1]

Roles of Reactive Oxygen Species in Promoting Seedling Growth and Pathogen Tolerance

Hydrogen peroxide, a reactive oxygen species, is toxic to cells because of its high oxidative ability. It has recently been highlighted as a signaling molecule for responses to both abiotic and biotic stresses. Furthermore, treatment of *Zinnia elegans* seeds with hydrogen peroxide enhances germination frequency and promotes seedling growth (Ogawa and Iwabuchi, 2001). This finding is true for many plants, including rice. Furthermore, seedlings that are treated with hydrogen peroxide (H_2O_2) rarely encounter diseases due to pathogen invasion. Antioxidant germination inhibitors exist in the pericarp and seed coat of seeds. Many antioxidants themselves have an inhibitory effect on germination and lose this effect after their oxidation by hydrogen peroxide (Ogawa and Iwabuchi, 2001). Taking into consideration the fact that H_2O_2 is generated in seedlings during germination, we suggest that H_2O_2 is required for seed germination, and that the antioxidant germination inhibitors present in the seed coat and pericarp block the promotional action of H_2O_2 during emergence from dormancy and seedling growth. Application of gibberellin abolishes the germination-inhibitory effect of the antioxidants. Furthermore, the H_2O_2-induced morphological alterations occur prominently in wild-type plants but are diminished in gibberellin-deficient rice mutants. These findings suggest that in seed germination, the pathways controlled by H_2O_2 are upstream of the gibberellin-regulated pathways.

The results of a rice cDNA microarray analysis of wild-type and gibberellin-deficient rice seedlings from H_2O_2-treated seeds have revealed how closely H_2O_2 and gibberellin are associated in the regulation of seed germination. In the gibberellin-deficient mutant, the H_2O_2-responsive genes can be categorized into 4 groups in light of their expression: upregulated in response to H_2O_2; downregulated after H_2O_2 treatment; and constitutively up- or downregulated regardless of H_2O_2. These results suggest that, in addition to its positive role in gene expression, gibberellin acts as a negative regulator of H_2O_2-induced responses. The gibberellin-regulated genes include those encoding the H_2O_2-scavenging enzyme ascorbate peroxidase and the H_2O_2-supplying enzyme superoxide dismutase. Therefore, it is likely that the regulatory mechanism for H_2O_2 and gibberellin constitutes a feedback loop. Furthermore, the H_2O_2-responsive genes are expressed similarly in seedlings germinated from H_2O_2-treated and untreated hulled seeds—a finding that strongly confirms that antioxidant germination inhibitors block the pathway controlled by H_2O_2 at the level of gene expression. The genes under the control of H_2O_2 include those for pathogenesis-related and systemic acquired resistance (SAR) DNA binding proteins, and this situation likely accounts for the low frequency of growth impairment due to onset of disease in the plants pretreated with H_2O_2. These genes are highly expressed in the gibberellin-deficient mutants, suggesting that gibberellin is a negative regulator for SAR.

Gibberellin regulates not only seed germination but also flowering. Moreover, we recently found that ascorbate peroxidase activity is associated with flowering in *Arabidopsis thaliana* ecotypes (Lokhande et al., 2002). Considering these facts, we hypothesize that the feedback-loop regulation involving hydrogen peroxide and gibberellin is a common development in plants.[2]

Chilling Tolerance at the Booting Stage Using Near-Isogenic Lines in Rice

In northern Japan, sterility caused by low temperature is the most serious problem in rice, because it inevitably leads to yield reduction (Saito et al., 2001). In the sterility type of cold injury, the growth stage most sensitive to low temperature is the booting stage—especially the young microspore stage spanning from the tetrad to the first contraction phase (Satake and Hayase, 1970). Low temperature during microsporogenesis causes degeneration of microspores, resulting in sterility.

We developed a cold-tolerant, near-isogenic line, NIL510-2 (B_5F_5), by backcross breeding, in which a cold-tolerant cultivar, *Hayayuki*, and a cold-sensitive cultivar, *Toyohikari*, were used as donor and recurrent parents, respectively. NIL510-2 produced sufficient pollen for successful pollination even after four days of 12°C treatment at the young microspore stage, whereas cold-sensitive *Toyohikari* failed to produce intact pollen grains because of the degeneration of microspores after cold stress.

We have used cDNA microarrays to monitor transcript abundance and expression patterns in spikelets of NIL510-2 and *Toyohikari* exposed to 12°C for four days at the microspore stage. Total RNA extracted from spikelets (length = 5.5-6.5 mm) of rice plants at the young microspore stage and exposed to 12 or 25°C for four days was analyzed. Transcripts corresponding to cDNA clones derived from callus or leaves were markedly upregulated in *Toyohikari* after four days of exposure to 12°C. Many of these cDNA clones showed sequence homology to stress-inducible genes or ESTs. In contrast, such genes were not upregulated in NIL510-2 after cold stress. Transcripts corresponding to clones derived from young panicles were more abundant in NIL510-2 than in *Toyohikari*. These data indicate that cold-sensitive *Toyohikari* expresses stress-inducible genes under cold stress, resulting in the degeneration of microspores, whereas cold-tolerant NIL510-2 responds less to cold stress and tends to continue expressing genes responsible for microsporogenesis even under these conditions.[3]

Cool-Temperature Stress During Rice Pollen Formation

In northern Japan, rice crop production is occasionally damaged severely by cool temperatures during the summer, and the development of rice subspecies with extremely high cold tolerance is desired. The most cold-temperature-sensitive stage of rice is the young microspore stage. At this stage, the tetrad-releasing enzyme callase is secreted from the tapetum, and its activity is upregulated after the acidification of the anther locule. To date, very little information is available regarding the molecular aspects of this low-temperature damage. Using the microarray system, we analyzed rice anthers cooled at the young microspore stage to identify and characterize the genes responsive to low temperature or related to the low-temperature damage process in rice plants. We used rice lines selected to be extremely cold tolerant, which we generated by crossing Japanese cold-tolerant cultivars with stocks originated from Chinese *Yunnan* or *Bhutan* cultivars. These lines are much more cold tolerant than is *Koshihikari*, the most cold-tolerant Japanese cultivar.

A growth chamber and ordinary greenhouse conditions were used for growing the cold-tolerant Japanese cultivar *Hayayuki* and the F_7 generation plants of four extremely cold-tolerant selected rice lines developed by the Miyagi Prefectural Furukawa Agricultural Experiment Station (line nos. 2, 4, 6, and 8; nos. 4 and 6 are progenitors of nos. 2 and 8, which are more cold-tolerant than are nos. 4 and 6). The anther stage was identified by panicle length and auricle distance. Rice plants were cooled at 12°C for five days at the young microspore stage, and 2,000 to 4,500 anthers were collected after cool treatment. About 2,000 to 4,500 uncooled anthers (controls) were collected at the early middle microspore stage, because during the cool treatment, the anthers progressed from the young microspore stage to the early middle microspore stage. Collected anthers were rapidly frozen in liquid nitrogen. Total RNA was extracted from the frozen samples by using cetyltrimethylammonium bromide and LiCl precipitation. Fluorescently labeled targets were made by reverse transcription, and we completed gene expression profiling by using a cDNA microarray comprising 8,987 rice cDNA clones. The fluorescence intensities of each spot in the Cy5-labeled images were quantified, and we subtracted and normalized the fluorescence levels of the local background. In the *Hayayuki* cultivar, 114 genes were downregulated and 20 were upregulated by cool-temperature treatment during the young microspore stage. Of these, one gene for polyamine biosynthesis was upregulated markedly in *Hayayuki*. Moreover, this upregulation clearly occurred in the most cold-tolerant line (no. 8), but this gene was rather downregulated in the less cold-tolerant line no. 4. These results suggest that polyamine synthesis is important in cold tolerance during rice pollen development. Furthermore, a gene related to jasmonate biosynthesis was downregulated by the cool-temperature treatment during the young microspore stage. This finding raises the possibility that jasmonate signaling is involved not only in pollen maturation and anther dehiscence, but also in early microspore development and cold-tolerance regulation in rice plants.[4]

Responses of Rice Roots to Moderately Low Temperatures

Because of its tropical origins, rice is much more sensitive to low temperatures than are other cereal crops, such as wheat and barley. Although moderately low temperatures (12-17°C) are not problematic for the viability of most *japonica* cultivars, they do inhibit certain stages of pollen development and cause male sterility (Satake and Hayase, 1970). Therefore, it is of great interest to reveal the molecular responses of rice to moderate to low temperatures.

We germinated seeds of *Oryza sativa* L. cv. *Yukihikari* for one day at 25°C in the dark. Germinated seeds were evenly placed onto a plastic mesh grid supported by a plastic container filled with water just to the base of the mesh grid. The container was kept in a growth chamber at 25°C under continuous illumination (256 μmol $\times$ m^{-2}s^{-1}) for seven days. Low-temperature treatment was conducted by transferring the mesh grid with seven-day-old seedlings onto a plastic container filled with water pre-equilibrated at 12°C in a growth chamber.

Total RNA was extracted from seedling roots and labeled with Cy5. Microarray plates containing 8,987 cDNA clones were screened with target cDNA from nonstressed (NS) plants or plants cold-stressed (CS) for 2 h or 24 h. Cold-induced clones were selected from among those with a CS/NS signal ratio of at least 2. We obtained 49 cDNA clones for genes that were induced by the 2-h cold treatment. Database searches indicated that putative functions had been assigned to 26 clones; these genes included *lip9* (Aguan et al., 1991), *OsCDPK7* (Saijo et al., 2000), which are known to be cold inducible. Moreover, all of the 12 clones that were analyzed by northern blots showed induction within 2 h of treatment at 12°C.

Collectively, these data suggest that the rice microarray system is highly reliable for the identification of stress-related genes. The newly identified 12°C-induced clones include NAC-family (Nitrogen Assimilation Control) transcription factors, a zinc finger protein, and a ring finger protein—data suggesting that these DNA-binding proteins may be involved in signal transduction after chilling. Genes for the biosynthesis of phospholipids, a disaccharide, polyamines, and jasmonate were also induced. These data suggest that metabolic changes occur as early responses to moderate- to low-temperature stress.

Exposure to these temperatures for a longer period (24 h) induced different responses in rice roots. Putative functions for 29 of the 49 clones identified were assigned after database searches; the corresponding genes included those for protein sorting and vesicle transport, translation, and protection from oxidative stress.[5]

Silicon Nutrition-Regulated Gene Expression

Silicon (Si) is the second most-abundant element in the Earth's crust. Unlike oxygen, Si is an inert element in biological systems; Si is not considered essential for most living organisms, including plants. Unlike most plant species, rice actively takes up Si and transports it to the leaves, where it accumulates to a relatively high concentration. In the leaves of rice plants grown with a sufficient supply of Si, Si content can reach more than 10% on a dry-weight basis. Si accumulates in the form of a gel in the apoplasts of the cuticles in the leaves of rice plants. Although Si is taken up actively, it is not an essential element for rice, because the rice plant can complete its life cycle in the absence of a detectable Si supply. However, Si application to rice plants has a broad range of beneficial effects, including increased photosynthetic activity, reduced water evaporation from the leaf surface, resistance to pests and insects, and mitigation of the toxic effects of Al and other metals.

Despite physiological evidence of the beneficial effects of Si, little information is available that addresses the effects of Si in rice plants at the molecular level. To our knowledge, no prior report has described genes that are regulated by Si nutrition. We carried out microarray analysis to determine the effects of Si at the level of transcript accumulation. Rice plants *(Oryza sativa* L. cv. *Nipponbare)* were grown hydroponically in a greenhouse under natural lighting at 30°C during the day and 25°C during the night for 4 weeks in hydroponic medium containing 0, 15, 150, or 1500 µM silicate. The plants were harvested, and RNA was extracted from the leaves and subjected to microarray analysis on glass slides with 8,987 cDNA clones. We fluorescently labeled 40 µg total RNA using the Atlas Glass Fluorescent Labeling Kit (Clontech). Hybridization was carried out using ExpressHyb Hybridization Buffer (Clontech). Images were obtained using Gene Pix 400a (Axon Instrument) and analyzed by Array Gauge (Fuji Film). Local background subtraction, normalization to the median values, and *t*-tests were used for data analysis. The Si content of the samples used for RNA extraction showed good correlation with the silicate concentration of the media, indicating proper treatment with Si during our experiments.

After two independent hybridizations, we identified about 80 clones that were markedly upregulated by Si application. Furthermore, about 80 other clones were downregulated after Si treatment. The extent of up- or downregulation was less than fivefold for most of the clones identified by using the RNA samples from plants grown with 0 and 1.5 mM silicate, suggesting that changes in Si nutrition do not have dramatic effects on the pattern of transcript accumulation among the 8,987 clones examined. Interestingly, a number of stress-related genes are represented among the downregulated clones, including *Xa21* (a resistance gene against *Xanthomonas*) and those for chitinase, copper amine oxidase, heat-shock protein, and metallothionein synthesis. These genes are known to be induced by pathogenic, heat, and heavy-metal stresses. Downregulation of these stress-related clones by

Si treatment is the first molecular evidence at the transcript level for the beneficial effects of Si supplementation.

We also found that asparagine synthase was upregulated by Si application. Amino acid analysis of leaf samples indicated that the relative content of asparagine was increased after this treatment. It is likely that the increased expression of asparagine synthase plays an important role in asparagine accumulation after Si treatment. Overall, microarray analysis revealed novel aspects of the effect of Si supplementation that support the physiological changes seen in response to this treatment.[6]

Responsiveness to N-Acetylchitooligosaccharide (Oligochitin) Elicitor Treatment

N-acetylchitooligosaccharide is a potent elicitor at subnanomolar concentrations in suspension-cultured rice cells, inducing a variety of defense reactions, such as the production of phytoalexins (Yamada et al., 1993). The expression of defense-related genes is thought to play important roles in defense reactions. In this suspension-cultured rice cell system, we have used subtractive hybridization to isolate and characterize novel elicitor-responsive genes that are rapidly induced after treatment with *N*-acetylchitooligosaccharide (Minami et al., 1996; Takai et al., 2001). To obtain further insight into the changes in gene expression during the elicitor response, we carried out DNA microarray analysis. Suspension-cultured rice cells *(Oryza sativa* L. cv. *Nipponbare)* were treated with *N*-acetylchitooctaose at 1 µg/mL for 15 min or 120 min, and poly(A)$^+$ RNA was isolated and used as a probe.

By using a 1,265-probe microarray system, we screened for elicitor-responsive genes upregulated within 15 min; RNA blot analysis confirmed 7 ESTs, representing genes upregulated by elicitor treatment (Day et al., 2002). These included ESTs encoding *OsMyb8, OsRac, OsCAM2,* and 2 *GRAS* family proteins. Detailed analysis of the structure and expression of the 2 *GRAS* genes is ongoing. These positive results prompted us to screen 8,987 ESTs using mRNAs from rice cells treated with *N*-acetylchitooctaose for 120 min. After repeating these experiments several times, we were able to identify 261 ESTs as elicitor-responsive genes. One-third of these were classified as unknown genes, and the rest included genes putatively involved in various ways in cellular structure and function. The levels of transcripts from cell-cycle-related genes such as histones and cyclins were decreased, consistent with the results of previous studies in which housekeeping genes related to cell-cycle progression or photosynthesis were repressed during the defense response (Kombrink and Hahlbrock, 1990; Logemann et al., 1995). As expected, defense-related genes such as chitinases and PR proteins were induced after elicitor treatment. Furthermore, genes involved in the shikimate and phenylpropanoid pathways were upregulated. Although we have not identified any phenylpropanoid compounds by using this experimental system, gas chromatography revealed the accumulation of numerous low-molecular-weight compounds in the media, consistent with the activation of these pathways.

Under stress conditions, ATP is generated mainly by anaerobic respiration, resulting in the formation of ethanol or lactic acid. It is noteworthy that two genes for trehalose metabolism—trehalose-6-phosphate synthase and trehalose-6-phosphate phosphatase—were included as upregulated genes, although the occurrence of trehalose in higher plants is not known. Recently, we showed that *EL5*, encoding a ring-H2 finger protein, is an elicitor-responsive gene; encodes ubiquitin ligase, which is in the ubiquitin/proteasome system; and interacts with the rice ubiquitin-conjugating enzymes OsUBC5a and OsUBC5b. OsUBC5b was shown to be induced by *N*-acetylchitooligosaccharide (Takai et al., 2002), consistent with the results of the present study. These results strongly suggest that a genome-wide change in gene expression occurs in rice cells in response to *N*-acetylchitooligosaccharide elicitor, resulting in dynamic changes in cellular metabolism and signaling.[7]

Blast Infection

The *gene-for-gene* theory proposed by Flor (1971) is realized well between rice cultivars and blast fungal strains. That is, depending on whether a rice cultivar has a specific resistance gene, it becomes resistant (or remains susceptible) to a particular race of blast fungus. We aim to isolate rice genes involved in blast resistance in addition to the resistance gene itself. For this purpose, we compared gene expression after blast infection between a resistant and a susceptible cultivar on a genome-wide scale.

In this study, we used two rice cultivars, *Sasanishiki* (the S-0 line, containing the *Pi-a* resistance gene) and its isogenic line (the S-1 line, with the *Pi-a* and *Pi-k* genes), which was developed in the Rice Breeding Section of the Miyagi Prefecture Furukawa Agricultural Experiment Station in Japan. Seedlings were grown in a greenhouse until the 6-leaf stage, when a conidial suspension of blast fungus strain Kyu89-246 (the S-0 is susceptible and the S-1 resistant to infection with this strain) was sprayed on the seedlings. Inoculated seedlings were kept in a chamber with 100% relative humidity at 25°C for 24 h; they were then grown in a greenhouse at 25 to 30°C. Every 24 h until 72 h after inoculation, leaf blades of the sixth leaves were sampled, and the total RNA was extracted. A series of RNA samples from mock-inoculated S-0 plants also was prepared. For each RNA sample, microarray hybridization experiments were conducted with the rice 8,987-clone micro-array. Dividing spot intensities by the median signal intensity normalized gene expression data among the different glass slides.

First, we compared the expression profile of S-0 inoculated plants with that of S-0 mock-inoculated seedlings and that of the S-1 inoculated seedlings with that of the mock-inoculated S-1 plants. These comparisons revealed that the number of clones whose expression levels were changed increased with the time after inoculation. In both types of plants, clones that were induced more than four times the levels in mock-inoculated controls included many novel genes encoding unknown proteins as well as genes for thaumatin-like proteins, PR-1 and PR-10 proteins, lipid transfer proteins, MAP kinases, and CDPKs. Most of these genes were generally induced by blast infection, and their expression levels did not correlate with whether the rice cells can inhibit fungal growth.

Next, a comparison of gene expression between S-1 and S-0 revealed that about 50 clones were upregulated more than tenfold specifically in the S-1 line. More than 50% of these clones have no functional annotation at present. Interestingly, the expression of some of these genes was increased in the S-1 line and in mock-inoculated controls but not in the S-0 line. This finding may indicate that blast fungus suppresses the activation of these genes during infection.

Our next challenges include determining whether the expression levels of the genes specifically induced in the S-1 line turn out to be increased during the interaction between this S-1 line and a virulent blast race, and whether other *Sasanishiki* isogenic lines that have different resistance genes also show increased expression of this group of genes, or whether a different set of genes is involved in the resistance via each true resistance gene.[8]

Response to Probenazole, a Chemical Inducer of Disease Resistance

Probenazole (3-allyloxy-1,2-benzisothiazole-1,1-dioxide; PBZ) is a chemical activator of disease resistance and has been used to protect rice plants against blast fungus infection (Watanabe et al., 1977). Several lines of evidence have shown that PBZ induced the expression of rice genes, including *pPB-1* (as a clone's name) (Minami and Ando, 1994), *PBZ1* (Midoh and Iwata, 1996) and *PRP1* (Sakamoto et al., 1999). More recently, PBZ was shown to induce systemic acquired resistance (SAR) in *Arabidopsis* and tobacco (Yoshioka et al., 2001). Another chemical compound, benzo(1,2,3) thiadiazole-7-carbothionic acid S-methyl ester (BTH) also induced SAR and the host genes *wc1-1* to *wc1-5* in barley and tobacco (Friedrich et al.,

1996; Goerlach et al., 1996). However, PBZ's mode of action is somewhat different from that of BTH (Yoshioka et al., 2001). We used a cDNA micro-array of 1,265 rice clones to obtain data on PBZ-induced changes in the gene expression profile; northern blotting and RT-PCR analyses confirmed these PBZ-induced changes in gene expression (Shimono et al., 2003).

At 12 days after sowing, rice seedlings *(Oryza sativa* L. cv. *Nipponbare)* were transferred to water containing PBZ (1.0 g/l Oryzemate, granular form containing 24% PBZ; Meiji Seika Kaisya Ltd., Tokyo) and incubated for 7 days. Then the leaves were harvested for total RNA extraction followed by mRNA preparation. The mRNA was labeled with Cy5-dCTP for use as a probe for both PBZ-treated and control plants. Probe preparation and hybridization conditions have been reported previously (Yazaki et al., 2000). The hybridization signals were compared between PBZ-treated and untreated (control) rice plants.

We compared the hybridization signals identified by using a probe from PBZ-treated rice with those from control rice. Ten cDNA clones were selected that showed more than a threefold change in their levels of gene expression. All of these clones showed induced expression in the PBZ-treated plants. Northern blot analyses confirmed the increased expression of seven of the ten clones we identified through microarray analysis (S12707, S03727, S16157, E02880, S10124, S10163, and S12429). The remaining three clones underwent further semi-quantitative RT-PCR analysis, and two showed increased expression in the PBZ-treated samples (S10962 and S02370). Thus, nine out of ten clones were confirmed to be similar in their expression patterns in response to PBZ treatment. Therefore, cDNA microarray analysis was shown to be a highly reliable method for monitoring a large number of gene expression profiles simultaneously.

Among the nine selected clones, six have putative identities with known genes for methylenetetrahydrofolate reductase (S12707), β-1,3 glucanase (S03727), S-adenosylmethione (S16157), caffeic acid 3-*o*-methyltransferase (E02880), ribosomal protein (S10124), and receptor kinase-like protein (S12429). Some of these genes are likely to be related to the defense reaction, including lignification and antimicrobial functions. Others may be disease-resistance-related genes. The putative functions of the remaining three clones are not known; it would be interesting to introduce these genes into plants to ascertain whether they have any effect on the defense reaction.

We further tested some of these clones as probes during rice blast fungus infection. Three clones (S03727, S10124, and S12429) showed increased expression under the conditions used (Shimono et al., 2003). This finding suggests that the SAR induced by PBZ shares some pathway with that of plant responses to pathogens. We performed additional experiments using the 8,987-clone and 2,884-clone cDNA microarrays with cDNA clones from PBZ-treated rice leaves. The expression profiles of more and different genes (but including *PBZ1* and *PRP1*) were changed (positively and negatively) after PBZ treatment (data not shown).[9]

Brassinolide-Enhanced Genes in Rice Seedlings

Brassinosteroids (BRs) are naturally occurring plant steroids with structural similarities to insect and animal steroid hormones. Exogenous application of BRs to plant tissues evokes various growth responses, such as cell elongation, proliferation, differentiation and organ bending, and a number of other physiological processes. The bending of the second leaf and its leaf sheath (lamina joint) in rice is very sensitive to the concentration of BRs. This unique characteristic of rice leaves has been used as a quantitative bioassay for BRs. To begin assessing brassinolide (BL)-induced changes in gene expression by means of the cDNA microarray, we adopted the lamina inclination system.

At first, the overall gene expression pattern in the lamina joint was assessed using a cDNA microarray containing 1,265 independent rice genes randomly selected from 9,600 ESTs. The array was hybridized with Cy5-labeled first-strand cDNA prepared from a lamina joint. Among the 1,265 clones, 30 (2.4%) are in the highly expressed group, and as expected, many of them are housekeeping genes. For example, the transcripts of genes homologous to those for ubiquitin or its precursor or RNA-binding proteins were detected at high levels in the lamina joints. The number of genes expressed at a modest level is 132, representing 10.4% of the total. Most of the expressed genes (84.3%) had low abundance, whereas the other 37 (2.9%) genes were undetectable in the lamina joint, indicating that these genes may be expressed at much lower levels in the lamina joint or in a limited number of other tissues.

To identify genes responsive to BL, the aforementioned cDNA microarray system was used to monitor changes of gene expression in the lamina joint after treatment with 1 μM BL for 6, 12, or 24 h. The arrays were hybridized with Cy5-labeled probes of BL-treated and control lamina joint, respectively. Data analysis revealed that the expression of 12 of the 1,265 arrayed cDNA clones was enhanced (more than twofold difference compared with control) by BL treatment. Among these 12 clones, five showed homology to known genes on the basis of a search of the GenBank database using the BLAST program. One of these five clones was a vacuolar H^+-transporting ATPase homologue, which showed increased expression in the BL-treated lamina joint, suggesting a role in BL-mediated cell expansion. The other four clones showed homology to an ACC oxidase-related protein, a putative kinetochore protein homologue (a protein involved in the mitosis machinery), the p23 chaperon, and a ubiquitin-conjugating enzyme involved with protein metabolism.

The remaining seven of the 12 clones we identified through the DNA microarray analysis had no noteworthy homology to any entry in the database. Among these seven, we have identified and cloned two novel BL-enhanced genes, designated as *OsBLE1* and *OsBLE2*. Northern blot results showed that these two genes were most responsive to BL in the lamina joint and leaf sheath in rice seedlings. Auxin and gibberellins also increased their expression. *In situ* hybridization revealed that these two genes were highly and similarly expressed in vascular bundles and root primordia, where the cells are actively undergoing division, elongation, and differentiation. Transgenic rice lines expressing antisense *OsBLE1* or *OsBLE2* exhibited various degrees of repressed growth. BL fails to enhance the expression of *OsBLE1* and *OsBLE2* in transgenic rice expressing antisense BRI1, a BR receptor. This finding suggests that BR signaling for the enhanced expression of *OsBLE1* and *OsBLE2* is mediated through BRI1. Together, our results suggest that *OsBLE1* and *OsBLE2* may be involved in BL-regulated growth and development processes in rice.[10]

Screening of Gibberellin/Brassinosteroid-Responsive Genes in Rice Seedlings

Our objective was to identify genes whose expressions are regulated by the interaction of multiple plant hormones and to reveal the functions of the encoded proteins. Each process of the plant life cycle is controlled by a complex series of plant hormone signal transduction reactions. Mainly gibberellin (GA) and auxin have been studied as regulators of shoot elongation. Recently, mutational analysis has demonstrated the additional importance of brassinosteroid (BR) for regulating stem elongation. The interaction of these hormones has been investigated only at the level of physiological response, not at the level of the crosstalk of their signal transduction. We focused on identifying genes whose expressions in rice seedlings are regulated by the combination of GA and BR, either additively or synergistically.

Rice seeds were sterilized in 0.1% Benlate for 15 min. After being washed with water, they were incubated in water for three days at 27°C under continuous light. The germinated seeds were planted on 0.8% agar containing uniconazole (Uni), a GA biosynthetic inhibitor; brassinazole (Brz), a BR biosynthetic inhibitor; or both. After they were incubated for two days at 27°C under continuous light, GA_3 or brassinolide (BL) or both were applied to the seedlings by the microdrop method. The optimal quantities of Uni and GA_3 were determined to be 0.1 μM and 10 ng/plant, respectively. Because neither Brz nor BL showed any effects on shoot elongation in this assay system, we used 1 μM Brz and 10 ng/plant BL, as has been reported for *Arabidopsis*. Total RNA was isolated from the aerial part of the rice seedlings 6 and 24 h after treatment with GA_3 or BL. The RNA samples were fluorescently labeled and used in microarray experiments.

We identified a total of 166 clones as GA/BR-responsive genes. Those clones could be categorized into many different groups according to their response patterns. One of the largest groups consisted of clones whose expression levels were upregulated only in the presence of both GA_3 and BL. When we allocated the 166 clones into two groups in light of their expression levels, clones whose expression levels were downregulated by GA_3 treatment predominated among those showing high expression after combined treatment with GA_3 and BL. In comparison, the group of clones with low expression after treatment with both hormones primarily comprised those clones whose expression levels were upregulated by GA_3 and those that were downregulated by either GA_3 or BL.

We used northern blotting to confirm the hormone responsiveness of some of the clones selected through the array analysis and roughly established the localization of the expression. The clones that were upregulated by combined treatment with GA_3 and BL included β-expansin, which has been reported to be GA-inducible in deepwater rice. This clone was more responsive to BL than to GA_3 in our system, and the effect of the two hormones was additive. Its expression level was high in the elongating leaf sheath and roots. Other clones in the same expression category showed different localizations. A clone with a farnesylation motif was expressed mainly in the elongation zone, and the expression of another clone with an unknown function was limited to the dividing part of the seedling. The clones whose expression levels were upregulated by GA_3 included genes involved in ATP synthesis and in the photosynthetic electron-transfer reaction.[11]

Profiles of Genes Expressed in a Rice Virescent Mutant

In a virescent mutant of rice, the seedling initially grows yellow leaves because of reduced chlorophyll accumulation. The leaves gradually become greener as the plant matures. A mutable virescent line (yl-v) was isolated from the progeny of a hybrid cross between *indica* and *japonica* rice cultivars, and subsequently a nearly isogenic line (yl-stb) in the genetic background of the *japonica* cultivar T-65 was established (Maekawa, 1995). The yl-v plant displayed yellow leaves with green sectors. An apparently stable virescent derivative (yl-stb), showing completely yellow leaves without green sectors spontaneously appeared among the progeny of the yl-v line (Maekawa et al., 1999). Although yl-stb lines typically cannot grow past the four-leaf stage under normal growth conditions, we were able to identify conditions under which yl-stb routinely grows and sets seeds. The yl-stb plants showed a weak cold-sensitive phenotype: yl-stb grown at 18°C bears yellow leaves, whereas at 33°C it produces pale green leaves. For profiling the genes expressed in yl-stb or T-65 using microarray analysis, the plants were grown under strictly controlled environmental conditions, and RNA was isolated from the upper part of four-leaf-stage seedlings. Analysis with the 8,987-clone microarray revealed that the expressions of genes related to photosynthesis, such as carbonic anhydrase-3 and rubisco activase, were reduced in yl-stb. Northern blot analysis confirmed the results obtained through the microarray analysis. About 330 genes showed more than three-fold increased expression in yl-stb compared with T-65 at 18°C, whereas only 95 genes showed increased expression in T-65 compared with yl-stb under the same growth conditions. Interestingly, many of the genes with increased expression in yl-stb were stress-related genes, including those for the Ras-related protein, germin-like protein, and salt-stress-induced protein. The results indicate that the stable virescent mutant yl-stb experiences strong stress conditions even if it can grow to set seeds.[12]

Cell Cycle-Regulated Genes
of Suspension-Cultured Rice Cells

We were interested in identifying genes that are expressed specifically during the S phase of the cell cycle in rice. Suspension-cultured rice cells were synchronized with aphidicolin, an inhibitor of DNA replication. Synchronization of the cells was confirmed by ^{3}H-thymidine incorporation into DNA (S phase) and by the mitotic index (10% of the total cell population, M phase). Cells at each stage of the cycle (S, G2, M, and G1) were collected, and their RNAs were extracted. Cy5-labeled cDNAs were used to hybridize to the cDNA microarray representing 8,987 full-length cDNA inserts. We identified the genes whose expression was more than threefold upregulated or downregulated in each cell cycle stage-specific population compared with the expression in nonsynchronized cells. We focused on the 50% of genes with the greatest signal intensity differences in each population. This choice yielded 48 genes that were upregulated and four that were downregulated during the S phase; 121 upregulated and 47 downregulated during G2; 73 and 19 at M; and 40 and 44 at G1. We found that the so-classified "S phase-regulated genes" included those induced by washing the cells to remove the inhibitor and transferring them to a new medium. The corrected numbers of S phase-upregulated and -downregulated genes were 31 and 20, respectively. Of the 31 upregulated clones, 11 were histones.

A BLAST homology search revealed that our 8,987 cDNA microarray contained several clones that we assumed to be involved in DNA replication, DNA repair, or cell cycle regulation in rice. Among the clones expressed most highly during S phase are genes for proliferating cell nuclear antigen (*OsPCNA*) (Suzuka et al., 1991), origin recognition complex 1 (*OsORC1*) (Kimura et al., 2000a), and the replication protein A 32-kDa protein (*OsRPA32*) (Ishibashi et al., 2001). However, others were expressed most highly at other phases or showed no distinct cell cycle-associated difference in expression. These clones include genes for flap endonuclease-1 (*OsFen-1*) (Kimura et al., 2000b), *cdc2Os-1* (Hashimoto et al., 1992), and cyclin-dependent kinase-activating kinase R2 (Hata, 1991).[13]

Overexpression of the Cyclin B Gene

The activation of cell division and transitions between the phases of the cell cycle are controlled by a family of cyclin-dependent serine-threonine protein kinases (CDKs) and their cyclin partners. We have shown previously that rice genes for B2-type cyclins (*cycB2;1* and *cycB2;2*) are expressed only from the G2 to M phase, indicating that they function as mitotic cyclins (Umeda et al., 1999). Doerner et al. (1996) have reported that ectopic expression of *cycB1;1* in *Arabidopsis* markedly accelerated root growth without inducing neoplasia. This finding prompted us to express B2-type cyclins in rice plants to activate cell division and to investigate the change in gene expressions associated with acceleration of the cell cycle.

Using the glucocorticoid-mediated transcriptional induction system, we generated transgenic plants that expressed mRNA for *cycB2;2*. In this system, the glucocorticoid derivative dexamethasone (DEX) activates the transcription factor GVG, which in turn induces *cycB2;2* expression (Aoyama and Chua, 1997). We isolated three transgenic lines that expressed *cycB2;2* in a DEX-dependent manner. When phenotypic changes were observed in the presence of 1 μM DEX, root growth accelerated, suggesting that increased expression of *cycB2;2* led to activation of cell division in the root meristem (M. Umeda, unpublished results).

In transgenic seedlings, the mRNA level of *cycB2;2* varied after induction with DEX; increased transcript levels were observed after 12 and 24 h. Therefore, we isolated total RNA from seedlings before (0 h) and 12, 20, 24, and 48 h after treatment with 1 μM DEX. The total RNA was labeled with

Cy5 and hybridized to the rice 8,987-clone microarray. In the glucocorticoid-inducible system, transgene expression is mediated by the transcription factor GVG; thus, fluctuations caused by GVG itself are false positive responses. Taking this point into consideration, we selected a vector control line (V7) in which GVG is constitutively expressed under the control of the cauliflower mosaic virus 35S promoter. First, we compared gene expression levels between wild-type and V7 seedlings to extract clones that were regulated by GVG in the absence of DEX. Second, we monitored gene expression in V7 12, 20, and 24 h after 1 μM DEX treatment and compared them with that at 0 h (no treatment). This comparison enabled us to identify clones that were up- or downregulated by GVG in the presence of DEX. Finally, we assessed the gene expression levels in the *cycB2;2* transgenic line as we did for the control line, and subtracted from these clones those regulated by GVG in the absence or presence of DEX.

Genes whose expression was upregulated by *cycB2;2* expression included several kinds of receptor protein kinases and cytosolic factors associated with the ubiquitin pathway and hormonal signaling. Although the mRNA level of *cycB2;2* increased 12 and 24 h after DEX treatment but decreased after 48 h, almost all of the clones that were most highly upregulated showed the highest transcription levels at 24 and 48 h. The list of downregulated genes included enzymes related to alcohol fermentation, such as pyruvate decarboxylase and alcohol dehydrogenase. This finding may indicate that the TCA cycle rather than alcohol fermentation functions to afford efficient generation of ATP during the activation of cell division. Additional rounds of hybridization using target RNA from other transgenic rice lines will lead us to further analysis of the coordinate expression of genes associated with the activation of cell division.[14]

Starch Biosynthesis

It is well known that many plants switch from vegetative to reproductive growth in proportion to various environmental or programmed stimuli. These stimuli induce the expression of genes involved in the initiation of floral meristem, flowering, and seed development, resulting in great changes in the physiological and morphological features of the plants.

Until now, mutants for various stages of floral transition have been isolated in dicotyledoneous plants, such as *Arabidopsis*. By use of these mutants, the molecular mechanisms underlying this process in dicotyls have been investigated extensively. In contrast, the mechanism of flower induction in monocotyledoneous plants remains to be clarified, although its similarity to those of dicotyledoneous plants is likely.

The molecular mechanisms of seed development have been well analyzed in monocotyledoneous plants. It has been shown that specific changes in gene expression occur when the embryo and endosperm are formed immediately after fertilization. In the endosperm, the expressions of a large number of genes involved in the biosynthesis of storage substances are induced. Sucrose synthetase is considered a key enzyme that catalyzes the conversion of sucrose and UDP to UDP-glucose and fructose, because sucrose transported from source organs is initially metabolized by the activity of this enzyme (Winter and Huber, 2000). It has also been reported that the activity of sucrose synthetase is controlled through phosphorylation by the calcium-dependent protein kinase SPK. When SPK activity is absent in transgenic plants harboring an antisense *SPK* gene, abnormal storage starch inevitably occurs in immature seeds (Asano et al., 2002).

The biosynthesis of storage starch in the rice endosperm reaches a maximum at two weeks after pollination. The product of the *flo2* gene is a regulatory factor involved in starch biosynthesis. A *flo2* mutant shows severely reduced expression of the starch branching enzyme gene and several related genes, resulting in alteration of the physical properties of storage starch (Kawasaki et al., 1996).

We analyzed gene expression during the stages of panicle formation and early and middle seed development by using a mutant of panicle formation, the antisense *SPK* transformant, and the *flo2* mutant, respectively. The gene expressions of these mutants were comprehensively compared with those of the wild-type plant using the microarray system.

The mutant of panicle formation showed infinite inflorescence branches on the panicles and set no flowers. Gene expression in immature panicles of the mutant was compared with that of wild-type plants. This analysis detected 14 genes whose expression was severely decreased in the mutant and 18 genes with increased expression. RT-PCR analyses of these genes confirmed the results of the microarray experiments. Among the identified clones was a homologue of a filamentous flower gene. Because the filamentous flower gene in *Arabidopsis* defines the fate of abaxial floral cells (Sawa et al., 1999), we assume that this rice gene is involved in this phenotype. This analysis also yielded eight novel genes showing remarkable changes in expression, whose functions are as yet unclear.

A microarray analysis on seeds in early seed development from the antisense *SPK* transformant and wild-type rice showed strongly reduced expression of 21 genes in the transformant and increased expression of 44 genes. RT-PCR analyses confirmed the microarray results for 11 of these 65 genes. Most of the genes exhibiting markedly altered expression in the transformant are involved in starch biosynthesis and sugar metabolism. Because it is evident that this phenotype is responsible for lacking SPK activity that may reg-ulate sucrose metabolism (Asano et al., 2002), there are probably other effects of this limited gene expression.

The microarray analysis of the *flo2* mutant was carried out using the im-mature seeds at ten days after flowering. Of the 224 genes whose expres-sion was decreased compared with the wild-type ones, 45 genes that are considered to be involved in the transcription, translation, and synthesis of storage products were further analyzed using RT-PCR. This analysis re-vealed reduced expression of genes for several key enzymes: starch branch-ing enzyme, which is involved in the accumulation of storage starch (Preiss, 1991); ADP-glucose pyrophosphorylase, which is involved in the storage of phosphate (Yoshida et al., 1999); and alanine aminotransferase, which is involved in nitrogen metabolism (Kikuchi et al., 1999). This analysis also revealed decreased expression of genes for several storage proteins. These results strongly suggest that the product of the *flo2* gene is a multifunctional regulatory factor that plays diverse important roles in the accumulation of various storage substances, such as storage starch, proteins, phosphate, and amino acids.[15]

Sink-Source Transition in Rice Leaf Sheaths

In rice (*Oryza sativa* L), the sheaths of the upper leaves accumulate a large amount of starch before heading, and the accumulated starch is con-verted to sucrose and translocated to the panicles after heading. To analyze the regulation of this sink-source transition, we performed large-scale mon-itoring of gene expression using microarray analysis. The sheaths of the first leaves were sampled from 14 days before heading to 14 days after and classified into seven developmental stages (stages 1 to 3, before heading; stage 4, heading; and stages 5 to 7, after heading) using 3 parameters (the length of the panicle, length of the neck internode, and fresh weight of the panicle). The starch content of these leaf sheaths increased during stages 1 to 3 and decreased during stages 5 to 7.

The levels of expression of the 8,987 rice ESTs in the microarray were compared among seven developmental stages (stages 1, 3, 5, and 7), and 93 genes showed more than a fivefold difference in mRNA abundance. Iden-tified genes included multiple genes involved in starch biosynthesis, cell division and expansion, or photosynthesis. All of the genes in these subsets showed preferential expression during early stages, probably reflecting head-ing-associated decreased starch biosynthesis, the completion of elongation, and the onset of photosynthesis in the first leaf sheaths. In contrast, we did not identify any developmentally regulated expression of genes related to starch degradation, sucrose synthesis, or sucrose transport.[16]

Studies on Grain Filling

The objective of our study was to identify the genes involved in grain filling in rice. For this purpose, we adopted a mutant derived from Nourin-8 (N8); the grain-filling ability of this mutant was very poor compared with that of N8. We also compared possible superior spikelets (SS), which were derived from upper primary rachis branches and may show better grain filling, with inferior spikelets (IS), which were derived from secondary rachis branches on lower primary rachis branches and may show poor grain filling, from both the mutant and N8 lines. Each spikelet was sampled 1, 7, and 13 days after flowering (DAF), mRNA was prepared from these samples, and the microarray system was used to identify genes whose expression differed according to spikelet properties (SS or IS), developmental stage (1, 7, or 13 DAF), and line (N8 or mutant). However, these 3-way analyses yielded an excessive number of genes whose expression varied with at least one parameter. To reduce the number of candidate genes, we first analyzed the expression of genes that were known to be involved in carbohydrate metabolism and that were definitely related in some way to grain filling. We then searched for other genes that showed similar expression patterns to those key genes of carbohydrate metabolism. This process yielded four groups of carbohydrate metabolism genes with differing expression patterns.

Group A included the genes for invertase, hexokinase, glucose-6-phosphate isomerase, granule-bound and soluble starch synthases, β-amylase, sucrose synthase, and a sugar transporter. For this group of genes, the level of expression typically decreased as spikelet development progressed. Furthermore, their expression in IS tended to be delayed compared with that in SS, resulting in higher signal levels in IS during later stages. However, some genes showed distinct expression patterns. For example, one putative sugar transporter gene was expressed much later than was the sucrose transporter gene *OsSUT1*. This novel transporter may be active mainly after *OsSUT1*.

Another interesting group A gene was ADP-glucose pyrophosphorylase (AGPase), for which the expression patterns of the large and small subunit genes differed. The large subunit gene showed the expression pattern typical of group A, in that expression in IS was delayed. However, the expression levels of the small subunit in IS were uniformly low. This finding suggests that the gene expression of the small subunit of AGPase in IS limits grain filling.

Group B genes included branching enzyme, for which expression increased according to spikelet development. Moreover, its expression levels were higher in SS than in IS. This difference may result in the different grain-filling abilities of the spikelets.

Group C genes included the gene for alpha amylase. The expression levels of Group C genes varied, but their most important characteristic was that the level of expression in IS was higher than that in SS. Alpha amylase catalyzes starch degradation; therefore, this enzyme may inhibit the accumulation of starch, resulting in the poor grain filling of IS.

We then classified additional genes according to this schema. Group A now has 117 candidate genes, Group B has 104, and Group C has 48. We are proceeding to analyze these genes as well as other interesting genes that show intriguing expression data.[17]

Anther- and Pistil-Specific Genes

The developmental programs of the male and female reproductive organs of rice are regulated by complicated genetic pathways. To identify the genes preferentially expressed during the development of anthers and pistils, we screened 1,265 EST clones using cDNA microarray techniques.

To monitor the gene expression patterns during the development of the sexual reproductive organs, we isolated anther and pistil tissues from two *japonica* cultivars, *Koshihikari* and *Hitomebore,* at different developmental stages according to the number of the nuclei per cell in the male gametophytes: unicellular microspore, bicellular pollen, and tricellular mature pollen stages (Tsuchiya et al., 1992). mRNAs from these six samples were directly isolated and used in first-strand cDNA synthesis. cDNAs derived from the anther and pistil tissues were labeled with Cy3 and Cy5, respectively, and used in hybridization reactions. The hybridization signals were used to identify genes whose expression was specific to sexual reproductive organs.

Of the 1,265 clones analyzed, 37 were abundantly expressed genes in anther tissues, and 6 were highly expressed in pistil tissues. The expression patterns of these 43 clones did not differ between the two *japonica* cultivars (*Koshihikari* and *Hitomebore*). This result suggests that the sequences and expression patterns of these reproductive organ-specific genes were relatively conserved among the cultivars.

Among the 37 clones highly expressed in anthers, the functions of 18 were not assigned, because they did not have any similarity to known sequences. Among the remaining 19 clones, genes similar to those for dihydroflaconol-4-reductase (DFR) and phenylalanine-ammonia-lyase (PAL) were expressed in anther tissues at the unicellular microspore stage. The known rice anther-specific genes expressed abundantly at this stage, such as *Osc4* and *Osc6* (Tsuchiya et al., 1994) and *YY1* and *YY2* (Hihara et al., 1996), have been characterized as tapetum-expressed genes. However, their functions have not been determined. The DFR and PAL genes are involved in the flavonoid synthetic pathway, which is essential for the development of viable pollen grains (Wiermann, 1970); and antisense inhibition of *PAL* causes male sterility in transgenic tobacco plants (Matsuda et al., 1996). Therefore, the two genes we identified might be expressed in the tapetum, and they also may be important for normal male gametogenesis in rice. Furthermore, two clones were highly expressed in anther tissues at the tricellular mature pollen stage; these clones had similarity to the genes for the *Cynd7* pollen allergen (calcium binding protein) from Bermuda grass and the *PhlpII* pollen allergen from timothy grass, respectively. *Cynd7* and *PhlpII* were highly expressed in these grasses, indicating that these two genes might have important roles in pollen grain maturation in the grass family (Poaceae). We were unable to assign putative functions to any of the six clones highly expressed in pistil tissues.

Using the cDNA microarray comprising 1,265 ESTs enabled us to identify only 43 clones (3.4%) that were abundantly expressed in the reproductive organs of rice. The reason for this low efficiency may be due to the fact that the vast majority of the EST clones represented in the microarray were derived from vegetative organs and calluses. This low efficiency suggests the importance of preparing a cDNA microarray whose clones are derived from specific (in this case, reproductive) organs, as has been reported for the legume *Lotus japonicus* (Endo et al., 2002).[18]

STUDIES USING THE MICROARRAY SYSTEM ON PLANTS OTHER THAN RICE

Iron Deficiency Stress in Barley

Fe is an essential nutrient for plant growth and crop productivity. Although Fe is abundant in mineral soils (>6%), under aerobic conditions at the physiological pH range, Fe is only sparingly soluble and not available to plants. Plants have therefore developed sophisticated and tightly regulated mechanisms for acquiring Fe from the soil (Mori, 1999). Grasses secrete Fe chelators, called mugineic acid family phytosiderophores (MAs), from their roots to solubilize Fe in the soil (Takagi, 1976). The resulting Fe^{3+}-MAs complexes are then reabsorbed into the roots through a specific transporter

(Curie et al., 2001) in the cell membrane. The production and secretion of MAs increase markedly in response to Fe deficiency, and tolerance to Fe deficiency in grasses is strongly correlated with the quantity and quality of the MAs secreted. Three molecules of methionine are used for the synthesis of 1 MA molecule (Mori and Nishizawa, 1987; Shojima et al., 1990). To date, we have isolated the genes encoding all the enzymes participating in the biosynthetic pathway of MAs (Higuchi et al., 1999; Kobayashi et al., 2001; Nakanishi et al., 2000; Takahashi et al., 1999). However, other genes that may participate in plants' tolerance to low Fe availability remain to be defined. For example, genes encoding factors that sense intracellular levels of Fe, transcriptional activators for regulating gene expression in response to Fe deficiency, and components of signaling pathways to monitor Fe status in the environment have not yet been identified. Furthermore, the molecular mechanism of MA secretion remains unclear. Neither the genes nor the proteins responsible for the transport of MAs to the outside of cells have been identified. MA secretion in barley shows a distinct diurnal rhythm (Takagi et al., 1984). A secretion peak occurs just after initial illumination and ceases in 2-3 h. It has been proposed that the particular vesicles that appeared in root cells of Fe-deficient barley are the sites of MA synthesis (Nishizawa and Mori, 1987). We are interested in elucidating the sequence of events that link the biosynthesis of MAs to their diurnal secretion.

We used a cDNA microarray containing 8,987 rice EST clones to analyze the gene expression profile in barley roots during Fe deficiency (Negishi et al., 2002). Genes of cereal crops tend to be highly conserved at the DNA sequence level (Devos and Gale, 2000), and this conservation allows the use of heterologous probes to identify orthologous DNA sequences in different species in DNA hybridization experiments. As a result, we were able to identify approximately 200 genes of which the expression levels were enhanced by Fe deficiency. To meet the increased demand for methionine to produce MAs, Fe deficiency enhances the expression of genes that participate in methionine synthesis, as well as recycling methionine through the Yang cycle. Of these 200 genes, approximately 50 exhibited different transcription levels in Fe-deficient roots at noon and at night. Northern blot analysis of time-course experiments confirmed that five of these genes exhibited a diurnal change in their level of expression. The diurnal changes in the expression of these genes suggest that polar vesicle transport is involved in the diurnal secretion of MAs.[19]

Gene Expression Regulated
by Cold Acclimatization in Wheat

Cold-tolerant or overwintering plants can induce freezing tolerance after a period of exposure to low but nonfreezing temperature. This cold acclimatization is one of the adaptive responses in plants against cold or freezing temperatures. We have analyzed cold-regulated genes to study cold acclimatization in winter wheat. During the acclimatization process, complex changes, including numerous modulations of gene expression, are mediated by independent signal transduction pathways. To analyze the processes involved in the adaptation to cold and freezing stresses, we have undertaken a large-scale analysis of the wheat transcriptome during cold acclimatization using a rice cDNA microarray.

We have isolated total RNA from control and cold-treated wheat and rice plants. After sowing, seeds of winter wheat (*Triticum aestivum* L. cv. *Valuevskaya*) were kept in the dark for two days at 25°C, and then at 20°C during the day and at 15°C during the night for seven days. Rice plants (*Oryza sativa* L. cv. *Nipponbare*) were grown at 28°C for ten days. Control plants were maintained under the same conditions. For cold treatment, the plants at the same stage as the controls were exposed to 4°C for one day. The RNA samples from the various plants were transcribed, labeled with Cy5-dCTP, and hybridized to rice cDNA microarrays containing 8,987 individual clones. The results of these heterologous experiments demonstrated that the hybridization with cDNA targets derived from wheat gave pronounced and informative signals, suggesting that the rice cDNA microarray is suitable for gene expression profiling in wheat. In control plants, approximately 3.6% of the total cDNA clones were differentially expressed between the two species, whereas in cold-treated plants these numbers were increased to approximately 7.6%. In the comparison between control and cold-treated plants, approximately 2.3% of the 8,987 clones in rice and 1.8% of those in wheat were markedly regulated by cold. It is interesting that many of the cDNAs showing cold-dependent expression revealed different regulation between rice and wheat under the cold conditions. Our results suggest that the rice cDNA microarray is useful for gene expression profiling of other grass species, and that those profiling data also provide comparative information on the regulation of gene expression in grass species.

In addition to *Valuevskaya* (which is highly tolerant to freezing), we used the spring wheat cultivar Haruyutaka (with poor freezing tolerance). Comparing the gene expression patterns of these cultivars showed clear differences after exposure to cold. As a result, 85 cDNA clones were differentially expressed between the two cultivars at a low temperature. Among these differentially expressed cDNAs, 36 clones were up- or downregulated by cold treatment. Of these, 22 were expressed at least fivefold higher in *Valuevskaya* than in *Haruyutaka,* and the rest were higher in *Haruyutaka.* Although approximately 40% of 36 selected cDNAs showed no significant homology to proteins with known function, this list included potential regulatory factors related to signal transduction, such as protein kinases, a calcium-binding protein, and transcription factors. These microarray analyses have provided the profiling data for the initial characterization of gene expression patterns involved in early events during cold acclimatization and will enable us to identify candidate genes that play pivotal roles in this process.[20]

Post-Anthesis in Wheat

Rice microarrays are potentially useful for monitoring the gene expression of other plants, particularly monocotyls. The first-stage rice microarray comprised only 1,265 clones, most of which came from callus cDNA libraries. We were interested in monitoring changes in mRNA expression in the developing wheat seed, and we could detect few differences in expression patterns when labeled cDNAs derived from such samples were hybridized to the rice microarray. Therefore, a library was constructed from wheat seed collected 0 to 40 days post-anthesis (DPA), and 1,200 randomly selected clones were amplified using amino-primers and bound to 3D-Link amine-binding slides (SurModics). PolyA$^+$ mRNA was extracted from 0, 2, 5, and 10 DPA immature seeds and from 10, 15, 20, 30, and 40 DPA embryos. First-strand cDNA was labeled with Cy3 or Cy5, and hybridization was conducted at 42°C overnight. After being washed, microarray slides were scanned by using a ScanArray scanner (Packard BioScience). Results were normalized by using five internal control genes or by global normalization methods, depending on the experiment.

Of the 1,200 clones arrayed, four represented β-amylase cDNAs. Wheat processes two types of β-amylase, one of which is expressed ubiquitously, including in pericarp tissue; the second is expressed chiefly in the endosperm. The two types have high sequence homology at the cDNA level, but the ubiquitous type lacks a glycine-rich C-terminal sequence. The expression of both cDNAs has been reported to increase during early seed development. Therefore, this gene is a good model for determining the accuracy of our microarray experiments.

The microarray data indicated that the expression of the gene increased more than twofold at each of the comparison points among 0 to 2 DPA, 2 to 5 DPA, and 5 to 10 DPA, and an approximately 30-fold higher expression level was detected in endosperm and pericarp tissue than in embryo tissue of 10 to 15 DPA immature seed. These data were consistent with those of previous reports, and all four clones showed the same expression pattern. From these results, we concluded that our array system monitored gene expression quite accurately. However, further experiments using northern blot or real-time PCR analysis demonstrated consistently larger changes in gene expression than those determined from microarray data. Although this finding suggests that our results are on the conservative side, choosing a cutoff of a twofold increase or decrease in expression likely will have eliminated false positives.

Our use of microarrays to monitor gene expression in developing embryo tissue clearly showed that the expression of stress-related, ABA-induced, pathogenesis-related, and translation initiation factor genes—as well as many transcripts with homology to rice genes encoding products with as yet unknown functions—increased dramatically during embryo development, especially after 30 DPA. Biologically, the gene expression patterns provided information on genes potentially involved in the acquisition of desiccation tolerance and preparation for dormancy. Although only 1,200 clones were screened, this was a sufficient number to provide new data on gene expression, largely because the clones came from the tissue we wished to assess.[21]

Crown Rust Resistance in Italian Ryegrass (Lolium Multiflorum *Lam.*)

Italian ryegrass (*Lolium multiflorum* Lam.) is an important grass species widely used for agricultural purposes in Japan, and is appreciated for its high nutritive value to herbivores. Crown rust, caused by the fungal pathogen *Puccinia coronata* f. sp. *lolii,* can reduce the yield, palatability, and nutritional quality of grasses and is recognized worldwide as one of the most serious diseases of plants of the genus *Lolium* (Kimbeng, 1999). *Yamaiku 130*, an Italian ryegrass line grown in Japan, shows a high level of resistance to crown rust. Although resistant cultivars are being bred in many countries, there is a paucity of information on the molecular characterization of crown rust resistance in Italian ryegrass. The microarray technique is a powerful tool for simultaneously analyzing the expression of thousands of genes. To elucidate the molecular events and genetic mechanism involved in the resistance to crown rust, we are using this technology to analyze the gene expression profile of the resistant line *Yamaiku 130* after crown rust infection.

We constructed three cDNA libraries using mRNA from *Yamaiku 130* leaves inoculated with spores of crown rust to obtain a comprehensive set of genes associated with disease resistance. Total RNA was prepared from leaves using Trizol (Gibco-BRL), and mRNA was prepared using the Oligotex-dT30-Super mRNA Purification kit (TaKaRa). A total of 11,616 clones from these libraries were purified using the Biomeck 2000 system (Beckman). cDNA fragments were amplified by PCR and purified by using MultiScreen FB (Millipore). We randomly selected 9,216 clones for micro-array analysis.

Spores of crown rust were mixed with talc and used to inoculate the plants by spraying. The plants then were placed in an inoculation box (20°C; high humidity). Control plants were treated in the same way but without the ad-dition of spores. Total RNA was extracted from leaves harvested 0, 0.5, 1, 2, 8, 12, and 24 h after treatment; the RNA samples were immediately frozen in liquid nitrogen. mRNA was purified and then labeled with Cy3, and the FLA-8000 (Fujifilm) was used for detection of signals. For the data analy-sis, spot intensities from scanned slides were quantified using ArrayGage software (Fujifilm).

A redundant 9,216-cDNA microarray was used for a time course analy-sis of the gene expression of the defense response after crown rust infection. Our results showed that the expression of 106 unique clones was induced by crown rust infection of *Yamaiku 130* at at-least one time point, and 125 clones showed reduced expression. Furthermore, 32 of the 106 upregulated clones and 58 of the 125 downregulated ones were differently expressed at 8 h after treatment. Some of the 231 clones identified included genes potentially in-volved in the hypersensitivity reaction. Other clones represented novel genes. Our findings suggest that time-course profiling using the microarray tech-nique provides valuable information on the molecular mechanism of crown rust resistance in Italian ryegrass. Comparison of the changes in the transcript levels of *Yamaiku 130* with those of susceptible or other resistant lines will be likely to yield additional interesting clones for more detailed analyses, such as transformation and linkage analysis.[22]

Apomixis in Guineagrass

Apomixis provides a method for cloning plants through the seed. The ad-vantage of apomixis for commercial seed production include the propaga-tion of hybrid seeds showing heterosis and the development of permanently useful genetic combinations through hybridization. Guineagrass (*Panicum maximum* Jacq.) is a leading forage grass in tropical and subtropical re-gions. Apomixis of guineagrass is classified as gametophytic aposporous apomixis—asexual reproduction through seeds that arise from a somatic cell of the nucellus. In nature, most apomictic plants of this species are tetraploid, whereas sexual plants are diploid. In our research, tetraploid sexual lines have been generated by chromosome-doubling methods. These tetraploid sexual lines play an important role in our research. Guineagrass is a useful species for mapping apospory using molecular markers and for cDNA analysis, because the reproductive methods are easily identified by embryo sac analysis. AFLP linkage analysis has indicated that apospory is controlled by a single gene. In guineagrass, the aposporous apomictic embryo sacs arise from somatic cells instead of the basal megaspore. To identify the gene(s) related to apomictic development, it is necessary to detect differences between apomictic and sexual plants by using gene profiling of initial embryo sac development (Ebina and Nakagawa, 2001; Nakagawa and Ebina, 2001).

We constructed 4,896 expressed cDNAs of immature inflorescences of apomictic guineagrass. Among these clones, 4,128 were cDNAs of apomixis progeny crossed between a sexual guineagrass and an apomictic cultivar, and 768 cDNAs came from an apomictic accession. Among these cDNAs, 288 came from libraries of the apomictic plants subtracted from the sexual plants.

During the first microarray screening, probes were prepared from immature inflorescences at five developmental stages and from pistils at three stages of an apomictic guineagrass accession. At the apomixis embryo initiation stage, we used the low-stringency criterion of a twofold change in expression to select 1,139 clones. During the second screening of the microarray, we used these 1,139 clones to compare mRNA expression levels from immature inflorescences between the two-sib apomixis progeny and the maternal sexual line. We then selected the clones whose expression levels were at least fivefold higher and that occurred commonly in the two-sib apomixis progeny. In our third analysis, we performed expression profiling among different tissues using the same 1,139-clone microarray as for the second screening and chose those clones especially expressed in immature inflorescences.

Finally, the results of microarray profiling of immature ovaries and inflorescences of apomictic and sexual progeny indicated that approximately 60 genes are specifically expressed at the developmental stage of the immature apomictic inflorescence. These genes included 29 unknown sequences that could not be characterized through a BLASTX homology search among all available databases. Another ten of these clones were homologous to signal transduction genes, including histidine kinase. These genes may play a role in the signal transduction for apomixis initiation in the nucellus cell. Furthermore, these candidate genes can be converted into linkage markers, such as CAPS or RFLP, by using the AFLP linkage frame-mapped apomixis locus of guineagrass.[23]

Isolation of Reproductive Organ-Specific Genes from Lotus japonicus

To gain understanding of the molecular mechanisms of sexual organ development in plants, we used cDNA microarray technology to isolate clones that were specifically expressed during the development of the sexual organs (anther and pistil) in the model legume *Lotus japonicus*. We first constructed two cDNA libraries derived from immature and mature flower buds and obtained about 1,000 ESTs from each library. Among these EST clones, we identified those that were homologous to pollen-specific genes (e.g., pectin methylesterase, ascorbate oxidase, polygalacturonase, and pollen allergen protein) in other plant species. Comparison of these EST sequences with those derived from the entire plant of *L. japonicus* revealed that more than 50% of the EST sequences from the immature and mature flower buds were not among those from the whole plant. This finding indicates that the EST clones from the flower buds were useful in evaluating patterns of gene expression in floral organ development in *L. japonicus* (Endo et al., 2000).

We then generated a cDNA microarray comprising 4,048 cDNA clones (1,817 EST clones and 2,231 non-EST clones) derived from the two flower-bud libraries. We separately collected samples of anther and pistil tissues according to bud length (stages 1, 2, and 3). After purification of mRNA from each sample, mRNAs derived from anther or pistil tissues at different developmental stages were labeled with Cy5. In contrast, mRNAs from leaf tissues (a vegetative organ) were labeled with Cy3. Using the cDNA microarray, we compared the expression patterns of anther and pistil (stages 1, 2, and 3) with that of leaves of *L. japonicus*. We selected those clones whose expression was at least sixfold upregulated in the anther or pistil tissues. We ultimately identified 22 independent cDNA clones that were specifically or predominantly expressed in immature anthers, and 111 independent cDNA clones that were specifically or predominantly expressed in mature anthers. Among the 111 clones from mature anthers, we identified clones homologous to genes related to cell wall reorganization (pectin methylesterase, ascorbate oxidase, and β-galactosidase), cytoskeleton (actin), and sugar metabolism (sucrose transporter, glucose transporter, and cell wall invertase). Approximately half of these cDNA clones showed no significant similarity to known sequences (BLASTX score < 100) or fell in the category of proteins with unidentified function. We also identified 23 independent cDNA clones that were specifically expressed in pistils. To determine the accuracy of the DNA microarray data, we performed RT-PCR analyses of several representative anther-specific clones. These results confirmed the accuracy and reproducibility of our approach. cDNA microarray technology is thus a powerful tool for the identification of novel reproductive organ-specific genes (Endo et al., 2002).[24]

Genes Expressed During Sexual Reproductive Organ Development in Ipomoea trifida

To understand the molecular mechanisms underlying the sexual reproductive organ development of plants, we isolated thousands of EST clones from the anthers and pistils/stigmas of rice *(Oryza sativa* L. cv. *Koshihikari)* and wild-type sweet potato *(Ipomoea trifida)* and characterized the expression patterns of these clones using cDNA microarray technology.

We first constructed cDNA libraries derived from anthers and pistils of several developmental stages, which were determined by the number of cells in the microspores or pollen grains. From *I. trifida,* we obtained 640 ESTs from mature anthers and 998 ESTs from mature stigmas; from *O. sativa,* we obtained and partially sequenced 1,002 ESTs from mature pistils, 2,016 anther-derived non-ESTs, and 2,304 pistil-derived non-ESTs from mixed libraries from the unicellular microspore stage to the tricellular pollen stage (Tsuchiya et al., 1992). Comparison of these ESTs with reported DNA sequences in the public DNA database revealed that less than 30% of the ESTs showed similarity to reported function-identified genes; we could not assign putative functions for the products of the remaining clones, because they showed either similarity only to reported EST or genomic sequences or no similarity to any known sequence. These results suggest that there are many genes still unidentified that are important to floral reproductive organ development.

Using these ESTs and non-ESTs as probes, we performed cDNA microarray analyses by using reproductive organ-derived targets. To prepare the targets, mRNAs were purified from anthers and stigmas *(I. trifida)* or pistils *(O. sativa)* at the unicellular microspore stage (stage 1), bicellular pollen stage (stage 2), and tricellular mature pollen stage (stage 3). mRNAs from these samples then were labeled with Cy5 during first-strand cDNA synthesis, and mRNAs from mature leaves (controls) were similarly labeled with Cy3. Hybridizations using three targets were carried out in six combinations for each plant species. To identify the genes abundantly expressed in the reproductive organs, the ratio of the level of transcript in the specific reproductive organ to that in leaves was calculated (Endo et al., 2002); we selected clones for which Cy5/Cy3 > 100. In this way, 190 of the 1,628 EST clones from *I. trifida* and 411 of the 5,322 combined EST and non-EST clones from *O. sativa* were identified as being specifically or predominantly expressed in reproductive organs. We classified the selected clones into six groups according to their expression patterns.

Regarding *I. trifida*, the putative functions of the products of more than 50% of the selected clones could not be assigned, because the clones had no similarity to reported genes. However, the analysis identified several genes related to polysaccharide metabolism, fatty acid metabolism, and cell wall reorganization. For *O. sativa*, in addition to those genes that were identified in *I. trifida*, there were several well-characterized reproductive organ-specific genes. For example, the chitinase gene was highly expressed in the pistil at stage 3, and the expression of this gene is known to be strong in pistils just before anthesis (Takei et al., 2000). In addition, the *Osc4* and *Osc6* genes were highly expressed in the anthers at stage 1, and the expression of these genes is limited in the tapetum (Tsuchiya et al., 1994). These two anther-specific genes were obtained by differential screening of 5×10^4 plaques (Tsuchiya et al., 1992). Together, these results show that cDNA microarray technology is a powerful tool for the screening of novel reproductive organ-specific genes.[25]

Pathogen Infection in Apple
(Malus × domestica *Borkh.*)

Although apple (*Malus × domestica* Borkh.) is one of the most widely cultivated temperate fruit crops, the gene function analyses and the genomic information available to date have been insufficient for their use in breeding programs. To identify useful genes and DNA markers for apple breeding, we have developed ESTs of apple cDNA clones obtained from cultured shoots and maturing fruits. We determined the partial sequences of about 2,000 clones and compared these ESTs with those in various databases. This effort revealed that half of our ESTs describe genes with as yet unknown functions.

To characterize the functions of variously expressed genes in apple, we used PCR-amplified cDNA fragments to generate a microarray, which we have used to monitor gene expression during the stress of pathogen infection. Alternaria leaf blotch of apple is caused by *Alternaria alternata*. During spore germination, this fungus produces a host-specific toxin (AM-toxin), which causes necrotic regions in the leaves of toxin-sensitive apple cultivars. It has been reported that the necrotic reaction requires de novo synthesis of proteins, and the primary sites for AM-toxin action are the chloroplast and plasma membrane (Otani et al., 1995).

To elucidate genes associated with AM-toxin-induced necrosis, we used the apple cDNA microarray to monitor gene expression in apple leaves treated with AM-toxin. Leaves of the AM-toxin-sensitive cultivar *Indo* were vacuum-infiltrated with the toxin solution and incubated in the dark for 1, 3.5, 5, or 10 h; leaves treated without toxin were used as controls. The mRNA isolated from treated leaves was labeled with Cy5 and used as the target. The fluorescence intensity of each spot was compared between the toxin-treated and control samples at each time point.

Almost all of the genes related to photosynthesis and CO_2 fixation were downregulated 1 h after treatment with AM-toxin, and the amounts of these mRNAs were reduced to 1/10 after 3.5 h. In particular, the mRNA level of sorbitol-6-phosphate dehydrogenase—an enzyme responsible for sorbitol synthesis—was rapidly reduced to 1/10 within 1 h. In the Rosaceae family, which includes the genus *Malus,* sorbitol is the major photosynthetic product and is translocated from leaves to growing tissues, such as fruits. We also observed that the mRNA level of the chalcone synthase gene, which is a known light-regulated gene, was depressed after 1 h. The inhibition of CO_2 fixation by AM-toxin has been reported (Otani et al., 1995). Our results showed that AM-toxin dramatically affects the transcription of photo-regulated genes in the dark. In contrast, clones upregulated by AM-toxin constituted approximately 10% of the microarray. Many of these upregulated genes were related to the stress response or defense mechanism.[26]

Fruit Development in Citrus Plants

Microarray technology is anticipated as a powerful tool for monitoring gene expression networks related to events of citrus fruit development, such as parthenocarpy, rind peeling, coloring, and maturation. The identification of the key genes regulating the onset of such physiological events would contribute to improving fruit productivity and quality through genetic and molecular biological approaches. To this end, the Citrus Genome Analysis Team (CGAT) at the National Institute of Fruit Tree Science (NIFTS) has been performing EST analyses of fruit development in citrus. So far, more than 5,000 ESTs have been catalogued from the various stages and tissues of citrus fruit, and these sequences represent a wide range of gene repertories with many as yet functionally unidentified ESTs. In a further contribution to the EST analysis program in citrus genomics, we have initiated profiling of the expression patterns of citrus ESTs using cDNA microarray analysis.

To date, the citrus cDNA microarray (Cit2.3k) comprises 2,300 cDNA sequences that were selected from nonredundant ESTs with or without functional annotation. This microarray yielded eightfold higher signal intensities (on average) for hybridization probes derived from citrus fruit than did cDNA microarrays from other plant sources, providing clear differences in signal strength among tissues and developmental stages. Using the Cit2.3k microarray, we are profiling the gene expression patterns between juvenile and adult phases of tree development, stages of fruit development, and citrus fruit tissues and organs. In addition to the comparisons between aspects of citrus fruit and tree development, the microarray analysis revealed the genes commonly expressed during citrus fruit maturation through comparative hybridizations with many other fruits, such as peach, apple, and banana.

In a rough comparison of the gene expression features of the vegetative stems and leaves of juvenile plants and the reproductive flower tissue of adult trees, 20 cDNA clones (<1%) showed marked differences in signal intensity. These clones were further investigated through RT-PCR and northern blot analyses to refine their expression patterns. Genes for chitinase and a seed protein homolog were highly expressed in floral organs and young fruitlets. ESTs expressed only during a specific stage were also detected in flowers at anthesis and in fruit at maturity stage. In addition to clones with specific expression patterns, we could classify many other EST clones into four groups in light of the gene's expression pattern during fruit development: (1) remains at the constitutive level, (2) peaks in middle season, (3) downregulated as maturation approaches, and (4) upregulated as maturation approaches. Of these clones, we have isolated a genomic sequence of an MT-like gene that is highly expressed during the middle and late phases of fruit development; this sequence affords a unique promoter for the genetic manipulation of fruit characteristics.

To identify key genes related to various physiological events during fruit development, we are constructing a 10,000-clone citrus cDNA microarray and refining the background levels of the analyses.[27]

NOTES

Contributors to research for the respective sections are listed as follows:

1. N. Kishimoto, J. Yazaki, and S. Kikuchi, National Institute of Agrobiological Science (e-mail: naoki@nias.affrc.go.jp).

2. K. Ogawa, T. Kawase, K. Henmi, and M. Iwabuchi, Research Institute for Biological Sciences (RIBS), Okayama (e-mail: ogawa_k@ns.bio-ribs.com).

3. Y. Sato, National Agricultural Research Center for the Hokkaido Region (e-mail: yutaka@affrc.go.jp).

4. T. Yamaguchi, K. Nakayama, T. Hayashi, and S. Koike, Plant Physiology Laboratory, National Agricultural Research Center for the Tohoku Region, National Agricultural Research Organization (e-mail: tomyamag@affrc.go.jp).

5. R. Imai, Winter Stress Laboratory, National Agricultural Research Center for the Hokkaido Region (e-mail: rzi@affrc.go.jp).

6. T. Fujiwara, S. Watanabe, and N. Ohkama, Department of Applied Biological Chemistry, The University of Tokyo (e-mail: atorufu@mail.ecc.u-tokyo.ac.jp).

7. C. Akimoto-Tomiyama, R. Bradley Day, R. Takai, and E. Minami, Department of Biochemistry, National Institute of Agrobiological Sciences (e-mail: eiminami@nias.affrc.go.jp).

8. Y. Nishizawa, National Institute of Agrobiological Sciences (e-mail: ynishi@nias.affrc.go.jp).

9. M. Nishiguchi[1,2] and M. Shimono[1], [1]National Institute of Agrobiological Sciences; [2]Present address: Faculty of Agriculture, Ehime University (e-mail: mnishigu@agr.ehime-u.ac.jp).

10. G. Yang and S. Komatsu, National Institute of Agrobiological Sciences (e-mail: skomatsu@affrc.go.jp).

11. T. Oka, Y. Suzuki, and I. Yamaguchi, Department of Applied Biological Chemistry, The University of Tokyo (e-mail: isomar@mail.ecc.u-tokyo.ac.jp).

12. K. Tsugane[1], M. Maekawa[1,2], and S. Iida[1], [1]National Institute for Basic Biology; [2] Okayama University (e-mail: tsugane@nibb.ac.jp).

13. T. Furukawa and J. Hashimoto, National Institute of Agrobiological Sciences (e-mail: junji@nias.affrc.go.jp).

14. M. Umeda, Institute of Molecular and Cellular Biosciences, The University of Tokyo (e-mail: mumeda@iam.u-tokyo.ac.jp).

15. K. Nishida, C. Mitsuhashi, K. Koizumi, and H. Shimada, Plant Molecular Biology and Molecular Genetics, Department of Biological Science and Technology, Tokyo University of Science (e-mail: shimadah@rs.noda.tus.ac.jp).

16. S. Takahashi and K. Ishimaru, Department of Plant Physiology, National Institute of Agrobiological Sciences (e-mail: sakiko@nias.affrc.go.jp).

17. H. Tabuchi[1] and T. Terao[2], [1]Rice Applied Genetics Laboratory; [2]Rice Applied Physiology Laboratory, Department of Rice Research, Hokuriku Research Center, National Agricultural Research Center, National Agricultural Research Organization (e-mail: htabuchi@affrc.go.jp).

18. T. Tsuchiya[1], M. Endo[2], T. Kokubun[2], and M. Watanabe[2], [1]Laboratory of Plant Molecular Genetics and Breeding, Faculty of Bioresources, Mie University; [2]Laboratory of Plant Breeding, Faculty of Agriculture, Iwate University (e-mail: tsuchiya@bio.mie-u.ac.jp).

19. T. Negishi, S. Mori, and N. Nishizawa, Laboratory of Plant Biotechnology, Department of Global Agricultural Sciences, The University of Tokyo (e-mail: annaoko@mail.ecc.u-tokyo.ac.jp).

20. E. Shimosaka and H. Handa, Plant Genecology Laboratory, National Agricultural Research Center for the Hokkaido Region (e-mail: etsuo@affrc.go.jp).

21. T. Nakamura[1], P. Vrinten[1], T. Shinbata[2], M. Saito[1], J. Iida[2], and J. Yonemaru[1], [1]Tohoku National Agricultural Research Center; [2]Nippon Flour Milling Company (e-mail: tnaka@affrc.go.jp).

22. M. Fujimori, Biotechnology Laboratory, Department of Forage Crop Breeding National Institute of Livestock and Grassland Science (NILGS), National Agricultural Research Organization (NARO) (e-mail: masafuji@affrc.go.jp).

23. M. Ebina, Department of Plant Breeding, National Grassland Research Institute (e-mail: triticum@affrc.go.jp).

24. M. Watanabe[1], M. Endo[1], H. Masuko[1], K. Hakozaki[1], H. Matsubara[1], T. Kokubun[1], H. Fukuda[2, 3], T. Demura[3], and T. Tsuchiya[4], [1]Faculty of Agriculture, Iwate University; [2]Department of Biological Sciences, Graduate School of Science, The University of Tokyo; [3]Plant Science Center, The Institute of Physical and Chemical Research; [4]Faculty of Bioresources, Mie University (e-mail: nabe@iwate-u.ac.jp).

25. T. Tsuchiya[1], A. Ando[1], C. Ogawa[1], K.Futagami[1], T. Hariyama[1], Y. Yano[1], T. Watase[1], H. Fukuda[2,3], T. Demura[2], M. Endo[4], and M. Watanabe[4], [1]Laboratory of Plant Molecular Genetics and Breeding, Faculty of Bioresources, Mie University; [2]Plant Science Center, Institute of Physical and Chemical Research; [3]Department of Biological Sciences, Graduate School of Science, The University of Tokyo; [4]Laboratory of Plant Breeding, Faculty of Agriculture, Iwate University (e-mail: tsuchiya@bio.mie-u.ac.jp).

26. Y. Hatsuyama, T. Fukasawa-Akada, M. Igarashi, and M. Suzuki, Aomori Green BioCenter (e-mail: yoshimichi_hatsuyama@ags.pref.aomori.jp).

27. T. Shimada, H. Fuji, T. Endo, and M. Omura, National Institute of Fruit Tree Science, National Agricultural Research Organization (e-mail: tshimada@affrc .go.jp).

REFERENCES

Aguan, K., K. Sugawara, N. Suzuki, and T. Kusano (1991). Isolation of genes for low-temperature-induced proteins in rice by a simple subtractive method. *Plant Cell Physiology 32:* 1285-1289.

Aoyama, T., and N.-H. Chua (1997). A glucocorticoid-mediated transcriptional induction system in transgenic plants. *The Plant Journal 11:* 605-612.

Arabidopsis Genome Initiative. (2000). Analysis of the genome sequence of the flowering plant *Arabidopsis thaliana. Nature 408:* 796-815.

Asano, T., N. Kunieda, Y. Omura, T. Ibe, T. Kawasaki, M. Takano, M. Sato, H. Furuhashi, T. Mujin, F. Takaiwa, and C. Y. Wu (2002). Rice SPK, a calmodulin-like domain protein kinase, is required for storage product accumulation during seed development: Phosphorylation of sucrose synthase is a possible factor. *Plant Cell 14:* 619-628.

Ballare C. L., M. C. Rousseau, P. S. Searles, J. G. Zaller, C. V. Giordano, T. M. Robson, M. M. Caldwell, O. E. Sala, and A. L. Scopel (2001). Impacts of solar ultraviolet-B radiation on terrestrial ecosystems of Tierra del Fuego (Southern Argentina). An overview of recent progress. *Journal of Photochemistry and Photobiology 62:* 67-77.

Curie, C., Z. Panaviene, C. Loulergue, S. L. Dellaporta, J. F. Briat, and E. L. Walker (2001). Maize yellow stripe1 encodes a membrane protein directly involved in Fe (III) uptake. *Nature 409:* 346-349.

Day, R., C. Akimoto, J. Yazaki, K. Nakamura, F. Fujii, K. Shimbo, K. Yamamoto, K. Sakata, T. Sasaki, N. Kishimoto, et al. (2002). Large-scale identification of elicitor-responsive genes in suspension-cultured rice cells by DNA microarray. *Plant Biotechnology 19:* 153-155.

Devos, K. M., and M. D. Gale (2000). Genome relationships: The grass model in current research. *Plant Cell 12:* 637-646.

Doerner, P., J.-E. Jørgensen, R. You, J. Steppuhn, and C. Lamb (1996). Control of root growth and development by cyclin expression. *Nature 380:* 520-523.

Ebina, M., and H. Nakagawa (2001). RAPD analysis of apomixis and sexual lines in guineagrass (*Panicum maximum* Jacq.). *Grassland Science 47:* 251-255.

Endo, M., T. Kokubun, Y. Takahata, A. Higashitani, S. Tabata, and M. Watanabe (2000). Analysis of expressed sequence tags of flower buds in *Lotus japonicus*. *DNA Research 7:* 213-216.

Endo, M., H. Matsubara, T. Kokubun, H. Masuko, Y. Takahata, T. Tsuchiya, H. Fukuda, T. Demura, and M. Watanabe (2002). The advantages of cDNA microarray as an effective tool for identification of reproductive organ-specific genes in a model legume, *Lotus japonicus*. *FEBS Letters 514:* 229-237.

Feng, Q., Y. Zhang, P. Hao, S. Wang, G. Fu, Y. Huang, Y. Li, J. Zhu. Y. Liu, X. Hu, et al. (2002). Sequence and analysis of rice chromosome 4. *Nature 420:* 316-320.

Flor, H. H. (1971). Current status of the gene-for-gene concept. *Annual Review of Phytopathology 9:* 275-296.

Friedrich, L., K. Lawton, S. Dincher, A. Winter, T. Staub, S. Uknes, H. Kessmann, and J. Ryals (1996). Benzothiadiazole induces systemic acquired resistance in tobacco. *The Plant Journal 10:* 61-70.

Goerlach, J., S. Volrath, G. Knauf-Beiter, G. Hengy, U. Beckhove, K. H. Kogel, M. Oostendorp, T. Staub, E. Ward, E. Kessmann, et al. (1996). Benzothiadiazole, a novel class of inducers of systemic acquired resistance, activates gene expression and disease resistance in wheat. *Plant Cell 8:* 629-643.

Goff, S. A., D. Ricke, T.-H. Lan, G. Presting, R. Wang, M. Dunn, J. Glazebrook, A. Sessions, P. Oeller, H. Varma, et al. (2002). A draft sequence of the rice genome *(Oryza sativa* L. *ssp. japonica). Science 296:* 92-100.

Hashimoto, J., T. Hirabayashi, Y. Hayano, S. Hata, Y. Ohashi, I. Suzuka, T. Utsugi, A. Toh-E, and Y. Kikuchi (1992). Isolation and characterization of cDNA clones encoding cdc2 homologues and cognate variants. *Molecular and General Genetics 233:* 10-16.

Hata, S. (1991). cDNA cloning of a novel cdc2+/CDC28-related protein kinase from rice. *FEBS Letters 279:* 149-152.

Higuchi, K., K. Suzuki, H. Nakanishi, H. Yamaguchi, N. K. Nishizawa, and S. Mori (1999). Cloning of nicotianamine synthase genes, novel genes involved in the biosynthesis of phytosiderophores. *Plant Physiology 119:* 471-479.

Hihara, Y., C. Hara, and H. Uchimiya (1996). Isolation and characterization of 2 cDNA clones for mRNAs that are abundantly expressed in immature anthers of rice (*Oryza sativa* L.). *Plant Molecular Biology 30:* 1181-1193.

Ishibashi, T., S. Kimura, T. Furukawa, J. Hashimoto, and K. Sakaguchi (2001). Two types of replication protein A 70 kDa subunit in rice, *Oryza sativa*: Molecular cloning, characterization, and cellular and tissue distribution. *Gene 272:* 335-343.

Kawasaki, T., K. Mizuno, H. Shimada, H. Satoh, N. Kishimoto, S. Okumura, N. Ichikawa, and T. Baba (1996). Coordinated regulation of the genes participating in starch biosynthesis by the rice floury-2 locus. *Plant Physiology 110:* 89-96.

Kikuchi, H., S. Hirose, S. Toki, K. Akama, and F. Takaiwa (1999). Molecular characterization of a gene for alanine aminotransferase from rice *(Oryza sativa)*. *Plant Molecular Biology 39:* 149-159.

Kimbeng, C. A. (1999). Genetic basis of crown rust resistance in perennial ryegrass, breeding strategies, and genetic variation among pathogen populations: Review. *Australian Journal of Experimental Agriculture 39:* 361-378.

Kimura, S., T. Ishibashi, M. Hatanaka, Y. Sakakibara, J. Hashimoto, and K. Sakaguchi (2000a). Molecular cloning and characterization of a plant homologue of the origin recognition. *Plant Science 158:* 33-39.

Kimura, S., T. Ueda, M. Hatanaka, M. Takenouchi, J. Hashimoto, and K. Sakaguchi (2000b). Plant homologue of flap endonuclease-1: Molecular cloning, characterization, and evidence of expression in meristematic tissues. *Plant Molecular Biology 42:* 415-427.

Kobayashi, T., H. Nakanishi, M. Takahashi, S. Kawasaki, N. K. Nishizawa, and S. Mori (2001). In vivo evidence that *Ids 3* from *Hordeum vulgare* encodes a dioxygenase that converts 2'-deoxymugineic acid to mugineic acid in transgenic rice. *Planta 212:* 864-871.

Kombrink, E., and K. Hahlbrock (1990). Rapid, systemic repression of ribulose 1,5-bisphosphate carboxylase small subunit mRNA synthesis in fungus-infected or elicitor-treated potato leaves. *Planta 181:* 216-219.

Logemann, E., S. C. Wu, J. Schröder, E. Schmelzer, I. E. Somssich, and K. Hahlbrock (1995). Gene activation by UV light, fungal elicitor, or fungal infection in *Petroselinum crispum* is correlated with repression of cell cycle-related genes. *The Plant Journal 8:* 865-876.

Lokhande, S. D., K. Ogawa, A. Tanaka, and T. Hara (2002). Effect of temperature on ascorbate peroxidase activity and flowering of *Arabidopsis thaliana* ecotypes under different light conditions. *Journal of Plant Physiology 160:* 57-64.

Maekawa, M. (1995). Irregular segregation of variegated chlorophyll deficiency derived from a cross between distantly related rice varieties in *Oryza sativa* L. In *Modification of gene expression and non-Mendelian inheritance,* K. Oono and F. Takaiwa (Eds.). Tsukuba, Japan: National Institute of Agrobiological Resources, pp. 379-388.

Maekawa, M., K. Rikiishi, T. Matsuura, and K. Noda (1999). A marker line, H-126, carries a genetic factor making chlorophyll mutation variegated. *Rice Genetics Newsletter 16:* 61-62.

Matsuda, N., T. Tsuchiya, S. Kishitani, Y. Tanaka, and K. Toriyama (1996). Partial male sterility in transgenic tobacco with antisense and sense *PAL* cDNA regulated by a tapetum-specific promoter. *Plant and Cell Physiology 37:* 215-222.

Midoh, N., and M. Iwata (1996). Cloning and characterization of a probenazole-inducible gene for an intracellular pathogenesis-related protein in rice. *Plant Cell Physiology 37:* 9-18.

Minami, E., and I. Ando (1994). Analysis of blast disease resistance induced by probenazole in rice. *Journal of Pesticide Science 19:* 79-83.

Minami, E., K. Kuchitsu, D. Y. He, H. Kouchi, N. Midoh, Y. Ohtsuki, and N. Shibuya (1996). Two novel genes rapidly and transiently activated in suspension-cultured rice cells by treatment with *N*-acetylchitoheptaose, a biotic elicitor for phytoalexin production. *Plant Cell Physiology 37:* 563-567.

Mori, S. (1999). Iron acquisition by plants. *Current Opinion in Plant Biology 2:* 250-253.

Mori, S., and N. K. Nishizawa (1987). Methionine as a dominant precursor of phytosiderophore in graminaceae plants. *Plant Cell Physiology 28:* 1081-1092.

Nagata, T., S. Todoriki, T. Hayashi, Y. Shibata, M. Mori, H. Kanegae, and S. Kikuchi (1999). Gamma-radiation induces leaf trichome formation in *Arabidopsis*. *Plant Physiology 120:* 113-119.

Nakagawa, H., and M. Ebina (2001). Development of molecular markers for the analysis of apomixis. In *Molecular breeding of forage crops,* G. Spangenberg (Ed.), pp. 161-173. Dordrecht, Netherlands: Kluwer.

Nakanishi, H., H. Yamaguchi, T. Sasakuma, N. K. Nishizawa, and S. Mori (2000). Two dioxygenase genes, *Ids3* and *Ids2,* from *Hordeum vulgare* are involved in the biosynthesis of mugineic acid family phytosiderophores. *Plant Molecular Biology 44:* 199-207.

National Institute of Agrobiological Sciences (2001). *RMOS: Rice microarray opening site.* http://cdna01.dna.affrc.go.jp/RMOS/index.html (last accessed February 25, 2007).

National Institute of Agrobiological Sciences (2002). *KOME: Knowledge-based Oryza molecular biological encyclopedia.* http://cdna01.dna.affrc.go.jp/cDNA/ (last accessed February 25, 2007).

National Institute of Agrobiological Sciences (2003). *Welcome to the Rice Expression Database.* http://red.dna.affrc.go.jp/RED/index.html (last accessed February 25, 2007).

Negishi, T., H. Nakanishi, J. Yazaki, N. Kishimoto, F. Fujii, K. Shimbo, K. Yamamoto, K. Sakata, T. Sasaki, S. Kikuchi, et al. (2002). cDNA microarray analysis of gene expression during Fe-deficiency stress in barley suggests that polar transport of vesicles is implicated in phytosiderophore secretion in Fe-deficient barley roots. *The Plant Journal 30:* 1-14.

Nishizawa, N. K., and S. Mori (1987). The particular vesicle appearing in barley root cells and its relation to mugineic acid secretion. *Journal of Plant Nutrition 15:* 695-713.

Ogawa, K., and M. Iwabuchi (2001). A mechanism for promoting the germination of *Zinnia elegans* seeds by hydrogen peroxide. *Plant and Cell Physiology 42:* 286-291.

Otani, H., K. Kohmoto, and M. Kodama (1995). *Alternaria* toxins and their effects on host plants. *Canadian Journal of Botany 73*(Supplement 1): S453-S458.

Preiss, J. (1991). Biology and molecular biology of starch synthesis and its regulation. In *Surveys of plant molecular and cell biology*, Vol. 7, B. J. Miflin (Ed.), pp. 59-114. Oxford, UK: Oxford University Press.

Rice Chromosome 10 Sequencing Consortium (2003). In-depth view of structure, activity, and evolution of rice chromosome 10. *Science 300:*1566-1569.

Rice Full-Length cDNA Consortium (2003). Collection, mapping, and functional annotation of over 28,000 full-length cDNA clones from *japonica*. *Rice Science 301:* 376-379.

Rousseaux, M. C., C. L. Ballare, C. V. Giordano, A. L. Scopel, A. M. Zima, M. Szwarcberg-Bracchitta, P. S. Searles, M. M. Caldwell, and S. B. Diaz (1999). Ozone depletion and UVB radiation: Impact on plant DNA damage in southern South America. *Proceedings of the National Academy of Sciences of the USA 96:* 15310-15315.

Saijo Y., S. Hata, J. Kyozuka, K. Shimamoto, and K. Izui (2000). Over-expression of a single Ca^{2+}-dependent protein kinase confers both cold and salt/drought tolerance on rice plants. *The Plant Journal 23:* 319-327.

Saito, K., K. Miura, K. Nagano, Y. Hayano-Saito, H. Araki, and A. Kato (2001). Identification of 2 closely linked quantitative trait loci for cold tolerance on chromosome 4 of rice and their association with anther length. *Theoretical and Applied Genetics 103:* 862-868.

Sakamoto, K., Y. Tada, Y. Yokozeki, H. Akagi, N. Hayashi, T. Fujimura, and N. Ichikawa (1999). Chemical induction of resistance in rice is correlated with the expression of a gene encoding a nucleotide binding site and leucine-rich repeats. *Plant Molecular Biology 40:* 847-855.

Sasaki, T., T. Matsumoto, K. Yamamoto, K. Sakata, T. Baba, Y. Katayose, J. Wu, Y. Niimura, Z. Cheng, Y. Nagamura, et al. (2002). The genome sequence and structure of rice chromosome 1. *Nature 420:* 312-316.

Satake, T., and H. Hayase (1970). Male sterility caused by cooling treatment at the young microspore stage in rice plants. V. Estimation of pollen developmental stage and most sensitive stage to coolness. *Proceedings of Crop Science Society of Japan 39:* 468-473.

Sawa, S., K. Watanabe, K. Goto, E. Kanaya, E. H. Morita, and K. Okada (1999). *FILAMENTOUS FLOWER*, a meristem and organ identity gene of *Arabidopsis*, encodes a protein with a zinc finger and HMG-related domains. *Genes and Development 13:* 1079-1088.

Shimono M., J. Yazaki, K. Nakamura, N. Kishimoto, S. Kikuchi, M. Iwano, K. Yamamoto, K. Sakata, T. Sasaki, and M. Nishiguchi (2003). cDNA microarray analysis of gene expression in rice plants treated with probenazole, a chemical inducer of disease resistance. *Journal of General Plant Pathology 69:* 76-82.

Shojima, S., N. Nishizawa, S. Fushiya, S. Nozoe, T. Irifune, and S. Mori (1990). Biosynthesis of phytosiderophores: In vitro biosynthesis of 2'-deoxymugineic acid from L-methionine and nicotianamine. *Plant Physiology 93:* 1497-1503.

Suzuka, I., S. Hata, M. Matsuoka, S. Kosugi, and J. Hashimoto (1991). Highly conserved structure of proliferating cell nuclear antigen (DNA polymerase delta auxiliary protein) gene in plants. *European Journal of Biochemistry 195:* 571-575.

Takagi, S. (1976). Naturally occurring iron-chelating compounds in oat- and rice-root washings. *Soil Science and Plant Nutrition 22:* 423-433.

Takagi, S., K. Nomoto, and S. Takemoto (1984). Physiological aspect of mugineic acid, a possible phytosiderophore of graminaceous plants. *Journal of Plant Nutrition 7:* 469-477.

Takahashi, M., H. Yamaguchi, H. Nakanishi, T. Shioiri, N. K. Nishizawa, and S. Mori (1999). Cloning 2 genes for nicotianamine aminotransferase, a critical enzyme in iron acquisition (strategy II) in graminaceous plants. *Plant Physiology 121:* 947-956.

Takai, R., K. Hasegawa, H. Kaku, N. Shibuya, and E. Minami (2001). Isolation and analysis of expression mechanisms of a rice gene, *EL5,* which shows structural similarity to ATL family from *Arabidopsis,* in response to *N*-acetylchitooligo-saccharide elicitor. *Plant Science 160:* 577-583.

Takai, R., N. Matsuda, A. Nakano, K. Hasegawa, C. Akimoto, N. Shibuya, and E. Minami (2002). EL5, a rice *N*-acetylchitooligosaccharide elicitor-responsive RING-H2 finger protein, is a ubiquitin ligase which functions in vitro in cooperation with an elicitor-responsive ubiquitin-conjugating enzyme, *Os*UBC5b. *The Plant Journal 30:* 447-455.

Takei, N., T. Nakazaki, T. Tsuchiya, Y. Kowyama, and H. Ikehashi (2000). Isolation and expression of a pistil-specific chitinase gene in rice (*Oryza sativa* L.). *Breeding Science 50:* 225-228.

Tsuchiya, T., K. Toriyama, S. Ejiri, and K. Hinata (1994). Molecular characterization of rice genes specifically expressed in the anther tapetum. *Plant Molecular Biology 26:* 1737-1746.

Tsuchiya, T., K. Toriyama, M. E. Nasrallah, and S. Ejiri (1992). Isolation of genes abundantly expressed in rice anthers at the microspore stage. *Plant Molecular Biology 20:* 1189-1193.

Umeda, M., N. Iwamoto, C. Umeda-Hara, and M. Yamaguchi (1999). Molecular characterization of mitotic cyclins in rice plants. *Molecular and General Genetics 262:* 230-238.

Watanabe, T., H. Igarashi, K. Matsumoto, S. Seki, S. Mase, and Y. Sekizawa (1977). The characteristics of probenazole (Oryzemate) for the control of rice blast. *Journal of Pesticide Science 2:* 291-296.

Wiermann, R. (1970). Die synthese von phenylpropanen während der pollenent-wicklung. *Planta 95:* 133-145.

Winter, H., and S. C. Huber (2000). Regulation of sucrose metabolism in higher plants: Localization and regulation of activity of key enzymes. *Critical Reviews in Plant Science 19:* 31-67.

Yamada, A., N. Shibuya, O. Kodama, and T. Akatsuka (1993). Induction of phyto-alexin formation in suspension-cultured rice cells by *N*-acetylchitooligosac-charides. *Bioscience, Biotechnology, and Biochemistry 57:* 405-409.

Yazaki, J., N. Kishimoto, M. Ishikawa, and S. Kikuchi (2002). Rice Expression Database: The gateway to rice functional genomics. *Trends in Plant Science 12:* 563-564.

Yazaki, J., N. Kishimoto, K. Nakamura, F. Fujii, K. Shimbo, Y. Otsuka, J. Z. Wu, K. Yamamoto, K. Sakata, and T. Sasaki (2000). Embarking on rice functional genomics via cDNA microarray: Use of 3' UTR probes for specific gene expression analysis. *DNA Research 7:* 367-370.

Yoshida, K. T., T. Wada, H. Koyama, R. Mizobuchi-Fukuoka, and S. Naito (1999). Temporal and spatial patterns of accumulation of the transcript of myo-inositol-1-phosphate synthase and phytin-containing particles during seed development in rice. *Plant Physiology 119:* 65-72.

Yoshioka, K., H. Nakashita, D. F. Klessig, and I. Yamaguchi (2001). Probenazole induces systemic acquired resistance in *Arabidopsis* with a novel type of action. *The Plant Journal 25:* 149-157.

Yu, J., S. N. Hu, J. Wang, G. K.-S. Wong, S. G. Li, B. Liu, Y. J. Deng, L. Dai, Y. Zhou, X. Q. Zhang, et al. (2002). A draft sequence of the rice genome *(Oryza sativa* L. *ssp. indica). Science 296:* 79-92.

Chapter 3

Rice Improvement:
Taking Advantage of New Technologies

D. J. Mackill

INTRODUCTION

Rice is one of the world's most important food crops, and certainly the main food of the world's poor. It is also a crop that has been well served by the international scientific community over the last four decades. Before 1960, rice was a scientifically neglected crop, aside from the work on the temperate *japonica* subspecies in Japan. This changed with the establishment of the International Rice Research Institute (IRRI) in the Philippines in 1960 and the international effort to develop input-responsive rice cultivars. Over the next decade, rice breeders at IRRI and other Asian locations laid the foundation for the Green Revolution cultivars that brought self-sufficiency to most tropical rice-growing countries (Khush, 1999).

Outside Japan, rice was also neglected as a system for basic biological studies. This changed again with the establishment of the Rockefeller Foundation's program on rice biotechnology (Toenniessen, 1992). This program focused worldwide attention on rice as a model for advanced genetic studies. The small size of the rice chromosomes, initially considered a nuisance for cytogenetic studies, was found to be a major advantage for biotechnological and molecular manipulation, and ultimately for genome sequencing. The draft sequences of the *indica* (Yu et al., 2002) and *japonica* (Goff et al., 2002) subspecies have already been published, and the completed sequence of the International Rice Genome Sequencing Program (Sasaki and Burr, 2000) will be available soon (see Chapter 1).

The completed sequence of both major subspecies offers unprecedented opportunities to scientists attempting to understand the genetic control of economically important traits in rice and related cereal crops. With the subsequent rush to decipher the functions of the tens of thousands of genes identified, it is appropriate to ask how this information will be of use in improving this important crop.

ACHIEVEMENTS AND LIMITATIONS
OF CONVENTIONAL RICE BREEDING

Conventional rice breeding programs are based on selection among local "land race" cultivars or selection following generation of genetic diversity through hybridization. Selection among local land races was practiced extensively in the early part of the twentieth century and, in fact, has been used even recently in developing cultivars for unfavorable rain-fed ecosystems. Superior pure breeding lines selected from heterogeneous traditional populations have achieved considerable success in providing farmers with improved cultivars. A prominent example is the cultivar *Khao Dawk Mali 105*, grown on several million hectares in Thailand and responsible for the jasmine rice exports from that country (Pushpavesa and Jackson, 1979). Farmers in Cambodia continue to request seeds of pure lines selected from local strains.

By far the most popular conventional breeding approach is hybridization followed by selection. There are many variations on this theme, and breeders have modified breeding methods to expedite their work. In most cases, breeders attempt to assemble various combinations of genes of major or minor (quantitative) effect from different parents into a single superior breeding line. These procedures have met with substantial success in terms of the adoption of modern cultivars by farmers. The cultivation of traditional rice cultivars is now the exception (Khush, 1995), and is mainly limited to areas with moderate to severe abiotic stresses or areas where farmers demand a premium grain quality that has not been obtained against a high-yield background.

The efforts to increase rice production in the tropics through breeding began in the middle of the twentieth century, with a primary focus on nitrogen responsiveness. Rice cultivars in temperate regions, belonging to the *japonica* subspecies, were seen as an early model for achieving higher N responsiveness. An FAO project centered in India attempted to transfer this characteristic into the tropical *indica* cultivars through intercrossing (Parthasarathy, 1960). This program was largely unsuccessful, and was overtaken by the discovery that the semidwarf genes inherited from Chinese cultivars such as *I-geo-tse* or *Dee-gee-woo-gen* could dramatically increase N responsiveness. The Green Revolution was initiated with the development and rapid adoption of cultivars such as IR8 at IRRI in the Philippines and Jaya in India (Khush, 1999). The rapid elevation of yield potential through increased N response was followed by a progression of improvements in other characteristics (Table 3.1). Breeders, with the assistance of pathologists and entomologists, began incorporating resistance to diseases and insects into modern cultivars. Most notable successes occurred for bacterial blight *(Xanthomonas oryzae* pv. *oryzae)*, blast *(Magnaporthe grisea)*, and grassy stunt virus diseases; and brown plant hopper (BPH; *Nilaparvata lugens)*, green leaf hopper *(Nepthotettix virescens)*, and gall midge *(Orseolia oryzae)* insect pests. Another major improvement has been the reduction in growth duration. Although it has been difficult to improve the overall yield potential established by IR8, the yield per day was improved remarkably through reducing the growth duration from 135 to 105 days or less.

TABLE 3.1. Plant traits improved in rice breeding programs in the years 1960-2000.

Trait	Traditional rices	Improved cultivars	Examples among IRRI cultivars
N responsiveness, harvest index	Lodged when fertilized (poor response)	Semidwarf varieties have higher harvest index	IR8
Resistance (biotic)	Most traditional rices susceptible to most diseases and insects, but many sources of resistance available	Cultivars combining resistance to several pathogens and/or insect pests	IR26
Early maturity	Most traditional rices photoperiod sensitive (late); first improved rices long duration	Progressively shorter duration to about 100-105 days duration in the tropics	IR36, IR50
Grain quality	Many high-quality traditional cultivars	Initial high-yield cultivars showed poor grain quality; improved in later cultivars	IR64
Resistance (abiotic)	Some traditional cultivars possess tolerance	Mostly susceptible; some early success with adverse soils tolerance	IR42 (adverse soils) IR49830-7 (submergence) PSBRc 102 (drought)

The first semidwarf cultivars were poor in palatability. This trait was gradually improved, and the IR64 cultivar, released in the Philippines in 1985, combined exceptional cooking quality with high yield and multiple resistance to diseases and insects. IR64 subsequently became the world's most widely cultivated rice variety (Khush, 1995).

The progress in breeding for tolerance of abiotic stress, such as drought, salinity, cold and submergence, has been less than for the biotic stresses aforementioned. Most of these stresses are under complicated genetic control. Drought resistance has been the most difficult trait to breed for, but some progress has been attained. Some rain-fed cultivars have been released that possess significant levels of drought resistance. PSBRc102 and Surin 1 are examples for rain-fed lowland rice. Some IRRI cultivars developed for irrigated conditions also possess tolerance to adverse soils (Ponnamperuma, 1979). Submergence tolerance is under relatively simple control, and this trait has been transferred into high-yield lines, such as IR49830-7 (Mackill et al., 1993), released as Popoul in Cambodia.

The breeding procedure preferred by rice breeders in tropical Asia and many other countries has been the standard pedigree method. The advantages of this method are well known and could certainly be responsible for the many successes of rice breeding, although it may also be to blame for some of its shortcomings in relation to more complex traits. Handling early generation nurseries through bulk breeding or its modifications has many advantages—in particular, its lower labor requirements, ability to handle larger population sizes, and postponement of selection for low-heritability traits to later generations. The early bias against this method due to selection for overly competitive plants (Jennings and Herrera, 1968) may not be applicable to populations of more uniform plant types or in more unfavorable environments. Once breeding lines become fixed for most traits, they are bulk-harvested and evaluated in field experiments at progressively higher levels of replication and number of locations, and larger plot sizes. The traditional breeding approach is characterized by an absence of farmer participation until varieties are released or in the last stage of evaluation. However, farmer participatory approaches can be incorporated into plant breeding programs without major modification of methods (Witcombe and Virk, 2001).

LINKAGE MAPS AND MARKER-ASSISTED SELECTION

Using DNA markers such as RFLPs and the more convenient PCR-based markers has served several purposes for rice improvement (Mackill, 1999). Mapping with molecular markers allowed the determination of how many genes controlled a trait (or the allelism of known genes) and could also serve as the starting point for cloning the genes through a map-based approach. For breeding, the use of linked markers must offer considerable advantages over visual selection to justify the added costs. The development of molecular maps of rice and other crops in the late 1980s generated considerable confidence that traits under complex genetic control could be handled similarly to those conditioned by major genes.

One of the main limitations of conventional marker-assisted selection (MAS) is finding markers that are close enough to the gene and that are polymorphic in the types of crosses being used in breeding programs. This problem is being solved through the development of a high-density microsatellite map (Temnykh et al., 2001). Microsatellite markers, sometimes called simple sequence repeats (SSR), have been shown to be highly polymorphic even in closely related rice germplasm (Ni et al., 2002; Saghai Maroof et al., 1994). The complete genome sequence will reveal all potential microsatellite markers. For example, the sequence published by Goff et al. (2002) indicated the presence on average of one SSR repeat (defined as at least eight repeats of 2-4 bp) every 8,000 bp, for a total of 48,351 SSRs. Although SSR markers are highly polymorphic, there are still drawbacks to their use in MAS. They may not be polymorphic in closely related varieties. There will also be recombination during MAS, because linkage is not close enough. Moreover, polyacrylamide gel electrophoresis is usually required for accurate phenotyping. Identifying the sequence change that causes a phenotype will overcome many of the difficulties in MAS.

The use of markers for tagging major gene traits in rice was reviewed by Mackill and Ni (2001). Examples of conditions where MAS would be advantageous include selection for traits that are difficult or expensive to measure (e.g., salt tolerance); pyramiding multiple genes that confer a similar or identical phenotype (e.g., multiple genes for resistance to blast or bacterial blight); or selecting against the donor chromosomal segments in backcrossing schemes.

Most molecular mapping work has concentrated on the identification of loci underlying quantitative traits (for reviews, see Li, 2001; McCouch and Doerge, 1995; Yano and Sasaki, 1997). The use of MAS to select for quantitative trait loci (QTLs) has lagged behind that for major gene traits, despite the fact that molecular markers were thought to offer great promise for the more complex quantitative traits. One reason is that QTL mapping is usually a highly inaccurate process, and the resolution of the QTL map is low. Fine mapping of a QTL involves developing advanced generations by selfing or backcrossing and replicated testing to accurately phenotype families (Monna et al., 2002; Paterson et al., 1990).

Another reason is that QTL mapping is usually done in special populations developed for this purpose, and that markers of the QTL itself may not
be appropriate for the populations used by breeders. This has led to the use
of backcrossing populations, where an elite cultivar is used as the recurrent
parent. This method was originally proposed to transfer QTLs from wild species into a cultivated variety (Tanksley and Nelson, 1996), but it is also being used to identify and transfer QTLs from diverse cultivated backgrounds
into elite rice varieties (Li, 2001). The use of such advanced backcrosses to
generate introgression lines is gradually replacing single-cross derivatives
as a means of QTL identification.

"Medalization" of quantitative traits can be achieved when a trait is controlled by one or two QTLs that are responsible for most of the variation in
segregating populations. Mackill and Ni (2001) defined a major QTL as
one controlling more than 25% of the phenotypic variation in a segregating
population. A number of QTLs have been found to account for 40-50% or
more of the phenotypic variation (Table 3.2). These QTLs could be the first
targets of MAS based on QTL mapping studies. In some cases, QTLs have
been shown to be composed of more than one gene, as was found for heading date at the *Hd3* locus in rice (Monna et al., 2002).

FUNCTIONAL GENOMICS

Research has intensified to identify the function of all plant genes, with a
focus on *Arabidopsis* and rice as models. Approaches such as microarray

TABLE 3.2. Examples of major quantitative trait loci (QTLs) responsible for more
than 40% of the genetic variation of particular quantitative traits in rice.

Trait	QTL	Chromosome	% Variation explained	Reference
Submergence tolerance	Sub1	9	69	Xu and Mackill, 1996
Partial resistance to blast	1 locus	11	46	Zenbayashi et al., 2002
Grain length	1 locus	3	64	Tan et al., 2000
P deficiency tolerance	1 locus	12	54-60	Ni et al., 1998
Salt tolerance	Saltol	1	39-44	Bonilla et al., 2002
Stem rot resistance	2 loci	2, 3	38-50	Ni et al., 2001

and insertional mutants will lead to the rapid identification of rice gene functions. However, an explanation of the contribution of the alleles of these loci to phenotype will require much painstaking research. Whereas the total number of rice genes may be up to 50,000 or more, the number of mapped genes and major QTLs is probably in the hundreds. Thus, the number of genes with allelic variation that contribute significantly to phenotype could be relatively low. Many of the most important genes will likely be regulatory genes involved in signal transduction or control of transcription.

The discovery of candidate genes underlying important traits will hasten the identification of alleles that are responsible for advantageous major gene or QTL effects. The sequence change can then be used to develop a marker that will identify that allele in segregating populations or even in germplasm collections or advanced breeding lines. Most allelic differences will be found to be single base changes or insertions/deletions (indels) in the coding region or promoter. These types of markers will lend themselves to high-throughput methods of genotyping large samples and will greatly facilitate the application of MAS. Using a sequence of the gene itself for MAS would be a great advantage over using linked markers. This would not only avoid the problem of recombination between the gene and the marker (or the need to use two flanking markers), but it should eliminate the problem of low polymorphism in closely related germplasm. Rice cultivars that differ in a number of agronomically important traits may be remarkably similar even in microsatellite alleles (Mackill et al., 1996; Ni et al., 2002). However, if the exact sequence change represented by a favorable allele is known, a primer could be designed that would detect this sequence change in any genetic background.

The identification of important candidate genes will be used to screen core germplasm sets to identify useful alleles of these loci. Allele mining of the gene bank can proceed through array-based technologies such as DarT (Jaccoud et al., 2001) or TILLING (Colbert et al., 2001).

Single nucleotide polymorphism (SNP) frequencies in maize are known to be quite high, up to 1 per 131 bp in coding regions (Rafalski, 2002). In rice, the availability of both an *indica* and a *japonica* sequence will provide an excellent means of identifying SNPs—at least those that are divergent between these two subspecies. Although SNPs are less polymorphic than microsatellites, their abundance makes up for this. Furthermore, SNPs are organized in a more limited number of haplotypes, regions of uniform distribution of markers bounded by hotspot regions of recombination (Yao et al., 2002). Association genetics using SNP-based methods will allow fine mapping and candidate gene discovery without the need for the large segregating populations required in conventional linkage studies (Rafalski, 2002).

Genes involved in important processes in the plant may not reveal allelic differences of direct use in breeding. Although these genes are instrumental in the expression of a trait, genotypic differences may be due to their regulation rather than to sequence differences in the gene itself. In this case, the use of transgenics will be an important avenue to the application of genomics. Gene expression can be altered by changing the promoter of the gene. A good example in rice is the overexpression of a pyruvate decarboxylase (*pdc*) gene to increase submergence tolerance (Quimio et al., 2000). Allelic variation at the *pdc* loci themselves has not been shown to be involved in submergence tolerance. Furthermore, the sequence of the gene can be altered to confer a new phenotype. In many cases, the changes in sequence needed to produce an improved phenotype may not be known, but orthologous genes in other species can be cloned and transformed into rice.

FUTURE PROSPECTS

Breeding programs can be thought of as including three parallel and overlapping objectives: (1) the development of elite lines with high yield and adaptation to a specific target environment, (2) base broadening to ensure future breeding progress, and (3) making incremental improvements to elite genotypes.

The development of elite lines will generally involve the conventional breeding methods, including the pedigree and bulk methods, as well as single seed descent. Traditionally, this method has relied on growing large populations of plants and plots with intense selection pressure for yield and quality among other traits. Crosses often focus on adapted parental material. The breeder attempts to accumulate QTLs controlling agronomic traits. Among the large number of crosses and large populations, the recombinants of superior alleles will be sought, and, with skill and a good dose of luck, obtained in very rare cases. Of these new varieties released to farmers, a few will distinguish themselves and spread to many farmers. In this way, some very productive and robust genotypes will spread to large areas.

Base broadening refers to the enhancement of breeding pools by the incorporation of genetic variation from exotic germplasm (Frey, 1996). This is thought of as a longer-term breeding process to ensure future progress from selection. This type of activity has been a feature of public sector breeding programs. Base broadening can be practiced without any knowledge of the underlying genetic control of the desirable traits that are being introduced.

The aforementioned approaches are largely dependent on conventional breeding; however, molecular tools can be applied in these efforts.

The development of introgression lines has become a common activity in rice breeding programs, and these lines can serve to introduce new variability into the gene pool. By hybridization between exotic cultivars or wild relatives and elite cultivars, several chromosomal fragments are introduced into breeding populations. Molecular markers can be used to associate specific fragments with agronomically useful genes. Each rice cultivar is a collection of chromosomal fragments, and multiple fragments will be segregating in the breeding populations. At some point, we will be able to determine the alleles present in these fragments and to follow them in the breeding program.

The success of rice breeding programs in most producing regions has resulted in the availability of elite cultivars that combine genes necessary for optimum productivity and product quality. It is certain that conventional breeding methods will continue to make improvements to these elite cultivars. But rice breeding is increasingly becoming a strategy of incremental improvements to the existing cultivars. It is in this area that genomics will have a major impact. Advantageous genes introduced through hybridization and marker-assisted backcrossing or novel genes introduced through genetic transformation will be used to develop the future generations of rice cultivars required to meet the challenge of feeding the world's rice consumers with minimum adverse effect on the environment.

REFERENCES

Bonilla, P., J. Dovorak, D. J. Mackill, K. Deal, and G. Gregorio (2002). RFLP and SSLP mapping of salinity tolerance genes in chromosome 1 of rice (*Oryza sativa* L.) using recombinant inbred lines. *Philippine Agricultural Scientist 87:* 68-76.

Colbert, T., B. J. Till, R. Tompa, S. Reynolds, M. N. Steine, A. T. Yeung, C. M. McCallum, L. Comai, and S. Henikoff (2001). High-throughput screening for induced point mutations. *Plant Physiology 126:* 480-484.

Frey, K. J. (1996). National plant breeding study. I. Human and financial resources devoted to plant breeding research and development in the United States in 1994. *Iowa Agriculture and Home Economics Special Report 98.* Ames: Iowa State University.

Goff, S. A., D. Ricke, T. H. Lan, G. Presting, R. L. Wang, M. Dunn, J. Glazebrook, A. Sessions, P. Oeller, H. Varma, et al. (2002). A draft sequence of the rice genome (*Oryza sativa* L. ssp. *japonica*). *Science 296:* 92-100.

Jaccoud, D., K. Peng, D. Feinstein, and A. Killian (2001). Diversity arrays: A solid state technology for sequence information independent genotyping. *Nucleic Acids Research 29:* No. 4 e25, pp. 1-7. Oxford, UK: Oxford University Press.

Jennings, P. R., and R. M. Herrera (1968). Studies on competition in rice II: Competition in segregating populations. *Evolution 22:* 332-336.

Khush, G. S. (1995). Modern varieties—Their real contribution to food supply and equity. *GeoJournal 35:* 275-284.

Khush, G. S. (1999). Green revolution: Preparing for the 21st century. *Genome 42:* 646-655.

Li, Z. (2001). QTL mapping in rice: A few critical considerations. In *Rice genetics IV,* G. S. Khush, D. S. Brar, and B. Hardy (Eds.), pp. 153-171. Los Baños, Philippines: International Rice Research Institute.

Mackill, D. J. (1999). Genome analysis and breeding. In *Molecular biology of rice,* K. Shimamoto (Ed.), pp. 17-41. New York: Springer-Verlag.

Mackill, D. J., M. M. Amante, B. S. Vergara, and S. Sarkarung (1993). Improved semidwarf rice lines with tolerance to submergence of seedling. *Crop Science 33:* 749-753.

Mackill, D. J., and J. Ni (2001). Molecular mapping and marker-assisted selection for major-gene traits in rice. In *Rice genetics IV,* G. S. Khush, D. S. Brar, and B. Hardy (Eds.), pp. 137-151. Los Baños, Philippines: International Rice Research Institute.

Mackill, D. J., Z. Zhang, E. D. Redona, and P. M. Colowit (1996). Level of polymorphism and genetic mapping of AFLP markers in rice. *Genome 39:* 969-977.

McCouch, S. R., and R. W. Doerge (1995). QTL mapping in rice. *Trends in Genetics 11:* 482-487.

Monna, L., H. X. Lin, S. Kojima, T. Sasaki, and M. Yano (2002). Genetic dissection of a genomic region for a quantitative trait locus, *Hd3,* into two loci, *Hd3a* and *Hd3b,* controlling heading date in rice. *Theoretical and Applied Genetics 104:* 772-778.

Ni, J., P. M. Colowit, and D. J. Mackill (2002). Evaluation of genetic diversity in rice subspecies using microsatellite markers. *Crop Science 42:* 601-607.

Ni, J., P. M. Colowit, J. J. Oster, K. Xu, and D. J. Mackill (2001). Molecular markers linked to stem rot resistance in rice. *Theoretical and Applied Genetics 102:* 511-516.

Ni, J. J., P. Wu, D. Senadhira, and N. Huang (1998). Mapping QTLs for phosphorus deficiency tolerance in rice (*Oryza sativa* L.). *Theoretical and Applied Genetics 97:* 1361-1369.

Parthasarathy, N. (1960). Final report on the International Rice Hybridization Project. *International Rice Commission Newsletter 9:* 12-23.

Paterson. A. H., J. W. De Verna, B. Lanini, and S. D. Tanksley (1990). Fine mapping of quantitative trait loci using selected overlapping recombinant chromosomes in an interspecific cross of tomato. *Genetics 124:* 735-742.

Ponnamperuma, F. N. (1979). IR42: A modern variety suited to the small rice farmer. *International Rice Research Newsletter 4*(3): 9-10.

Pushpavesa, S., and B. R. Jackson (1979). Photoperiod sensitivity in rainfed rice. In *Rainfed lowland rice: Selected papers from the 1978 International Rice Research Conference,* pp. 139-147. Manila, Philippines: International Rice Research Institute.

Quimio, C. A., L. B. Torrizo, T. L. Setter, M. Ellis, A. Grover, E. M. Abrigo, N. P. Oliva, E. S. Ella, A. L. Carpena, Q. Ito, et al. (2000). Enhancement of submergence tolerance in transgenic rice overproducing pyruvate decarboxylase. *Journal of Plant Physiology 156:* 516-521.

Rafalski, J. A. (2002). Novel genetic mapping tools in plants: SNPs and LD-based approaches. *Plant Science 162:* 516-521.

Saghai Maroof, M. A., R. M. Biyashev, G. P. Yang, Q. Zhang, and R. W. Allard (1994). Extraordinarily polymorphic microsatellite DNA in barley: Species diversity, chromosomal locations, and population dynamics. *Proceedings of the National Academy of Sciences of the USA 91:* 5466-5470.

Sasaki, T., and B. Burr (2000). International Rice Genome Sequencing Project: The effort to completely sequence the rice genome. *Current Opinion in Plant Biology 3:* 138-141.

Tan, Y. F., Y. Z. Xing, J. X. Li, S. B. Yu, C. G. Xu, and Q. F. Zhang (2000). Genetic bases of appearance quality of rice grains in Shanyou 63, an elite rice hybrid. *Theoretical and Applied Genetics 101:* 823-829.

Tanksley, S. D., and J. C. Nelson (1996). Advanced backcross QTL analysis: A method for the simultaneous discovery and transfer of valuable QTLs from unadapted germplasm into elite breeding lines. *Theoretical and Applied Genetics 92:* 191-203.

Temnykh, S., G. DeClerck, A. Lukashova, L. Lipovich, S. Cartinhour, and S. McCouch (2001). Computational and experimental analysis of microsatellites in rice (*Oryza sativa* L.): Frequency, length variation, transposon associations, and genetic marker potential. *Genome Research 11:* 1441-1452.

Toenniessen, G. (1992). Building plant science research capacity in developing countries. *Plant Cell 4:* 5-6.

Witcombe, J. R., and D. S. Virk (2001). Number of crosses and population size for participatory and classical plant breeding. *Euphytica 122:* 451-462.

Xu, K., and D. J. Mackill (1996). A major locus for submergence tolerance mapped on rice chromosome 9. *Molecular Breeding 2:* 219-224.

Yano, M., and T. Sasaki (1997). Genetic and molecular dissection of quantitative traits in rice. *Plant Molecular Biology 35:* 145-153.

Yao, H., Q. Zhou, J. Li, H. Smith, M. Yandeau, B. J. Nikolau, and P. S. Schnable (2002). From the cover: Molecular characterization of meiotic recombination across the 140-kb multigenic al-sh2 interval of maize. *Proceedings of the National Academy of Sciences of the USA 99:* 6157-6162.

Yu, J., S. N. Hu, J. Wang, G. K. S. Wong, S. G. Li, B. Liu, Y. J. Deng, L. Dai, Y. Zhou, X. Q. Zhang, et al. (2002). A draft sequence of the rice genome (*Oryza sativa* L. ssp. *indica*). *Science 296:* 79-92.

Zenbayashi, K., T. Ashizawa, T. Tani, and S. Koizumi (2002). Mapping of the QTL (quantitative trait locus) conferring partial resistance to leaf blast in rice cultivar Chubu 32. *Theoretical and Applied Genetics 104:* 547-552.

Chapter 4

Haploid Breeding in Rice Improvement

S. K. Datta
L. B. Torrizo
G. B. Gregorio
H. P. Moon

INTRODUCTION

Rice is undoubtedly one of the world's most important crops, as it supplies the caloric requirement of 40-50% of the world's population, the majority of who come from the developing countries. If no drastic measures to control the world's population are implemented, by the year 2020, the population will be 7.5 billion people (Food and Agriculture Organization of the United Nations, 2004). This translates to 40-50% more rice having to be produced for food. Food production is further held back by plant pests and diseases, unpredictable changes in weather, increased pressure on agricultural land in use for increased productivity diversification, and encroaching urbanization (Khachatourians et al., 2002). There are numerous strategies to realize increased food production, and one of them is breeding for varieties with better performance than those now in cultivation.

Breeding for self-pollinated crops such as rice is a very lengthy process, especially in locations where the climate permits only one cropping per year. The speedy production of homozygous lines that could immediately undergo field evaluation, by skipping at least six generations of inbreeding needed to attain homozygosity, could spell tremendous savings in time, labor, and money. This is possible through the various methods of haploid breeding, by inducing the gametes to proceed to sporophytic instead of gametophytic development, and eventually produce dihaploid plants (Figure 4.1). There are several methods to achieve doubled haploid (DH) production, including anther culture, culture of whole spikes, shed pollen culture (Figure 4.2) and microspore culture, as described by Datta (2001). Anther culture is the most popular of these four methods.

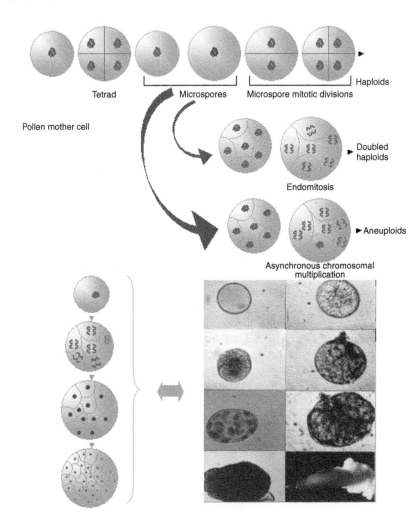

FIGURE 4.1. Schematic representation of microspore androgenesis. *Source:* Datta, 2001.

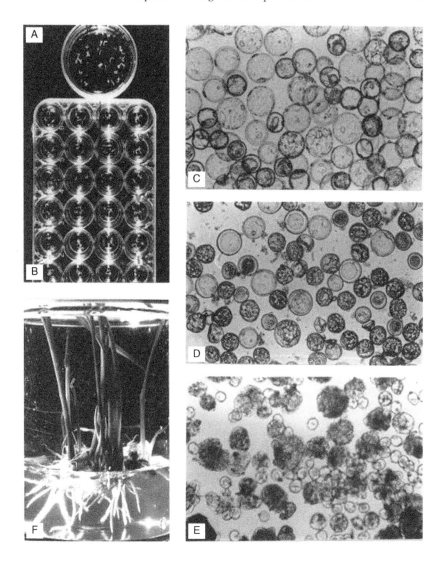

FIGURE 4.2. Shed pollen culture of rice. A and B: anther cultures in a 10-ml petri plate and a 24-well dish, respectively; C: microspores at the initial stage of culture; D: a culture showing several dividing microspores and plasmolyzed degenerating microspores; E: microspore-derived calli; and F: dihaploid plants regenerated from microspore calli. *Source:* Datta et al. (1990a).

BREEDING OBJECTIVES AND ACCOMPLISHMENTS

Doubled haploid production boosts the efficiency of a breeding program in two ways: (1) time saving—this system is the quickest method of advancing heterozygous breeding lines to homozygosity and (2) increase in selection efficiency (Snape, 1982). Realizing the potential facility of using anther culture (AC) in the fast production of varieties, rice breeding programs have increasingly used this method in various countries, with China, Korea, and Japan having substantial numbers of varieties produced (Table 4.1). Certainly a lot more varieties are not included in the list for lack of documented reports. For example, China has released 38 rice varieties as of 1998 (Zhu et al., 1998), and this number should have significantly increased by now. PSBRc50 is the first AC-derived variety produced by the International Rice Research Institute (IRRI). The suitability of this line in saline conditions has been shown (Zapata et al., 1991), leading to its release as a variety after passing rigorous multilocational tests by the National Rice Cooperative Testing for Saline Prone Areas (Bonilla and Desamero, 1996). From the same cross (IR51500), sister lines of PSBRc50 have been tested in India, and two varieties (CSR21 and CSR28) have been released in 1998. An AC-derived line, IR61673-AC201-1-1-3-3 (Cheolweon 31//Hua110/Osok), which has been tested in five experimental locations in the Philippines, is being considered a prerelease variety for cool, elevated areas. Anther culture has also been widely used in hybrid rice programs (Balachandran et al., 2003; Datta, 2005; Zhu et al., 1998).

IN VITRO SELECTION

Whether dihaploids represent a random array or hybrid gametic selection has always been the subject of debate. This should not be confused with gametoclonal variation, which results from variability displayed by tissue culture-derived plants due to the influence of artificial conditions to which the cells or tissues have been subjected during in vitro culture.

Studies have been conducted to address such issues as comparing AC-derived plants and plants produced through conventional breeding. Isozyme markers are advantageous in the study of segregation and recombination, as these permit assessment of the germinal origin (Guiderdoni et al., 1989). This method has been used in comparing gametic selection in sexual and AC derivatives from distant rice hybrids (Guiderdoni, 1989, 1991). Using the segregation of heterozygous isozyme markers among selfed and AC derivatives, there was a significant departure from the expected 1/1 isozyme phenotype

TABLE 4.1. Varieties of rice developed through anther culture.

Variety name	Parentage	Purpose/Desirable characteristics	Scientists involved in development/ Reported by
Aya (Norin 309)	Eikei 84271/Kitaake	Good eating quality, early maturing, intermediate lodging resistance, and cold tolerance	Kikuchi and Kunihiro, 1991; Kunihiro et al., 1993; Suzuki, 1992
CSR21	IR5657-33-2/IR4630-22-2-5-1-3	Salinity tolerance	
CSR28	IR5657-33-2/IR4630-22-2-5-1-3	Salinity tolerance	
Dama	—	—	Heszkyand Simon-Kiss, 1992
Fusaotome	Tohoku 143 (Hitomebore)/ Etsunan 146 (Hanaechizen)	Early maturity, cold resistance, high grain quality, excellent eating quality	Watanabe et al., 1998
Ganzhaoxian 11	Indica hybrid Shanyou 2	Early maturity, high and stable yield, high cold tolerance at seedling stage, good grain quality	Zhu et al., 1995
Guan 18	Zhen-Shan 97/IR24	High yield ability, improved milling, grain and cooking quality	Zhu and Pan, 1990
Hashiriaji	Niigatawase/ Hananomai	Early maturity, medium heat and cold tolerance, lodging resistance, good grain quality, medium yield potential	Hoshi et al., 1998
Hirohikari	Fukei 130/Akihikari	Early maturity, lodging resistance, medium floral impotency at low temperature, high yield ability, medium eating and milling quality	Maeda et al., 1990; Tsuchiya, 1992
Hirohonami	Fukei 130/Yoneshiro	Early maturity, resistance to floral impotency under low temperature, resistance to blast and brown spot, high yield ability, good eating and milling quality	Maeda et al., 1990; Tsuchiya, 1992

TABLE 4.1 *(continued)*

Variety name	Parentage	Purpose/Desirable characteristics	Scientists involved in development/ Reported by
Hiroshima 21 (Brand name: Koimomiji)	Sachiizumi/Fukei 141	Early maturity, blast resistance, cold tolerance, lodging resistance	Maeda et al., 2000
Hua 1A	Liao Geng 10/Zheng Gang B	WA-CMS line	Li et al., 1998
Hua Han Zao	Jia Nong 485/Labalate/ Tainan 13	Early maturity, low temperature tolerance, stable yield	Zhang et al., 1988
Hua Yu I Hua Yu II	Nipponbare/Ch'ien Chun Pang	High yield, improved bacterial blight resistance, wide adaptability to different growing conditions	Tiensin Agricultural Research Institute, 1976
Hwacheonbyeo	Suweon 298/Milyang 64	BPH, blast, BLB and stripe virus resistance, good eating quality, medium-late maturity	Lee et al., 1989
Hwadongbyeo	Daegwanbyeo/SR13345-20-1	Early maturity, cold tolerance	Moon et al., 1998
Hwajinbyeo	Milyang 64/Iri353	Medium-late maturity, resistant to BLB and stripe virus, resistance to leaf blast, stable to cold weather, high quality and yield	Moon et al., 1989a
Hwajoongbyeo	SasanishikiCheonmabyeo	High quality, lodging resistant, tolerant to stripe virus	Moon et al., 1994b
Hwajungbyeo	Sasanishiki/Cheonmanbyeo	High quality, lodging resistant, resistant to stripe virus, excellent cooking quality	Moon, 1995
Hwamyeongbyeo	M101/SR14779-HB234-32	Lodging tolerance, direct seeding	Kim et al., 1998

Hwasambyeo	Milyang 101/Iri389	High yield, good eating quality, short plant height, resistant to bacterial blight and stripe virus, moderate resistance to blast	Oh et al., 1997
Hwaseonchalbyeo	Milyang 64/Iri 355 (Shinseonchalbyeo)	Medium maturity and height, resistant to stripe virus and BLB, stable in cold weather	Moon et al., 1993
Hwaseongbyeo	Aichi 37/Samnambyeo (Suweon 295)	Medium height, medium maturity, stripe virus resistance, cold tolerance	Moon et al., 1986
Hwashinbyeo	Iri390/Milyang 110	good quality, high yield	Lee et al., 1996
Joiku No.394	–	–	Sasaki et al., 1988
Kokoromachi	–	–	Hayasaka et al., 1993
Kouiko 55	Kouiko 27/H3A$_3$70	Earliness, good eating quality, high yield	Kameshima et al., 1998
Kouikuka 37	Koganibare/Hieri	Scented rice, excellent quality	Nakamura et al. 1996
Kouikusake 54	Hamadanisiki/Hinohikari	Sake brewing, high yield	Mizobuchi et al., 1998
Manamusume	–	–	Hayasaka et al., 1997
Marianna	Belozem/Plovdiv 22	Short-stemmed, lodging resistant	Boyadjiev, 1991
Mocoi	H342/H161-28-2-2-1	Short stature, lodging resistant, intermediate threshability, high yield potential	Marassi et al., 2000
Parag 401	Prabhavati/Basmati 370	Superior grain quality, resistance to iron chlorosis	Patil et al., 1997
Petei	Quella/Guayquiraro	Early maturity, cold tolerance, salinity tolerance	Marassi et al., 2000
PSBRc50 (Bicol)	IR5657-33-2/IR4630-22-2-5-1-3	Salinity tolerance, high yield	Zapata et al., 1991

TABLE 4.1 (continued)

Variety name	Parentage	Purpose/Desirable characteristics	Scientists involved in development/ Reported by
Senbonnishiki	Nakate-Shinsenbon/ Yamadnishiki	Sake making, medium maturity, short stature, lodging resistance	Tsuchiya et al., 2001
Shan Hua 369	Zhen Shan 97A/IR24	Early maturity	Zhu et al., 1990
Shirayukihime	–	–	Sugimoto et al., 1991
Tan Feng No. 1	–	–	Heilungkiang Academy Agricultural Experiment Station in Rice, 1975
Xin xiou	Keng Gui/Ke Qing 3	High yield, lodging resistance, tolerance of high N fertilizer	Zhang et al., 1988
Yamagata 54	–	Cold tolerance	Chuba et al., 1997
Yamagata No. 54	Sho 389/Tuohoku No. 143	Early maturity	Sano et al., 1998

segregation in both plant populations. Such deviation could be due to hybrid sterility breakdown, which interferes with the androgenetic process, and hence, with the nature of the AC derivatives (Guiderdoni et al., 1989). In another study, it was observed that the segregation of the AC-derived plant populations fitted the expected 1/1 ratio in two crosses, whereas distorted segregation in at least one locus was observed in the F_2s (Guiderdoni, 1989). It seems that the morphogenic pathway of androgenesis seems to avoid in vivo pollen selection occurring during subsequent stages of the microsporogenesis, and in vitro gametic selection. Random assortment patterns on several hundred AC plants derived from three hybrids, each bearing two heterozygous morphological markers, have also been noted by Chen et al. (1982, 1983). They, however, used *japonica/japonica* crosses in their study, which are not prone to the hybrid breakdown phenomenon.

The performance of DH and conventionally bred lines has also been assessed by field evaluation. In one such study by Martinez et al. (1996), pedigree method (PM) and AC-derived lines coming from the same crosses were planted together and evaluated under rain-fed conditions, with the principal objective of comparing the efficiency of these two methods in producing rice lines resistant to blast. Stable blast resistance was assessed based on field performance over three years. Martinez et al. found that AC was superior to PM in getting lines that showed stable resistance to blast in some cross types and, with a few exceptions, with savings in time and labor. Despite the overall slight advantage of AC over PM, in terms of the numbers of blast-resistant progeny, more crosses generated stable resistant lines through PM than by AC. This result could simply be a reflection of the poor response to AC in several crosses.

Anther culture actually has a comparative advantage for selection of blast-resistant lines compared to the pedigree method, as shown by Perez-Almeida et al. (1995). Variability range for panicle blast reaction was higher in AC-derived rice than in pedigree-derived populations, indicating that AC offers a greater variability spectrum to the breeder and provides a better chance to select extreme genotypes.

In another study comparing AC-derived and conventional breeding-derived lines for grain yield and palatability, no difference was found between the mean heading date and the palatability of the DH and conventionally bred lines (Oosato et al., 1999). The mean yield of the DH was lower, however, and this could be due to selection for good plant types in the conventionally derived lines resulting in a higher yield. Likewise, the necessity of increasing the efficiency of anther culture and selecting DH lines from a large population was recognized.

Comparison of single seed descent (SSD) and AC-derived lines based on the distribution parameters (mean, variance, skewness, and kurtosis) of 12 quantitative traits was done on three single crosses (Courtois, 1993). For most of the traits, data were normally distributed, and the means and variances were found to be identical for SSD and AC-derived lines of the same hybrids. Differences were observed for some other traits, mainly in the lines from the intra-*japonica* crosses, probably due to a sampling problem, but the results suggested that SSD and AC are equally effective breeding techniques for producing agronomically useful lines of rice.

Based on our experience, the expression of a wide range of recombination from F1 crosses is directly related to the number of plants regenerated, or to a sampling problem (Courtois, 1993). If a sufficient number of dihaploid lines are produced, the higher variability of regenerated phenotypes is evident where it is possible to select for lines having superior phenotypes (Figure 4.3).

CULTURE CONDITIONS

The genotype and physiology of donor plants are two factors that influence anther culturability, but they will not be dealt with at length in this

FIGURE 4.3. Superior phenotype of anther culture-derived plant of IR72976 (IR631 33-49-1/IR67962-84-2-2), a cross for the new plant type, irrigated ecosystem.

chapter. Suffice to mention, however, the long-accepted fact that genotype plays a major role in determining anther culture response, with the *japonica* type generally more responsive than the *indica* lines (Datta et al., 1990a; Guha-Mukherjee, 1973; International Rice Research Institute, 1980; Quimio and Zapata, 1990). *Japonica*s are cultivated in only about 10% of the area planted with rice, making the *indica*s economically more important, but the former are good materials not only in the study of doubled haploids but also for tissue culture and transformation in general. Since the first report on the use of the *japonica* variety *Taipei 309* as a model system (Zapata et al., 1981), it has been used in establishing the protocols of protoplast culture (Abdullah et al., 1986) and in demonstrating the possibility of developing totipotent embryogenic cell suspension (Datta et al., 1990a) and the generation of transgenic *indica* (Datta et al., 1990b) and *japonica* (Peterhans et al., 1990) rice.

Likewise, the conditions for the growth of donor plants are critical, and experience-wise, we observed a better response from plants grown during the summer than during cold months. Raina (1997) observed that plants of IR43 (an *indica*) give the highest response when the conditions during panicle emergence were (1) long days (>12 h), (2) high solar radiation (18 Mjm^{-2}), (3) sunshine (>8 h), and (4) day/night temperatures of 34/24°C. Similarly, *japonica* rice grown at 23.5-25.7°C with 13-14 h of sunshine gave better response (Sun, 1999).

Anther culture is a laborious process. Attempts to simplify the technique have included devising a simple apparatus for mass-collection of anthers (Moon et al., 1994a) and one-step procedure from callus induction to plant regeneration (Moon et al., 1988b, 1991).

Optimal conditions are believed to play a great role in increasing the efficiency of double haploid production. There are quite a number of reports about the culture conditions for increasing the amenability of rice to anther culture.

Preculture Conditions

The most popularly employed preculture treatment is temperature shock, with cold shock being considered a routine procedure. Cold shock induces synchronized sporophytic development of the microspores. At IRRI, panicles are cold-shocked at 8-10°C for 7-8 days. A number of studies reported optimum temperature at around 10°C with the duration varying from a few days to almost a month (Kim et al., 1991; Matsushima et al., 1988). It was also found that the treatment temperature is more critical than the duration

(Moon et al., 1989b). Other preculture conditions that were found to increase AC efficiency are heat treatment (Zapata and Torrizo, 1986), irradiation (Aldemita and Zapata, 1991), and treatment of rice plants with male gametocide (Beaumont and Courtois, 1990).

Basal Culture Medium

The two most popular basal culture media used in rice anther culture are –6 (Chu et al., 1975, 1978) and MS (Murashige and Skoog, 1962). These two media put emphasis on the type and amount of nitrogen source. Modifications of these basal media have been made, mainly in the amounts of macronutrients. NH_4^+ has been found to be critical for anther culture (Iwai et al., 1990), and other nitrogen sources considered were amino acids such as proline and glutamine (Cho and Zapata, 1988; Habayashi et al., 1987).

Carbon Source

Sucrose has been the most often used carbohydrate in the culture medium, and optimized concentrations usually range from 30 to 60 g/l (Chen, 1978). However, this energy source rapidly breaks down to glucose and fructose, and the toxicity of sucrose in androgenesis is due to the sensitivity of the microspores to fructose, leading to studies on the substitution or addition of other carbon sources, like 2-deoxyglucose (Zeng, 1985) and maltose. Maltose yields two glucose units upon hydrolysis, so that it has become attractive and has been shown to be a superior carbohydrate source, replacing sucrose usually in equimolar amounts. Maltose has been demonstrated to stimulate androgenesis of *japonica* (Sun et al., 1993; Xie et al., 1995; Zhang et al., 1992) and *indica* cultivars (Lentini et al., 1995) and even of rice protoplasts (Ghosh Biswas and Zapata, 1993).

Media Additives and Gelling Agents

Various organic additives have been added to culture medium to enhance anther culturability. Rice and potato extracts (Kim et al., 1989), activated charcoal (Ji et al., 1998; Kim et al., 1989), and rice root extract (Su and Huang, 1979) have been found to stimulate anther culturability. Other tested unconventional additives to the culture medium include bleeding sap of towel gourd (Zang and Zhang, 1991), ginseng (Chen and Tsay, 1986), and silver nitrate (Lentini et al., 1995).

Other culture conditions investigated include gamma irradiation at different stages of anther culture (Kazuhiro and Hattori, 1997), sucrose starvation (Ogawa et al., 1994), osmotic stress (Raina and Irfan, 1998), addition of abscisic acid to the medium in the presence of optimum auxin-cytokinin ratio (Torrizo and Zapata, 1986), and use of magnetized water (Cho et al., 1996). The type and amount of gelling agent have also been considered as factors in anther culture (Feng and Li, 1999; Gau et al., 1990). A concentration of 0.8% agar is routinely used, but increasing agar concentration to 1.2-1.6% has been shown to remarkably increase regeneration frequency and callusing ability (Moon et al., 1988a).

PHENOTYPIC SELECTION

Plants derived from in vitro cultures exhibit phenotypic and physiological deviations from the parental type. These phenotypic variations may be induced by genomic, chromosomal, genic, and cytoplasmic mutations and by physiological effects (Oono, 1983). The variability displayed by anther culture- or microspore culture-derived plants has been termed *androclonal variation* (Raina, 1989). Although the anthers of F1 hybrids are routinely cultured to produce segregating populations of doubled haploids, the culture conditions are chosen to minimize the frequency of drastic genetic changes in the recombinants (Pramanik and Mandal, 2000). Be it a product of androclonal variation or recombination of mutations of microspore-derived cells, various studies have shown that haploid breeding is a rich source of plants with desirable phenotypes.

Oono (1983) reported variability of AC-derived plants in terms of heading date, plant height, panicle length, and seed fertility. Similar variations in agronomic characteristics have been noted (Pramanik and Mandal, 2000; Zapata et al., 1986, 1991). The general tendency is for AC-derived plants to have a shorter plant height, less tiller and lower panicle numbers per plant, and lower fertility, although improved characteristics, such as an increase in culm length and grains per panicle have been also observed in AC-derived progenies of tetraploid rice (Kishikawa et al., 1986). Nevertheless, it is still possible to select for phenotypes that could exceed the parental material in at least one trait, as, for example, in the breeding of PSBRc50, an anther culture-derived line for saline areas, where the dihaploid progeny outyielded both parents under both saline and nonsaline conditions (Zapata et al., 1991). Through anther culture, desirable phenotypes have been produced for cold tolerance (Zapata et al., 1986), salinity tolerance (Miah et al., 1996; Zapata et al., 1991), bacterial blight (Zhang et al., 1991) and rice blast resistance (Perez-Almeida et al., 1995), and photoperiod insensitivity in the aromatic

Thailand rice variety *KDML 105* (Sripichitt et al., 1994) and the aromatic Philippine variety *Dinorado.*

MARKER-ASSISTED SELECTION OF DIHAPLOIDS

Biotechnological tools complement breeding programs in many ways, one of which is to be able to identify target genes (or mapped genes of agronomic importance) with the assistance of DNA markers, a process called *marker-assisted selection* or MAS (Zheng et al., 1995).

Anther culturability is a quantitative trait, controlled by nuclear-encoded genes (Miah et al., 1985; Quimio and Zapata, 1990; Yamagishi et al., 1998). Earlier genetic studies on haploidy, however, merely determined whether there were differences in response among varieties and whether traits such as callus induction and plant regeneration were heritable. With the development of the MAS system, these characteristics can now be detected at the molecular level.

Quantitative trait loci (QTLs) responsible for anther culturability were surveyed and analyzed with the molecular map constructed from a population resulting from anther culture of a DH line (He et al., 1998a, 1998b). Parameters for four traits were callus induction, green plant differentiation frequency, albino plant differentiation frequency, and green plantlet yield frequency. All four traits displayed continuous distributions among the DH lines. For callus induction frequency, five QTLs were identified on chromosomes 6, 7, 8, 10, and 12. Two QTLs for green plantlet differentiation frequency were located on chromosomes 1 and 9, whereas there was a major QTL for albino plantlet differentiation on chromosome 9. No independent QTL was found for green plantlet yield frequency. These results may be useful in the selection of parents with high response to anther culture for rice haploid breeding and in the establishment of permanent DH populations for molecular mapping.

To clarify the association between chromosomal regions showing distorted segregation and anther culturability, the anther culturability of DH lines derived from a *japonica/indica* cross having distorted segregation on chromosomes 1, 3, 7, 10, and 11 was examined (Yamagishi et al., 1998). One region on chromosome 1 was found to control callus formation from microspores, and one region on chromosome 10 appeared to control the ratio of green to albino regenerated plants. In both regions, the *Nipponbare* (*japonica* parent) allele had a positive effect. Three regions on chromosomes 3, 7, and 11, however, showed no significant effect on anther culturability.

Likewise, using recombinant inbred lines from a cross between *Milyang 23* and *Gihobleo,* QTLs associated with green plant regeneration were mapped to chromosomes 3 and 10 (Kwon et al., 2002a, 2002b). The QTL on chromosome 10 was detected repeatedly using three AC methods and was tightly linked to three markers. One of these three markers, RZ400, was able to effectively identify genotypes with good (>10) and poor (<3) regenerability based on the cultivars and two F2 populations. This marker enables the screening of rice germplasm for anther culturability and introgression into elite lines in breeding programs.

The growth and development of plants from callus cultures is influenced by genes controlling the production of certain enzymes essential for the metabolism of differentiating cells, tissues and organs. Peroxidases are metalloprotein enzymes containing porphyrin-bound iron and are found to be associated with many physiological processes, including morphogenesis. Subhadra and Reddy (1998) quantified the amount of peroxidase present in four *indica* cultivars and all possible F1 combinations. They found that calli with high regeneration capacity showed high values of peroxidases, whereas those with low amounts of peroxidases gave only albinos or predominantly albinos with few green plants, suggesting that peroxidases play a role in the morphogenesis of anther calli. By employing isozyme markers like peroxidase, the embryogenic and regeneration potentials of calli can be identified and used for selecting high-regenerating calli.

On the other hand, investigation of albino plants by Southern blotting showed that seven of the ten albino plants investigated contained plastid DNA that had suffered large-scale deletions. The size and location of these deletions differed among plants. The results indicated that some albino plants lack the region coding the *rbc*L gene in the plastid genome. The deleted plastid DNA molecules were retained in calluses derived from the roots of each albino plant (Harada et al., 1991).

DIHAPLOIDS IN GENOMICS

Genomics implies DNA sequencing, the routine use of DNA microarray technology to analyze the gene expression profile at the mRNA level, and improved informatic tools to organize and analyze such data (Datta, 2000). It is expected to become the engine that drives the discovery of traits and helps solve intractable problems in crop production. Through genomics, every functional diversity of rice genes in germplasm and the overall architecture of genetic, biochemical, and physiological systems in rice will be discovered (Zaid et al., 1999).

Doubled haploid lines are useful for genetic analysis, particularly of quantitative traits (Snape, 1982). Quantitative trait loci (QTLs) affect some important agronomic traits in cultivated rice. QTL studies have been facilitated by the development of molecular markers using segregating populations, F2, or backcross populations, but these studies have been difficult to replicate in order to obtain accurate phenotypic values for precise QTL mapping. The use of recombinant inbred lines (RILs) provides many advantages in QTL studies, but it will take a long time to develop such populations. Recently, many studies have employed DH populations to construct genetic maps and locate QTLs. Because DH lines are homozygous, they can be propagated without further segregation. This characteristic allows for the precise measurement of quantitative traits by repeated trials and for a reduction in the environmental component of the total phenotypic variance (Lu et al., 1996). However, one hazard in the use of AC-derived materials is the possible distorted segregation of RFLP markers in AC-derived DH populations. Yamagishi et al. (1996) observed that 10 and 11 of the 50 markers in two AC-derived populations showed distorted segregation ratios from the theoretical ratio of 1/1. Parental alleles were not randomly transmitted from the F1 plant to the AC-derived plants. Furthermore, the segregation ratios of seven and six RFLP markers, respectively, were distorted both from the 1/1 ratios and from the observed ratios in the F2 population. The chromosomal regions involving these markers were on chromosomes 1, 3, 7, 10, 11, and 12. The percentage of the markers showing segregation distortion in the AC-derived populations was, however, similar to that in the F2 population. Thus, distortion in segregation does not appear to be a major drawback in the use of AC populations for rice breeding and genetics. The importance of doubled haploid populations in the study of quantitative traits was confirmed by Chen et al. (1996), who demonstrated that most gametoclonal variations among DH plants involve quantitative traits, and that the frequency of distinct variations was not high. Biochemical and molecular analysis proved a high degree of genetic stability of gametoclones, concluding that although AC may to some extent modify the performance of microspore-derived plants, it will not dramatically affect their use in plant breeding and genetic engineering programs.

Extensive research involving dihaploids in the study of QTLs can be found in the literature. One such trait is tiller angle, which has great significance in the high-yield breeding of rice; too small a tiller angle reduces the resistance to disease, whereas too large a tiller angle is undesirable for high yield (Qian et al., 1999a). Based on the constructed linkage map of a DH population from a female parent with a spreading plant type and a male parent having a compact plant type, two major QTLs were detected on chromosomes 9 and 11, and one minor QTL on chromosome 9. Similarly, doubled haploids have been used in the study of QTLs for length of top internodes, plant height and days to heading (Tan et al., 1997a), ratooning ability and grain yield traits (Tan et al., 1997b), and seedling cold tolerance (Qian et al., 1999b).

The employment of molecular genetic markers is particularly useful as an alternative strategy to phenotype selection for rice root traits. Breeding varieties with increased root penetration ability through hardpans and other root traits is very difficult, as screening numerous genotypes under field conditions is laborious and time consuming. Furthermore, the fact that soil compaction is not uniform and inconsistent throughout rice fields makes the evaluation of root traits difficult (Ali et al., 2000). Studies of QTLs for rice root characteristics such as root vitality (Teng et al., 2000); constitutive root morphology, such as deep root morphology and root thickness (Kamoshita et al., 2002); and osmotic adjustment, root penetration index, basal root thickness, penetrated root thickness, root pulling force, total root dry weight, penetrated root dry weight, and penetrated root length (Zhang et al., 2001) have all used doubled haploid populations.

A DH population is a kind of permanently segregating population. Its genetic structure is fixed, so it can be grown at different times and locations for detecting QTLs and evaluating the interactions between genotype and environment—that is, the phenotypic expression level of QTLs in different environments. This technique has been applied in the identification of 22 QTLs for six agronomic traits of rice in three different locations (environments). QTLs for spikelets and grains per panicle were common across environments, whereas traits like heading date and plant height were more sensitive to environment (Lu et al., 1996).

Doubled haploid rice populations have also been used in QTL studies on rice grain quality (Bao et al., 2002), grain shape (Huang et al., 1997), paste viscosity characteristics (Bao and Xia, 1999), aromatic traits (Lorieux et al., 1996), and brown planthopper resistance (Huang et al., 1997).

Accumulation and fixation of marker genes was achieved using genetic male sterile composite crosses and then employing anther culture techniques (Suh and Song, 1991). The dihaploid plants induced from the AC of the composite crossed plants showed the segregation ratio for male sterility as well as five to six marker genes generated through this method.

Aneuploids (plants with extra chromosomes in addition to the normal haploid chromosome complement) are useful for genetic research—for example, for investigating genetic imbalance caused by extra chromosomes at the haploid and diploid levels and for studying chromosome behavior in meiosis and rice genome construction (Wang and Iwata, 1995). It has been difficult to produce aneuploids in rice, but through anther culture, haploid plants with one extra chromosome ($n + 1$), aneuploids and tetrasomics have been derived from anther culture of trisomic rice plants (Wang and Iwata, 1991, 1995). These aneuploids could be used to assign DNA markers to individual chromosomes. Meiotic behavior and morphological features of auto-pentaploid rice plants derived from anther culture have also been investigated for genetic and cytological studies (Watanabe, 1974).

SALT TOLERANT LINES DEVELOPED THROUGH ANTHER CULTURE OF F1s

The development of a promising salinity-tolerant line through hybridization and selection could take eight to ten years. This is aggravated by the fact that, in areas where salinity or other abiotic stresses exist, only one cropping season per year is possible, and thus the generation of stable lines takes longer. This period could be substantially reduced by applying the F1 anther culture technique.

In 1996, some high-yield, salt-tolerant, AC-derived lines were generated at IRRI. In most cases, it only took three years for the tolerant lines to be isolated. These AC-derived lines were IR51500-AC11-1, IR51500-AC17, IR51485-AC6534-4, IR72132-AC6-1, IR69997-ACC1, IR69997-AC2, IR69997-AC3, and IR69997-AC4. Most of these lines have been used as donor parents in breeding programs in Bangladesh, the Dominican Republic, Egypt, Mexico, Myanmar, the Philippines, and Thailand. IR51500-AC11-1 was released as a salt-tolerant variety in the Philippines with the name PSBRc50 or *Bicol* (Figure 4.4). This is the first time that an AC-derived line from an *indica/indica* cross has been released as a variety, and this is also the first variety to be recommended for cultivation in adverse environments. IR51500-AC17 and IR51485-AC6534-4, named CSR21 and CSR28, respectively, were identified as commercial varieties in India. Both varieties were derived from F1 anther culture and were recommended for cultivation on saline-alkaline soils (International Rice Research Institute, 1997). Other promising AC-derived salt-tolerant lines are now undergoing further evaluation (Table 4.2).

FIGURE 4.4. Field evaluation of PSBRc50 (IR51500-AC11-1) under saline condition. IR29 and IR42 are tolerant and susceptible checks, respectively.

USE OF ANTHER CULTURE IN TRANSGENICS

There are several methods of using haploid breeding in transgenics. One is to transform pollen by gene delivery with two objectives: (1) to produce genetically modified seed by pollinating plants with transformed pollen, and (2) to produce haploid plants from transformed pollen and microspores (Sawhney, 2002). Another is to use haploidy techniques to fix the homozygosity of segregating transgenic populations.

There are only a few reports on the production of transgenics using haploid cells as starting materials or explants. This is mainly due to the difficulty of regenerating plants from haploid materials as compared to somatic cells. Nevertheless, the regeneration of plants from protoplasts derived from haploid cell suspensions has been done in preliminary studies to transformation (Datta et al., 1990a; Su et al., 1992). Transgenic homozygous genetically engineered *indica* rice has been produced from microspore-derived cell suspensions (Datta et al., 1990b). Likewise, fertile transgenic plants have been regenerated from anther-derived cell suspensions of *Zhonghua No. 11* transformed via electroporation with maize transposable element *Ac/Ds* to be used in the transposon tagging of the mutated gene (Liu et al., 1995).

TABLE 4.2. New salt-tolerant lines developed at IRRI using some improved donor sources.

Improved variety/line	Reaction to salinity[a]	Source of tolerance
IR65185-3B-8-3-2	3	TCCP266-2-49-2B-3
IR65185-3B-2-3	3	TCCP266-2-49-2B-3
IR5713-2B-8-2B-1-2	5	Pokkali
IR63291-B-3R-8-3	3	TCCP266-2-49-2B-3
IR61920-3B-22-2-1	3	Wagwag
IR72132-AC6-1	3	TCCP266-2-49-2B-3, IR51491-AC10
IR58443-6B-10-3	3	AT69-5
IR69997-AC1-2-3-4	3	TCCP266-2-49-2B-3
IR63307-4B-4-3	3	TCCP266-2-49-2B-3
PSBRc50 (Bicol)[b]	3	IR4630-22-2-5-1-3

[a]According to IRRI's *Standard Evaluation System for Rice* (International Network for Genetic Evaluation of Rice, 1996) with slight modification (Gregorio et al., 1997). On a scale of 1 to 9, where 1 = *highly tolerant* and 9 = *highly susceptible*.

[b]Released as a salt-tolerant variety in the Philippines.

Using pollen-derived callus, transgenic homozygous plants containing the maize transposable element *Ds* and using hygromycin resistance and *PPt* resistance genes are markers were produced via *Agrobacterium tumefaciens* (Fu et al., 2001). Haploid as well as diploid plants were regenerated. All transgenic lines showed resistance to the herbicide, and the T_1 plants from the diploid lines also resisted BASTA without segregation within each line, indicating the homozygosity of the foreign gene in diploid transgenic lines.

The rapid fixation of homozygosity in segregating populations derived from the transformation of somatic tissues has also benefited from haploid techniques. Herbicide-resistant homozygous lines were obtained through anther culture of transgenic Chinese cultivars derived via particle bombardment (Si et al., 1999; Sun and Si, 1996). Other studies used anther culture to fix homozygosity of the inherited transgenes in hybrids between a transgenic and a nontransgenic variety (Feng and Xue, 2001) or between two transgenic materials (Chen and Li, 2000). In the latter study, stable AC-derived hybrids expressing both transgenes resulted from the hybridization of a line containing a rice borer resistance gene and one with a herbicide resistance gene.

Pathogens can mutate easily to overcome resistance genes in the plants and render them susceptible again. Single-gene resistances, when widely deployed in monoculture, often break down in a few years (Leach et al., 1995). For example, cultivars carrying the *Xa-4* gene for resistance to bacterial blight, which were widely deployed in the Philippines from the early 1970s, became susceptible to the predominant race of *Xanthomonas oryzae* pv. *oryzae* within five years (Mew et al., 1992). This means that to be able to sustain high productivity for the ever-burgeoning world population, rice breeders should produce new or improved varieties in a much shorter time.

There are no available donor parents for sheath blight resistance in the rice gene pool, giving rise to a genetic engineering strategy involving high-level expression of PR (pathogenesis related) proteins with different modes of action against target organisms with broad-spectrum resistance to rice varieties (Datta et al., 1999). It has been demonstrated that homozygous dihaploid transgenics of an *indica* rice cultivar harboring the *chitinase* gene for enhanced sheath blight resistance could be produced in less than a year through anther culture of primary transgenic plants (Baisakh et al., 2001). This result demonstrates the possibility of producing and making available to rice farmers resistant lines much faster than biotic resistance breaks down.

One of the most significant breakthroughs in rice genetic engineering in this millennium was the incorporation of three genes required to function concurrently in the biosynthesis of β-carotene in the grains of *japonica* variety *Taipei 309* (Ye et al., 2000). As these genes were introduced through cotransformation and possibly integrated in different loci, the fixation of homozygosity for lines having all three genes through the conventional method would have been lengthy. But through anther culture of primary regenerants, it was possible to produce double haploids that were homozygous for all three genes, including their rearranged copies, in only one step (Baisakh et al., 2002). Elite *indica* cultivars (IR64, BR29, *Mot Bui, Nang Huong Cho Dao,* etc.) have now been transformed with genes for ß-carotene biosynthesis (Datta et al., 2003). Homozygous transgenic BR29 has now been developed using anther culture (Datta et al., 2005). These plants could now be used as donor parents in transferring these genes into other locally adapted *indica* backgrounds through hybridization.

CONCLUSIONS AND FUTURE PROSPECTS

Starting with the first successful haploid production using anther culture techniques in rice (Niizeki and Oono, 1968) until the early 1980s, persistent problems in anther culture were inadequate regenerating techniques, inadequate techniques for generating diploid plants from haploids, high rates of mutation (abnormality) in pollen-derived dihaploid plants, high incidence of albino plants, and possible dedifferentiation of calli from cells of somatic tissue anthers (Niizeki, 1983). Notwithstanding these constraints, the production of AC-derived varieties has been enormous, especially in China, Korea, and Japan. The aforementioned problems have not been totally eradicated; however, growing the explants in proper conditions and optimization of preculture and culture conditions and media have seemed to help. Production of a dihaploid population with a wide range of variability or recombination is a game of numbers, and the low production of regenerants can be a drawback of the haploidy method relative to conventional techniques. On the assumption that the percentage response (callus induction and plant regeneration) for a given genotype has reached a plateau, the only alternative is to plate as many anthers or calli as possible, until such time as a technique or procedure becomes available to break the barrier of genotype dependency, not only of anther culture, but of tissue culture in general.

Haploid breeding has not only proven its utility in the creation of varieties but also in the production of homozygous lines that are necessary in the identification of genes encoding important agronomic traits—especially those that are recessive in character. The combination of transgenic and haploid breeding methods has been successful, and the prospect of being able to immediately obtain transgenic doubled haploids by the efficient transformation of highly regenerable haploid explants will overcome breeding constraints even more.

REFERENCES

Abdullah, R., E. C. Cocking, and J. A. Thompson (1986). Efficient plant regeneration from rice protoplasts through somatic embryogenesis. *Bio/Technology 4:* 1087-1090.

Aldemita, R. R., and F. J. Zapata (1991). Anther culture of rice: Effects of radiation and media components on callus induction and plant regeneration. *Cereal Research Communications 19:* 9-32.

Ali, M. L., M. S. Pathan, J. Zhang, G. Bai, S. Sarkarung, and H. T. Nguyen (2000). Mapping QTLs for root traits in a recombinant inbred population from two *indica* ecotypes in rice. *Theoretical and Applied Genetics 101:* 756-766.

Baisakh, N., K. Datta, N. Oliva, I. Ona, G. J. N. Rao, T. W. Mew, and S. K. Datta (2001). Rapid development of homozygous transgenic rice using anther culture harboring rice *chitinase* gene for enhanced sheath blight resistance. *Plant Biotechnology 18:* 101-108.

Baisakh, N., K. Datta, M. Rai, S. Rehana, P. Beyer, I. Potrykus, and S. K. Datta (2002). Development of dihaploid transgenic "golden rice" homozygous for genes involved in the metabolic pathway for β-carotene biosynthesis. *Rice Genetics Newsletter 18:* 91-94.

Balachandran S., G. Chandel, M. F. Alam, J. Tu, S. S. Virmani, K. Datta, and S. K. Datta (2003). Improving hybrid rice through anther culture and transgenic approaches. In *Proceedings of the 4th international symposium on hybrid rice for food security, poverty alleviation, and environmental protection,* S. S. Virmani, C. X. Mao, and B. Hardy (Eds.), pp. 105-118. Hanoi, Vietnam and IRRI, Philippines.

Bao, J. S., Y. R. Wu, B. Hu, P. Wu, H. R. Cui, and Q. Y. Shu (2002). QTL for rice grain quality based on a DH population derived from parents with similar apparent amylose content. *Euphytica 128:* 317-324.

Bao, J. S., and Y. W. Xia (1999). Genetic control of paste viscosity characteristics in *indica* rice (*Oryza sativa* L.). *Theoretical and Applied Genetics 98:* 1120-1124.

Beaumont, V., and B. Courtois (1990). Anther culturability of rice plants treated with male gametocide chemicals. *International Rice Research Newsletter 15*(1): 9.

Bonilla, P. S., and N. V. Desamero (1996). Development of salt-tolerant varieties at PhilRice: Past, present and future. In *Proceedings: Workshop on soil salinity in rice,* pp. 5-14. Philippine Rice Research Institute, Maligaya, Muñoz, Nueva Ecija. January 24-26, 1996.

Boyadjiev, P. (1991). New rice cultivar *Marianna* obtained through anther culture. *International Rice Research Newsletter 16*(6): 13-14.

Chen, C. C. (1978). Effects of sucrose concentration on plant production in anther culture of rice. *Crop Science 18:* 905-906.

Chen, C. C., W. L. Chiu, L. J. Yu, S. S. Ren, W. J. Wu, and M. M. H. Lin (1983). Genetic analysis of anther-derived plants of rice, independent assortment of unlinked genes. *Canadian Journal of Genetics and Cytology 25:* 18-24.

Chen, C. M., C. C. Chen, and M. H. Lin (1982). Genetic analysis of anther-derived plants of rice. *Journal of Heredity 73:* 49-52.

Chen, J. J., and H. S. Tsay (1986). Effects of ginseng on rice anther culture. *Journal of Agricultural Research of China 35:* 139-144.

Chen, W., and N. T. Li (2000). Di-transgenic rice was developed in Zhejiang University. *Chinese Rice Research Newsletter 8*(1): 16.

Chen, Y., C. F. Lu, Y. B. Xu, P. He, and L. H. Zhu (1996). Gametoclonal variation of microspore derived doubled haploids in *indica* rice: Agronomic performance, isozyme and RFLP analysis. *Chinese Journal of Genetics 23:* 97-105.

Cho, E. G., S. J. Kweon, D. Y. Suh, D. J. Choi, B. S. Choi, C. Y. Kim, J. K. Sohn, and D. U. Kim (1996). Effects of magnetized water on callus formation and plant regeneration in rice anther culture. *Korean Journal of Crop Science 41:* 650-655.

Cho, M. S., and F. J. Zapata (1988). Callus formation and plant regeneration in isolated pollen culture of rice (*Oryza sativa* L. cv. Taipei 309). *Plant Science 58:* 239-244.

Chu, C. C. (1978). The −6 medium and its applications to anther culture of cereal crops. In *Proceedings: Symposium on plant tissue culture,* pp. 43-50. Peking. May 25-30, 1978. Science Press.

Chu, C. C., C. C. Wang, C. S. Sun, C. Hsu, K. C. Yin, C. Y. Chu, and F. Y. Bi (1975). Establishment of an efficient medium for anther culture of rice through comparative experiments on the nitrogen sources. *Scientia Sinica 18:* 659-668.

Chuba, R., S. Satoh, T. Sano, H. Sakurada, N. Yokoo, K. Yuhi, E. Kikuchi, T. Kuroki, M. Chuba, A. Yanakawa, and Y. Tanifugi (1997). Characteristics of a new rice variety, *Yamagata 54. Tohoku Agricultural Research 50:* 37-38 (in Japanese).

Courtois, B. (1993). Comparison of single seed descent and anther culture-derived lines of three single crosses of rice. *Theoretical and Applied Genetics 85:* 625-631.

Datta, K., N. Baisakh, N. Oliva, L. Torrizo, E. Abrigo, J. Tan, M. Rai, S. Rehana, S. Al-Babili, P. Beyer, et al. (2003). Bioengineered "golden" *indica* rice cultivars with ß-carotene metabolism in the endosperm with hygromycin and mannose selection systems. *Plant Biotechnology Journal 1:* 81-90.

Datta, K., S. Muthukrishnan, and S. K. Datta (1999). Expression and function of PR-protein genes in transgenic plants. In *Pathogenesis-related proteins in plants,* S. K. Datta and S. Muthukrishnan (Eds.), pp. 261-277. Boca Raton, FL: CRC Press.

Datta, S. K. (2000). Potential benefit of genetic engineering in plant breeding: Rice, a case study. *Agricultural and Chemical Biotechnology 43:* 197-206.

Datta, S. K. (2001). Androgenesis in cereals. In *Current trends in the embryology of angiosperms,* S. S. Bhojwani and W. Y. Soh (Eds.), pp. 471-488. Dordrecht, Netherlands: Kluwer.

Datta, S. K. (2005). Androgenic haploids: Factors controlling development and its application in crop improvement. *Current Science 89:*1870-1878.

Datta, S.K., K. Datta, V. Parkhi, M. Rai, N. Baisakh, G. Sahoo, S. Rehana, A. Bandyopadhyay, M. Alamgir, M. S. Ali, E. Abrigo, N. Oliva, and L. Torrizo (2007). Golden rice: Introgression, breeding, and field evaluation. *Euphytica 154:* 271-278.

Datta, S. K., K. Datta, and I. Potrykus (1990a). Embryogenesis and plant regeneration from microspores of both *indica* and *japonica* rice *(Oryza sativa). Plant Science 67:* 83-88.

Datta, S. K., A. Peterhans, K. Datta, and I. Potrykus (1990b). Genetically engineered fertile *indica* rice plants recovered from protoplasts. *Bio/Technology 8:* 736-740.

Feng, S. H., and Y. S. Li. (1999). Effect of agar concentration in media on green plantlet regeneration frequency in rice anther culture. *Chinese Rice Research Newsletter 7*(2): 7-8.

Feng, Y., and Q. Z. Xue (2001). Homozygous basta resistance rice *(Oryza sativa* L.) lines obtained via anther culture. *Journal of Agricultural Biotechnology 9:* 330-333.

Food and Agriculture Organization of the United Nations *(FAOSTAT)* (2004). http://faostat.fao.org/ (last accessed February 25, 2007).

Fu, Y. A., H. M. Si, Z. G. Zhu, G. C. Hu, and Z. X. Sun (2001). Homozygous transgenic plants regenerated from rice pollen derived callus via *Agrobacterium tumefaciens*. *Journal of Zhejiang University (Agriculture and Life Sciences) 27:* 407-410.

Gau, S. L., C. C. Yeh, J. Y. Shu, and H. S. Tsay (1990). Effects of agar brand and gelling agent on rice anther culture. *Journal of Agricultural Research in China 39:* 14-20.

Ghosh Biswas, G. C., and F. J. Zapata (1993). High-frequency plant regeneration from protoplasts of *indica* rice (*Oryza sativa* L.) using maltose. *Journal of Plant Physiology 141:* 470-475.

Gregorio, G. B., D. Senadhira, and R. D. Mendoza (1997). Screening rice for salinity tolerance. *IRRI Discussion Paper Series No. 22.* Manila, Philippines: International Rice Research Institute.

Guha-Mukherjee, S. (1973). Genotypic differences in the in vitro formation of embryoids from rice pollen. *Journal of Experimental Botany 24:* 139-144.

Guiderdoni, E. (1989). Comparative gametic selection in sexual and anther culture derivatives from five distant F1 hybrids of rice. *Rice Genetics Newsletter 6:* 91-93.

Guiderdoni, E. (1991). Gametic selection in anther culture of rice (*Oryza sativa* L.). *Theoretical and Applied Genetics 81:* 406-412.

Guiderdoni, E., B. Courtois, and J. C. Glaszmann (1989). Use of isozyme markers to monitor recombination and assess gametic selection among anther culture derivatives of remote crosses of rice (*Oryza sativa* L.). In *Review of advances in plant biotechnology 1985-88:2nd International Symposium on Genetic Manipulation in Crops,* A. Mujeeb-Kazi and L. A. Sitch (Eds.), pp. 43-55. Mexico and Manila: CIMMYT and International Rice Research Institute.

Habayashi, T., S. Misoo, O. Kamijima, and M. Sawano (1987). Effects of several amino acids in the medium on the anther culture of rice. *Japan Journal of Breeding 37*(Suppl.1): 16-17.

Harada, T., T. Sato, D. Asaka, and I. Matsukawa (1991). Large-scale deletions of rice plastid DNA in anther culture. *Theoretical and Applied Genetics 81:* 157-161.

Hayasaka, H., B. Chiba, K. Nagano, and K. Matsunaga (1997). Characteristics of a new rice cultivar, *Manamusume. Tohoku Agricultural Research 50:* 29-30 (in Japanese).

Hayasaka, H., K. Matsunaga, K. Nagano, H. Takizawa, and T. Sasaki (1993). Characteristics of a new rice cultivar, *Kokoromachi,* developed by anther culture. *Tohoku Agricultural Research 46:* 9-10 (in Japanese).

He, P., J. Z. Li, and L. H. Zhu (1998a). The interaction between quantitative trait loci for anther culturability in rice. *Acta Genetica Sinica 26:* 524-528.

He, P., L. S. Shen, C. F. Lu, Y. Chen, and L. H. Zhu (1998b). Analysis of quantitative trait loci which contribute to anther culturability in rice (*Oryza sativa* L.). *Molecular Breeding 4:* 165-172.

Heilungkiang Academy Agricultural Experiment Station in Rice (1975). Some experiences of breeding the new cultivar *Tan Feng No. 1* of rice by means of anther culture in vitro. *Acta Botanica Sinica 17:* 268-272 (in Chinese).

Heszky L. E., and I. Simon-Kiss (1992). *Dama,* the first plant variety of biotechnology origin in Hungary, registered in 1992. *Hungarian Agricultural Research 1*(1): 30-32.

Hoshi, T., S. Abe, K. Ishizaki, S. Azuma, K. Kobayashi, K. Higuchi, T. Tamura, A. Harada, M. Ozeki, H. Kanayama, et al. (1998). On the new rice cultivar *Hashiriaji. Journal of Niigata Agricultural Experiment Station 43:* 1-13.

Huang, N., A. Parco, T. Mew, G. Magpantay, S. McCouch, E. Guiderdoni, J. C. Xu, P. Subudhi, E. R. Angeles, and G. S. Khush (1997). RFLP mapping of isozymes, RAPD and QTLs for grain shape, brown planthopper resistance in a double haploid rice population. *Molecular Breeding 3:* 105-113.

International Network for Genetic Evaluation of Rice (1996). *Standard evaluation system for rice* (4th ed.). Los Baños, Philippines: International Rice Research Institute.

International Rice Research Institute (1980). *The International Rice Research Institute annual report for 1979.* Los Baños, Philippines: IRRI.

Iwai, M., S. Yoshida, and K. Watanabe (1990). Effect of NH_4^+ in callus formation medium and regeneration medium on anther culture of rices. *Kinki Chugoku Agricultural Research 80:* 14-17.

Ji, B. J., S. Y. Jiang, Q. F. Chen, W. M. Li, J. M. Qi, and D. M. Mao (1998). Effects of activated charcoal on rice anther culture. *Journal of Fujian Agricultural University 27*(1): 16-19.

Kameshima, M., Y. Nakamura, M. Mizobuchi, and H. Uga (1998). A new rice extremely early line (Kouiko 55). *Bulletin of Kochi Agricultural Research Center 7:* 71-79.

Kamoshita, A., J. X. Zhang, J. Siopongco, S. Sarkarung, H. T. Nguyen, and L. J. Wade (2002). Effects of phenotyping environment on identification of quantitative trait loci for rice root morphology under anaerobic conditions. *Crop Science 42:* 255-265.

Kazuhiro, N., and K. Hattori (1997). Effect of ^{60}Co gamma-ray irradiation at different culture stages on rice anther culture. *Breeding Science 47:* 101-105.

Khachatourians, G. G., A. McHughen, R. Scorza, W. K. Nip, and Y. H. Hui (Eds.). (2002). *Transgenic plants and crops.* New York: Marcel Dekker.

Kikuchi, H., and F. Kunihiro (1991). A new paddy rice variety, *Aya. Nogyo Gijutsu 46:* 472.

Kim, H. S., Y. T. Lee, T. S. Kim, and S. Y. Lee (1989). Effects of rice and potato extracts and activated charcoal on callus formation and plant regeneration in rice anther culture. *Research Report of Rural Development Administration (Biotechnology) 31*(1): 1-5.

Kim, H. S., Y. T. Lee, S. Y. Lee, and T. S. Kim (1991). Anther culture efficiency in different rice genotypes under different cold pretreatment durations and culture temperatures. *Research Report of Rural Development Administration (Biotechnology) 33*(1): 5-13.

Kim, H. Y., H. P. Moon, I. S. Choi, K. H. Kang, S. Y. Cho, H. W. Park, H. G. Hwang, H. C. Choi, N. K. Park, Y. G. Choi, et al. (1998). An anther-derived high quality and lodging tolerant variety adaptable to direct seeding, *Hwamyeongbyeo. Korean Journal of Breeding 30:* 382.

Kishikawa, H., Y. Takagi, M. Agashira, H. Yamashita, and Y. Takamori (1985). Morphological variations in the progenies of plants regenerated from anther culture of tetraploid rice. *Bulletin of the Faculty of Agricultural Saga University 60:* 37-48.

Kunihiro, Y., Y. Ebe, N. Shinbashi, H. Kikuchi, H. Tanno, and K. Sugawara (1993). A new paddy rice variety *Aya* with good eating quality due to low amylose content developed by anther culture breeding. *Japanese Journal of Breeding 43:* 155-164.

Kwon, Y. S., K. M. Kim, M. Y. Eun, and J. K. Sohn (2002a). QTL mapping and associated marker selection for the efficacy of green plant regeneration in anther culture of rice. *Plant Breeding 121:* 10-16.

Kwon, Y. S., K. M. Kim, D. H. Kim, M. Y. Eun, and J. K. Sohn (2002b). Marker-assisted introgression of quantitative trait loci associated with plant regeneration ability in anther culture of rice (*Oryza sativa* L.). *Molecules and Cells 14:* 24-28.

Lakshmini Narayanan, S., S. M. Ibrahim, S. Ashok, M. M. Mohan, and K. P. Kalidhasan (2000). Effect of abscisic acid in enhancing plant regenerability in rice. *Rice Biotechnology Quarterly 41:* 16.

Leach, J. E., H. Leung, R. J. Nelson, and T. W. Mew (1995). Population biology of *Xanthomonas oryzae* pv. oryzae and approaches to its control. *Current Opinion in Biotechnology 6:* 298-304.

Lee, K., S. Y. Lee, H. J. Kang, Y. T. Lee, T. H. Noh, H. T. Shin, S. Y. Lee, S. Y. Cho, C. H. Kim, M. S. Shin, et al. (1996). An anther-derived good quality and high yielding rice variety, *Hwashinbyeo. Korean Journal of Breeding 28:* 480.

Lee, Y. T., M. S. Lim, H. S. Kim, H. T. Shin, C. H. Kim, S. H. Bae, and C. I. Cho (1989). An anther-derived new high quality rice variety with disease and insect resistance, *Hwacheonbyeo. Research Report of Rural Development Administration (Rice) 31*(2): 27-34.

Lentini, Z., P. Reyes, C. P. Martinez, and W. M. Roca (1995). Androgenesis of highly recalcitrant rice genotypes with maltose and silver nitrate. *Plant Science 110:* 127-138.

Li, W. M., Q. F. Chu, R. S. Pan, S. Zhen, J. M. Qi, G. L. Lin, S. Y. Jiang, L. H. Lin, B. J. Ji, and X. A. Zheng (1998). Breeding of *Hua 1A,* a rice WA-CMS line produced by anther culture. *Journal of Fujian Agricultural University 27:* 129-132.

Liu, M. H., X. M. Lou, Z. Y. Wang, J. L. Zhang, and M. M. Hong (1995). Introduction of maize transposable element Ac/Ds into rice anther-derived suspension cells and plant regeneration. *Acta Phytophysiologica Sinica 21:* 195-205.

Lorieux, M., M. Petrov, N. Huang, E. Guiderdoni, and A. Ghesquiere (1996). Aroma in rice: Genetic analysis of a quantitative trait. *Theoretical and Applied Genetics 93:* 1145-1151.

Lu, C., L. Shen, Z. Tan, Y. Xu, P. He, Y. Chen, and L. Zhu (1996). Comparative mapping of QTLs for agronomic traits of rice across environments using a doubled haploid population. *Theoretical and Applied Genetics 93:* 1211-1217.

Maeda, H., T. Tsuchiya, N. Hiraoka, M. Maeshige, S. Uemoto, and M. Nakayabu (1990). New cold resistant rice varieties *Hirohikari* and *Hirohonami*. *Bulletin of Hiroshima Agricultural Experiment Station 58:* 9-20 (in Japanese with English summary).

Maeda, M., O. Ito, M. Nakayabu, S. Nakazawa, H. Maeda, T. Hoshina, T. Tsuchiya, and Y. Doi (2000). A new paddy rice variety *Hiroshima 21* (brand name: *Koimomiji*). *Bulletin of Hiroshima Prefecture Agricultural Research Center 68:* 1-11.

Marassi, M. A., J. J. Marassi, J. E. Marassi, and L. A. Mroginski. (2000). *Petei* and *Mocoi:* Two rice cultivars developed through anther culture in Argentina. *International Rice Research Notes 25*(1): 10-11.

Martinez, C. P., F. C. Victoria, M. C. Amezquita, E. Tulande, G. Loema, and R. S. Zeigler (1996). Comparison of rice lines derived through anther culture and the pedigree method in relation to blast (*Pyricularia grisea* Sacc.) resistance. *Theoretical and Applied Genetics 92:* 583-590.

Matsushima, T., S. Kikuchi, F. Takaiwa, and K. Oono (1988). Regeneration of plants by pollen culture in rice (*Oryza sativa* L.). *Plant Tissue Culture Letters 5:* 78-81.

Mew, T. W., C. M. Vera Cruz, and E. S. Medalla (1992). Changes in race frequency of *Xanthomonas oryzae* pv. oryzae in response to rice cultivars planted in the Philippines. *Plant Disease 76:* 1029-1032.

Miah, M. A. A., E. D. Earl, and G. S. Khush (1985). Inheritance of callus formation ability in anther cultures of rice, *Oryza sativa* L. *Theoretical and Applied Genetics 70:* 113-116.

Miah, M. A. A., M. S. Pathan, and H. A. Quayum (1996). Production of salt tolerant rice breeding line via doubled haploid. *Euphytica 91:* 285-288.

Mizobuchi, M., Y. Nakamura, M. Kameshima, H. Uga, M. Uehiyashi, and H. Moriyama (1998). A new rice line for sake brewing *(Kouikusake 54)*. *Bulletin of Kochi Agricultural Research Center 7:* 57-69.

Moon, H. P. (1995). An anther-derived high quality and lodging-tolerant rice variety, *Hwajungbyeo*. *RDA (Rural Development Administration) Journal of Agricultural Science (Rice) 37*(1): 75-82.

Moon, H. P., S. Y. Cho, H. Y. Kim, K. H. Kang, H. C. Choi, N. K. Park, I. S. Choi, Y. G. Choi, Y. H. Son, R. K. Park, and C. Y. Cho (1993). An anther-derived disease resistance and high quality glutinous rice variety, *Hwaseonchalbyeo*. *RDA (Rural Development Administration) Journal of Agricultural Science (Rice) 35*(2): 55-63.

Moon, H. P., S. Y. Cho, R. K. Park, Y. H. Son, H. R. Kim, B. T. Jun, and K. H. Kang (1989a). An anther-derived and colchicine-doubled new high quality and high yielding rice variety, *Hwajinbyeo*. *Research Report of the Rural Development Administration (Rice) 31*(3): 1-10.

Moon, H. P., S. Y. Cho, Y. H. Son, B. T. Jun, M. S. Lim, H. C. Choi, N. K. Park, R. K. Park, and G. S. Chung (1986). An anther-derived new high quality and high yield rice variety, *Hwaseongbyeo*. *Research Report of the Rural Development Administration (Korea) Crops 28*(2): 27-33.

Moon, H. P., S. H. Choi, S. Y. Cho, and Y. H. Son (1988a). Effect of high agar medium on plant regeneration in rice (*Oryza sativa* L.) anther culture. *Korean Journal of Breeding 20:* 335-340.

Moon, H. P., S. H. Choi, K. H. Kang, and S. Y. Cho (1989b). Low-temperature pretreatment on callus induction and plant regeneration in rice anther culture. *Korean Journal of Breeding 21:* 47-54.

Moon, H. P., S. H. Choi, Y. H. Son, S. Y. Cho, and R. K. Park (1988b). Studies on one-step procedure in rice *(Oryza sativa)* anther culture. I. Direct generation of rice plantlets on NAA and 2,4-D containing N6 medium. *Korean Journal of Breeding 20:* 186-190.

Moon, H. P., K. H. Kang, and S. Y. Cho (1991). Studies on one-step procedure in rice *(Oryza sativa)* anther culture. II. Effects of 2,4-D, kinetin and ABA on direct plant generation. *Korean Journal of Breeding 23:* 153-160.

Moon, H. P., K. H. Kang, and S. Y. Cho (1994a). Aseptic mass collection of anthers for increasing efficiency of anther culture in rice breeding. *International Rice Research Notes, 19* (1), 30.

Moon, H. P., H. Y. Kim, S. Y. Cho, K. H. Kang, H. C. Choi, I. S. Choi, Y. S. Shin, K. W. Kim, Y. G. Choi, R. K. Park, and Y. S. Kim (1994b). An anther-derived high quality rice variety, *Hwajoongbyeo. Korean Journal of Breeding 26:* 451.

Moon, H. P., H. Y. Kim, K. H. Kang, I. S. Choi, S. Y. Cho, H. W. Park, N. G. Park, H. G. Hwang, H. C. Choi, J. D. Yea, et al. (1998). An anther-derived, early-maturing and cold-tolerant rice variety, *Hwadongbyeo. Korean Journal of Breeding 30:* 382.

Murashige, T., and F. Skoog (1962). A revised medium for rapid growth and bioassays with tobacco tissue cultures. *Physiologia Plantarum 15:* 473-497.

Nakamura, Y., M. Kameshima, M. Mozubuchi, H. Uga, and Y. Morita (1996). A new scented rice variety, *Kouikuka 37. Bulletin of Kochi Agricultural Research Center 5:* 38-49.

Niizeki, H. (1983). Uses and application of anther and pollen culture in rice. In *Cell and tissue culture techniques for cereal crop improvement.* Proceedings of a Workshop cosponsored by the Institute of Genetics, Academia Sinica, and the International Rice Research Institute, pp. 165-171. Beijing: Science Press.

Niizeki, H., and K. Oono (1968). Induction of haploid rice plant from anther culture. *Proceedings of the Japan Academy 44:* 554-557.

Ogawa, T., H. Fukuoka, and Y. Ohkawa (1994). Induction of cell division of isolated pollen grains by sugar starvation in rice. *Breeding Science 44:* 75-77.

Oh, B. G., S. J. Lim, S. J. Yang, H. G. Hwang, G. H. Yi, U. S. Yeo, N. B. Park, D. Y. Kwak, S. C. Kim, H. Y. Kim, et al. (1997). An anther derived high eating quality and high yielding rice variety with resistance to bacterial leaf blight, *Hwasambyeo. RDA (Rural Development Administration) Journal of Crop Science 39*(2): 94-100.

Oono, K. (1983). Genetic variability in rice plants regenerated from cell culture. In *Cell and tissue culture techniques for cereal crop improvement,,* pp. 95-104. Proceedings of a Workshop cosponsored by the Institute of Genetics, Academia Sinica, and the International Rice Research Institute. Beijing: Science Press.

Oosato, K. F., Y. Hamachi, and S. Imabayashi (1999). Comparison of the anther culture and conventional breeding lines in rice and a new elite line, *Chikushi 26*, developed by the anther culture method. *Japanese Journal of Crop Science 68:* 440-443.

Patil, V. D., Y. S. Nerkar, M. B. Misal, and S. R. Harkal (1997). *Parag 401*, a semidwarf rice variety developed through anther culture. *International Rice Research Notes 22*(2): 19.

Perez-Almeida, I., Z. Lentini, and E. P. Guimaraes (1995). Anther culture for developing rice blast *(Pyricularia grisea)* resistant germplasm. *Fitopatologia Venezuela 8:* 11-14.

Peterhans, A., S. K. Datta, K. Datta, G. H. Godall, I. Potrykus, and J. Paszkowski (1990). Recognition of efficiency of Dicotyledoneae-specific promoter and RNA processing signals in rice. *Molecular and General Genetics 221:* 362-368.

Pramanik, S. C., and A. B. Mandal (2000). Androclonal variation in major agronomic traits in an *indica* rice. *SABRAO Journal of Breeding and Genetics 32:* 119-120.

Qian, Q., P. He, S. Teng, D. Zeng, and L. Zhu (1999a). QTL analysis of rice *(Oryza sativa* L.) tiller angle in a double haploid population derived from anther culture of *indica/japonica. Chinese Rice Research Newsletter 7*(4): 1-2.

Qian, Q., D. Zhen, F. Huang, P. He, X. W. Zheng, Y. Chen, and L. H. Zhu (1999b). The QTL analysis of seedling cold tolerance in a double haploid population derived from anther culture of *indica/japonica. Chinese Rice Research Newsletter 7*(2): 1-2.

Quimio, C. A., and F. J. Zapata (1990). Diallel analysis of callus induction and green-plant regeneration in rice anther culture. *Crop Science 30:* 188-192.

Raina, S. K. (1989). Tissue culture in rice improvement: Status and potential. *Advances in Agronomy 42:* 339-398.

Raina, S. K. (1997). Doubled haploid breeding in cereals. *Plant Breeding Research 15:* 141-186.

Raina, S. K., and S. T. Irfan (1998). High-frequency embryogenesis and plantlet regeneration from isolated microspores of *indica* rice. *Plant Cell Reports 17:* 957-962.

Sano, T., H. Satoh, H. Sakurada, T. Kikuchi, M. Chuba, Y. Tanifuzi, T. Kuroki, R. Chuba, A. Yamakawa, N. Yokoo, and Y. Kazuhiro (1998). On the breeding of the new paddy rice variety, *Yamagata No. 54. Bulletin of Yamagata Prefecture Agricultural Experiment Station 32:* 1-19.

Sasaki, K., N. Shinbashi, K. Sasaki, M. Aikawa, T. Yanagawa, and Y. Numao (1988). New rice variety *Joiku No. 394. Bulletin of Hokkaido Prefecture Agricultural Experiment Station 58:* 13-23 (in Japanese with English summary).

Sawhney, V. K. (2002). Pollen biotechnology. In *Transgenic plants and crops,* G. G. Khachatourians, A. McHughen, R. Scorza, W. K. Nip, and Y. H. Hui (Eds.), pp. 99-107. New York: Marcel Dekker.

Si, H. W., Y. P. Fu, H. Xiao, G. C. Hu, J. P. Cao, D. N. Huang, and Z. X. Sun (1999). Homozygous lines of transgenic rice *(Oryza sativa* L.) obtained via anther culture. *Chinese Journal of Rice Science 13*(1): 19-24.

Snape, J. W. (1982). The use of doubled haploids in plant breeding. In *Induced variability in plant breeding,* pp. 52-58. Wageningen, the Netherlands: Center for Agricultural Publishing and Documentation.

Sripichitt, P., P. Pongtongkam, and A. Vanavichit (1994). Breeding of aromatic rice (*Oryza sativa* L.) variety KDML 105 for photoperiod insensitivity through anther culture of F1 hybrids. *Kasetsart Journal 28:* 499-511.

Su, R. V., M. L. Rudert, and T. K. Hodges (1992). Fertile *indica* and *japonica* rice plants regenerated from protoplasts isolated from embryogenic haploid suspension cultures. *Plant Cell Reports 12:* 45-49.

Subhadra, V. V., and G. M. Reddy (1998). Peroxidase, a marker for regeneration potential in anther culture of *indica* rice. *Oryza 35:* 363-364.

Sugimoto, K., S. Nohara, S. Otsubo, S. Suda, M. Muto, Y. Kawase, and H. Kirii (1991). New rice variety *Shirayukihime* developed by anther culture. *Bulletin of Gifu Agricultural Research Center 4:* 36-41 (in Japanese).

Suh, H. S., and Y. C. Song (1991). Accumulation of marker genes by using genetic male sterility and anther culture technique in rice. *Korean Journal of Breeding 23:* 64-67.

Sun, Z. X., and H. M. Si (1996). Anther culture of transgenic rice plants. *Chinese Rice Research Newsletter* 4(4): 2-3.

Sun, Z. X., H. M. Si, S. H. Cheng, and X. Y. Zhan (1993). Effect of maltose on efficiency of anther culture of rice. *Chinese Journal of Rice Science 7:* 227-231.

Suzuki, M. (1992). New summer crop varieties registered by the Ministry of Agriculture, Forestry and Fisheries in 1991. Paddy rice, soybean, peanut, buckwheat, sugarcane. *Japanese Journal of Breeding 42:* 167-175 (in Japanese with English summary).

Tan, Z. B., L. S. Shen, H. C. Kuang, C. F. Lu, Y. Chen, K. D. Zhou, and L. H. Zhu (1997a). Identification of QTLs for lengths of the top internodes and other traits in rice and analysis of their genetic effects. *Chinese Journal of Genetics 24:* 15-22.

Tan, Z. B., L. S. Shen, Z. L. Yuan, C. F. Lu, Y. Chen, K. D. Zhou, and L. H. Zhu (1997b). Identification of QTLs for rationing ability and grain yield traits of rice and analysis of their genetic effects. *Acta Agronomica Sinica 23:* 289-295.

Teng, S., D. Zeng, X. W. Zheng, K. Yasufumi, Q. Qian, and L. H. Zhu (2000). QTL analysis of rice (*O. sativa* L.) root vitality in a double haploid population derived from anther culture of an *indica/japonica* cross. *Chinese Rice Research Newsletter* 8(1): 3-4.

Tiensin Agricultural Research Institute, Rice Research Laboratory. (1976). New rice varieties *Hua Yu I* and *Hua Yu II* developed from anther culture. *Acta Genetica Sinica* 3(1): 19-24.

Torrizo, L. B., and F. J. Zapata (1986). Anther culture in rice. IV. The effect of abscisic acid on plant regeneration. *Plant Cell Reports 5:* 136-139.

Tsuchiya, K. (1992). Development of new rice varieties, *Hirohikari* and *Hirohonami,* used haploid method of breeding by anther culture. In *Plant tissue culture and gene manipulation for breeding and formation of phytochemicals,* K. Oono, T. Hirabayashi, S. Kikuchi, H. Handa, K. Kajiwara, and K. Tsukuba (Eds.), pp. 191-196. Tsukuba, Japan: National Institute of Agrobiological Resources.

Tsuchiya, T., Z. Katsuba, Y. Doi, M. Urano, O. Ito, Y. Sakai, S. Uemoto, M. Imoto, M. Kurao, K. Nishida, et al. (2001). A new rice variety *Senbonnishiki* for making sake. *Bulletin of Hiroshima Prefecture Agricultural Research Center 69:* 25-36.

Wang, Z. X., and N. Iwata (1991). Production of n+1 plants and tetrasomics by means of anther culture of trisomic plants in rice (*Oryza sativa* L.). *Theoretical and Applied Genetics 83:* 12-16.

Wang, Z. X., and N. Iwata (1995). Aneuhaploids and tetrasomics in rice (*Oryza sativa* L.) derived from anther culture of trisomics. *Genome 38:* 696-705.

Watanabe, T., K. Wada, Y. Nishikawa, T. Nagashima, R. Hayashi, Y. Ito, M. Ohara, and A. Fujie (1998). Breeding of a new rice cultivar *Fusaotome* with early maturity, high cold resistance, high grain quality and excellent eating quality. *Bulletin of the Chiba Prefecture Agricultural Experiment Station 39:* 15-26.

Watanabe, Y. (1974). Meiotic chromosome behaviours of auto-pentaploid rice plant derived from anther culture. *Cytologia 39:* 283-288.

Xie, J. H., M. W. Gao, Q. H. Kai, X. Y. Chen, Y. W. Shen, and Z. Q. Liang (1995). Improved isolated microspore culture efficiency in medium with maltose and optimized growth regulator combination in *japonica* rice (*Oryza sativa*). *Plant Cell, Tissue and Organ Culture 42:* 245-250.

Yamagishi, M., M. Otani, M. Higashi, Y. Fukuta, K. Fukui, M. Yano, and T. Shimada (1998). Chromosomal regions controlling anther culturability in rice (*Oryza sativa* L.). *Euphytica 103:* 227-234.

Yamagishi, M., M. Yano, Y. Fukuta, K. Fukui, M. Otani, and T. Shimada (1996). Distorted segregation of RFLP markers in regenerated plants derived from anther culture of an F1 hybrid of rice. *Genes and Genetic Systems 71:* 37-41.

Ye, X., S. Al-Babili, A. Kloti, J. Zhang, P. Lucca, P. Beyer, and I. Potrykus (2000). Engineering the provitamin A (β-carotene) biosynthetic pathway into (carotenoid-free) rice endosperm. *Science 287:* 303-305.

Zaid, A., H. G. Hughes, E. Porceddu, and F. W. Nicholas (1999). *Glossary of biotechnology and genetic engineering.* Rome: Food and Agriculture Organization of the United Nations.

Zang, Z. M., and Z. H. Zhang (1991). Studies on the effect of bleeding sap of towel gourd (*Luffa cylindrica* Roem) on rice anther culture. *Acta Agronomica Sinica 17:* 352-361.

Zapata, F. J., M. S. Alejar, L. B. Torrizo, A. U. Novero, V. P. Singh, and D. Senadhira (1991). Field performance of anther-culture-derived lines from F1 crosses of *indica* rices under saline and nonsaline conditions. *Theoretical and Applied Genetics 83:* 6-11.

Zapata, F. J., and L. B. Torrizo (1986). Heat treatment to increase callus induction efficiency in anther culture of IR42. *International Rice Research Newsletter 11*(4): 25-26.

Zapata, F. J., L. B. Torrizo, R. R. Aldemita, A. U. Novero, A. M. Mazaredo, R. V. Visperas, M. S. Lim, H. P. Moon, and M. H. Heu (1986). Rice anther culture: A tool for production of cold tolerant lines. In *Rice genetics*, pp. 773-780. Los Baños, Laguna, Philippines: International Rice Research Institute.

Zeng, J. H. (1985). Effect of 2-deoxyglucose on rice anther culture. *Acta Phytophysiologica Sinica 11:* 328-335.

Zhang, C. M., J. A. Lu, and Z. H. Zhang (1991). Genetic analysis of bacterial blight (BB) resistance in rice anther culture progenies. *International Rice Research Newsletter 16*(6): 7-8.

Zhang, J., H. G. Zheng, A. Aarti, G. Pantuwan, T. T. Nguyen, J. N. Triparthy, A. K. Sarial, S. Robin, R. C. Babu, B. D. Nguyen, et al. (2001). Locating genomic regions associated with components of drought resistance in rice: Comparative mapping within and across species. *Theoretical and Applied Genetics 103:* 19-29.

Zhang, X. Y., Z. X. Sun, H. M. Si, and H. S. Cheng (1992). Effect of maltose on efficiency of anther culture of rice (*Oryza sativa* L.). In *Agricultural biotechnology,* C. B. You and Z. L. Chen (Eds.), pp. 668-670. Beijing: China Science and Technology.

Zhang, Z. H., Z. L. Zheng, Y. H. Gao, and H. X. Cao (1988). Breeding, evaluation and utilization of anther-cultured varieties *Xin xiou* and *Hua Han Zao* in rice (*Oryza sativa* L.). In *Genetic manipulation in crops,* pp. 36-37. London: International Rice Research Institute, Philippines, and Cassell Tycooly.

Zheng, K., N. Huang, J. Bennett, and G. S. Khush (1995). PCR-based marker-assisted selection in rice breeding. *IRRI Discussion Paper Series No. 12.* Los Baños, Philippines: International Rice Research Institute.

Zhu, D. Y., X. G. Pan, C. Y. Chen, X. H. Ding, J. H. Yi, and Z. Y. Chang (1995). *Ganzhaoxian 11,* a product of anther culture in *indica* rice. *Chinese Rice Research Newsletter 3*(1): 10-11.

Zhu D. Y., and X. G. Pan (1990). Rice (*Oryza sativa* L.): *Guan 18,* an improved variety through anther culture. In *Haploids in crop improvement I,* Y. P. S. Bajaj (Ed.), pp. 204-211. Berlin: Springer-Verlag.

Zhu, D. Y., C. Y. Pen, and X. G. Pan (1990). The development and evaluation of a new variety *Shan Hua 369* through anther culture in *indica* rice. *Genetic Manipulation in Plants 6*(2): 7-14.

Zhu, D. Y., Z. X. Sun, X. G. Pan, X. H. Ding, X. H. Shen, Y. Wan, H. Pan, J. H. Yin, M. S. Alejar, L. B. Torrizo, and S. K. Datta (1998). Use of anther culture in hybrid rice breeding. In *Advances in hybrid rice technology,* S. S. Virmani, E. A. Siddiq, and K. Muralidharan (Eds.), pp. 265-281. Proceedings of the 3rd International Symposium on Hybrid Rice, Hyderabad, India.

Chapter 5

Hybrid Rice Technology

S. S. Virmani
I. Kumar

INTRODUCTION

In the past two to three decades, many rice-growing countries in South and Southeast Asia have been able to produce sufficient rice for their ever-burgeoning populations. By 2020, the world must produce 754 million tons of paddy rice, compared to 579 million tons produced in 2002. This challenge is more serious in the irrigated rice fields of South and Southeast Asia, particularly in rice bowl areas, where yield gains through conventional high-yield varieties (HYVs) have plateaued because most of the exploitable yield gaps between researchers' yield and farmers' yield have more or less been bridged. Among the various technological options currently available to increase rice yield beyond the levels of presently grown high-yield varieties, hybrid rice has proven to be the most successful. The superiority of rice hybrids has been amply demonstrated in China for the past 25 years. China currently covers 50% of its total rice area (30 million ha) with hybrid rice varieties, which contribute 60% of its total rice production (Ma and Yuan, 2002) due to their higher yield (6.9 tons/ha) than that of the inbred HYVs (5.4 tons/ha). The higher yield potential of hybrid rice contributes to food security, saves land and water for diversified farming, and reduces the cost of rice production per unit area.

Since 1979, the International Rice Research Institute (IRRI) has been involved in developing hybrid rice technology in countries outside China and has helped about 20 countries in developing this technology. During the past 10 years, this technology has been commercialized in Vietnam, India, the Philippines, Bangladesh, and the United States. In 2002, about 800,000 ha were covered with hybrid rice varieties in these countries, resulting in a yield advantage of about 1-1.5 tons/ha over inbred HYVs. During the next five years, several other countries are likely to commercialize this technology. This chapter highlights the progress made in developing and using hybrid rice technology around the world.

EXTENT AND PHYSIOLOGICAL BASIS OF HETEROSIS

Virmani (1994) reviewed the literature on the subject of hybrid rice up to 1990 and reported the presence of significant standard heterosis for yield and yield components. Growth duration did not correlate with the expression of heterosis. Maruyama et al. (1982, 1984, 1987) reported heterosis for photosynthetic efficiency. Other traits reported to increase include leaf area, shoot, dry weight, cumulative growth rate (Blanco et al., 1986; Sunohara et al., 1985; Yamauchi and Yoshida, 1985), and leaf area index (Ponnuthurai et al., 1984). Akita (1988) reported that heterosis in leaf area development during the seeding stage was the key to vigorous growth after transplanting. Heterosis for root characteristics, such as a stronger and active root system, number of roots, root thickness, root dry weight, root density, and so forth, has been reported by Lin and Yuan (1980), Wang and Yoshida (1984), Namuco et al. (1988), and Sarawagi and Srivastava (1988). Increased yield of rice hybrids has been attributed to their increased dry matter, resulting from a higher leaf area index (LAI), and their higher crop growth rate and increased harvest index, resulting from their increased spikelet number and to some extent increased grain weight (Agata, 1990; Akita et al., 1986; Ponnuthurai et al., 1984).

Recently, Wang, X. et al. (2001, 2002) compared the physiological basis of super-high-yield hybrid rice *Xieyou 9308* (11-12 tons/ha) to that of another high-yield hybrid, *Xieyou 63* (9.8 tons/ha). Biomass per stem, dry weight per stem during the panicle initiation stage, and the dry matter accumulated after the full heading stage were higher by 40% in the super-high-yield hybrid than in the normal high-yield hybrid. The tillering in the seedbed stage reached 3 to 4, and the effective tillering process terminated earlier (about two weeks after transplanting) in super-high-yield hybrids. The latter also had a compact plant type, and the light intensity at quarter top canopy was 50% that of total canopy. The peak grain filling rate was observed two weeks after flowering, whereas in normal high-yield hybrids, it occurred as late as three to four weeks after flowering. The photosynthetic rate of flag leaf decreased slowly, and the leaves still had 5-10 μmol CO_2, whereas in other high-yielding hybrids, it varied from 10 to 20 μmol. The root exudate intensity per stem was higher than 1 gm at the yellow ripening stage in super-high-yielding hybrid *Xieyou 9308*, whereas it was <0.5 gm in hybrid *Xieyou 63*. China has now characterized super-high-yielding hybrids as those that give a yield of 100 kg/ha/day and possess the following desirable features:

- Long erect flag leaf (50 cm) that is about 20 cm above the panicle top, second leaf from the top 10% longer than the flag leaf, and the third leaf should reach the middle of the panicle. Plant has narrow and V-shaped leaves, which maintain erect leaf angles until maturity.
- Dry weight of upper three leaves to be 0.55 gm/100 cm^2, compared to 0.4-0.45 gm/100 cm^2 for common HYVs.
- Moderately compact plant type, with panicle top about 60 cm from the ground.
- Average panicle weight nearly 5 gm, with 2.73 million panicles/ha.
- Leaf area index higher than 6.0 based on upper three leaves, with ratio of leaf area to grain weight of 100/2.2-2.3.
- Harvest index about 0.55.

Studies at IRRI (International Rice Research Institute, 1993) and in Vietnam (Guong et al., 1995) have indicated that hybrid rice has a higher N use efficiency than inbred lines. Shi et al. (2002) reported higher efficiency of N uptake in hybrid rice compared to their parents under different N levels. Superiority of hybrid rice over their parents was evident in the stronger ability of roots to absorb N nutrition, higher number, higher weight, and higher activity of ATPase and oxidation ability of roots. However, hybrids were not superior to parents in N use efficiency. In their study on the relative N use efficiency of rice hybrids and inbreds, Reddy et al. (2002a) concluded that the superior yield performance of the hybrids was partly attributable to their N use efficiency but also to N-responsive characteristics across locations, as indicated by their better agronomic efficiency, physiological efficiency, and nitrogen harvest index. Reddy et al. (2002b) further concluded that, as hybrid rice has more sink size in the form of bigger panicles and more spikelets than inbred varieties, it could show a decrease in grain filling unless the crop is properly fertilized. The adequate nutritional status of the late growth stage—by virtue of its implication in the translocation of photosynthates to the sink and both loading and unloading of photosynthates— is expected to be maintained to better regulate the relation between source and sink. Thus, the full potential of higher sink size in hybrids can be realized through the synchronization of K supply at the late growth stage. Hybrids performed better than inbreds, with a mean yield increase of 22% due to basal K application and 26% due to split K application. It is therefore evident that hybrids require different strategies for nutrient management than inbreds to maximize the expression of their yield advantage.

DEVELOPING AND COMMERCIALIZING RICE HYBRIDS IN CHINA AND ELSEWHERE

The first rice hybrid was commercialized in China in 1976. Since then, much progress has been made in developing new parental lines of hybrids. More than 100 hybrids, which were developed at more than 50 provincial and prefectural research centers, have been released in different provinces of China. The major cytoplasmic male sterile (CMS) lines developed and used in the initial stages were V20A, *Er Jiu Nan 1A,* ZS 97A, V41A, and others. Among restorers, IRRI lines such as IR24R, IR26R, and IR9761-19R were frequently used in the first and second-generation rice hybrids. In the 1970s, rice hybrids such as *Wei You 2, Wei You 6, Shan You 2,* and *Shan You 6* were popular, and later on, hybrids like *Wei You 64, Shan You 63, Xie You 63, Nan You 2,* and *Wei You 49,* played a key role in developing further hybrids using female lines from China and male lines both from China and IRRI. *Wei You 64* was the first short-duration hybrid developed in China in collaboration with IRRI (Yuan et al., 1985). The area under hybrid rice cultivation increased from 8.43 million ha in 1985 to 15.6 million ha in 2002. In the mid-1980s, based on the country's average, the hybrids yielded 6.5 tons/ha compared to inbred varieties yielding 4.8 tons/ha. Rice hybrid *Shan You 63* covered the largest area at one time and is still a popular three-line hybrid. Currently, rice hybrids in China yield 6.9 tons/ha against 5.4 tons/ha by commercial HYVs, thus having an advantage of 1.5 tons/ha. In Sichuan and Jiangsu provinces, the average yield of 2.7 million ha of hybrid rice was 7.5 tons/ha. In Hunan province, the average yield of about 1 million ha covered with rice hybrids was 6 tons/ha, whereas the conventional high-yield varieties yielded 4.5 tons/ha. For smaller areas, an average yield of 7.6 tons/ ha was recorded in 1983 for *Wei You 6* planted on 24,000 ha. In 1984, hybrid *Shan You 2* averaged 11.3 tons/ha on a 62-ha area. The maximum yield of *japonica* hybrid *Li You 57* was 13.7 tons/ha in Fujieng province. In an experiment under special ecological conditions at Yougsheng county, Yunan province, the highest yield of a rice hybrid *(Shan You 63)* was recorded to be 15.27 tons/ha, with a biological yield of 28-29 tons/ha and a harvest index of 0.54 (Cheng and Shaokai, 2001).

Since the 1990s, significant progress has been made in China in developing parental lines and two-line hybrids based on the environment-sensitive genetic male sterility (EGMS) system. These lines have yielded either the same or 5-10% higher than the three-line hybrids (based on the cytoplasmic male sterility system) of the same maturity duration. The major two-line hybrids released in China were *Peiai 64S/Tequing, Peiai 64S/9311,* and *Peiai 64S/E-32.* The average yield of hybrid *Peiai 64S/9311* was 9.6 tons/ha (average of 232,000 hectares). A maximum yield of 17.1 tons/ha was obtained in a two-line hybrid on a small area of 487 m^2 in Yougshang, Yunnan Province, in 1999 (Luo, 2000); this was about 2 tons/ha higher than the highest yielding common variety, IR64. So far, nearly 20 two-line hybrids have been released in China. The area under two-line hybrids increased from 73,000 ha in 1995 to 2.67 million ha in 2002, covering a total of 17% hybrid rice area. The major emphasis in China now is to develop super-high-yield hybrids, with a yield of 100 kg/day/ha (Wang, X. et al., 2001, 2002).

Outside China, the development and spread of hybrid rice has been rather slow in the initial stages on account of the lack of commercially usable CMS lines. Chinese CMS lines (viz., V20A, Z S 97A, V41A, etc.), when introduced from China into tropical areas, were short in duration and suffered heavily from tropical rice diseases and insect pests. CMS lines suitable for tropical conditions were later developed by IRRI and shared freely with different collaborating countries. The hybrids outside China were developed primarily using one CMS line (IR58025A) that is adapted to tropical conditions, has long slender grains, good combining ability, and good outcrossing capability. Since the late 1990s, hybrid rice technology has been commercialized in India, Vietnam, and the Philippines. Recently, three-line hybrids have also been developed and released in Bangladesh and Indonesia. These hybrids yield higher than the local inbred HYVs, but their grain quality is not widely accepted due to their low head rice recovery, aroma, and stickiness on cooking. In the Philippines, however, these hybrids are preferred over common varieties such as IR64. New CMS lines have now been developed by IRRI that have no aroma and intermediate to high amylose content. These hybrids are being shared with different countries. Some of the first generation hybrids (e.g., IR69690H) have been widely adopted and released in India *(Sahyadari),* Bangladesh (as BRRI Dhan Hybrid 1), Vietnam (as HYT 57) and Indonesia *(Rokan).* The hybrids (viz., ProAgro 6201, 6444) developed by Hybrid Rice International, Inc., in India were adopted in the Philippines and Myanmar and yielded 1-2 tons/ha more than the best local common HYVs of similar growth duration. Yields up to 12 tons/ha or more have been obtained at various sites in India and the Philippines with three-line *indica* hybrids.

The Chinese hybrids have been found to adapt well in North Vietnam. However, their grain size, shape, and quality were not accepted in South Vietnam. In other countries like Bangladesh, Indonesia, India, the Philippines, Sri Lanka, Myanmar, and Thailand, the Chinese hybrids have not been accepted due to their relatively low adaptability, poor grain quality, and susceptibility to pests and diseases. Some new Chinese hybrids are now showing adaptability to tropical conditions, although their grain quality is still a problem.

The slower spread of hybrid rice technology outside China can be attributed to (1) inadequate experience with hybrid seed production; (2) inadequate seed industry infrastructure; (3) inadequate coordination between research, seed production, and extension agencies; (4) inadequate policy support; and (5) inadequate action plans and partnerships between the public and private sectors for the promotion of hybrid rice. Nevertheless, the area covered by hybrid rice has reached 0.75-0.8 million ha, and during the next 5 years, it may reach 1.5-2.0 million ha.

GENETIC TOOLS FOR DEVELOPING RICE HYBRIDS

Rice, being a self-pollinated crop, requires male sterility systems in order to develop hybrids. As a result of male sterility, rice spikelets do not produce functional anthers or viable pollen, but their ovaries function normally. Such spikelets cannot self-pollinate, but they can set hybrid seed through cross-pollination from a pollen parent grown along with the male sterile line in isolation. The most effective male sterility systems used so far to develop commercial rice hybrids are cytoplasmic-genetic (CMS) and environment-sensitive genetic male sterility (EGMS) systems.

Cytoplasmic-Genetic Male Sterility System

In rice, the role of cytoplasm in causing male sterility was first reported by Sampath and Mohanty (1954), Katsuo and Mizushima (1958), and Kitamura (1962). However, CMS lines using sterile cytoplasm were developed by Shinjyo and Omura (1966), Erickson (1969), Watanabe (1971), Carnahan et al. (1972), and Athwal and Virmani (1972). China took the lead in developing commercially usable CMS lines to develop commercial rice hybrids (Lin and Yuan, 1980). The CMS lines in China were developed in 1972, after the discovery of a male sterile plant of a wild rice (*O. sativa* f. *spontanea*) population on Hainan Island in 1970 (Yuan, 1977). The male sterility of this plant was maintained by several Chinese rice varieties. By 1973, the first set of restorers was identified, and the first rice hybrid was released by 1976 (Lin and Yuan, 1980). Subsequently, IRRI used this so-called wild abortive (WA) cytoplasm to develop *indica* CMS lines for developing tropical rice hybrids and distributed these to many countries. Recently, some private seed companies in India (viz., Hybrid Rice International, Pioneer) and the United States (viz., Rice Tec) have also developed proprietary CMS lines possessing WA cytoplasm for developing commercial hybrids.

The degree of pollen abortion in a CMS line depends on the interaction of cytoplasmic and nuclear genes. The extent of pollen sterility is higher in lines where pollen abortion occurs at the uninucleate stage (i.e., before the first pollen mitosis) and the morphology of pollen grain is irregular. Such pollen sterility is usually stable and does not break down. If the pollen abortion occurs at a later stage, the pollen grains are nearly round and sometimes even lightly stained with IKI solution.

CMS lines suitable for developing commercial rice hybrids should possess (1) stable and complete male sterility, (2) easy fertility restoration, (3) cytoplasmic and genetic diversity, (4) acceptable grain and cooking quality, (5) improved outcrossing rate, (6) resistance to diseases and insects, and (7) high general combining ability.

Cytoplasm Diversification

In China, nearly 85% of the CMS lines used in developing *indica* hybrids have WA cytoplasm (Yuan, 1998), whereas outside of China, nearly 100% of the CMS lines used in developing commercial hybrids belong to the WA type. Although concerns have been shown about the use of only one type of cytoplasm, so far, WA cytoplasm has presented no problem in spite of its cultivation on a very large scale. Studies conducted at IRRI (Faiz, 2000) did not show any association between the WA-based CMS system and resistance or tolerance to several biotic and abiotic stresses.

Efforts are being made to identify CMS systems that would overcome the risks of a sudden outbreak of some disease or insect pest associated with the widely used WA CMS system. Various CMS sources discovered in rice, compiled by Virmani and Shinjyo (1988), Li and Zhu (1988), Virmani (1994), and Pradhan et al. (1990a,b), also identified new CMS sources. Although many CMS lines have been developed in China and by IRRI (Li and Zhu, 1988; Lin and Yuan, 1980; Virmani, 1996; Virmani et al., 1986; Virmani and Wan, 1988), only a few possess features desirable for commercial use. Chinese scientists are currently using CMS-G, CMS-D, and CMS-ID systems in addition to the WA CMS system (Zeng et al., 2000). A new cytoplasm from DISSI D52/37, a West African cultivar, has been reported by Xie and Chen (2000). It has been used for commercial hybrids like *Hongyou 5355* and *Hongyou 22* in China. A new CMS line with Y-type cytoplasm has been developed by Cai (1997, 2001, 2002) in China. Improved CMS lines possessing diverse CMS systems have also been developed at IRRI.

Chen (1992) reported that two different cytoplasms had significant effects on plant height, panicle weight, and heterosis (varying from 3.9-9.1%), indicating the need to identify CMS lines showing stronger heterosis. Wang et al. (1996) reported that K-type cytoplasm showed stigma exsertion, longer duration of flowering, and higher 1,000-grain weight than WA and D cytoplasms, but it had similar effects on fertility restoration and on some quantitative traits. RAPD analysis showed differences in mitochondrial DNA for K-type CMS compared to WA and D cytoplasms. You et al. (1997) crossed CMS line *Fuji A* having WA cytoplasm and CMS line *Fuji A* having KV-type cytoplasm with six restorers and found KV-type cytoplasm to be superior to WA-type cytoplasm for various agronomic traits, including yield.

You et al. (1998) reported that heterosis with KV cytoplasm was higher than that of WA cytoplasm. WA-type cytoplasm was reported to be earlier than that of KV-type cytoplasm. You et al. further indicated that the KV-type CMS line showed higher hormone content than the WA-type CMS line. Jiang et al. (1999) reported a fine quality CMS line, *Yue 4A*, with a new type of cytoplasm with high outcrossing rate and high combining ability. Xie and Chen (2000) reported *Hong'ai A* as a new CMS line with a type of cytoplasm with high outcrossing rate and high combining ability. Different male sterile cytoplasms identified in different countries are listed in Table 5.1.

Fertility Restoration

Certain fertility restorer lines, when crossed with CMS lines, result in commercial rice hybrids. The frequency of restorers was found to be higher (64%) in Asian cultivars, compared to only 4% in American strains (Shinjyo, 1975). Work at IRRI also indicated a 30-40% frequency of fertility restorers among elite inbred lines developed in the tropics. The maintainer frequency was high among elite *indica* lines developed in China, temperate and tropical *japonica* rices, and basmati rices. The restorer and maintainer frequency was intermediate in *indica-japonica* derivative lines. Hundreds of restorer lines have been identified for WA and other CMS systems in China, India, the Philippines, Vietnam, Bangladesh, Indonesia, the United States, Egypt, and so forth. The genetics of fertility restoration have been studied by many researchers (Bharaj et al., 1991; Govinda Raj and Virmani, 1988; Hu and Li, 1985; Li and Yuan, 1986; Teng and Shen, 1994; Virmani et al., 1986; Wang, 1980; Young and Virmani, 1984; Zhou et al., 1983). Govinda Raj and Virmani (1988) reported two fertility restoring genes, the action of one of which was stronger than that of the other. Teng and Shen (1994) also

TABLE 5.1. Cytoplasmic-genetic male sterility sources identified in different countries.

Cytoplasm type	Country	Reference
O. sativa f. spontanea	Japan	Sasahara and Katsuo, 1965
Chinsurah Boro (BT)[a]	Japan	Shinjyo and Omura, 1966
Khaiboro cytoplasm	Japan	Nagamine et al., 1995
WA[a]	China	Lin and Yuan, 1980
ARC	Philippines	International Rice Research Institute, 1986
Gambiaca[a]	China	Lin and Yuan, 1980
Kalinga	India	Pradhan et al., 1990b
Dissi D52	China	Xie and Chen, 2000
WA Dwarf[a]	China	Virmani and Wan, 1988
MS Ma-Wei-Zhan[a]	China	Virmani and Wan, 1988
Red awned wild rice[a]	China	Lin and Yuan, 1980
TN1	Philippines	Athwal and Virmani, 1972
V20B	India	Pradhan et al., 1990a
Mutagenized IR62829B	Philippines	International Rice Research Institute, 1995
TI* (E-Shan Ta Bai)	China	Virmani and Wan, 1988
O. glaberrima	USA	Carnahan et al., 1972
O. perennis (Acc. 104823)	Philippines	Dalmacio et al., 1995
O. glumaepetula (Acc. 100969)	Philippines	Dalmacio et al., 1996
O. rufipogon (W 1080)	Japan	Shinjyo et al., 1981
O. rufipogon (W 1092)	Japan	Shinjyo and Motomura, 1981
O. rufipogon	India	Unpublished
O. nivara	India	Unpublished
K[a]	China	Wang et al., 1996
Y[a]	China	Cai, 1997
?[a]	China	Jiang et al., 1999
ID	China	Virmani and Wan, 1988
KV[a]	China	You et al., 1997

[a]Commercially used cytoplasm.

reported the role of two fertility restoring genes, one dominant and the other incompletely dominant for WA cytoplasm. Bharaj et al. (1995) located the stronger restorer gene on chromosome 7 and the weaker restorer gene on chromosome 10. Fertility restoration in CMS-Bo and CMS-D male sterile lines was found to be controlled by a single dominant gene, *Rf1* (Hu and Li,

1985; Shinjyo, 1969, 1975; Shinjyo et al., 1974; Teng and Shen, 1994). In both these CMS lines, the effect of the restorer gene was gametophytic, as the hybrids showed partial pollen fertility but normal seed setting. Cheng and Xue (1998) reported that sterility of CMS-Bo and CMS-Di could most easily be restored, followed by CMS-DA; fertility restoration of CMS-WA was reported to be relatively more difficult than for the other cytoplasms.

Although many *indica* varieties have genes for fertility restoration of WA cytoplasm, temperate *japonica* and tropical *japonica* varieties lack restorer genes. These lines, therefore, cannot be used directly for hybrid rice development using the CMS system. However, some NPT lines derived from *indica*/tropical *japonica* crosses have been found to have fertility restoration ability.

Hybrid rice breeders have also developed new restorer lines from crosses. Restorer genes can be incorporated in non-restorer genotypes by crossing restorer and non-restorer genotypes, followed by backcrossing to the non-restorer genotype. An ISSR marker, UCB-835, was reported to be closely linked to a restorer gene of the WA system with a distance of 3.58 cM (Jing et al., 2000). He et al. (2000) reported that the polymorphic segment AP 1 generated from primer E-AGC/M-CAA is associated with a restorer gene; the distance between AP 1 and the restorer gene was reported to be 4.76 cM. Molecular markers for gene *Rf3* have been identified and used in the IRRI breeding program with about 80% efficiency. Efforts are underway to identify molecular markers for restorer gene *Rf4*. Screening of a population with both molecular markers should further increase restorer breeding efficiency.

Restorers for other cytoplasms (viz., CMS-ARC, CMS-Dissi, CMS-Gambiaca) have been identified at IRRI (Virmani et al., 1994), and in China (CMS-Di, CMS-Dissi, CMS-Gambiaca, CMS-DA, CMS-GA, etc.). So far, no restorers have been identified for CMS-*O. perennis* or CMS-*O. glumaepetula,* developed by Dalmacio et al. (1996), and CMS-TN, developed by Athwal and Virmani (1972).

Environmental Genetic Male Sterility System

This male sterility system is controlled by the expression of nuclear genes that are affected by environmental conditions at the critical plant growth stage (15-20 days after panicle initiation). EGMS plants can become completely male sterile at a specific day length, temperature, or combination of both, and the same plants show fertility if grown under other environmental conditions. Depending on their response to day length or temperature, such lines are known as photoperiod-sensitive (PGMS) or thermosensitive genetic male sterile (TGMS) lines. The genes controlling PGMS or TGMS traits have been found to be recessive. Because the PGMS and TGMS lines can be multiplied through self-pollinating by growing them under suitable day length or temperature regimes, there is no need for a maintainer line to maintain their sterility. Any fertile line can restore the fertility of the hybrids derived from them. Such hybrids are known as *two-line* hybrids compared to the three-line (A, B, R) hybrids derived from the CMS system.

Major advantages of the two-line over the three-line hybrid breeding system are

- The choice of parental lines (especially male lines) becomes wider; therefore, there is a better chance to exploit genetic diversity to breed heterotic hybrids.
- Cytoplasm has no role to play in this system; therefore, there are no negative effects (if any) of the cytoplasm inducing male sterility.

PGMS was first discovered in 1973 by Shi while working in the Hubei province of China, where he developed the first male sterile line in *japonica* variety *Nongken 58* (Shi, 1981). The male sterile mutants showed complete pollen sterility at long day lengths and normal pollen fertility at short day lengths. Soon after, efforts were initiated in China to transform other elite lines into PGMS lines. During these studies, it was observed that the fertility alteration in some genotypes was not only controlled by day length but also by temperature at the critical growth stage. As rice is grown at different latitudes and altitudes, the EGMS system is practically usable in this crop to develop hybrids.

In temperate zones, the PGMS system is considered more important, whereas in tropical areas, where the variation of the photoperiod is small, the TGMS system is considered of greater significance. IRRI's hybrid rice breeding program, therefore, emphasizes developing TGMS lines (Virmani and Ilyas Ahmed, 2001). Mou et al. (1998) described the procedure for breeding TGMS and PGMS lines. TGMS lines developed in China initially had a high critical temperature point (CTP) and were not commercially usable because of their unstable male sterility in the field. Subsequently, TGMS lines with low CTP (24°C) have been developed (e.g., W91607S, W9451S, W9461S, and W9593S) and are being used widely to breed two-line commercial hybrids (Xiao et al., 2002a). Initially, TGMS lines developed by IRRI also had a high CTP (30-22°C, mean 26°C), but more recently, TGMS lines with low CTP (27-19°C, mean 23-24°C), which appear to be more stable, have been developed.

At IRRI, the TGMS genes *tms2* (from Norin PL 12) and *tms3* (IR32364S) and ID24 (a TGMS line from India) have been deployed for breeding new

TGMS lines. TGMS lines derived from the ID24 mutant (viz., IR73827-23S and IR73834S) have stable sterility and are being used in developing experimental hybrids.

In order to characterize PGMS/TGMS lines, the selected lines are grown so that their sensitive growth stage falls under the critical photoperiod or temperature conditions (Mou et al., 1998; Virmani and Ilyas Ahmed, 2001; Zeng and Zhang, 2001). The development of near isogenic lines of *Peiai 64S* (viz., P2364S, P3464S, P2664S, and P2864S, with CTPs for male sterility of 23°C, 24°C, 26°C, and 28°C, respectively) has been reported. Line P2364S carries a special significance, as it has a low critical temperature and is stable with complete sterility in high-temperature conditions. This line was reported to have stable fertility at low temperature (<22°C) and short day length and thus can be multiplied under such climatic conditions.

Lu et al. (1998) have reviewed elite PGMS and TGMS lines commonly used in China. *Peiai 64S* is indeed the first practical thermosensitive (TGMS) line used in commercial hybrid production. It recombined PGMS genes from *japonica* line *Nongken 58S* with wide compatibility genes from *Peiai 64*.

Genetics of EGMS Traits

The PGMS trait has been reported to be controlled by a single recessive gene (Jin and Li, 1991) or by two recessive loci, *pms1* on chromosome 7 and *pms2* on chromosome 3 (Zhang et al., 1994). The effect of *pms1* is two to three times stronger than that of *pms2*. Gui et al. (2002) reported three PGMS genes, *pms1, pms2,* and *pms3,* located on chromosome 7, 3, and 12, respectively. Zhang et al. (1994) identified positive markers from three regions located on chromosomes 1, 3, and 7 that are positively linked to PGMS.

Three TGMS mutants (viz., 5460S, H89-1, and IR32364) were reported to carry a single recessive gene (Borkakati and Virmani, 1996; Maruyama et al., 1991; Yang et al., 1992). Wang et al. (1995) reported that the TGMS gene was located on chromosome 8 near RFLP markers RZ 667 and RG 648. Subudhi et al. (1997) reported 4 RAPD markers to be linked to TGMS gene *tms3*. Gui et al. (2002) reported five TGMS genes, *tms1, tms2, tms3, tms4,* and *tms5,* to be located on chromosomes 8, 7, 6, 9, and 2, respectively. The location of the various *tms* and *pms* genes in the rice genome is given in Figures 5.1 and 5.2.

The instability of sterility expression (ISE) in many photo/thermosensitive genetic male sterile lines (PTGMS) has also been observed (Liao and Yuan, 2003). This instability may be caused by (1) minor genes controlling the trait (critical sterility-inducing temperature point); (2) genetic heterozygosity; and (3) quantitative loci controlling the sterility expression.

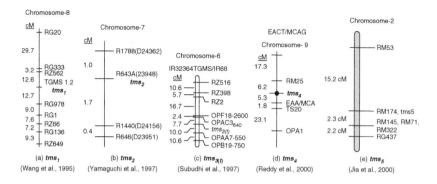

FIGURE 5.1. Location of *tms* genes in the rice genome.

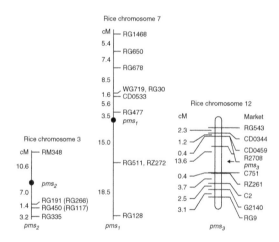

FIGURE 5.2. Location of *pms* genes in the rice genome.

Techniques such as anther culture can be helpful in reducing the effect of heterozygosity and increase the efficiency of breeding PTGMS lines with stable sterility.

Duan et al. (2003) observed a drift in the critical sterility-inducing temperature (CSIT) of PTGMS lines (viz., *Peiai 64S* and *An Xiang S*) possessing different male sterility genes. This conclusion was drawn on the basis of the change in average pollen sterility of different generations of PTGMS lines from core seed to foundation seed level grown under long day length and low temperature conditions in a phytotron. It was inferred that (1) the phenomenon of CSIT drift existed commonly among PTGMS lines, and (2) the degree of drift and frequency of CSIT varied among lines. Extra care, therefore, must be taken to maintain and monitor the purity of PTGMS lines during multiplication of their nucleus—breeder and foundation seeds—in different seasons and years.

HYBRID RICE SEED PRODUCTION

The commercial success of hybrid rice depends not only on its yield superiority over the popular high-yield varieties but also on the ease and economy of hybrid seed production. Even a superior hybrid can only become popular if its seed production is economically viable. As the male and female line of every hybrid differ in their flowering behavior, the production technology has to be developed for every hybrid. Indeed, improved hybrid seed production techniques are the keys to the success of any hybrid-breeding program.

In the initial commercialization phase of hybrid rice technology in China in 1976, hybrid seed yield was very low (376 kg/ha). Therefore, special attention was paid to seed production technology, and by 1980, seed yield had increased to nearly 0.71 tons/ha, further rising to more than double (1.74 tons/ha) in 1983, and nearly tripling by 1985 (2.07 tons/ha). Currently, the average yield of hybrid rice seed production in China has increased to 2.75 tons/ha (Mao et al., 1998). The recorded maximum yield obtained in a hybrid rice seed production plot was 7.386 tons/ha on a small plot of 0.113 ha at Zixing City, Hunan Province, in 1993 (Mao et al., 1998). In tropical countries, mean hybrid rice seed yields are about 1.2 tons/ha (range 0.5-4.0 tons/ha).

Hybrid rice seed production involves many technical skills to get high quality F_1 seed at economical prices (Lin and Yuan, 1980; Virmani and Sharma, 1993; Virmani et al., 2002; Yuan, 1985). These include production of pure seed of parental (A, B, R, PGMS, and TGMS) lines and seeds of the derived three- and two-line hybrids. Detailed guidelines, procedures, and practices for hybrid rice seed production have been laid down by Yuan (1985), Virmani and Sharma (1993), and Virmani et al. (2002). The guidelines include (1) selection of seed and pollen parents with synchronized time of anthesis; (2) selection of seed parents with long, exserted stigma and longer duration and wider range of floret opening; (3) selection of pollen parent with high percentage of residual pollen per anther after anther exsertion;

(4) high pollen shedding potential, attained by getting 2,000-3,000 spike-lets/m^2 bloomed per hour during the flowering period; (5) synchronization of flowering time of the two parents by seeding them at different dates, depending on their growth duration or estimated accumulated temperature requirements for initiation of flowering; (6) use of optimum seed-pollen parent ratio such that the ratio of spikelet number per unit area of seed parent and pollen parent is about 3.5/1; (7) selection of seed and pollen parent possessing small and horizontal flag leaves or cutting long and erect flag leaves; (8) use of gibberellic acid (GA_3) to improve panicle exsertion and prolong duration of floret opening and stigma receptivity; (9) planting of seed parent and pollen parent rows across the prevailing wind direction at the time of flowering, and use of supplementary pollination with a rope or stick when wind velocity is below 2.5 m/sec; (10) selection of optimum time of flowering of parental lines in seed production plots; and (11) observing strict time and distance isolation standards to avoid contamination from the adjoining fields and produce seed of acceptable purity.

With experience, hybrid rice seed yields are continuously improving in China and elsewhere. Consequently, the cost per kilogram of seed production is decreasing, making hybrid rice economically viable in different situations. Hybrid rice seed production is labor intensive and requires more labor than normal, nonmechanized rice cultivation. It therefore generates additional employment opportunities, which has a great benefit especially for landless women in countries (e.g., India, Bangladesh, Vietnam, the Philippines, and Indonesia) that have a labor surplus. If six million hectares were planted with hybrid rice by the year 2010, hybrid rice seed production would generate about three to four million man-days of rural employment in Asia.

BIOTECHNOLOGICAL TOOLS FOR HYBRID RICE BREEDING AND SEED PRODUCTION

The development of biotechnological tools has opened new avenues in hybrid rice breeding. Anther culture is useful for developing homozygous male lines quickly from a cross between two restorer lines or from superior crosses involving restorer and non-restorer lines. Many new varieties have been developed in China using this technique (Brar and Khush, 1993).

Molecular marker-aided selection (MAS) is now technically mature, following the development of high-density molecular linkage maps (Causse et al., 1994; Kurata et al., 1994). The use of molecular markers has provided a unique opportunity for hybrid rice programs to expedite the breeding process. Examples include genes for fertility restoration, wide compatibility, TGMS, disease and insect resistance, and QTL for heterosis. Fertility restoration of WA cytoplasm has been reported to be governed by two independent loci on chromosomes 7 and 10 (Bharaj et al., 1995). Zhang et al. (1997) found three RFLP markers (RG 532, RG 140, and RG 458) to be closely linked with *Rf3* on chromosomes 1 and 10. The locus on chromosome 10 had a much stronger effect than the one on chromosome 1. There are other reports using *Rf* genes (Akagi et al., 1996; Huang et al., 1999; Shen et al., 1998; Tan et al., 1998; Yang et al., 1997). MAS has also been successfully used as a tool for transferring the *Xa21* gene (imparting resistance to bacterial blight) into PGMS line 3418S. A wide compatibility-linked marker, RG213, has been reported by Yanagihara et al. (1995). Markers have been used to pyramid several desirable genes. Reports are available on the successful transfer of the *Xa21* gene into eight hybrids and their parental lines in China (Chen and Qian, 2000) using transformation procedures. Success was achieved in transferring genes for intermediate amylose from *Minghui 63* to ZS 97.

Using the transformation approach, methods have been developed for disrupting normal pollen development (male sterility) and restoring normal pollen development (fertility restoration) in hybrids. The barnase-barstar system uses the tapetal specific transcriptional activity of the tobacco TA29 gene and an RNAse *(barnase)* paired with an RNAse inhibitor *(barstar)* from the bacterium *Bacillus amyloliquefaciens*. The toxin gene *barnase* was made to express itself in tapetum tissue of developing anthers, where it destroys tapetal cells, inhibits pollen formation, and causes male sterility. The introduction and expression of a restorer gene construct (TA29; tapetum specific promoter-barstar) into another plant leads to the development of a restorer line, which is then crossed with the male sterile line for hybrid seed production. As barstar is the specific inhibitor of barnase, the tapetal development in the hybrid plants is not affected, and the F_1 plants are male fertile. The dominant nature of male sterility due to *barnase* is maintained by crossing the male sterile line *(barnase/-)* with a normal, untransformed (without *barnase; -/-)* male fertile parental line (Williams and Leemans, 1993). As the progeny of the cross of sterile *(barnase/-)* and normal fertile plants (-/-) will produce 50% sterile *(barnase/-)* and 50% fertile (-/-) plants, the fertile plants have to be removed before flowering for hybrid seed production. The identification of fertile plants has been achieved by linking the *barnase* gene with the *bar* gene that confers resistance to the herbicide BASTA. All the male sterile plants *(barnase/bar)* are resistant to herbicide, whereas male fertile plants (-/-) are susceptible, and thus killed by herbicide spraying. This system has been successfully used in canola and maize and can possibly be used in hybrid rice development.

In rice, Wang et al. (2001) reported transfer of the gene for male sterility, T-29-*barnase*, into rice embryo calli by *Agrobacterium tumefaciens*. Hu et al. (2000) reported *Agrobacterium*-mediated transformation for the *bar* gene into restorer lines R 187 and *Xiushui 04*. Ling et al. (1998) transferred the Barnase *ps1* gene into rice genomic DNA and successfully engineered male sterile plants in *Taipei 309* and *Qui Guang* (*japonica* rice). The pollen grains of completely male sterile plants of Barnase *ps1* were shrunken and irregular in shape and not stained by IKI solution. The female organ of these plants was normal, as these plants could set seed upon crossing.

Restorer lines of popular hybrids (viz., *Ce64, Teqing, Minghui 63*) in different cytoplasms have been converted into herbicide-resistant male lines from *bar*-transgenic rice varieties by cross breeding. The physiological and agronomic characteristics of these homotypic herbicide-resistant restorer lines were similar to those of the original restorer lines. Their hybrids shared similar or higher heterosis compared to their original non-herbicide-resistant combinations. The impurity problem faced in hybrid rice production can thus be solved through such a herbicide resistance gene.

Success has already been achieved in developing bacterial leaf blight (BLB)-resistant (*Xa21*) transgenic restorer lines to breed BLB-resistant hybrids. At IRRI, the *Xa21* gene has most successfully been transformed into a very popular CMS line, IR58025A (Laha et al., 2003). Being an elite line, this line would be very helpful in breeding BLB-resistant rice hybrids. In China, Zhao et al. (2002) and Zhou et al. (2002) studied hybrids involving BLB-resistant transgenic restorer lines *Minghui 63-Xa21, Yanhui 559-Xa21*, and C418-*Xa21* with different CMS lines. The transgenic hybrids with transgenes from the male parent showed a wide spectrum of resistance and a similar yield to their controls.

Bacillus thuringiensis (*Bt*) genes have been introduced into the maintainer line IR68899B, which showed a very high resistance to yellow stem borer (Alam et al., 1999). *Minghui 63*, a well-known restorer line of the Chinese hybrid rice *Shanyou 63*, was transformed with a *Bt* fusion gene derived from *cryIA(b)* and *cryIA(c)* under the control of the rice *actinI* promoter and showed resistance to all major lepidopteran insect pests in China (Datta et al., 1998; Tu et al., 1998). Later, *Minghui 63-Bt* was used to develop hybrid *Bt-Shanyou 63* rice. The *Bt-Shanyou 63* was field-tested in natural and repeated heavy manual infestation of two lepidopteran insects, leaf folder and yellow stem borer. The transgenic hybrid plants showed high protection against both insect pests without reduced yield (Tu et al., 2000, 2003).

ECONOMIC CONSIDERATIONS
IN HYBRID RICE TECHNOLOGY

For the large-scale adoption of any hybrid, two major factors are very important: yield superiority, and seed production at economical prices. Normally, hybrid seed is priced at seven to ten times the cost of high-yield varieties; therefore, the economic benefit to farmers is the major factor for its repeat demand. For a hybrid rice cultivator, the F_1 seed cost is the decisive factor, as other recommendations are more or less the same as for traditional high-yield varieties. Normally, hybrid seed is sold at $2-2.50/kg, and a farmer has to buy 15 kg seed/ha, giving an extra cost of $30-37.50/ha. This cost can be compensated if the farmer gets an additional yield of 250-330 kg/ha. As the farmer gets 1-1.5 tons/ha additional yield, worth $115-170/ha, this results in a cost-benefit ratio of ⅓ - ¼ for a farmer's investment on hybrid rice seed.

For a hybrid rice seed producer, the additional costs are in parent seed, labor for transplanting and supplementary pollination (about 70-80 man-days), and GA_3. Although GA_3 is an expensive input, some seed agencies in India obtain it cheap directly from China. It has been seen that the seed-producing farmer (at a yield level of 1.5 tons/ha) earns much more than the hybrid rice cultivator, after deducting the extra cost for GA_3 and labor (Virmani et al., 2002).

CURRENT APPROACHES TO IMPROVE YIELD HETEROSIS

Exploitation of Inter-Subspecific Diversity

Current hybrids have been developed primarily from *indica/indica* crosses. After a steady increase in productivity of three-line hybrids, their yield is now plateauing in China (Wu, 2000). In India, progress has been made in yield improvement by new three-line *indica* rice hybrids by about 1-2 tons/ha over that of traditional HYVs.

Tropical *japonica* germplasm has been shown to be genetically diverse relative to *indica* types. Significant heterosis for various agronomic traits has been observed in crosses involving *indica* and tropical *japonica* (TJ) lines (Khush et al., 1998). Heterosis studies at IRRI (Khush et al., 1998; Yuan, 1994), conducted using different ecotypes, showed the order of heterosis to be *indica*/TJ > *indica/indica* > TJ/TJ. In temperate countries, the order of heterosis is likewise *indica/japonica* > *indica/indica* > *japonica/japonica*. Commercial exploitation of heterosis from *indica*/TJ crosses (in the tropics) and *indica/japonica* crosses (in temperate countries) is constrained by (1) lack of a restorer gene in tropical *japonica* and *japonica* parents to restore fertility in three-line hybrids involving *indica* CMS lines; (2) lack of a wide compatibility (WC) gene (Ikehashi and Araki, 1984) in either male or female parent, resulting in partial fertility in such crosses; (3) Differences in the anthesis time of *japonica* and *indica* lines, resulting in poor outcrossing and seed setting in seed production plots due to late availability of pollen; and (4) absence of a semidwarf gene (*sd1*) in the desired tropical *japonica* and *japonica* types, so that hybrids with an *indica* parent become tall. These constraints are being overcome in hybrid rice breeding programs in China and at IRRI. Restorer genes are being incorporated in *indica*/TJ-derivative lines showing new plant types and higher yield potential. Similarly, WC genes have been identified in some tropical *japonica* rice cultivars, and these are being incorporated into lines derived from *indica*/TJ or TJ/*japonica* crosses. Several tropical *japonica* lines and some NPT lines having a WC gene at the S^5 locus have been identified at IRRI. With specifically developed parental lines carrying WC and *Rf* genes, Ren et al. (1999) and Xiao et al. (2002b) reported high fertility and high heterosis (i.e., up to 40% in specifically developed *indica/japonica* crosses compared to *indica/indica* crosses).

To improve tropical *japonica* lines, IRRI and Chinese rice breeders have been crossing these with *indica* lines and selecting for new plant types (low to intermediate tillering, long panicle). A number of progenies derived from such crosses can possess a significant proportion of tropical *japonica* genome to maintain genetic distance from *indica* lines (e.g., *Peiai 64S*, developed in China) and also possess WC genes. The crosses of elite *indica* lines with such derived lines would help to enhance heterosis. In fact, such crosses have resulted in the so-called "super rice hybrids" in China, which have significantly outyielded *indica/indica* rice hybrids in China (Yuan, 2002). Rice breeders at IRRI have developed several elite lines possessing the *sd1*, gene, low to intermediate tillering, higher grain number and long panicles, thick stems, and resistance to lodging. Such lines are now available with restoration and maintainer ability and are being used in crosses with elite *indica* CMS and TGMS lines to develop three-line and two-line rice hybrids for the tropics.

Pyramiding Yield Genes from Wild Rices Using Molecular Markers

Wild species of rice possessing different genomes are an important reservoir of useful genes not only for biotic and abiotic stresses, but also for enhancing yield, which can be tagged with molecular markers (McCouch et al., 2001). A feasible approach to enhance rice heterosis would be the pyramiding of favorable QTL and heterotic gene blocks in wild species and traditional rice varieties using molecular marker-assisted reciprocal recurrent selection (Wu, 2000; Li et al., 2002). Two QTLs from the wild rice species *O. rufipogon* have been reported to improve yield by a margin of 18% each (Xiao et al., 1996). Through marker-assisted selection and backcrossing, several near-isogenic lines carrying these QTLs have been developed. Using these lines as parents (e.g., in cross J23A/NIL611), rice hybrids were reported to outyield controls by a margin of 35% (Xiao et al., 1998).

At IRRI, many introgression lines involving wild species like *O. glaberrima, O. australiensis, O. minuta, O. officinalis,* and *O. rufipogon* have been developed. These lines have one or the other favorable yield attribute (Brar and Khush, 2002). Such lines could be useful as parents to enhance heterosis in rice. Zhong et al. (2000) reviewed the use of wild species for transferring elite genes into cultivated rice through molecular mapping of alien introgressed genes from wild species.

Use of Alien Genes in Parental Lines

Wang, F. et al. (2002) suggested the incorporation of valuable alien genes into rice plants as another means of enhancing genetic diversity and, hence, heterosis. He reported nine introgression lines with small segments from corn DNA or barnyard grass DNA through the pollen tube pathway method into a rice restorer line, R122. Analysis using SSR markers showed barnyard grass DNA to be incorporated on different chromosomes, with a segment length of 7.1 cM. Similarly, corn DNA was found to be inserted on chromosomes 6, 11, and 12; heterosis for yield in the crosses involving these introgressed lines ranged from 11-32% over the original restorer, suggesting the usefulness of such an approach.

Use of the C4 Gene

Transgenic rice plants expressing maize phosphoenol pyruvate carboxylase (PEPC) and pyruvate orthophosphate dikinase (PPDK) have been reported to exhibit higher photosynthetic capacity than untransformed ones (Ku et al., 2001). Preliminary field data indicate that grain yield was higher by 10-20% in PEPC and 30-35% in PPDK transgenic plants due to increased tillers. It is therefore important to study heterosis in C4/C3 crosses vis-à-vis C3/C3 and C4/C4 crosses.

CURRENT APPROACHES FOR GRAIN
QUALITY IMPROVEMENT IN HYBRID RICE

In addition to high yield, rice hybrids must also possess acceptable grain and cooking quality; otherwise, millers and consumers would not pay prices for hybrid grains comparable to what they pay for popular HYV grains. Therefore, more attention should be paid to physical quality parameters (viz., milling and head rice recovery, grain appearance, grain chalkiness, including white belly, white core, or white center) and eating quality parameters (viz., amylose content, gelatinization temperature, and gel consistency of endosperm). Amylose content is the major factor determining rice cooking quality. Rices with low amylose cook soft, glossy, and sticky, whereas the ones with high amylose content cook dry, fluffy, and separate. The intermediate amylose rices remain moist and soft upon cooking. Various CMS lines and genetically divergent male parents, which differed widely in quality parameters, have been used so far in developing commercial hybrids in China and elsewhere. Although the F_1 plants were uniform for grain length, width, and chalkiness score, their F_2 grains (i.e., grains on F_1 plants) differed widely if the parent lines differed widely for endosperm traits like amylose content, gelatinization temperature, and gel consistency (Kumar and Khush, 1986, 1987, 1988).

In China, the grain quality of most CMS lines developed before 1990 was poor; therefore, the grain quality of hybrids derived from them was also poor (Hu et al., 2002). Even today, about 90% of the hybrid rice area in China is occupied by hybrids developed from CMS lines like *Zhen Shan, II You A, Gangyou A, Xie You A, Wei You A, Te You A, Dyon You A*, and others that have poor grain quality (Hu et al., 2002). A few hybrids developed from CMS lines (viz., *Bo You A, Ji You A*) have superior grain quality, but these currently occupy only 11-12% of the hybrid growth area. Outside China, a widely used CMS line (IR58025A) possesses better grain quality than the widely used Chinese CMS lines, but its low amylose content and strong aroma is not liked in some countries.

Recently, some new CMS lines with improved grain quality have been reported in China. Wang et al. (2000) bred a fine quality CMS line, 45A, with cytoplasm of Indonesian paddy number 6. Yang et al. (2001) bred a new CMS fine quality line, *Deshan A*, using WA cytoplasm. The hybrids derived from this CMS line have been reported to have good quality along with high yield. Hu et al. (2001) developed a fine quality rice hybrid, *Jin You 928*, from fine quality CMS line *Jin 23A* and fine quality restorer line 928. Considerable attention has been paid at IRRI to developing new CMS lines (e.g., IR68897A, IR68888A, IR73728A) possessing intermediate amylose content, desirable grain type and good milling ability. Recently bred IRRI hybrids *Mestizo 2* and *Mestizo 3,* released in the Philippines have used the CMS lines with intermediate amylose content. Likewise, in India, a superior quality *Basmati* type hybrid, *Pusa* RH10, of which both the parents have basmati grain quality features, has been commercialized. Hybrid Rice International, a private sector seed company, has developed new superior quality, intermediate amylose, nonaromatic CMS lines and restorers and commercialized a *Sona Mahsuri* type hybrid, 6516, which is nonaromatic and has superior cooking and eating quality. Rice Tec has also developed and commercialized rice hybrids XL7 and XL8, which have grain quality comparable to the popular long grain rice varieties cultivated in the southern United States (Walton and Tsuchiya, 2002).

Breeding strategies for developing new superior quality hybrid rices have been proposed by Khush and Virmani (1988) and Virmani et al. (2002). Amylose content of rice has been reported to be controlled by a single gene (Kumar and Khush, 1986, 1987, 1988) or by two genes, $Amy^{i(t)}$ and $Amy^{s(t)}$ (Chen et al., 2002). Similarly, aroma has been reported to be controlled by two recessive genes. For breeding a nonaromatic hybrid, both the parents should be nonaromatic. If one of the parents is aromatic, the grains of an F_1 hybrid would have a mild aroma due to segregation within a panicle. The quality parameters of *indica/indica* or *indica/*TJ hybrids can be improved either by breeding both parental lines with desirable quality or by appropriately choosing parental lines with desirable and similar grain quality features from the available germplasm. The genetics of head rice recovery have not yet been clearly understood, but variation for head rice recovery has been recorded, and the trait is known to be highly influenced by environment. Newly bred CMS lines in China, India, at IRRI, and elsewhere offer opportunity to breed good-quality hybrids.

FUTURE OUTLOOK

Currently, hybrid rice technology is widely used in China, but in the near future it will be used extensively in India, Vietnam, the Philippines, Bangladesh, Indonesia, Sri Lanka, Myanmar, Thailand, the United States, and Egypt. The extent of heterosis in the future rice hybrids may be higher due to the use of new approaches such as intersubspecific crosses, heterotic gene blocks, yield genes from wild rices, and C4 rice genes. Although currently rice hybrids are being cultivated exclusively in irrigated rice ecosystems in China, in India they have shown adaptability to the favorable rain-fed lowland conditions of Maharashtra. Future hybrids would also be cultivated in aerobic rice and certain rain-fed lowland ecosystems. Grain quality and disease/insect resistance of future rice hybrids can be improved by the choice of appropriate parents and critical evaluation of derived hybrids before their release.

Sufficient cytoplasmic diversity in rice has been identified to develop future rice hybrids with a broader cytoplasmic base. This would minimize the risk of potential genetic vulnerability of CMS-based hybrid rices. Fortunately, there is no evidence so far that the widely used WA-type CMS is associated with any disease or insect susceptibility. Deployment of the EGMS system in hybrid rice breeding would increase, and private companies would be coming up with genetically engineered male sterility systems. These systems would not only help in increasing hybrid breeding efficiency but would also minimize the potential genetic vulnerability associated with the CMS system. Future hybrid rice breeding programs would not only use the conventional breeding procedures but also selectively use biotechnological tools (e.g., anther culture selection, genetic transformation) to increase hybrid breeding efficiency.

Similar to the experience in China, hybrid rice seed production technology outside China should also improve over time with the development of improved male sterile lines possessing higher outcrossing potential and improved seed production management systems. The economic viability of hybrid technology will improve with the development of good grain quality hybrids and improved, cost-effective hybrid seed production technology. This would encourage more private seed companies to invest in hybrid rice, and consequently, the hybrid seed industry would be instrumental in generating increased rural employment opportunities. The higher seed yields and competition in the seed industry should reduce the cost of hybrid seed, and the technology may also be used for direct seeding conditions where seed rate is high and the current price of hybrid seed makes it prohibitive to use hybrid rice. If apomixis in rice is successfully identified or developed, this would make hybrid rice technology affordable even for resource-poor farmers.

Hybrid rice technology holds great promise, as the increased yield through rice hybrids would not only increase farmers' income but also contribute toward national and global food security. The increased yield per unit area would save land and water and protect the environment. The saved land would help in increasing crop diversification and farmers' household income.

REFERENCES

Agata, W. (1990). Mechanisms of high yield achievements in Chinese F_1 compared with cultivated rice varieties. *Japanese Journal of Crop Science 59:* 270-273.

Akagi, H., Y. Yokazaki, A. Inagaki, A. Nakamura, and T. Fujimura (1996). A codominant DNA marker closely linked to the rice nuclear restorer gene, *Rf-1,* identified with inter-SSR fingerprinting. *Genome 39:* 1205-1209.

Akita, S. (1988). Physiological bases of heterosis in rice. In *Hybrid rice,* pp. 67-77. Manila, Philippines: International Rice Research Institute.

Akita, S., L. Blanco, and S. S. Virmani (1986). Physiological analysis of heterosis in rice plant. *Japanese Journal of Crop Science 55*(Special issue 1): 14-15.

Alam, M. F., K. Datta, A. Vasquez, J. Tu, S. S. Virmani, and S. K. Datta (1999). Transgenic insect resistant maintainer line (IR68899B) for improvement of hybrid rice. *Plant Cell Reports 18:* 572-575.

Athwal, D. S., and S. S. Virmani (1972). Cytoplasmic male sterility and hybrid breeding in rice. In *Rice breeding,* pp. 615-620. Manila, Philippines: International Rice Research Institute.

Bharaj, T. S., S. S. Bains, G. S. Sidhu, and M. R. Gagneja (1991). Genetics of fertility restoration of wild abortive cytoplasmic male sterility in rice, *Oryza sativa* L. *Euphytica 56:* 199-203.

Bharaj, T. S., S. S. Virmani, and G. S. Khush (1995). Chromosomal location of fertility restoring genes for "wild abortive" cytoplasmic male sterility using primary trisomic in rice. *Euphytica 83:* 169-173.

Blanco, L., S. Akita, and S. S. Virmani (1986). Growth and yield of F_1 rice hybrids in different levels of nitrogen. *Japanese Journal of Crop Science 55*(Special issue 1): 12-13.

Borkakati, R., and S. S. Virmani (1996). Genetics of thermosensitive male sterility in rice. *Euphytica 88:* 1-7.

Brar, D. S., and G. S. Khush (1993). Application of biotechnology in integrated pest management. *Journal of Insect Science 6:* 7-14.

Brar, D. S., and G. S. Khush (2002). Transferring genes from wild species into rice. In *Quantitative genetics, genomics and plant breeding,* M. S. Kang (Ed.), pp. 197-217. Wallingford, UK: CAB International.

Cai, S. X. (1997). Effects of male sterile cytoplasms of Yegong (*Oryza sativa* L. subsp. *indica*) and abortive wild rice on agronomic characters of hybrid rice. *Chinese Journal of Rice Science 11:* 211-214.

Cai, S. X. (2001). Breeding of CMS line Y Huanong A with Y-type cytoplasm in rice. *Hybrid Rice 16*(6): 9-10.

Cai, S. X. (2002). Breeding of male sterile lines with the cytoplasm of Yeqong (Y type) in Hunan late *indica* rice. *Chinese Journal of Rice Science 16:* 185-188.

Carnahan, H. L., J. R. Erickson, S. T. Tseng, and J. N. Rutger (1972). Outlook for hybrid rice in USA. In *Rice breeding,* pp. 603-607. Manila, Philippines: International Rice Research Institute.

Causse, M. A., T. M. Fulton, Y. G. Cho, S. N. Ahn, J. Chunwongse, K. S. Wu, J. H. Xiao, P. C. Ronald, S. E. Harrington, G. Second, et al. (1994). Saturated molecular map of the rice genome based on an interspecific backcross population. *Genetics 138:* 1251-1274.

Chen, P. (1992). A study on effect of different cytoplasms on heterosis. *Hybrid Rice 3:* 42-44.

Chen, P., Z. H. Huang, W. Liu, and Q. L. Huang (2002). Breeding of new *indica* CMS line Bo II in rice. *Hybrid Rice 17*(3): 9-10.

Chen, W. H., and Q. Qian (2000). Bacterial blight resistant gene was transferred into hybrid rice. *Chinese Rice Research Newsletter 8*(3): 15.

Cheng, C. Y., and Q. Z. Xue (1998). Analysis of restoring genes to different type of cytoplasmic male sterility. *Acta Agronomica Sinica 24:* 361-367.

Cheng, S. H., and M. Shaokai (2001). Super rice breeding in China. *Chinese Rice Research Newsletter 9*(2): 13-15.

Dalmacio, R., D. S. Brar, T. Ishii, L. A. Sitch, S. S. Virmani, and G. S. Khush (1995). Identification and transfer of a new cytoplasmic male sterility source from *Oryza perennis* into *indica* rice *(O. sativa). Euphytica 82:* 221-225.

Dalmacio, R., D. S. Brar, S. S. Virmani, and G. S. Khush (1996). Male sterility in rice *(O. sativa) developed with O. glumaepetula* cytoplasm. *International Rice Research Newsletter 21*(1): 22-23.

Datta, K., A. Vasquez, J. Tu, L. Torrizo, M. F. Alam, N. Oliva, E. Abrigo, G. S. Khush, and S. K. Datta (1998). Constitutive and tissue-specific differential expression of *cryIA(b)* gene in transgenic rice plants conferring resistance to rice insect pest. *Theoretical and Applied Genetics 97:* 20-30.

Duan, M. J., D. Y. Yuan, Q. Y. Deng, and X. Q. Li (2003). Studies on fertility stability of PTGMS rice. IV. Rule of drift in critical sterility-inducing temperature. *Hybrid Rice 18*(2): 62-64.

Erickson, J. R. (1969). Cytoplasmic male sterility in rice *(Oryza sativa* L.). *Agronomy Abstracts 1969:* 6.

Faiz, F. A. (2000). *Effect of wild abortive cytoplasm including inducing male sterility on biotic/abiotic resistance/tolerance and agronomic and grain quality in some basmati rice hybrids.* PhD thesis, Central Luzon State University, Philippines.

Govinda Raj, K., and S. S. Virmani (1988). Genetics of fertility restoration of WA type cytoplasmic male sterility in rice. *Crop Science 28:* 787-792.

Gui, L. X., T. M. Mou, N. Hoan, and S. S. Virmani (2002). *Two line hybrid rice breeding in and outside China.* Paper presented at the 4th International Symposium on Hybrid Rice, May 14-17, 2002, Hanoi, Vietnam.

Guong, V. T., T. T. Lap, N. M. Hoa, E. G. Castillo, J. L. Padilla, and U. Singh (1995). Nitrogen use efficiency in direct seeded rice in Mekong river delta: Varietal and phosphorus response. In *Vietnam and IRRI: A partnership in rice research,* G. L. Denning and V. T. Xuan (Eds.), pp. 151-159. Proceedings of a Conference held in Hanoi, Vietnam, May 4-7, 1994.

He, G. H., Y. Pei, G. W. Yang, M. Tang, R. Xie, L. Hou, Z. L. Yang, and Y. H. Li (2000). AFLP markers of restoring genes of the wild-abortive hybrid rice. *Acta Genetica Sinica 27:* 304-310.

Hu, G. C., H. Xiao, Y. H. Yu, Z. G. Zhu, H. M. Si, Y. P. Fu, and Z. X. Sun (2000). *Agrobacterium*-mediated transformation of the restorer lines of two-line hybrid rice with *bar* gene. *Chinese Journal of Applied and Environmental Biology 6:* 511-515.

Hu, J., and Z. Li (1985). A preliminary study on the inheritance of male sterility of rice male sterile lines with four different kinds of cytoplasm. *Journal of Huazhong Agricultural University 4:* 15-22.

Hu, P., H. Zhai, S. Tang, and J. Wan (2002). Rice quality improvement in China. *China Rice Research Newsletter 10*(1): 13-15.

Hu, X., B. H. Hu, B. G. Duan, H. M. Xu, T. Z. Wang, J. Tu, G. P. Yuan, and B. Shu (2001). Breeding and application of fine quality *indica* hybrid rice *Jinyou 928*. *Hybrid Rice 16:* 13-14.

Huang, Q., Y. He, R. Jin, H. Huang, and Y. Zhu (1999). Tagging of the restorer gene for rice HL-type CMS using microsatellite markers. *Rice Genetics Newsletter 16:* 75-77.

Ikehashi, H., and H. Araki (1984). Varietal screening of compatibility types revealed in F_1 fertility of distant crosses in rice. *Japanese Journal of Breeding 34:* 304-313.

International Rice Research Institute (1986). *Annual report for 1985.* Manila, Philippines: IRRI.

International Rice Research Institute (1993). *Program report for 1992.* Manila, Philippines: IRRI.

International Rice Research Institute (1995). *Program report for 1994.* Manila, Philippines: IRRI.

Jiang, X. P., X. Y. Xie, and G. H. Ou (1999). Breeding of fine quality CMS line *Yue 4A* with a new type of cytoplasm. *Hybrid Rice 14*(2): 3-5.

Jin, D., and Z. Li (1991). Inheritance behavior of photoperiod sensitive genic male sterility (PGMS) in the crosses between *japonica* PGMS lines and *japonica* varieties. *Journal of Huazhong Agricultural University 10:* 136-144.

Jing, R. C., Q. P. He, Q. Y. Huang, and Y. G. Zhu (2000). Analysis of the fertility restorer gene in the wild abortive (WA) type cytoplasmic male sterility (CMS) system with the ISSR and SSLP markers. *Scientia Agricultura Sinica 33*(2): 10-15.

Katsuo, K., and U. Mizushima (1958). Studies on cytoplasmic differences among rice varieties, *Oryza sativa* L. *Japanese Journal of Breeding 8:* 1-5.

Khush, G. S., R. C. Aquino, S. S. Virmani, and T. S. Bharaj (1998). Using tropical *japonica* germplasm to enhance heterosis in rice. In *Advances in hybrid rice technology,* S. S. Virmani, E. A. Siddiq, and K. Muralidharan (Eds.), pp. 59-66. Proceedings of the 3rd International Symposium on Hybrid Rice, November 14-16, 1996, Hyderabad, India. Manila, Philippines: International Rice Research Institute.

Khush, G. S., and S. S. Virmani (1988). *Breeding strategies for high yielding rices at IRRI.* Paper presented at Japan-IRRI seminar on development and physiological characteristics of high yielding rice varieties, Tsukuba, Japan.

Kitamura, E. (1962). Studies on cytoplasmic sterility of hybrids in distantly related varieties of rice *O. sativa* L. I. Fertilities of F_1 hybrids between strains derived from certain Philippines × Japanese varietal crosses and Japanese varieties. *Japanese Journal of Breeding 12:* 81-84.

Ku, M. S. B., D. Cho, X. Li, D. M. Jiao, M. Pinto, M. Miyao, and M. Matsuoka (2001). Introduction of genes encoding C4 photosynthesis enzyme into rice plant. Physiological consequences. In *Rice biotechnology: Improving yield, stress tolerance and grain quality,* J. Goode and D. Chadwick (Eds.), pp. 100-110. Los Baños, Novartis Foundation Symposium.

Kumar, I., and G. S. Khush (1986). Genetics of amylose content in rice (*Oryza sativa* L.). *Journal of Genetics* 65(142): 1-11.

Kumar, I., and G. S. Khush (1987). Genetic analysis of different amylose levels in rice. *Crop Science 27:* 1167-1172.

Kumar, I., and G. S. Khush (1988). Inheritance of amylose content in rice (*Oryza sativa* L.). *Euphytica 38:* 261-269.

Kurata, N., Y. Nagamura, K. Yamamoto, Y. Harushima, N. Sue, J. Wu, B. A. Antonio, A. Shomura, T. Shimizu, S. Y. Lin, et al. (1994). A 300-kilobase interval genetic map of rice including 883 expressed sequences. *Nature Genetics 8:* 365-372.

Laha, G. S., M. V. R. Rao, E. Abrigo, N. Oliva, K. Datta, and S. K. Datta (2003). Evaluation of durable resistance of transgenic hybrid maintainer line IR58025B for bacterial blight disease of rice. *Rice Genetics Newsletter 20:* 88-91.

Li, Y. C., and L. P. Yuan (1986). Genetic analysis of fertility restoration in male sterile lines of rice. In *Rice genetics,* pp. 617-632. Manila, Philippines: International Rice Research Institute.

Li, Z., and Y. Zhu (1988). Rice male sterile cytoplasm and fertility restoration. In *Hybrid rice,* pp. 85-102. Manila, Philippines: International Rice Research Institute.

Liao, F. M., and L. P. Yuan (2003). On genetic mechanism and cause of instability in sterility expression in photoperiod and thermosensitive genic male sterile rice. *Hybrid Rice 18*(2): 1-6.

Lin, S. C., and L. P. Yuan (1980). Hybrid rice breeding in China. In *Innovative approaches to rice breeding,* pp. 35-51. Manila, Philippines: International Rice Research Institute.

Ling, D. H., L. Z. Tao, Z. R. Ma, S. P. Zhang, and S. K. Datta (1998). Engineered male sterile transgenic plants of rice (*Oryza sativa* L.) with *psl 1*-barnase gene transformation by particle bombardment. *Chinese Journal of Genetics 25:* 289-297.

Lu, J. F., and M. H. Gu (1998). Effect of LH422 cytoplasm on the expression of wide compatibility genes. *Chinese Journal of Rice Science 12:* 121-124.

Lu, X., S. S. Virmani, and Y. Ren Cui (1998). Advances in two-line hybrid rice breeding. In *Advances in hybrid rice technology,* S. S. Virmani, E. A. Siddiq, and K. Muralidharan (Eds.), pp. 89-98. Proceedings of the 3rd International Symposium on Hybrid Rice, November 14-16, 1996, Hyderabad, India. Manila, Philippines: International Rice Research Institute.

Luo, R. L. (2000). Hybrid rice research and development in China and its new progress. *Hunan Agricultural Science and Technology Newsletter 1*(3): 5-7.

Ma, G. M., and L. P. Yuan (2002). *Achievement and development of hybrid rice in China.* Paper presented at the 4th International Symposium on Hybrid Rice, May 14-17, 2002, Hanoi, Vietnam.

Mao, C. X., S. S. Virmani, and I. Kumar (1998). Technological innovation to lower the cost of hybrid rice seed production. In *Advances in hybrid rice technology,*

S. S. Virmani, E. A. Siddiq, and K. Muralidharan (Eds.), pp. 111-128. Proceedings of the 3rd International Symposium on Hybrid Rice, November 14-16, 1996, Hyderabad, India. Manila, Philippines: International Rice Research Institute.

Maruyama, K., K. Araki, and H. Kato (1991). Thermosensitive genetic male sterility induced by irradiation. In *Rice genetics II,* pp. 227-232. Manila, Philippines: International Rice Research Institute.

Maruyama, S., K. Miyazato, and A. Nose (1987). Studies on dry matter production of F_1 hybrid in rice. I. Heterosis in the single leaf photosynthetic rate. *Japanese Journal of Crop Science 56:* 198-203.

Maruyama, S., Y. Norishama, S. Miyasato, and A. Nose (1982). Studies on dry matter production of F_1 rice hybrids. I. Heterosis in photosynthetic activity. *Japanese Journal of Crop Science* (Special issue 2): 85-86.

Maruyama, S., N. Ogasawara, N. S. Miyasato, and H. Nose (1984). Dry matter production of F_1 rice hybrids. III. Heterosis in photosynthetic activity. *Japanese Journal of Crop Science 53*(Special issue 2): 100-101.

McCouch, S. R., S. Temnykh, A. Lukashova, J. Coburn, G. Clerck, S. Cartinhour, S. Harrington, M. Thomson, E. Septiningsih, M. Semon, et al. (2001). Microsatellite markers in rice: Abundance, diversity and applications. In *Rice genetics IV,* G. S. Khush, D. S. Brar, and B. Hardy (Eds.), pp. 117-136. Proceedings of the Fourth International Rice Genetics Symposium, October 22-27, 2000, Los Baños, Philippines. New Delhi, India: Science Publishers.

Mou, T., C. Li, G. Young, and X. Lu, (1998). Breeding and characterizing *indica* PGMS and TGMS lines in China. In *Advances in hybrid rice technology,* S. S. Virmani, E. A. Siddiq, and K. Muralidharan (Eds.), pp. 78-88. Proceedings of the 3rd International Symposium on Hybrid Rice, November 14-16, 1996, Hyderabad, India. Manila, Philippines: International Rice Research Institute.

Nagamine, T., K. Kodawaki, A. Goto, and T. Aiioka (1995). A new source of cytoplasmic male sterility found in Bangladesh boro rice. *Rice Genetics Newsletter 12:* 232-233.

Namuco, O. S., K. T. Ingram, M. A. Maguling, and S. S. Virmani (1988). Hybrid rice in rainfed environments. *International Rice Research Newsletter 13*(5): 9-10.

Nguyen, T., T. Quang, V. Tran, and N. B. Thong (2002). *The fertility alternation of TGMS line: Peiai 64S and the ability of seed multiplication and hybrid seed production in Vietnam.* Paper presented at the 4th International Symposium on Hybrid Rice, May 14-17, 2002, Hanoi, Vietnam.

Ponnuthurai, S., S. S. Virmani, and B. S. Vergara (1984). Comparative study on growth and grain yield of some F_1 rice hybrids. *Philippine Journal of Crop Science 9:* 183-193.

Pradhan, S. B., S. N. Ratho, and P. J. Jachuck (1990a). Development of new cytoplasmic genetic male sterility in cultivated *indica* rice. *Euphytica 51:* 127-130.

Pradhan, S. B., S. N. Ratho, and P. J. Jachuck (1990b). New sources of cytoplasm genetic male sterility in cultivated *indica* rice. *Euphytica 48:* 215-218.

Reddy, M. N., K. Surekha, and V. Balasubramanyan (2002a). *Relative N-use efficiency of rice hybrids and inbred cultivars.* Poster presented at the 4th International Symposium on Hybrid Rice, May 14-17, 2002, Hanoi, Vietnam.

Reddy, M. N., K. Surekha, and V. Balasubramanyan (2002b). *Synchronization of potassium supply for rice hybrids.* Poster presented at the 4th International Symposium on Hybrid Rice, May 14-17, 2002, Hanoi, Vietnam.

Ren, G., X. Lu, Q. Li, and C. Zhang (1999). Breeding and utilization of wide compatibility rice restorer line *Chenghai 448. Chinese Journal of Rice Science 13:* 120-122.

Sampath, S., and H. K. Mohanty (1954). Cytology of semi-sterile rice hybrid. *Current Science 23:* 182-183.

Sarawagi, A. K., and M. N. Srivastava (1988). Heterosis in rice under irrigated and rainfed situations. *Oryza 25:* 10-15.

Sasahara, T., and K. Katsuo (1965). Studies on the cytoplasmic difference among rice varieties *O.sativa* L. III. On the abortive pollen of *Fujisaka No.5*-type plants with cytoplasm of Chinese wild variety *O. sativa* f. *spontanea. Japanese Journal of Breeding 15:* 291-296.

Shen, Y. W., Z. Q. Guan, J. Lu, J. Y. Zhuang, K. L. Zheng, M. W. Gao, and X. M. Wang (1998). Linkage analysis of a fertility restoring mutant generated from CMS rice. *Theoretical and Applied Genetics 97:* 261-266.

Shi, M. S. (1981). Preliminary report of breeding and utilization of late *japonica* natural double purpose line. *Journal of Hubei Agriculture 7:* 1-3.

Shi, Q. H., M. Y. Li, and Q. H. Tu (2002). Studies on efficiency of N-nutrition and physiological factors in roots of hybrid rice. *Hybrid Rice 17*(4): 45-48.

Shinjyo, C. (1969). Cytoplasmic genetic male sterility in cultivated rice, *Oryza sativa* L. II. The inheritance of male sterility. *Japanese Journal of Genetics 44:* 149-156.

Shinjyo, C. (1975). Genetical studies of cytoplasmic male sterility and fertility restoration in rice, *Oryza sativa* L. *Science Bulletin of the College of Agriculture University of Ryukyu 22:* 1-57.

Shinjyo, C., Y. Ishimura, and M. Tanaki (1981). Inheritance of male sterility in isogenic lines of Taichung 65 possessing male sterile cytoplasm and fertility restoring genes of *O. perennis* W 1080 strain. *Japanese Journal of Breeding 31:* 238-239.

Shinjyo, C., and K. Motomura (1981). Inheritance of male sterility in isogenic lines of Taichung 65 possessing male sterile cytoplasm and fertility restoring genes from *Oryza perennis* W1080 strain. *Japanese Journal of Breeding 31:* 240-241.

Shinjyo, C., R. Nishima, and Y. Watanabe (1974). Inheritance of fertility restoring gene *Rfx* and *Rf* in male sterile cytoplasm derived from variety Lead rice. *Japanese Journal of Breeding 24:* 130-131.

Shinjyo, C., and T. Omura (1966). Cytoplasmic genetic male sterility in cultivated rice *Oryza sativa* L.: Fertilities of F_1, F_2 and offsprings obtained for their mutual reciprocal backcrosses and segregation of completely male sterile plants. *Japanese Journal of Breeding 16:* 179-180.

Subudhi, P. K., R. P. Borkakati, S. S. Virmani, and N. Huang (1997). Molecular mapping of thermosensitive genic male sterility gene in rice using bulk segregant analysis. *Genome 40:* 188-194.

Sunohara, Y., M. Yajima, M. Suzuki, and H. Seki (1985). Dry matter production of F_1 rice hybrids. Heterosis in initial growth and sink size. *Japanese Journal of Crop Science 54* (Special issue 1): 132-133 (in Japanese).

Tan, X. L., A. Vanavichit, S. Amornsilpa, and S. Trangoonrung (1998). Genetic analysis of rice CMS-WA fertility restoration based on QTL mapping. *Theoretical and Applied Genetics 97:* 994-999.

Teng, L. S., and Z. T. Shen (1994). Inheritance of fertility restoration for cytoplasmic male sterility in rice. *Rice Genetics Newsletter 11:* 95-97.

Tu, J., K. Datta, M. F. Alam, G. S. Khush, and S. K. Datta (1998). Expression and function of a hybrid *Bt* toxin gene in transgenic rice conferring resistance to insect pests. *Plant Biotechnology (Japan) 15:* 183-191.

Tu, J., K. Datta, N. Oliva, G. Zhang, C. Xu, G. S. Khush, Q. Zhang, and S. K. Datta (2003). Site-independently integrated transgenes in the elite restorer rice line Minghui 63 allow removal of a selectable marker from the gene of interest by self-segregation. *Plant Biotechnology Journal 1:* 155-165.

Tu, J., G. Zhang, K. Datta, C. Xu, Y. He, Q. Zhang, G. S. Khush, and S. K. Datta (2000). Field performance of transgenic elite commercial hybrid rice expressing *Bacillus thuringiensis* δ-endotoxin. *Nature Biotechnology 18:* 1101-1104.

Virmani, S. S. (1994). Heterosis and hybrid rice breeding. In *Theoretical and Applied Genetics Monograph No. 22.* Berlin: Springer-Verlag, pp. 1-189.

Virmani, S. S. (1996). Hybrid rice. *Advances in Agronomy 57:* 378-462.

Virmani, S. S., K. Govinda Raj, C. Casal, R. D. Dalmacio, and P. A. Aurin (1986). Current knowledge of and outlook on cytoplasmic genetic male sterility and fertility restoration in rice. In *Rice genetics,* pp. 633-647. Manila, Philippines: International Rice Research Institute.

Virmani, S. S., and M. Ilyas Ahmed (2001). Environment-sensitive genic male sterility (EGMS) in crops. *Advances in Agronomy 72:* 139-195.

Virmani, S. S., C. X. Mao, R. S. Toledo, M. Hossain, and A. Janaiah (2002). Hybrid rice seed production technology and its impact on seed industries and rural employment opportunities in Asia. *Technical Bulletin 156, Food and Agriculture Technology Center,* pp. 1-13.

Virmani, S. S., and H. L. Sharma (1993). *Manual for hybrid rice seed production.* Los Baños, Philippines: International Rice Research Institute.

Virmani, S. S., and C. Shinjyo (1988). Current status of analysis and symbols for male-sterile cytoplasms and fertility restoring genes. *Rice Genetics Newsletter 5:* 9-15.

Virmani, S. S., and B. H. Wan (1988). *Development of CMS lines in hybrid rice breeding. I Hybrid rice,* pp. 103-114. Manila, Philippines: International Rice Research Institute.

Walton, M., and T. Tsuchiya (2002). *Hybrid rice breeding and seed production under mechanized farming situations.* Paper presented at the 4th International Symposium on Hybrid Rice, May 14-17, 2002, Hanoi, Vietnam.

Wang, C. Y., H. Q. Hong, and H. X. Qi (1995). Breeding of two-line *japonica* hybrid rice N5088S/R187. *Hybrid Rice 2:* 4-6.

Wang, F., W. Liu, S. Li, Y. Liao, and H. Peng (2002). *Studies on enhancing heterosis of indica hybrid rice through introgression lines of corn and barnyard grass DNA.* Paper presented at the 4th International Symposium on Hybrid Rice, May 14-17, 2002, Hanoi, Vietnam.

Wang, P. H., X. C. Zhang, T. Q. Liu, Y. C. Zhang, and Y. J. Luo (2000). Breeding of fine quality *indica* CMS line 45A. *Hybrid Rice 15*(5): 7-8.

Wang, S. (1980). Inheritance of R genes in rice and methods of selection of new R-lines. *Agricultural Science and Technology Hunan 4:* 1-4.

Wang, X., L. X. Tao, M. Y. Yu, and X. L. Huang (2001). Physiological model of super hybrid rice *Xie You 9308*. *Chinese Rice Research Newsletter 9*(3): 11-12.

Wang, X., L. X. Tao, M. Y. Yu, and X. L. Huang (2002). Physiological model of super hybrid rice *Xie You 9308*. *Chinese Journal of Rice Science 16:* 38-44.

Wang, W. M., H. C. Weu, G. L. Yuan, X. Q. Wan, and Y. C. Zhu (1996). Breeding of and studies on K-type hybrid rice. *Hybrid Rice 6:* 11-13.

Wang, Y., and S. Yoshida (1984). Studies on heterosis in physiological characters and grain yield of F_1 hybrid rice (V20A × IR54). *Acta Scientiarum Naturalium Universitatis Sunyatseni 4:* 115-121.

Watanabe, Y. (1971). Establishment of cytoplasmic and genetic male sterile lines by means of *indica-japonica* crosses. *Oryza 8:* 9-16.

Williams, M. E. and J. Leemans, J. (1993). Maintenance of male sterile plants. *Patent application No. WO 93/25695.*

Wu, X. J. (2000). Possible approaches to improve rice heterosis. *Chinese Journal of Rice Science 14:* 61-64.

Xiao, G. Y., X. X. Deng, and L. P. Yuan (2002a). Effect of water temperature on male fertility alteration of the sensitive TGMS lines in rice under the simulated low air temperature in high summer. *Hunan Agricultural Science and Technology Newsletter 1*(3): 7-12.

Xiao, G., L. P. Yuan, X. Deng, and L. Tang (2002b). *Studies on heterosis of Peiai 64S/Javanica rice and breeding of 770, a Javanica restorer line.* Paper presented at the 4th International Symposium on Hybrid Rice, May 14-17, 2002, Hanoi, Vietnam.

Xiao, J. H., S. Grandillo, S. N. Ahn, S. R. McCouch, S. D. Tanksley, J. Li, and L. P. Yuan (1996). Genes from wild rice improve yield. *Nature 384:* 223-224.

Xiao, J. H., J. Li, S. Grandillo, S. N. Ahn, L. P. Yuan, S. D. Tanksley, and S. R. McCouch (1998). Identification of trait improving quantitative trait allele from a wild rice relative *O. rufipogon*. *Genetics 150:* 899-909.

Xie, C. H., and Y. J. Chen (2000). Breeding and utilization of new CMS line *Hong'ai A*. *Hybrid Rice 15*(5): 9-10.

Yamauchi, M., and S. Yoshida (1985). Heterosis in photosynthetic rate, leaf area, tillering and some physiological characters of 35 F_1 rice hybrids. *Journal of Experimental Botany 36:* 274-280.

Yanagihara, S., S. R. McCouch, J. Ishikawa, Y. Ogi, K. Maruyama, and H. Ikehashi. (1995). Molecular analysis of the inheritance of the *S-5* locus conferring wide compatibility in *indica/japonica* hybrids of rice (*O. sativa* L.). *Theoretical and Applied Genetics 90:* 182-188.

Yang, D. C., G. B. Magpantay, M. Mendoza, N. Huang, and D. S. Brar (1997). Construction of a contig for a fertility restorer gene, *Rf 2*, in rice using BAC library and its sequence with *Rf 2* gene of maize. *Rice Genetics Newsletter 14:* 116-117.

Yang, N. C., S. P. Xia, Y. L. Li, and Z. Z. Yuan (2001). Breeding of fine quality *indica* CMS line Deshan A and its characteristics. *Hybrid Rice 16*(2): 7-9.

Yang, R. C., K. J. Liang, N. Y. Wang, and S. H. Chen (1992). A recessive gene in *indica* rice 5460S for thermosensitive genic male sterility. *Rice Genetics Newsletter 9:* 56-57.

You, N. S., J. C. Lu, L. X. Huang, X. H. Zheng, S. P. Li, and X. T. Zhu (1998). Cytoplasmic effect of KV type sterile lines in rice. *Chinese Journal of Rice Science 12*(3): 181-184.

You, N. S., J. Lu, X. Zheng, L. Houng, X. Zhu, and S. Lei (1997). Studies on KV type cytoplasmic effects on heterosis in rice. *Hybrid Rice 12*(3): 29-31.

Young, J. B., and S. S. Virmani (1984). Inheritance of fertility restoration in a rice cross. *Rice Genetics Newsletter 1:* 102-103.

Yuan, L. P. (1977). The execution and theory of developing hybrid rice. *Zhonggue Nongye Kexue (Chinese Agricultural Science) 1:* 27-31.

Yuan, L. P. (1985). *A concise course in hybrid rice.* Hunan, China: Hunan Science and Technology Press.

Yuan, L. P. (1994). Increasing yield potential in rice by exploitation of heterosis. In *Hybrid rice technology: New development and future prospects,* S. S. Virmani (Ed.), pp. 1-6. Selected papers from the International Rice Research Conference. Los Baños, Philippines: International Rice Research Institute.

Yuan, L. P. (1998). Hybrid rice breeding for super high yield. *Hunan Agricultural Research Newsletter 5:* 7-10.

Yuan, L. P. (2002). *Future outlook on hybrid rice research and development in China.* Paper presented at the 4th International Symposium on Hybrid Rice, May 14-17, 2002, Hanoi, Vietnam.

Yuan, L. P., S. S. Virmani, and G. S. Khush (1985). Wei You 64—an early duration hybrid for China. *International Rice Research Newsletter 10:* 11-12.

Zeng, H. L., and D. P. Zhang (2001). Developing near isogenic lines at different critical temperature for thermo-photoperiod sensitive male sterile rice Peiai 64S. *Acta Agronomica Sinica 27:* 351-355.

Zeng, Q. C., K. D. Zhou, Z. Zhu, and Q. Luo (2000). Current status in use of hybrid rice heterosis in China. *Chinese Journal of Rice Science 14:* 243-246.

Zhang, G., T. S. Bharaj, Y. Lu, S. S. Virmani, and N. Huang (1997). Mapping of *Rf₃* nuclear fertility-restoring gene for WA cytoplasmic male sterility in rice using RAPD and RLFP markers. *Theoretical and Applied Genetics 94:* 27-33.

Zhang, Q., Y. T. Gao, S. H. Yang, R. A. Ragab, M. A. Saghai Maroof, and Z. B. Li (1994). A diallel analysis of heterosis in elite hybrid rice based on RFLPs and microsatellites. *Theoretical and Applied Genetics 89:* 185-192.

Zhao, X. F., W. X. P. Li, C. L. Wang, X. B. Pan, Q. Qian, S. G. Li and L. H. Zhu (2002). Field tests and analyses of different *Xa21*-transgenic hybrid rice combinations. *Acta Agronomica Sinica 28:* 521-527.

Zhong, D. B., L. J. Luo, and C. S. Ying (2000). Advances on transferring elite gene from wild rice species into cultivated rice. *Chinese Journal of Rice Science 14:* 103-106.

Zhou, T. L., J. H. Shen, and F. C. Ye (1983). Analysis of R genes in hybrid *indica* rice of WA type. *Acta Agronomica Sinica 9:* 241-247.

Zhou, Y. L., Q. Zhang, C. L. Wang, Q. D. Xing, W. X. Zhai, S. P. Pan, and L. H. Zhu (2002). Resistance of *Xa 21* transgenic hybrid rice Shan You 63 and Shan You 559 to bacterial blight. *Chinese Journal of Rice Science 16:* 93-95.

Chapter 6

Stem Borer Resistance:
Bt (Bacillus thuringiensis) Rice

N. Baisakh
S. K. Datta

INTRODUCTION

The rising world population demands food sufficiency and food security in terms of continuing increase in global agricultural output. The basis for this increase must be improving the harvest yields of major cereal crops from existing cultivated land, as there is no scope for the horizontal or vertical expansion of arable land. Rice alone accounts for almost two thirds of the world's food and has been a prime target for improvement, higher yield, and plant protection. One practical means of achieving this increased yield would be to prevent more crops from suffering losses due to pests—particularly insect pests—thereby attaining the maximum yield potential (also referred to as *yield stability*).

Rice experiences insect predation, whose severity and economic importance vary with genotypes, time, and space. In the man-made environments of agricultural crop production for food, this predation almost always decreases the yield of the agricultural system. In fact, the development of *high-yield varieties* (HYVs) grown as monocultures managed with intensive fertilizer inputs and, in many cases, irrigation to produce luxuriant vegetative growth has altered the co-evolutionary balance between this crop and its insect predators and increased the potential for epizootics. Insects are not only responsible for massive direct losses of productivity as a result of their herbivory, but they also cause a great deal of indirect losses because of their role as vectors for various plant pathogens, especially viruses.

This review focuses on the rationale, present status, and future of biotechnology research with particular reference to transgenic *Bt* research for insect pest management in rice, an economically important crop that fills the food bowl of most Asians.

MAJOR INSECT PESTS OF RICE

Knowledge of the major regular and sporadic insect pests of economic importance and the crop losses specific to them is useful in formulating strategies to minimize losses and in estimating the gains from research on specific pests. Rice is the crop that suffers most losses from insect pests, followed by maize, sorghum, wheat, pearl millet, oats, barley, and rye. Table 6.1 lists the major insect pests of rice that are prevalent in Asian countries.

Yield losses from insect pests vary with geographical area, with or without the extensive use of insecticides. Available information from a case study on rice has shown a 10-35% loss in rice production in different Asian rice-growing countries (Table 6.2) and a 2-35% loss in rice productivity worldwide (Table 6.3) because of major insect pests. Lepidopteran insects are known to cause a majority of the loss to rice yield, of which the loss due to stem borer is the most devastating.

CROP PROTECTION STRATEGIES: PROS AND CONS

Integrated pest management (IPM) can be used to develop a package of practices for insect management in rice and other major cereals. Nevertheless, crop protection relies primarily on synthetic chemical pesticides

TABLE 6.1. Major insect pests of rice.

Major insect pest	Scientific name
Stem borer	
Yellow	*Scirpophaga incertulas*
Striped	*Chilo suppressalis*
Pink	*Sesamia inferens*
White	*Scirpophaga innotata*
Brown planthopper	*Nilaparvata lugens*
Green planthopper	*Nephotettix virescens*
Leaffolder	*Cnaphalocrocis medinalis*
Gall midge	*Orseolia oryzae*

Sources: Carozzi and Koziel (1997); Prakash and Rao (1999).

TABLE 6.2. Losses in rice production due to major insect pests in Asian countries.

Country	Loss (kg ha⁻¹)	Loss (%)
India	352	28.8-35
Sri Lanka	–	10-20
Philippines	–	16-30 (M = 18.3)
Bangladesh	135	–
Indonesia	399	–
Thailand	83	–
Nepal	376	–
East and South Asia	–	23.7

Source: Ramaswamy and Jatileksono (1996).

TABLE 6.3. Estimated rice yield losses caused by insect pests globally.

Region	Yield loss (%)
Asia	31.5
China	15.0
Africa	14.4
South America	3.5
North and Central America	3.4
Europe	2.0

Source: Cramer (1967).

(agrochemicals)—the basis of an approximately U.S. $10 billion per annum global pesticide market. However, pressure is increasing against the indiscriminate use of pesticides because of their large negative consequences. The extensive use of agrochemicals for the sake of plant protection is now seriously questioned, as it makes the environment unsustainable (Pingali et al., 1992). The limitations of agrochemicals are manifold. First and foremost is the huge cost in nonrenewable resources. For example, the market value of the insecticides used by Indian farmers for rice alone is $51 million, which is quite high. It is estimated that, despite the investment of $10 billion in management practices and chemical control, crop losses reach 20-30% across the globe (Estruch et al., 1997). Table 6.4 gives comprehensive information on the expenses incurred on pesticides for major crops and the volume and value of crop losses. This table also substantiates the inefficiency and inefficacy of agrochemicals in terms of the proportion that actually reaches the intended target. Chemical insecticides have not been successful in controlling stem borers, because the insect larvae remain for a very short time on the outer surface of the plant and feed deep inside the stem pith, beyond the reach of spray-ons (Teng and Revilla, 1996). Furthermore, slow pesticide decomposition rates (Cengel and Saatci, 1982) and the consequent long persistence of chemicals in the soil (Awasthi et al., 1984), leading to contamination of food chains and water sources and human health hazards (Pingali et al., 1992), are unacceptable environmental consequences. More important, the indiscriminate use of and predominant reliance on pesticides often leads to pest resurgence, frequent large-scale infestations, and disruption of the pest-predator balance, when the pesticides become hazardous to non-target, beneficial organisms, such as predators and parasites (Banerjee, 1996). Although the tremendous benefits accrued to agriculture from agrochemicals cannot be belittled, demand is urgent for developing partial substitution technologies that would allow a more limited use of synthetic pesticides and still provide adequate protection of crops within a sustainable agricultural framework such as IPM.

The use of biopesticides consisting of formulations of predators, paraitoids, and pathogens of insect pests is considered an alternative, environment-friendly component of IPM farming systems. However, biopesticides account for only 3% of the pesticide market, for the following reasons: (1) the difficulty in simultaneous management of three different biological populations, namely, predators, prey (pest), and the crop; (2) the difficulty in using

TABLE 6.4. Crop losses due to insect pests despite the use of chemical insecticides.

Crop	Crop volume loss (%)	Crop loss (million US$)	Insecticide used (% of total used on all crops)	Expenditure on insecticides (million US$)
Rice	27	45,000	1	1,190
Maize	12	8,000	17	620
Wheat	–	–	1	–

Source: Krattiger (1997).

them in annual field crops; and, above all, (3) the unavailability of predators or parasites for many major insect pests.

Apart from these possible pest control strategies, the inherent resistance of crop plants forms an integral and by far the most important and safe component of IPM (van Emden, 1987). Conventional breeding in the past has played a key role in developing insect pest-resistant varieties in an array of crop species. However, the success of such a conventional crossbreeding program is seriously handicapped for developing crops that are resistant to stem borers due to the unavailability of resistant donors in the interbreeding gene pool.

Nevertheless, *Bacillus thuringiensis (Bt)* formulations (spores and crystal mixtures) have been used as insecticidal sprays since the 1930s, but large-scale production started only with the production of Thuricide in the late 1950s, followed by similar products from several companies (Beegle and Yamamoto, 1992). Such an approach, though environment-friendly, did not have a large share of the insecticide market, and its use was largely limited to organic farmers, gardeners, and in forestry. The factors responsible for the lack of popularity of the *Bt* spray were

1. lack of stability;
2. failure to penetrate tissues, and therefore reach insects, in all parts of the plant;
3. narrow specificity;
4. rapid degradation by UV light and quick loss of activity;
5. nonsystemic, and thus ineffective against insects that do not come into direct contact with the crystals, such as the sap-sucking and piercing insects, root-dwelling pests, or larvae that, after hatching, rapidly burrow or bore into plant tissues.

Hence, an alternative strategy is the development of *built-in* protection by genetic engineering for the production of proteins with insecticidal activity by crop plants. The first such protein that has been used in transgenic plants is the crystalline delta-endotoxin from the soil-borne bacterium *Bacillus thuringiensis*. The toxicity and mode of action of *Bt* is described in greater detail later. There are manifold advantages in the use of transgenic plants to produce such insecticidal proteins. First, the transfer of genes across taxa—for example, from bacterium to plant—is possible. Second, insecticidal genes are often determined by a single dominant gene, unlike similar traits in varieties related to the crop of interest, which are often encoded by multiple genes. Therefore, the direct transfer of a single gene through the transgenic approach rules out the possibility of linkage drag associated with multiple-gene transfer through conventional breeding.

Bt *INSECTICIDAL PROTEIN*

Bacillus thuringiensis is a ubiquitous, Gram-positive, spore-forming bacterium that produces crystalline bodies during sporulation. These parasporal crystals usually contain potent insecticidal delta-endotoxins, classified as either crystal toxins (Cry) or cytolytic toxins (Cyt). The potency of *Bt* Cry and Cyt toxins with their highly specific insecticidal activity (Höfte and Whiteley, 1989) makes them ideal for controlling certain insect species in the orders Lepidoptera, Diptera, and Coleoptera (Beegle and Yamamoto, 1992). As a result of intensive screening programs, new strains producing delta-endotoxins active against representatives from the orders Hymenoptera, Homoptera, Orthoptera, and Mallophaga, and against nematodes, mites, and protozoa (Feitelson, 1993), have also been isolated. *B. thuringiensis* is already a useful alternative or supplement to synthetic chemical pesticide application in commercial agriculture, forest management, and mosquito control. It is also a key source of genes for transgenic expression to provide pest resistance in plants.

Diversity and Classification of Crystal Proteins

Since the first cloning of an insecticidal crystal gene from *B. thuringiensis* (Schnepf and Whiteley, 1981), nearly 200 such genes have been isolated. Initially, each newly characterized gene or protein received an arbitrary designation from its discoverers: icp (McLinden et al., 1985); cry (Donovan et al., 1988; Ward and Ellar, 1987); kurhdl (Geiser et al., 1986); Bta (Sanchis et al., 1989); bt1, bt2, etc. (Hofte et al., 1986); typB and typC (Höfte et al., 1988); and 4.5, 5.3, and 6.6 kb (Kronstad and Whiteley, 1986).

Höfte and Whiteley (1989) reviewed 42 crystal proteins known by then and proposed a systematic nomenclature for 14 distinct crystal protein genes based on both the structural similarities and insecticidal spectra of the encoded proteins. According to their proposal, 13 of these genes—the so-called *cry* genes—specify a family of related insecticidal proteins (Cry proteins) with four major classes and several subclasses. The four major classes are Lepidoptera-specific *(cryI)*, Lepidoptera- and Diptera-specific *(cryII)*, Coleoptera-specific *(cryIII)*, and Diptera-specific *(cryIV)* genes. One crystal protein gene of *B. thuringiensis* subsp. *israelensis* that codes for a 27-kDa protein that exhibits cytolytic activity against a variety of invertebrate and vertebrate cells and is totally unrelated to the *cry* genes was classified in the *cyt* gene family and is beyond the scope of the present review. Separate strains of *Bt* produce a variety of crystal toxins with distinct host ranges (Table 6.5).

This primary nomenclature system provided a useful framework for classifying the set of known genes. The protein named CryIC, for example, was reported to be toxic to both Diptera and Coleoptera (Smith and Ellar, 1994), while the protein designated CryIB was reported to be toxic to both Lepidoptera and Coleoptera (Bradley et al., 1995). Another example is that the *cryIIB* gene received a place in the Lepidoptera-Diptera class with *cryIIA*, even though toxicity against Diptera could not be demonstrated for the toxin designated CryIIB (Crickmore et al., 1998). Furthermore, the new genes encoding a diverse set of proteins without a common insecticidal activity were all classified as *cryV* (Shin et al., 1995; Sick et al., 1994).

TABLE 6.5. Specificity of different *Bt* toxins used in producing transgenic crops.

Crystal proteins	Insecticidally active on	Transformed plants
Cry1Aa	Lepidoptera	Cranberry, poplar
Cry1Ab	Lepidoptera	Apple, cotton, maize, poplar, potato, rice, tobacco, tomato, white clover, white spruce
Cry1Ac	Lepidoptera	Apple, broccoli, cabbage, cotton, grapevine, oilseed rape, peanut, maize, rice, soybean, tobacco, tomato, walnut
Cry1B	Lepidoptera	Rice, maize
Cry1Ba	Lepidoptera	White clover
Cry1C	Lepidoptera	Broccoli, alfalfa, tobacco
Cry1Ca	Lepidoptera	Alfalfa, *Arabidopsis*, tobacco
Cry1H	Lepidoptera	Maize
Cry2A	Lepidoptera	Rice
Cry2Aa	Lepidoptera	Cotton
Cry2a5	Coleoptera	Tobacco
Cry3A	Coleoptera	Eggplant, potato, tobacco
Cry3C	Coleoptera	Tobacco
Cry6A	Coleoptera	Alfalfa
Cry9C	Lepidoptera	Maize

Currently, 168 *cry* and *cyt* genes and 28 groups of *cry* genes have been identified (Crickmore, 1999). Among these, the Cry1 proteins, primarily active against lepidopteran pests, are the most extensively studied for structure and mode of action (Luo et al., 1998; Rajamohan et al., 1998; Schnepf et al., 1998).

Crystal Structure

The Cry proteins generally form a crystalline inclusion in the mother cell compartment. Depending on their protoxin composition, the crystals have various shapes: bipyramidal (Cry1), cuboidal (Cry2), flat rectangular (Cry3A), irregular (Cry3B), spherical (Cry4A and Cry4B), and rhomboidal (Cry11A; Schnepf et al., 1998). This ability of protoxins to crystallize allows them to rapidly and efficiently solubilize in the gut of insect larvae to become biologically active. Some reports, however, indicate that the solubility of crystals also depends on factors such as the secondary structure of protoxins (Bernhard, 1986), the energy of disulfide bonds (Bietlot et al., 1990), and the presence of additional, *Bt*-specific components (Agaisse and Lereclus, 1995; Baum and Malvar, 1995; Kostichka et al., 1996).

Höfte and Whiteley (1989) first reported the concept of five blocks of amino acids conserved among most of the Cry toxins known by then. Complete amino acid sequence alignment of the Cry proteins also revealed the same five tracts, or conserved blocks, in most of them. However, comparison of the carboxyl-terminal halves of sequences with more than 1,000 residues suggested the presence of an additional three blocks lying outside the active toxic core (Schnepf et al., 1998) (Figure 6.1).

The recent major breakthroughs in solving the three-dimensional structure of different classes of crystal proteins, such as Cry3A (Li et al., 1991) and Cry1Aa (Grochulski et al., 1995), by X-ray crystallography provide the molecular basis for a better interpretation of the toxicity mechanism. An analysis of primary sequences demonstrates that Cry3A and Cry1Aa show about 36% amino acid identity (Crickmore et al., 1998). This primary sequence similarity results in their high overall three-dimensional structural similarity. Both Cry3A and Cry1Aa are built from three domains, and the corresponding domains have the same topological folds (Grochulski et al., 1995). Domain I consists of a bundle of seven antiparallel α-helices, in which helix 5 is encircled by the remaining helices. Domain II makes up three antiparallel β-sheets that are packed around a central hydrophobic core, forming a so-called β-prism structure (Sankaranarayanan et al., 1996; Shimizu and Morikawa, 1996). The β-ribbons of each sheet terminate in loops in a small region of the molecule. These have similarity to the comple-

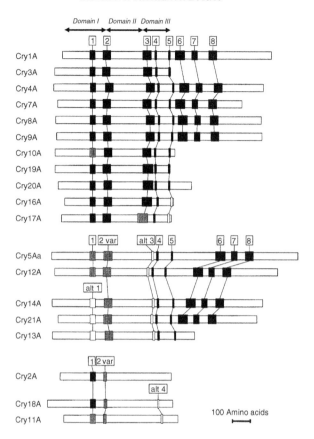

FIGURE 6.1. Positions of conserved blocks among Cry proteins. The figure shows the sequence arrangement for each holotype toxin (e.g., Cry1Aa1) having at least one of the conserved blocks. Sequence blocks are shown as dark gray, light gray, or white to indicate high, moderate, or low degrees of homology, to the consensus sequence for each conserved block. var = variant; alt = alternate. The lengths of each protein and the conserved blocks within them are drawn to scale. *Source:* Schnepf et al. (1998).

mentary determining regions of immunoglobulins and have been proposed as receptor binding sites (Li et al., 1991). Domain III consists of two twisted, antiparallel β-sheets forming a β-sandwich with a "jelly roll" topology.

Mechanism of Action

The mechanism of action of the *B. thuringiensis* Cry proteins involves solubilization of the crystal in the insect midgut, proteolytic processing of the protoxin by midgut proteases, binding of the Cry proteins to midgut receptors, and insertion of the Cry proteins into the apical membrane to create ion channels or pores (Figure 6.2). The Cry proteins are initially synthesized as inactive protoxins during sporulation of *B. thuringiensis* strains and are deposited in the parasporal crystal. Upon ingestion by susceptible insects, the crystals are solubilized in the alkaline environment of the insect gut (Hofmann et al., 1988a). Differences in the extent of solubilization sometimes explain differences in the degree of toxicity among Cry proteins (Aronson et al., 1991; Du et al., 1994). Reduced solubility could be one po-

FIGURE 6.2. Development and field evaluation of transgenic *Bt* rice. (a) Dead heart symptom; (b) production of putative transgenic rice in greenhouse with *Bt* gene and cut-stem bioassay showing larval mortality (Datta et al., 1998); (c) Southern blot showing integration of 1.8-kb cry gene in the genome of rice cv. *Minghui* 63 (Tu et al., 2003); (d) expression of Cry protein in transgenics of *indica* rice cv. *Tulasi;* (e) resistance reaction of transgenic IR72 *Bt* rice (T) against yellow stem borer; and (f) leaffolder infestation under natural field conditions at Wuhan, China (Ye et al., 2001). See corresponding Plate 6.2 in the central gallery.

tential mechanism for insect resistance to *Bt* toxins (McGaughey and Whalon, 1992).

Soon afterward, the solubilized protoxins are processed by insect midgut proteases (Milne and Kaplan, 1993) to become activated toxins. The Cry1A protoxins are digested to 55-65 kDa toxin proteins by clipping off nearly 500 amino acids from the C-terminus and 28 amino acids from the N-terminus (Choma et al., 1990). The mature Cry1Aa toxin is cleaved at R28 at the amino-terminal end (Nagamatsu et al., 1984), whereas Cry1Ac is cleaved at K623 at the carboxy-terminal end (Bietlot et al., 1989). An interesting and unexpected finding is that a DNA fragment of 20 kb is intimately involved in the proteolytic processing of the protoxin (Bietlot et al., 1993).

Activated Cry toxins then kill the midgut cell by inserting themselves into the brush border membrane to create a transmembrane leakage pore. This membrane penetration occurs in two steps: binding to a specific receptor exposed on the membrane surface is followed by insertion of the activated toxins into the membrane, leading to pore formation. The activated toxin binds readily to specific receptors on the apical brush border of the midgut microvillae of susceptible insects (Hofmann and Lüthy, 1986; Hofmann et al., 1988a, 1988b). In *Manduca sexta,* a cadherin-like 210-kDa membrane protein was identified as the specific receptor for Cry1Ab toxin (Francis and Bulla, 1997; Keeton and Bulla, 1997; Vadlamudi et al., 1993), whereas aminopeptidase N (APN) proteins with molecular masses of 120 kDa and 106 kDa proved to be the receptors for Cry1Ac and Cry1C, respectively (Knight et al., 1994; Luo et al., 1996; Sangadala et al., 1994). Cry1Ac toxin was found to attach to the membrane surface via a glycosyl phosphatidyl inositol (GPI) anchor (Garczynski and Adang, 1995). Incorporation of purified 120-kDa APN into the planar lipid bilayer catalyzed ion channel formation by Cry1Aa, Cry1Ac, and Cry1C (Schwartz et al., 1997). Alkaline phosphatase has also been proposed to be a Cry1Ac receptor (Sangadala et al., 1994). Furthermore, some evidence indicates that domain II from either Cry1Ab or Cry1Ac can promote binding to the larger protein, whereas domain III of Cry1Ac promotes binding to the presumed APN (de Maagd et al., 1996).

The recent cloning of the putative 210-kDa (Vadlamudi et al., 1995) and 120-kDa (Knight et al., 1995) Cry1Ac receptors opens possibilities for studies on toxin-receptor interactions. In *Heliothis virescens,* three aminopeptidases bound to Cry1Ac on toxin affinity columns. One of them, a 170-kDa APN, bound to Cry1Aa, Cry1Ab, and Cry1Ac, but not to Cry1C or Cry1E columns. Another experiment on modeling studies showed that binding of the Cry1Ac toxin, but not binding of Cry1Aa and Cry1Ab, to the APN receptor from *M. sexta* (Knight et al., 1994) is specifically prevented by the presence of the sugar *N*-acetyl galactosamine (GalNAc). *N*-acetyl gluco-samine, although having the identical charge and size as GalNAc, does not inhibit this binding activity of Cry1Ac (Ellar, 2005). These results provide an opportunity to verify that the *Bt* toxin could incorporate two lectin-binding sites (see the role of domain III). The three Cry1Ac toxins each recognized a high-affinity and low-affinity binding site on this 170-kDa APN (Luo et al., 1997a). In gypsy moth *(Lymantria dispar)*, the cry1Ac receptor also seems to be APN, whereas Cry1Aa and Cry1Ab bind to a 210-kDa brush border membrane vesicle (BBMV) protein (Valaitis et al., 1997). In *Plutella xylostella* (Luo et al., 1997b) and *Bombyx mori* (Yaoi et al., 1997), APN appears to function as a Cry1Ac-binding protein as well.

Binding is a two-stage process: a reversible binding event (Hofmann and Lüthy, 1986; Hofmann et al., 1988a,b), followed by an irreversible mem-brane association. It has been generally assumed that irreversible binding is exclusively associated with membrane insertion (Ihara et al., 1993; Rajamohan et al., 1995; van Rie et al., 1989). Certainly, the recent report that truncated Cry1Ab molecules containing only domains II and III can still bind to midgut receptors, but only reversibly, supports the notion that irreversible binding requires the insertion of domain I (Flores et al., 1997). Yet at least some published data are consistent with the notion of tight binding to puri-fied receptors. Tight binding of Cry1Aa and Cry1Ab to purified *M. sexta* APN has been observed (Masson et al., 1995), and Cry1Ac may also show some degree of irreversible binding to *M. sexta* as well as to purified *L. dispar* APN. Finally, the evidence from studying the interaction of toxin with the purified receptor by surface plasmon resonance (SPR) indicates that the receptor is reusable—that is, upon toxin insertion into the monolayer, APN is released and can bind more toxin (Cooper et al., 1998a,b).

After binding to the specific receptor, the toxin inserts itself into the apical membrane of the midgut epithelial cells to create transmembrane leakage pores (Knowles and Dow, 1993). The formation of the toxin-induced pores in the columnar cell apical membrane allows rapid fluxes of ions. The pores are K^+ selective (Sacchi et al., 1986), permeable to cations (Wolfersberger, 1989), permeable to anions (Hendrickx et al., 1989), or permeable to solutes such as sucrose, irrespective of the charge (Schwartz et al., 1991). There is evidence that midgut permeability was altered for monovalent cations, neu-tral solutes, and water in the presence of Cry1Ac (Carroll and Ellar, 1993). Knowles and Dow (1993) suggested that the inserted toxins lead to cessa-tion of the K^+ pump, which leads to swelling of the columnar cell and os-motic lysis, and eventually to the death of the insect. These pores possess both selective (only K^+ pass through) and nonselective (Na^+ and anions pass through) properties, depending on the pH (Schwartz et al., 1993). The lepidopteran insect midgut is alkaline, and the pores probably permit K^+ leakage (Sharma et al., 2000). The size of the toxin pore in susceptible insect cells was recently measured as 1 nm in diameter using osmotic protection experiments (Carroll and Ellar, 1997).

Pore Formation: The Role of Domain I

Planar lipid bilayer studies as well as protein cleavage studies have suggested that the helical domain I is responsible for ion channel formation (von Tersch et al., 1994). The isolated domain I alone can be partitioned into artificial phospholipid membranes (Grochulski et al., 1995). In fact, the notion of a pore-forming role for domain I is strongly supported by the striking similarities to the pore-forming or membrane translocating domains of two other bacterial protein toxins, colicin A and diphtheria toxin (Parker and Pattus, 1993). The membrane insertion domain I of colicin A consists of a bundle of ten antiparallel helices, with two central helices, $\alpha8$-$\alpha9$, embedded between them (Parker et al., 1989, 1992). Membrane insertion, followed by pore formation, is triggered by low pH, an environmental parameter causing the tertiary structure of domain I to change from a transition state to a molten globule state (van der Goot et al., 1991). Subsequently, the helical bundle unfolds, with retention of its secondary structure, so that the hydrophobic $\alpha8$-$\alpha9$-formed helical hairpin can insert itself into the membrane.

Like colicin A and diphtheria toxin, domain I of Cry proteins contains a relatively hydrophobic helical hairpin that initiates pore formation. As mentioned before, domain I of Cry proteins is composed of a 7-helix bundle, and five of them ($\alpha3$-$\alpha7$) are the external helices, long enough to span a 30-Å membrane (Grochulski et al., 1995). To form a membrane-spanning pore, the external helices are supposed to become inverted, so that their hydrophobic faces make contact with the membrane lipid bilayer, while their hydrophilic faces are able to form a nonspecific aqueous pore. Two mechanisms have been proposed for this inversion. In one proposal, the so-called penknife model, the strongly hydrophobic helices $\alpha5$ and $\alpha6$ that were joined by a loop at the top of the structure open in penknife fashion and insert themselves into the membrane, with oligomerization of several toxin molecules, thus forming an aqueous pore (Hodgman and Ellar, 1990). The rest of the molecule remains at the membrane surface or on the receptor. The second hypothesis, known as the "umbrella" model, proposes that helices $\alpha4$ and $\alpha5$ form a helical hairpin that initiates toxin insertion into the target membrane to create packing disorder, which allows other domain I helices to participate in pore formation (Li et al., 1991). Direct evidence to support this hypothesis comes from Schwartz et al (1997), who created a disulfide bond within domain I, and between domain I and domain II, to restrict intramolecular movements. Their results showed that helices $\alpha 4$ and $\alpha 5$ insert themselves into the membrane, while the rest of domain I flattens out on the membrane surface in an umbrella-like molten globule state. The lack of protein structural analysis in this work, however, leaves open the possibility that the disulfide bonds blocked the ability of these mutant proteins to penetrate the membrane. That domain I is responsible for pore formation is also confirmed by site-directed mutation studies (Wu and Aronson, 1992).

Receptor Binding: The Role of Domain II

For several Cry proteins, the receptor-binding regions have been mapped primarily to domain II via a set of in vitro-constructed reciprocal recombinants between closely related toxins of differing specificities (Masson et al., 1994; Schnepf et al., 1990; Widner and Whiteley, 1989). Van Rie et al. (1989) demonstrated that the receptor binding correlated with insect specificity, and Lee et al. (1992) demonstrated that the specificity and binding domains were collinear for Cry1Aa against *B. mori*. Examination of the crystal structure of Cry3A (Li et al., 1991) and Cry1Aa (Grochulski et al., 1995) suggested a molecular basis for receptor binding by the loops of domain II. These suggestions have now been substantiated by site-directed mutagenesis.

Lu et al. (1994) found that residues 365-371 of Cry1Aa are essential for binding to the membrane of midgut cells of *B. mori*. Smith and Ellar (1994) demonstrated that mutations in two predicted loop regions of Cry1C—residues 317-320 and 374-377—were able to modulate both toxicity and specificity to two different target insects. Rajamohan et al. (1995, 1996) used Cry1Ab, Smedley and Ellar (1996) used Cry1Ac, and Wu and Dean (1996) used Cry3A to repeat mutations at these loops of domain II. Consistent with the results mentioned before, most of these mutations led to dramatic changes in the kinetics of binding to the receptor of the insect midgut membrane (see review by Schnepf et al., 1998, for details). Therefore, all of these data strongly support the notion that these loops of domain II of Cry proteins are a primary determinant of insect specificity.

Receptor Binding and Toxicity Stabilization: Possible Role of Domain III

The β-sandwich structure of domain III has also been implicated in receptor binding. It was originally suggested that domain III functions in maintaining the structural integrity of the toxin molecules, perhaps by protecting them against gut proteolysis, but in fact all three domains would have to share this characteristic. Recently, experiments involving the reciprocal exchange of domain segments between Cry1Aa and Cry1Ac have shown that the recombining protein containing Cry1Aa domain III bound to a 210-kDa receptor, whereas the recombining protein containing Cry1Ac domain III bound to a 120-kDa receptor in gypsy moth, strongly suggesting a role of domain III in receptor binding (de Maagd et al., 1996). This role of domain III was also confirmed by domain switching experiments in which Cry1Ab domain III bound to the receptor in *Spodoptera exigua* (de Maagd et al., 1996). Because of the close structural similarity between the Cry1Ac domain III fold and the cellulose-binding domain of a protein from the bacterium *Cellulomonas fimi*, a series of mutations were made in the putative carbohydrate-binding site to monitor the effect on the ability of GalNAc to inhibit toxin binding to the receptor (Burton et al., 1999). The results obtained in a triple Cry1Ac mutant involving three residues located by sequence alignment on the outer sheet of the jelly roll fold of domain III showed that GalNAc recognition and receptor binding were completely abolished, and this confirmed that domain III is also a lectin-like domain that recognizes GalNAc on the epithelial APN (Burton et al., 1999). These results also indicate that the insecticidal specificity of these toxins could be determined by two lectin-like domains—domain II and domain III—that act either in concert or independently.

Is Bt *Hazardous to Human Beings, Other Mammals, and Nontarget Pests?*

As of today, there has been no definitive *yes* answer to this question. As is known, insecticidal crystal proteins have been used in agriculture for more than 50 years; so their safety and efficacy have been well established. Their mode of action—unlike the insect resistance trait in conventional breeding—has been characterized for protein structure (Li et al., 1991), specificity of receptor binding (Hofmann et al., 1988a,b; van Rie et al., 1989), solubility, and proteolytic activation only in the alkaline midgut fluids of insects (Höfte and Whiteley, 1989)—not in acidic ones, as in the case of human beings and other mammals. The specificity of *Bt* is such that it was expected to have no direct effects on predator populations. Initial observations support these expectations (e.g., Hofmann et al., 1988a,b). Hilbeck et al. (1998) suggested that there might be a reduction in fitness of predator chrysopid larvae directly attributable in part to preying on caterpillars that fed on *Bt* maize. However, it should be noted that there was high control mortality in these experiments, and there is expected to be a vast difference between forced laboratory experiments and natural field conditions.

In sum, *Bt* toxin, with 40 years of toxicological data, by virtue of its monogenic nature, activation through ingestion, with high activity at low dose, specificity, and safety for consumers and nontarget organisms, can be effectively deployed in genetic engineering to develop insect-resistant transgenic crops that might be a component of integrated pest management (IPM).

HISTORICAL DEVELOPMENT
IN GENETIC ENGINEERING FOR Bt TRANSGENICS

At least 12 genes encoding different *Bt* toxins that provide protection against mostly lepidopteran and coleopteran insects have been engineered into more than 30 different plant species, including rice (Table 6.6).

Although engineering plants with *Bt* genes has had priority in plant biotechnology, it has taken ten years to make transgenic plants and seeds carrying *Bt* genes available to farmers, growers, and end users. The year 1996 marked a milestone in agricultural biotechnology: for the first time, varieties of potato, cotton, and corn containing modified *cry* genes were sold to farmers and growers. The early transformation experiments with native bacterial *cry* genes (wild type; WT) under different promoters using the *Agrobacterium tumefaciens* method were successful in developing transgenic tobacco and tomato (Adang et al., 1987; Barton et al., 1987; Fischhoff et al., 1987; Vaeck et al., 1987). However, the expression level of this foreign protein was too low (less than 0.001% of leaf soluble protein) to provide adequate protection from the target insect pests. Subsequently, a substantial increase in the expression level of *Bt* protein in several crops was achieved through truncated or synthetic genes developed by the removal of potential mRNA processing (improper splicing sites) and polyadenylation signals during resynthesis of *Bt* genes (Estruch et al., 1997) and optimization of codon usage (G+C rich in place of A+T residues) without altering the amino acid sequences. Such codon-optimized *Bt* genes driven by a constitutive promoter are expressed at levels of 0.1-0.6% of total soluble protein (Armstrong et al., 1995; Kumar et al., 1998; Perlak et al., 1993; Stewart et al., 1996a,b; Strizhov et al., 1996; Williams et al., 1992). Higher expression up to 3-5% has also been achieved through chloroplast transformation of the native WT and truncated *Bt* gene (McBride et al., 1995) and even up to 45.3% of the total soluble protein with cry2Aa2 operon (De Cosa et al., 2001).

TABLE 6.6. Transgenic rice plants expressing crystal protein genes from *Bacillus thuringiensis*.

Bt gene	Target insect	Efficacy (% larval mortality or crop protection level)	Endotoxin (% of total soluble protein)	References
cry1A(b)	YSB	80-100	0.01-0.2	Datta et al., 1998
cry1A(b)/ *cry1A(c)*	YSB Leaffolder	71.1-100	0.002-0.2	Tu et al., 1998, 2000
cry1A(b)	YSB	88-100	–	Alam et al., 1999
cry1A(b)	YSB	100	0.01-0.1	Alam et al., 1998
cry1A(b)/ *cry1A(c)*	YSB	88.9-100	0.01-0.1	Baisakh, 2000
cry1A(b)/ *cry1A(c)*	YSB	67-100	–	Wu et al., 1997
cry1A(b)	Lepidoptera	–	–	Wünn et al., 1996
		10-40	0.05	Fujimoto et al., 1993
cry1A(c)	YSB	70-85	0.015-0.024	Nayak et al., 1997
cry1A(c)	YSB	–	–	Riazuddin et al., 1996
cry1A(b)	YSB/SSB	70-100	0.1	Ghareyazie et al., 1997
cry1A(b)/ *cry1A(c)*	YSB/SSB	97-100	up to 3	Cheng et al., 1998
cry1B	SSB	90-100	0.01-0.4	Breitler et al., 2000
cry2A	YSB Leaffolder	60-100	0.01-5	Maqbool et al., 1998
cry2Aa2	–	–	45.3	De Cosa et al., 2001

Note: YSB = yellow stem borer, SSB = striped stem borer.

Promoters Used in Developing Transgenic Bt *Crops*

The level of expression of the *Bt* toxin in different tissues (tissue specific) or throughout the plant in transgenic crops has largely depended on the relative efficiencies and advantages of different promoters. In the majority of cases, 35S, originating from cauliflower mosaic virus, has been used, which shows constitutive gene expression in most plant tissues. However, variable levels of toxin proteins have been reported among plant species and among different tissues of the same plant (Datta et al., 1998; Nilsson et al., 1996; Pauk et al., 1995). Continuous expression of the gene in all plant tissues poses a risk of pests developing resistance and also results in a yield penalty and tradeoff as the plant redirects more resources than is required for its defense. Considerable research is now directed toward concentrating the expression of *cry* genes in plant parts targeted by insects—for instance, a phloem-specific promoter for sap-sucking insects such as aphids. Among other kinds of promoters, wound-inducible promoters drive gene expression only when the plant is actually attacked (Finch, 1994). Other constitutive and specific promoters that have been widely used in the production of transgenic *Bt* cereals and other crops are comprehensively reviewed in Table 6.7.

Selectable Marker Genes Used in Bt Crops

A majority of the insect-resistant transgenic plants contain genes with resistance to bacterial antibiotics, such as *hph* coding for hygromycin phosphotransferase and the *nptII* gene for neomycin phosphotransferase, as selectable marker genes. Because there was more concern about the marker genes than about the insect resistance gene itself, research efforts were oriented toward using herbicide tolerance genes (e.g., *bar* for phosphinothricin-*N*-acetyl transferase) as alternative marker genes. The current trend is therefore

TABLE 6.7. Promoters used in developing *Bt* transgenic rice.

Promoter	Expression	Insecticidal protein	References
Cauliflower mosaic virus 35S promoter (CaMV 35S)	Most plant tissues (constitutive)	Most Cry proteins listed to date	Alam et al., 1998; Alam et al.,1999; Datta et al., 1998; Wu et al., 2000
Maize phosphoenol pyruvate-carboxylase (PEPC)	Green tissue	Cry1A(b)	Datta et al., 1998; Ghareyazie et al.; 1997
Maize pith-specific	Pith cells	Cry1A(B)	Datta et al., 1998
Maize ubiquitin-1 (Ubi-1)	All plant tissues (constitutive)	Cry1A(c)	Cheng et al., 1998; Khanna and Raina, 2002; Nayak et al., 1997; Riazuddin et al., 1996
		Cry1B Cry1B-Ab	Breitler et al., 2000; Ho et al., 2006
Rice actin-1 (Act-1)	All plant tissues (constitutive)	Cry1A(b)/ Cry1A(c)	Baisakh, 2000; Datta et al., 1998; Tu et al., 1998; Wu et al., 1997

not to use antibiotic resistance genes at all, but to use the Positech system with phosphomannose isomerase *(pmi)* or to deliver the marker gene in a different locus through cotransformation, so as to remove the marker gene later in the segregating generation (Datta, 2000). However, there remains a dearth of efficient selectable marker genes, thus leaving an opportunity for a more positive selection system that would make transgenic *Bt* plants safer and more acceptable to the public.

Field Trials and Commercialization of Transgenic Bt Crops

The total global area of transgenic *Bt* crops in 2003 was 12.2 million hectares, which is 18% of the total area (67.7 million hectares) devoted to transgenic crops (James, 2003). Commercially released or approved for release are *Bt* cotton (Australia, China, India, South Africa, United States), *Bt* potato (Canada, Japan, United States), and *Bt* maize (Argentina, Canada, Japan, the European Community, United States). The commercialization of other *Bt* crops such as canola and rice is in progress in several countries, including Asian countries such as India and China. *Bt* maize and rice with resistance to different target insect pests are under field evaluation in several countries (Table 6.8). Especially among the developing countries, Thailand was the

TABLE 6.8. Field trials of important transgenic crops expressing *Bt* toxin (Published reports only).

Crop	Bt gene	Target pest	References
Rice	*cry 1A(c)*	Stem borer	Altossar et al., 1999
	cry1A(b)/	Stem borer Stem borer/leaffolder	Datta et al., 1999; Tu et al., 2000
	cry1A(c)	Stem borer/leaffolder Stem borer Stem borer	Ye et al., 2001; Khanna and Raina, 2002; Bashir et al., 2004
Maize	*cry1A(b)*	European corn borer	Koziel et al., 1993; Armstrong et al., 1995
		Helicoverpa zea	Sims et al., 1996
		Lepidoptera	Pilcher et al., 1997
Tomato	*cry1A(b)*	Pinworm	Delannay et al., 1989
Potato	*cry 1A(b)*	Tuber moth	Peferoen, 1992
	cry 3A	Colorado potato beetle	Perlak et al., 1993
Cotton	*cry 1A(b)*	Pink boll worm	Wilson et al., 1992
Tobacco	*cry 1A(b)*	Heliothines	Warren et al., 1992

first Asian country to field-test *Bt* cotton in 1996, followed by Indonesia for *Bt* cotton and *Bt* maize, the Philippines for *Bt* maize, China for *Bt* rice, maize, and poplar, and India for *Bt* cotton, rice, and canola. This list is expected to expand as encouraging results from field trials on transgenic *Bt* crops accumulate each cropping season. These field-testing programs have resulted in a significant reduction in insecticide use, increased and more stable yield, effectiveness and stability of *Bt* on target pests, increased net returns per hectare, and substantial social, environmental, and human health benefits.

Deployment of Fused Bt or Two or More Bt Genes in Combination

After the report of the evolution of insect resistance to *Bt* toxins (Tabashnik, 1994), arguments have increased recently for developing two-toxin *Bt* crops, as two-toxin cultivars require smaller refuges to achieve successful resistance management and sustainable field release (Cohen et al., 2000). However, any two *Bt* toxins that are used in combination must not be too similar to each other; otherwise, a single mutation could lead to a breakdown in host resistance. Another suggested strategy is to use combinations of a *Bt* toxin gene and a gene encoding an unrelated toxin, because some insect mutations can confer cross-resistance to multiple, distantly related *Bt* toxins (Frutos et al., 1999). Based on earlier studies (Fiuza et al., 1996; Lee et al., 1997), combinations of *Bt* toxins such as Cry1Aa or Cry1Ac with Cry1C or Cry2A have been suggested as the most effective for yellow and striped stem borer in rice. Transgenic crops have been produced with combinations of different *Bt* genes or the fused hybrid *Bt* gene in rice and potato (Datta et al., 1998; Perlak et al., 1993; Tu et al., 1998, 2000).

Development of Transgenic Bt Rice

The first *japonica* rice engineered with the synthetic, codon-optimized *cry1A(b)* gene through electroporation was reported more than a decade ago (Fujimoto et al., 1993). The R_2 generation transgenic rice was reported to be more resistant to striped stem borer and leaffolder than the wild types. Transgenic *indica* rice cultivar IR58 expressing a synthetic *cry1A(b)* gene under the control of the 35S promoter showed resistance to yellow stem borer (YSB) and striped stem borer (SSB) and feeding inhibition of two leaffolder species (Wünn et al., 1996). However, this transformation event did not proceed further due to gene silencing. With the objective of increasing the expression level of *Bt* toxin, Nayak et al. (1997) transformed IR64 with reconstructed *cry1A(c)*. They claimed resistance of the transgenic plants to YSB damage even at a low level of toxin. Since then, several *indica* and *japonica Bt* rice varieties have been developed in laboratories worldwide (Breitler et al., 2000; Cheng et al., 1998; Ghareyazie et al., 1997; Maqbool et al., 1998; Marfà et al., 2002; Nayak et al., 1997).

Here we present the progress of our transgenic *Bt* rice research at the International Rice Research Institute (IRRI) in the Philippines, which is instrumental in the development and evaluation of *Bt* rice in greenhouse and in field conditions. Figure 6.2 presents the development of transgenic *Bt* rice at IRRI and its deployment under natural field conditions in collaboration with China as a case study. At IRRI, using all three established transformation methods *(Agrobacterium,* biolistic, and protoplast-mediated), a large number of homozygous transgenic *Bt* rice plants have been produced in the background of a number of cultivars—IR72, IR64, IR68899B, NPT, *Chinsurah Boro II, Basmati 370, Tulasi, Vaidehi, Azucena, Nang Huong Cho Dao, Mot Bui, Dinorado, Minghui 63,* and BPT-5204—adapted to the ecogeographic regions of different countries with various synthetic and modified *Bt* genes under different promoters (Datta et al., 2003).

Integration and Expression of Single or Fusion Cry Genes with Different Promoters

The efficiency of various synthetic and modified *cryIA(b)* and *cryIA(c)* genes (individually or fused) driven by constitutive (CaMV 35S, Rice Actin-1, and Maize Ubi-1 with its intron) as well as tissue-specific (green tissue-specific PEPC and pith-specific) promoters has been studied in detail (Datta et al., 1998) (see Table 6.7). Apart from the individual *cry* genes (Alam et al., 1998, 1999; Datta et al., 1998, 1999), a fused gene, *cryIA(b)-cryIA(c),* under the control of rice actin-1 promoter has been used to transform *indica* rice—IR72 and *Ming Hui 63* (Tu et al., 1998); BPT 5204 and BR827R (Balachandran et al., 2003); *Azucena, Dinorado* (Rao et al., 2003); *Vaidehi* and *Tulasi* (Alam et al., 1998; Baisakh, 2000). Another translational fusion gene, *cryIB-cryIA(b),* driven by maize ubiquitin promoter (in collaboration with Roger Frutos, CIRAD, Montpellier, France) has been introduced into elite *indica* rice lines *(Nang Huong Cho Dao* and *Mot Bui)* from Vietnam (Ho et al., 2006).

The stable integration of the transgenes was studied through polymerase chain reaction (PCR) and Southern blot analysis in primary (T_0) and subsequent selfed progenies. In some cases, the transgenics developed through the biolistic method carried rearranged fragments showing hybridization signals with *cry* genes. However, the frequency of such transgenics was very low in the case of transgenics developed through the *Agrobacterium*-mediated transformation method. By and large, the inheritance of the *Bt* genes followed a typical single-locus Mendelian segregation of 3 *Bt*(+)/1 *Bt*(−) with few exceptions, indicating that the *Bt* transgene was integrated in a single locus in the rice genome. Through careful selection of progenies at T_1 generation, homozygous lines were developed either in T_2 or T_3 in most of the transgenics.

The *Bt* protein expression was analyzed through western blot analysis and quantified either by enzyme-linked immunosorbent assay (ELISA) or by comparing the *Bt* protein bands with known amounts of purified *Bt* protein in immunoblots. The presence of a 65-kDa, 60-kDa, and 66-kDa proteins indicated the successful transcription and subsequent translation of *cry1A(b)*, *cry1A(c)*, *cry1A(b)-cry1A(c)*, and *cry1B-cry1A(b)* genes, respectively. Based on immunoblot assay and ELISA, the toxin level of the transgenic lines varied from a minimum of 0.01% to a maximum of 1% of the total soluble protein in the case of biolistically derived transgenics (Alam et al., 1998; Datta et al., 1998; Tu et al., 1998) and from 0.813 to 1.241 ppm in leaf soluble protein (Chandel et al., unpublished data). However, expression of up to 3% and 5% of total soluble protein has been reported by Cheng et al. (1998) with modified *cry1A(b)* or *cry1A(c)* under the Ubi promoter and Maqbool et al. (1998) with the *cry2A* gene.

Functional Efficacy of Bt Protein

The functionality of the transgenes was assessed by challenging the transgenic plants growing in the containment greenhouse against the neonate larvae of yellow stem borer using the cut-stem as well as whole-plant bioassay method. The bioassay results showed variation in the levels of protection of the transgenics among the transgenic lines of different cultivars and also among the progenies of a single homozygous line depending on the genes and the promoters used. However, most of the transgenics showed larval mortality in the range of 94.4-100%. The transgenic lines were identical in morpho-agronomic features. This established the fact that the amount of protein expressed was optimal to effect complete protection of the transgenics against the target insect without involving any phenotypic cost. Moreover, an investigation into the DNA fingerprinting of a few representative transgenics with a set of AFLP marker combinations did not show detectable polymorphisms, which further evidenced that the transgenics and their counterpart control cultivars were isolines with no change in their genetic make up except for the *Bt* transgenes (Chandel et al., unpublished data). Furthermore, the studies involving a feeding experiment with *Bt* transgenics of a nontarget insect pest, the brown planthopper *(Nilaparvata lugens),* and a secondary exposure of mirid bug *(Cyrtorrhinus lividipenis),* a predator of brown planthopper, showed no significant effect of *Bt* toxin on five fitness characters studied (Chandel et al., unpublished data).

Isolation of Selectable Marker-Free Transgenics (MFT) and Development of Bt *Hybrid Rice*

Selectable marker genes are one of the prerequisites for selecting transformed tissues. Antibiotic resistance genes, especially *hph* (hygromycin phosphotransferase), conferring resistance against hygromycin, are widely used in the development of transgenic plants. However, due to the public concern over the use of transgenics with such antibiotic marker genes, strategies are being developed to use alternative systems like PosiTech using *pmi* (phosphomannose isomerase) or to obtain marker-free transgenics (MFT). Among the various strategies (see recent review by Puchta, 2003), cotransformation still seems to be the simplest and can be accomplished through different delivery systems.

Using cotransformation with plasmids containing the genes of interest mixed with plasmids carrying a selectable marker gene *(hph),* we have been successful in identifying the transgenics from the progenies of two different lines *(Minghui 63* and *Azucena)* without selectable marker genes. Detailed molecular analysis of the integration pattern of the transgenes revealed differential integration patterns among the progenies of a single transgenic event; moreover, the segregation ratio did not fit that expected for a single Mendelian locus. This gave us a clue that the transgenes were clustered and integrated into the rice genome in more than one loci. Further analysis showed that *hph* was tightly linked to the rearranged fragments of the *Bt* transgenes in a single locus that was genetically unlinked with another locus carrying only the expected fragment of the Bt gene. This allowed the meiotic segregation of these two unlinked loci in the subsequent segregating generation, thereby resulting in recovery of transgenics free from the *hph* gene (Rao et al., 2003; Tu et al., 2003).

We extended our *Bt* research to hybrid rice programs by introducing the *Bt* genes into both maintainer (IR68899B; Alam et al., 1999) and restorer *(Minghui 63;* Tu et al., 2003; BR827-35R; Balachandran et al., 2003) lines. A hybrid rice *Bt-Shanyou 63* was derived from a cross between a marker-free *Bt*-transgenic Chinese restorer line, *Minghui 63* (Tu et al., 1998) and a CMS line, *Zhenshan97A.* The transgenics from other maintainer and restorer lines are presently being used in the development of hybrids.

Field Evaluation of Transgenic Bt Rice

Field evaluation of the transgenics is an important and essential component of transgenic breeding for studying the efficacy of the transgenes and assessing their agronomic features under natural conditions. Under a collaborative IRRI-China research program, transgenic IR72 (an IRRI-bred elite *indica* variety), *Minghui 63* (a Chinese restorer line), and a hybrid, *Shanyou 63,* are being field-evaluated for more than four years. The field trial results with *Bt*-IR72 clearly demonstrated the resistant reaction of the transgenic lines, which showed very high protection against four lepidopteran insects—yellow stem borer (YSB), pink stem borer (PSB), striped stem borer (SSB), and leaffolder—under natural and artificial infestations in field conditions at Zhezhang University in China (Ye et al., 2001b). In Wuhan province of China, hybrid *Bt-Shanyou 63* outyielded the non-*Bt-Shanyou 63* by 28.9% due to increased protection from leaffolder and stem borer (Tu et al., 2000). This is the first ever commercial hybrid rice with *Bt* gene under field trials in Wuhan, China. Moreover, as mentioned earlier, the *Bt* hybrid is free of the selectable marker gene that involved a marker-free *Bt-Minghui 63,* the restorer parent of the hybrid. This could ease the process in obtaining approval for commercializing *Bt* hybrid rice.

In India, with the approval from the Department of Biotechnology (DBT), field trials in both dry and wet seasons (over two years) have shown higher resistance of our transgenic *Bt*-IR72 cultivar to YSB and SSB at the Directorate of Rice Research, Hyderabad. Recent field testing of *Bt* rice has been also reported in China (Ye et al., 2001a), India (Khanna and Raina, 2002), and Pakistan (Bashir et al., 2004).

Developmental and Promoter-Dependent Variation in Bt Protein Expression

The transgenic *Bt* rice with *cry1A(b)* using the PEPC promoter was highly resistant to stem borers and leaf-feeding pests at the vegetative stage. However, the toxin titer and insect resistance dropped substantially at the flowering stage (Alinia et al., 2000). Our results with *Bt* rice carrying *cry1A(b)* driven by 35S, Actin-1, and pith-specific promoter showed these plants to retain resistance to stem borer and leaffolder at the flowering stage (Aguda et al., 2001). Variations in the expression of Cry1A(b) protein has been also found in Basmati rice in leaves, stems, and panicles under the control of PEPC, ubiquitin, or pollen-specific promoters (Husnain et al., 2002).

Nevertheless, many questions still need to be addressed, including the minimum requirement of *Bt* protein needed to protect a given rice variety from insect pests. We have observed that a minimum range of 0.01-0.2 μg/g *Bt* protein is capable of causing 100% YSB larvae mortality in the cut-stem method (Datta et al., 1998).

CONCERNS ABOUT INTRODUCING Bt CROPS IN THE FIELD AND MARKET

Controversial reports on genetically modified organisms (GMOs) are increasing on issues relating to their safe environmental release. One of the important concerns in the development and field release of transgenic crops is the horizontal gene flow to wild relatives. Preliminary experiments showed the possibility of cross-pollination among members of the Brassicaceae family and an increased survivorship of *Brassica napus* with a *Bt* transgene under natural field conditions (Stewart et al., 1996b). However, the stable inheritance and expression of the insect-resistant phenotype in the offspring of such natural hybrids needs to be confirmed to determine the likelihood and impact of such a transfer. Nonetheless, such a possibility could be taken care of by deployment strategies, such as tissue-specific or wound-inducible expression of the *Bt* gene, extending expression of the transgene in the chloroplast genome, and so forth.

Another concern is the development of insect resistance to the toxin genes. Some laboratory experiments have shown such instances. It is not surprising that, given the multiple steps involved in the processing of the crystal to an active toxin, the insect population might develop various means of resisting intoxication. Several laboratory experiments to select for *Bt* resistance in diamondback moths failed, although this is the only known insect reported so far to have developed resistance to *Bt* in the field. However, it is important to note that under laboratory conditions, insect populations are generally isolated; dilution of resistance by mating with susceptible insects, as observed in field populations, is excluded. Furthermore, the natural environment may contain factors that affect the viability and fecundity of the resistant insects. Resistance mechanisms may be associated with certain fitness costs that can be deleterious under natural conditions. Natural enemies such as predators and parasites can influence the development of resistance to *Bt* by preferring intoxicated susceptible or healthy resistant insects. Nevertheless, selection experiments in the laboratory are valuable, as they reveal the possible resistance mechanisms and make genetic studies of resistance possible.

Resistance management strategies were developed to prevent or diminish the selection of the rare individuals carrying resistance genes, thereby keeping the frequency of resistant alleles sufficiently low for insect control. Proposed strategies include using multiple toxins (stacking/pyramiding), crop rotation, high or ultrahigh doses, and spatial or temporal refugia (McGaughey and Whalon, 1992; Tabashnik, 1994). However, a recent retrospective analysis of resistance development does not support the use of refugia (Tabashnik, 1994). Among the different modes of resistance management strategies, a valuable option is the expression in crops of multiple Cry proteins with different modes of action. Cry toxins that recognize different receptors in the same target species could be deployed, because they are less prone to cross-resistance. This would still control even insects that are homozygous for one or two resistance genes yet heterozygous for another gene. The assumption in the multiple toxin strategy is that within a population, if insects homozygous for one gene are rare, then insects homozygous for multiple resistance genes will be extremely rare.

Another important consideration in the management strategy is to make the plants express crystal protein genes in certain specific tissues, such as limiting *cry* gene expression in cotton to young bolls. Crystal protein gene expression could be triggered by the feeding of the insect itself on the transgenic plants, when *cry* genes are controlled by wound-inducible promoters (Peferoen, 1992). If plants were to express *Bt* toxin only in response to specific damage thresholds, this might provide a mechanism to diminish toxin exposure to insects. Alternatively, toxin expression could be induced by applying chemicals (Williams et al., 1992). Farmers would then have the option of having Cry toxin present in the crop only when insect densities exceeded an economic threshold.

CONCLUSION

Biotechnology could be used to prevent pest problems and thus reduce the need for pest management and pesticide use. Since the beginning of agriculture, plant breeders have developed crop varieties that were resistant to or tolerant of particular pests. The tools of biotechnology could be used to make such plant breeding easier and quicker. Transgenic breeding for development of *Bt* rice with resistance to insect pests is quite important because of the lack of resistant donors for conventional breeding. Deployment of *Bt* rice varieties in the field with all precautionary measures of biosafety is an environment-friendly and cost-effective component of IPM. Moreover, the development of *Bt* rice free from the selectable marker gene, such rice could enter directly into the farmers' field or could be used as the parent for further breeding in the national programs. *Bt* rice will have a tremendous impact in Asia in controlling insect pests and reducing the cost and excessive use of environmentally burdensome and health-damaging pesticides.

REFERENCES

Adang, M. J., E. Firoozabady, J. Klein, D. Deboer, V. Sekar, J. D. Kemp, E. Murray, T. A. Rocheleau, K. Rashka, and G. Staffield (1987). Application of a *Bacillus thuringiensis* crystal protein for insect control. In *Molecular strategies for crop protection*, C. J. Arntzen and C. Ryan (Eds.), pp. 345-353. New York: Alan R. Liss.

Agaisse, H., and D. Lereclus (1995). How does *Bacillus thuringiensis* produce so much insecticidal crystal protein? *Journal of Bacteriology 177:* 6027-6032.

Aguda, R. M., K. Datta, J. Tu, S. K. Datta, and M. B. Cohen (2001). Expression of *Bt* genes under control of different promoters in rice at vegetative and flowering stages. *International Rice Research Notes 26:* 26-27.

Alam, M. F., K. Datta, E. Abrigo, N. Oliva, J. Tu, S. S. Virmani, and S. K. Datta (1999). Transgenic insect-resistant maintainer line (IR68899B) for improvement of hybrid rice. *Plant Cell Reports 18:* 572-575.

Alam, M. F., K. Datta, E. Abrigo, A. Vasquez, D. Senadhira, and S. K. Datta (1998). Production of transgenic deep water *indica* rice plants expressing a synthetic *Bacillus thuringiensis cryIA(b)* gene with enhanced resistance to yellow stem borer. *Plant Science 135:* 25-30.

Alinia, F., B. Ghareyazie, L. Rubia, J. Bennett, and M. B. Cohen (2000). Effect of plant age, larval age, and fertilizer treatment on resistance of a *cry1Ab*-transformed aromatic rice to lepidopterous stem borers and foliage feeders. *Journal of Economic Entomology 93:* 484-493.

Altossar, I., Z. Ali, R. Sardana, N. Muthukrishnan, X. Cheng, Q. Shu, and H. Kaplan (1999). Insect control technology in transgenic rice using *Bt* "actives": Field trial performance of synthetic genes and constitutive promoters. In *Abstracts of the Rockefeller Foundation general meeting of the International Program on Rice Biotechnology,* p. 126. Phuket, Thailand.

Armstrong, C. L., G. B. Parker, J. C. Pershing, S. M. Brown, P. R. Sanders, D. R. Duncan, T. Stone, D. A. Dean, D. L. DeBoer, and J. Hart (1995). Field evaluation of European corn borer control in progeny of 173 transgenic corn events expressing an insecticidal protein from *Bacillus thuringiensis*. *Crop Science 35:* 550-557.

Aronson, A. I., E. S. Han, W. McGaughey, and D. Johnson (1991). The solubility of inclusion proteins from *Bacillus thuringiensis* is dependent upon protoxin composition and is a factor in toxicity to insects. *Applied Environmental Microbiology 57:* 981-986.

Awasthi, M. D., L. Anand, P. L. Krishna-Moorthy, and S. V. Sarode (1984). Vertical movement and persistence of granular insecticides in soil. In *Pesticides and envi-*

ronment, K. Raghupathy, K. Rajukkanu, and S. Chelliah (Eds.), pp. 128-133. Tamilnadu Agricultural University, Coimbatore, India.

Balachandran, S., G. Chandel, M. F. Alam, J. Tu, S. S. Virmani, K. Datta, and S. K. Datta (2003). Improving hybrid rice through anther culture and transgenic approaches. In *Proceedings of the 4th international symposium on hybrid rice for food security, poverty alleviation, and environmental protection,* S. S. Virmani, C. X. Mao, and B. Hardy (Eds.), pp. 105-118. Hanoi, Vietnam and IRRI, Philippines.

Banerjee, A. K. (1996). Fungicides and their interactions with non-target organisms. *Journal of Mycopathological Research 34*(1): 7-11.

Barton, K., H. Whiteley, and N. S. Yang (1987). *Bacillus thuringiensis* δ-endotoxin in transgenic *Nicotiana tabacum* provides resistance to lepidopteran insects. *Plant Physiology 85:* 1103-1109.

Baisakh, N. (2000). Improvement of rainfed lowland *indica* rice through in vitro culture and genetic engineering. PhD dissertation, Utkal University, India.

Bashir, K., T. Husnain, T. Fatima, Z. Latif, S. A. Mehdi, and S. Riazuddin (2004). Field evaluation and risk assessment of transgenic *indica* basmati rice. *Molecular Breeding 13:* 301-312.

Baum, J. A., and T. Malvar (1995). Regulation of insecticidal crystal protein production in *Bacillus thuringiensis. Molecular Microbiology 18:* 1-12.

Beegle, C. C., and T. Yamamoto (1992). History of *Bacillus thuringiensis* Berliner research and development. *The Canadian Entomologist 124:* 587-616.

Bernhard, K. (1986). Studies on the delta-endotoxin of *Bacillus thuringiensis* var. *tenebrionis. FEMS Microbiology Letters 33:* 261-265.

Bietlot, H. P., P. R. Carey, M. Pozsgay, and H. Kaplan (1989). Isolation of carboxyl-terminal peptides from proteins by diagonal electrophoresis: Application to the entomocidal toxin from *Bacillus thuringiensis. Analytical Biochemistry 181:* 212-215.

Bietlot, H. P., J. P. Schernthaner, R. E. Milne, F. R. Clairmont, R. S. Bhella, and H. Kaplan (1993). Evidence that the CryIA crystal protein from *Bacillus thuringiensis* is associated with DNA. *Journal of Biological Chemistry 268:* 8240-8245.

Bietlot, H. P., I. Vishnubhatla, P. R. Carey, M. Pozsgay, and H. Kaplan (1990). Characterization of the cysteine residues and disulfide linkages in the protein crystal of *Bacillus thuringiensis. Biochemical Journal 267:* 309-316.

Bradley, D., M. A. Harkey, M. K. Kim, D. Biever, and L. S. Bauer (1995). The insecticidal CryIB protein of *Bacillus thuringiensis* ssp. *thuringiensis* has dual specificity to coleopteran and lepidopteran larvae. *Journal of Invertebrate Pathology 65:* 162-173.

Breitler, J. C., V. Marfà, M. Royer, D. Meynard, J. M. Vassal, B. Vercambre, R. Frutos, J. Messeguer, R. Gabarra, and E. Guiderdoni (2000). Expression of a *Bacillus thuringiensis Cry1B* synthetic gene protects Mediterranean rice against striped stem borer. *Plant Cell Reports 19:* 1195-1202.

Burton, S. L., D. J. Ellar, J. Li, and D. J. Derbyshire (1999). N-acetylgalactosamine on the putative insect receptor aminopeptidase N is recognized by a site on the domain III lectin-like fold of a *Bacillus thuringiensis* insecticidal toxin. *Journal of Molecular Biology 287:* 1011-1022.

Carozzi, N., and M. Koziel (1997). *Advances in insect control: The role of transgenic plants.* London: Taylor and Francis.

Carroll, J., and D. J. Ellar (1993). Proteolytic processing of coleopteran specific δ-endotoxin by *Bacillus thuringiensis* var. *tenebrionis. European Journal of Biochemistry 214:* 771-778.

Carroll, J., and D. J. Ellar (1997). Analysis of the large aqueous pores produced by a *Bacillus thuringiensis* protein insecticide in *Manduca sexta* midgut-brush-border membrane vesicles. *European Journal of Biochemistry 245:* 797-804.

Cengel, M., and F. Saatci (1982). Effect of pesticides and herbicides on soil microorganisms and biochemical turnover in the soil. *Ege-Universitesi-Zirrat-Fakultesi Derigisi 19*(1): 251-259.

Cheng, X., R. Sardana, H. Kaplan, and I. Altossar (1998). *Agrobacterium*-transformed rice plants expressing synthetic *cryIA(b)* and *cryIA(c)* genes are highly toxic to striped stem borer and yellow stem borer. *Proceedings of the National Academy of Sciences of the USA 95:* 2767-2772.

Choma, C. T., W. K. Surewicz, P. R. Carey, M. Pozsgay, and H. Kaplan (1990). Secondary structure of the entomocidal toxin from *Bacillus thuringiensis* subsp. *kurstaki* HD-73. *Journal of Protein Chemistry 9:* 87-94.

Cohen, M. B., F. Gould, and J. S. Bentur (2000). Bt rice: Practical steps to sustainable use. *International Rice Research Newsletter 25*(2): 4-10.

Cooper, M. A., J. Carroll, E. R. Trabis, D. H. Williams, and D. J. Ellar (1998a). *Bacillus thuringiensis* Cry1Ac toxin interaction with *Manduca sexta* aminopeptidase N in a model membrane environment. *Biochemical Journal 333:* 677-683.

Cooper, M. A., A. C. Try, J. Carroll, D. J. Ellar, and D. H. Williams (1998b). Surface plasmon resonance analysis at a supported lipid monolayer. *Biochimica et Biophysica Acta 1373:* 101-111.

Cramer, H. H. (1967). Plant protection and world crop production. *Pflanzenschutz Nachrichten Bayer 20.*

Crickmore, N. (1999). *Dr. Neil Crickmore.* http://www.lifesci.sussex.ac.uk/home/Neil_Crickmore/Bt/ (last accessed February 25, 2007).

Crickmore, N., D. R. Zeigler, J. Feitelson, E. Schnepf, J. van Rie, D. Lereclus, J. Baum, and D. H. Dean (1998). Revision of the nomenclature for the *Bacillus thuringiensis* pesticidal proteins. *Microbiology and Molecular Biology Reviews 62:* 807-813.

Datta, K., A. Vasquez, G. S. Khush, and S. K. Datta (1999). Development of transgenic new plant type rice for stem borer resistance. *Rice Genetics Newsletter 16:* 143-144.

Datta, K., A. Vasquez, J. Tu, L. Torrizo, M. F. Alam, N. Oliva, E. Abrigo, G. S. Khush, and S. K. Datta (1998). Constitutive and tissue-specific differential expression of *CryIA(b)* gene in transgenic rice plants conferring enhanced resistance to insect pests. *Theoretical and Applied Genetics 97:* 20-30.

Datta, S. K. (2000). Potential benefit of genetic engineering in plant breeding: Rice, a case study. *Agricultural Chemistry and Biotechnology 43:* 197-206.

Datta, S. K., N. Baisakh, S. Balachandran, L. B. Torrizo, G. Chandel, M. D. Arboleda, E. M. Abrigo, N. P. Oliva, J. Tu, and K. Datta (2003). Development of transgenic *Bt* rice and its deployment in Asia. In *4th Pacific Rim Conference on the biotechnology of Bacillus thuringiensis and its environmental impact,* R. J. Akhrust, C.E. Beard, and P.A. Hughes (Eds.), pp. 151-153. Australian National University, Canberra, Australia, November 11-15, 2001.

De Cosa, B., W. Moar, S. B. Lee, M. Miller, and H. Daniell (2001). Overexpression of the *Bt* Cry2Aa2 operon in chloroplasts leads to formation of insecticidal crystals. *Nature Biotechnology 19:* 71-74.

Delannay, X., B. J. LaVallee, R. K. Proksch, R. L. Fuchs, S. R. Sims, J. T. Greenplate, P. G. Marrone, R. B. Dodson, J. J. Augustine, J. G. Layton, and D. A. Fischhoff (1989). Field performance of transgenic tomato plants expressing the *Bacillus thuringiensis* var. *kurstaki* insect control protein. *Bio/Technology 7:* 1265-1269.

de Maagd, R. A., H. van der Klei, P. L. Bakker, W. J. Stiekema, and D. Bosch (1996). Different domains of *Bacillus thuringiensis* δ-endotoxins can bind to insect midgut membrane proteins on ligand blots. *Applied Environmental Microbiology 62:* 2753-2757.

Donovan, W. P., C. Dankocsik, and M. P. Gilbert (1988). Molecular characterization of a gene encoding a 72-kilodalton mosquito-toxic crystal protein from *Bacillus thuringiensis* subsp. *israelensis*. *Journal of Bacteriology 170:* 4732-4738.

Du, C., P. A. W. Martin, and K. W. Nickerson (1994). Comparison of disulfide contents and solubility at alkaline pH of insecticidal and noninsecticidal *Bacillus thuringiensis* protein crystals. *Applied Environmental Microbiology 60:* 3847-3853.

Ellar, D. J. (2005). *Summary of research programmes.* http://www.bioc.cam.ac. uk/~dje1/anchor919325 (last accessed February 25, 2007).

Estruch, J. J., N. B. Carozzi, N. Desai, N. B. Duck, G. W. Warren, and M. G. Koziel (1997). Transgenic plants: An emerging approach to pest control. *Nature Biotechnology 15:* 137-141.

Feitelson, J. S. (1993). The *Bacillus thuringiensis* family tree. In *Advanced engineered pesticides,* L. Kim (Ed.), pp. 63-71. New York: Marcel Dekker.

Finch, R. P. (1994). An introduction to molecular technology. In *Molecular biology in crop protection,* G. Marshall and D. Walters (Eds.), pp. 1-37. London: Chapman & Hall.

Fischhoff, D. A., K. S. Bowdish, F. J. Perlak, P. G. Marrone, S. M. McCormick, J. G. Niedermeyer, D. A. Dean, K. Kusano-Kretzmer, E. J. Mayer, and D. E. Rochester (1987). Insect tolerant transgenic tomato plants. *Bio/Technology 5:* 807-813.

Fiuza L. M., C. Nielsen-Leroux, E. Goze, R. Frutos, and J. F. Charles (1996). Binding of *Bacillus thuringiensis* Cry1 toxins to the midgut brush border membrane vesicles of *Chilo suppresalis* (Lepidoptera: Pyralidae): Evidence of shared binding sites. *Applied and Environmental Microbiology 62:* 1544-1549.

Flores, H., X. Soberon, J. Sanchez, and A. Bravo (1997). Isolated domain II and III from the *Bacillus thuringiensis* Cry1Ab delta-endotoxin binds to lepidopteran midgut membranes. *FEBS Letters 414:* 313-318.

Francis, B. R., and L. B. Bulla, Jr. (1997). Further characterization of BT-R$_1$, the cadherin-like receptor for Cry1Ab toxin in tobacco hornworm *(Manduca sexta)* midguts. *Insect Biochemistry and Molecular Biology 27:* 541-550.

Frutos, R., C. Rang, and M. Royer (1999). Managing insect resistance to plants producing *Bacillus thuringiensis* toxins. *Critical Reviews in Biotechnology 19:* 227-276.

Fujimoto, H., K. Itoh, M. Yamamoto, J. Kyozuka, and K. Shimamoto (1993). Insect resistant rice generated by introduction of a modified delta-endotoxin gene of *Bacillus thuringiensis*. *Bio/Technology 11:* 1151-1155.

Garczynski, S. F., and M. J. Adang (1995). *Bacillus thuringiensis* CryIA (c) δ-endotoxin binding aminopeptidase in the *Manduca sexta* midgut has a glycosyl-phosphatidylinositol anchor. *Insect Biochemistry and Molecular Biology 25:* 409-415.

Geiser, M., S. Schweitzer, and C. Grimm (1986). The hypervariable region in the genes coding for entomopathogenic crystal proteins of *Bacillus thuringiensis*: Nucleotide sequence of the *kurhd1* gene of subsp. *kurstaki* HD1. *Gene 48:* 109-118.

Ghareyazie, B., F. Alinia, C. A. Menguito, L. G. Rubia, J. M. de Palma, E. Liwanag, M. B. Cohen, G. S. Khush, and J. Bennett (1997). Enhanced resistance to two stem borers in an aromatic rice containing a synthetic *CryI(b)* gene. *Molecular Breeding 3:* 401-414.

Grochulski, P., L. Masson, S. Borisova, M. P. Carey, J. L. Schwartz, R. Brousseau, and M. Cygler (1995). *Bacillus thuringiensis* CryIA(a) insecticidal toxin: Crystal structure and channel formation. *Journal of Molecular Biology 254:* 447-464.

Hendrickx, K., A. De Loof, and H. van Mellaert (1989). Effects of *Bacillus thuringiensis* delta-endotoxin on the permeability of brush border membrane vesicles from tobacco hornworm *(Manduca sexta)* midgut. *Comparative Biochemistry and Physiology 95C:* 241-245.

Hilbeck, A., M. Baumgartner, P. M. Fried, and F. Bigler (1998). Effects of transgenic *Bacillus thuringiensis* corn-fed prey on mortality and development time of immature *Chrysoperla carnea* (Neuroptera: Chrysopidae). *Environmental Entomology 27:* 480-487.

Ho, N. H., N. Baisakh, N. Oliva, K. Datta, R. Frutos, and S. K. Datta (2006). Translational fusion hybrid *Bt* genes confer resistance against yellow stem borer *(Scirpophaga incertulas* Walker) in transgenic elite Vietnamese rice *(Oryza sativa* L.) cultivars. *Crop Science 46:* 781-789.

Hodgman, T. C., and D. J. Ellar (1990). Models for the structure and function of the *Bacillus thuringiensis* δ-endotoxins determined by compilational analysis. *DNA Sequence 1:* 97-106.

Hofmann, C., and P. Lüthy (1986). Binding and activity of *Bacillus thuringiensis* delta-endotoxin to invertebrate cells. *Archives of Microbiology 146:* 7-11.

Hofmann, C., P. Lüthy, R. Hütter, and V. Pliska (1988a). Binding of the delta-endotoxin from *Bacillus thuringiensis* to brush-border membrane vesicles of the cabbage butterfly *(Pieris brassicae)*. *European Journal of Biochemistry 173:* 85-91.

Hofmann, C., H. Vanderbrüggen, H. Höfte, J. van Rie, S. Jansens, and H. van Mellaert (1988b). Specificity of *Bacillus thuringiensis* δ-endotoxins is correlated with the presence of high-affinity binding sites in the brush-border membrane of target insect midguts. *Proceedings of the National Academy of Sciences of the USA 85:* 7844-7848.

Höfte, H., H. de Greve, J. Seurinck, S. Jansens, J. Mahillon, C. Ampe, M. Vandekerckhove, M. van Montagu, M. Zabea, and M. Vaeck (1986). Structural and functional analysis of a cloned delta endotoxin of *Bacillus thuringiensis berliner* 1715. *European Journal of Biochemistry 161:* 273-280.

Höfte, H., J. van Rie, S. Jansens, A. van Houtven, H. Vanderbriggen, and M. Vaeck (1988). Monoclonal antibody analysis and insecticidal spectrum of three types of lepidopteran-specific insecticidal crystal proteins of *Bacillus thuringiensis*. *Applied and Environmental Microbiology 54:* 2010-2017.

Höfte, H., and H. R. Whiteley (1989). Insecticidal crystal proteins of *Bacillus thuringiensis*. *Microbiological Reviews 53:* 242-255.

Husnain, T., A. Jan, S. B. Maqbool, S. K. Datta, and S. Riazuddin (2002). Variability in expression of insecticidal *Cry*1AB gene in *indica* basmati rice. *Euphytica 128:* 121-128.

Ihara, H., E. Kuroda, A. Wadano, and M. Himeno (1993). Specific toxicity of δ-endotoxins from *Bacillus thuringiensis* to *Bombyx mori. Bioscience, Biotechnology, and Biochemistry 57:* 200-204.

James, C. (2003). *Preview: Global status of commercialized transgenic crops: 2003.* ISAAA Briefs No. 30. Ithaca, NY: International Service for the Acquisition of Agri-biotech Applications.

Keeton, T. P., and L. A. Bulla, Jr. (1997). Ligand specificity and affinity of BT-R$_1$, the *Bacillus thuringiensis* Cry1A toxin receptor from *Manduca sexta,* expressed in mammalian and insect cell cultures. *Applied Environmental Microbiology 63:* 3419-3425.

Khanna, H. K., and S. K. Raina (2002). Elite *indica* transgenic rice plants expressing modified Cry1Ac endotoxin of *Bacillus thuringiensis* show enhanced resistance to yellow stem borer *(Scirpophaga incertulas). Transgenic Research 11:* 411-423.

Knight, P. J. K., N. Crickmore, and D. J. Ellar (1994). The receptor for *Bacillus thuringiensis* CryIA(c) delta-endotoxin in the brush border membrane of the lepidopteran *Manduca sexta* is aminopeptidase N. *Molecular Microbiology 11:* 429-436.

Knight, P. J. K., B. H. Knowles, and D. J. Ellar (1995). Molecular cloning of an insect aminopeptidase N that serves as a receptor for *Bacillus thuringiensis* CryIA(c) toxin. *Journal of Biological Chemistry 270:* 17765-17770.

Knowles, B. H., and J. A. T. Dow (1993). The crystal endotoxin of *Bacillus thuringiensis:* Models for their mechanism of action on the insect gut. *BioEssays 15:* 469-476.

Kostichka, K., G. W. Warren, M. Mullins, A. D. Mullins, J. A. Craig, M. G. Koziel, and J. J. Estruch (1996). Cloning of a *cry*V-type insecticidal protein gene from *Bacillus thuringiensis:* The *cry*V-encoded protein is expressed early in stationary phase. *Journal of Bacteriology 178:* 2141-2144.

Koziel, M. G., G. L. Beland, C. Bowman, N. B. Carozzi, R. Crenshaw, L. Crossland, J. Dawson, N. Desai, M. Hill, S. Kadwell, et al. (1993). Field performance of elite transgenic maize plants expressing an insecticidal protein derived from *Bacillus thuringiensis. Bio/Technology 11:* 194-200.

Krattiger, A. F. (1997). *Insect resistance in crops: A case study of Bacillus thuringiensis (Bt) and its transfer to developing countries.* Ithaca, NY: International Service for the Acquisition of Agri-biotech Applications.

Kronstad, J. W., and H. R. Whiteley (1986). Three classes of homologous *Bacillus thuringiensis* crystal protein genes. *Gene 43:* 29-40.

Kumar, P. A., A. Mandaokar, K. Sreenivasu, S. K. Chakrabarti, S. Bisaria, S. R. Sharma, S. Kaur, and R. P. Sharma (1998). Insect-resistant transgenic brinjal plants. *Molecular Breeding 4:* 33-37.

Lee, M. K., R. Aguda, M. B. Cohen, F. L. Gould, and D. H. Dean (1997). Determination of receptor binding properties of *Bacillus thuringiensis* δ-endotoxins to rice stem borer midguts. *Applied and Environmental Microbiology 63:* 1453-1459.

Lee, M. K., R. E. Milne, A. Z. Ge, and D. H. Dean (1992). Location of a *Bombyx mori* receptor binding region on a *Bacillus thuringiensis* endotoxin. *Journal of Biological Chemistry 267:* 3115-3121.

Li, J., J. Carroll, and D. J. Ellar (1991). Crystal structure of insecticidal δ-endotoxin from *Bacillus thuringiensis* at 2.5 Å resolution. *Nature 353:* 815-821.

Lu, H., F. Rajamohan, and D. H. Dean (1994). Identification of amino acid residues of *Bacillus thuringiensis* δ-endotoxin CryIA(a) associated with membrane binding and toxicity to *Bombyx mori. Journal of Bacteriology 176:* 5554-5559.

Luo, K., D. Banks, and M. J. Adang (1998). Toxicity, binding and permeability analyses of four *Bacillus thuringiensis* Cry1 δ-endotoxins using brush border membrane vesicles of *Spodoptera exigua* and *Spodoptera frugiperda. Applied and Environmental Microbiology 65:* 457-464.

Luo, K., Y. J. Lu, and M. J. Adang (1996). A 106 kDa form of aminopeptidase is a receptor for *Bacillus thuringiensis* CryIC delta-endotoxin in the brush border membrane of *Manduca sexta. Insect Biochemistry and Molecular Biology 26:* 783-791.

Luo, K., S. Sangadala, L. Masson, A. Mazza, R. Brousseau, and M. J. Adang (1997a). The *Heliothis virescens* 170 kDa aminopeptidase functions as "receptor A" by mediating specific *Bacillus thuringiensis* Cry1A δ-endotoxin binding and pore formation. *Insect Biochemistry and Molecular Biology 27:* 735-743.

Luo, K., B. E. Tabashnik, and M. J. Adang (1997b). Binding of *Bacillus thuringiensis* Cry1Ac toxin to aminopeptidase in susceptible and resistant diamondback moths *(Plutella xylostella). Applied and Environmental Microbiology 63:* 1024-1027.

Maqbool, S. B., T. Husnain, S. Riazuddin, and P. Christou (1998). Effective control of yellow rice stem borer and rice leaf folder in transgenic rice *indica* varieties Basmati 370 and M7 using novel δ-endotoxin *Cry2A Bacillus thuringiensis* gene. *Molecular Breeding 4:* 501-507.

Marfà, V., E. Melé, R. Gabarra, J. M. Vassal, E. Guiderdoni, and J. Messeguer (2002). Influence of the developmental stage of transgenic rice plants (cv. Senia) expressing the *cry1B* gene on the level of protection against the striped stem borer *(Chilo suppressalis). Plant Cell Reports 20:* 1167-1172.

Masson, L., Y. J. Lu, A. Mazza, R. Brousseau, and M. J. Adang (1995). The CryIA(c) receptor purified from *Manduca sexta* displays multiple specificities. *Journal of Biological Chemistry 270:* 20309-20315.

Masson, L., A. Mazza, J. L. Gringorten, D. Baines, V. Anelunias, and R. Brousseau (1994). Specificity domain localization of *Bacillus thuringiensis* insecticidal CryIA toxins is highly dependent on the bioassay system. *Molecular Microbiology 14:* 851-860.

McBride, K. E., Z. Svab, D. J. Schaaf, P. Hogan, D. M. Stalker, and P. Maliga (1995). Amplification of a chimeric *Bacillus* gene in chloroplasts leads to an extraordinary level of an insecticidal protein in tobacco. *Bio/ Technology 13:* 362-365.

McGaughey, W. H., and M. E. Whalon (1992). Managing insect resistance to *Bacillus thuringiensis* toxins. *Science 258:* 1451-1455.

McLinden, J. H., J. R. Sabourin, B. D. Clark, D. R. Gensler, W. E. Workman, and D. H. Dean (1985). Cloning and expression of an insecticidal k-73 type crystal protein gene from *Bacillus thuringiensis* var. *kurstaki* into *Escherichia coli. Applied and Environmental Microbiology 50:* 623-628.

Milne, R. E., and H. Kaplan (1993). Purification and characterization of a trypsin-like digestive enzyme from spruce budworm *(Choristoneura fumiferana)* responsible for the activation of δ-endotoxins from *Bacillus thuringiensis. Insect Biochemistry and Molecular Biology 23:* 663-673.

Nagamatsu, Y., Y. Itai, C. Hatanaka, G. Funatsu, and K. Hayashi (1984). A toxic fragment from the entomocidal crystal protein of *Bacillus thuringiensis. Agricultural and Biological Chemistry 48:* 611-619.

Nayak, P., D. Basu, D. Das, A. Basu, D. Ghosh, N. A. Ramakrishna, M. Ghosh, and S. K. Sen (1997). Transgenic elite *indica* rice plants expressing CryIA(c) δ-endotoxin of *Bacillus thuringiensis* are resistant against yellow stem borer *(Scirpophaga incertulas). Proceedings of the National Academy of Sciences of the USA 94:* 2111-2116.

Nilsson, O., C. H. A. Little, G. Sandberg, and O. Olsson (1996). Expression of two heterologous promoters, *Agrobacterium rhizogenes* rolC and cauliflower mosaic virus 35S, in the stem of transgenic hybrid aspen plants during the annual cycle of growth and dormancy. *Plant Molecular Biology 31:* 887-895.

Parker, M. W., and F. Pattus (1993). Rendering a membrane protein soluble in water: a common packing motif in bacterial protein toxins. *Trends in Biochemical Sciences 18:* 391-395.

Parker, M. W., F. Pattus, A. D. Tucker, and D. Tsernoglou (1989). Structure of the membrane-pore forming fragment of colicin A. *Nature 337:* 93-96.

Parker, M. W., J. P. M. Postma, F. Pattus, A. D. Tucker, and D. Tsernolgou (1992). Refined structure of pore-forming domain of colicin A at 2.4 Å resolution. *Journal of Molecular Biology 224:* 639-657.

Pauk, J., I. Stefanov, S. Fekete, L. Bögre, I. Karsai, A. Fehér, and D. Dudits (1995). A study of different (CaMV and mas) promoter activities and risk assessment of field use of transgenic rapeseed plants. *Euphytica 85:* 411-416.

Peferoen, M. (1992). Engineering of insect-resistant plants with *Bacillus thuringiensis* crystal protein genes. In *Plant genetic manipulation for crop protection,* A. M. R. Gatehouse, V. A. Hilder, and D. Boulter (Eds.), pp. 135-153. Wallingford, UK: CAB International.

Perlak, F. J., T. B. Stone, Y. M. Muskopf, L. J. Petersen, G. B. Parker, S. A. McPherson, J. Wyman, S. Love, G. Reed, D. Biever, and D. A. Fischhoff (1993). Genetically improved potatoes: Protection from damage by Colorado potato beetles. *Plant Molecular Biology 22:* 313-321.

Pilcher, C. D., J. J. Obrycki, M. E. Rice, and L. C. Lewis (1997). Preimaginal development, survival, and field abundance of insect predators on transgenic *Bacillus thuringiensis* corn. *Environmental Entomology 26:* 446-454.

Pingali, P. L., C. B. Marquez, and F. G. Palis (1992). *Farmer health impact of long term pesticide exposure: A medical and economic analysis for the Philippines.* Paper presented at the workshop on measuring the health and environment effects of pesticides, Bellagio, Italy.

Prakash, A., and J. Rao (Eds.). (1999). *Insect pests of cereals and their management. Applied Entomology Vol. 1.* Cuttack, India: Applied Zoologists Research Association.

Puchta, H. (2003). Marker-free transgenic plants. *Plant Cell Tissue and Organ Culture 74:* 123-134.

Rajamohan, F., E. Alcantara, M. K. Lee, X. J. Chen, A. Curtiss, and D. H. Dean (1995). Single amino acid changes in domain II of *Bacillus thuringiensis* CryIAb δ-endotoxin affect irreversible binding to *Manduca sexta* midgut membrane vesicles. *Journal of Bacteriology 177:* 2276-2282.

Rajamohan, F., J. A. Cotrill, F. Gould, and D. H. Dean (1996). Role of domain II, loop 2 residues of *Bacillus thuringiensis* CryIAb δ-endotoxin in reversible and irreversible binding to *Manduca sexta* and *Heliothis virescens. Journal of Biological Chemistry 271:* 2390-2396.

Rajamohan, F., M. K. Lee, and D. H. Dean (1998). *Bacillus thuringiensis* insecticidal proteins: Molecular mode of action. *Progress in Nucleic Acid Research and Molecular Biology 60:* 1-27.

Ramaswamy, C., and T. Jatileksono (1996). Intercountry comparison of insect and disease losses. In *Rice research in Asia: Progress and priorities,* R. E. Evenson, R. W. Herdt, and M. Hossain (Eds.), pp. 305-316. Wallingford, UK: CAB International, in association with International Rice Research Institute.

Rao M. V. R., E. Abrigo, M. Rai, N. P. Oliva, K. Datta, and S. K. Datta (2003). Marker-free transgenic Bt rice conferring resistance to yellow stem borer. *Rice Genetics Newsletter 20:* 98-101.

Riazuddin, S., T. Husnain, E. Khan, S. Karim, F. Khanum, R. Makhdoom, and I. Altossar (1996). Transformation of *indica* rice with *Bt* pesticidal genes. In *Rice genetics III: Proceedings of the Third International Rice Genetics Symposium,* G. S. Khush (Ed.), pp. 730-734. Los Baños, Philippines: International Rice Research Institute.

Sacchi, V. F., P. Parenti, B. Giordana, G. M. Hanozet, P. Luthy, and M. G. Wolfersberger (1986). *Bacillus thuringiensis* inhibits K^+ gradient dependent amino acid transport across the brush border membrane of *Pieris brassicae* midgut cells. *FEBS Letters 204:* 213-218.

Sanchis, V., D. Lereclus, G. Menou, J. Chaufaux, S. Guo, and M. M. Lecadet (1989). Nucleotide sequence and analysis of the N-terminal coding region of the *Spodoptera*-active delta-endotoxin gene of *Bacillus thuringiensis aizawai.* 7.29. *Molecular Microbiology 3:* 229-238.

Sangadala, S., F. S. Walters, L. H. English, and M. J. Adang (1994). A mixture of *Manduca sexta* aminopeptidase and phosphatase enhances *Bacillus thuringiensis* insecticidal CryIA(c) toxin binding and ^{86}Rb$^+$-K$^+$ efflux in vitro. *Journal of Biological Chemistry 269:* 10088-10092.

Sankaranarayanan, R., K. Sekar, R. Banerjee, V. Sharma, A. Surolia, and M. Vijayan (1996). A novel mode of carbohydrate recognition in jacalin, a *Moraceae* plant lectin with a β-prism fold. *Nature Structural Biology 3:* 596-603.

Schnepf, H. E., N. Crickmore, J. van Rie, D. Lereclus, J. Baum, J. Feitelson, D. R. Zeigler, and D. H. Dean (1998). *Bacillus thuringiensis* and its pesticidal crystal proteins. *Microbiology and Molecular Biology Reviews 62:* 775-806.

Schnepf, H. E., K. Tomczak, J. P. Ortega, and H. R. Whiteley (1990). Specificity-determining regions of lepidopteran-specific insecticidal proteins produced by *Bacillus thuringiensis. Journal of Biological Chemistry 265:* 20923-20930.

Schnepf, H. E., and H. R. Whiteley (1981). Cloning and expression of the *Bacillus thuringiensis* crystal protein gene in *Escherichia coli. Proceedings of the National Academy of Sciences of the USA 78:* 2893-2897.

Schwartz, J. L., L. Garneau, L. Masson, and R. Brousseau (1991). Early response of cultured lepidopteran cells to exposure to δ-endotoxin from *Bacillus thuringiensis*: Improvement of calcium and anionic channels. *Biochimica et Biophysica Acta Biomembranes 1065:* 250-260.

Schwartz, J. L., L. Garneau, D. Savaria, L. Masson, R. Brousseau, and E. Rousseau (1993). Lepidopteran specific crystal toxins from *Bacillus thuringiensis* form cation and anion-selective channels in planar lipid bilayer. *Journal of Membrane Biology 132:* 53-62.

Schwartz, J. L., Y. L. Lu, P. Sohnlein, R. Brousseau, R. Laprade, L. Masson, and M. J. Adang (1997). Ion channels formed in planar lipid bilayers by *Bacillus thuringiensis* toxins in the presence of *Manduca sexta* midgut receptors. *FEBS Letters 412:* 270-276.

Sharma, H. C., K. K. Sharma, N. Seetharama, and R. Ortiz (2000). Prospects for using transgenic resistance to insects in crop improvement. *Electronic Journal of Biotechnology 3*(2): 1-20.

Shimizu, T., and K. Morikawa (1996). The β-prism: A new folding motif. *Trends in Biochemical Sciences 21:* 3-6.

Shin, B. S., S. H. Park, S. K. Choi, B. T. Koo, S. T. Lee, and J. I. Kim (1995). Distribution of *cryV*-type insecticidal protein genes in *Bacillus thuringiensis* and cloning of *cryV*-type genes from *Bacillus thuringiensis* subsp. *kurstaki* and *Bacillus thuringiensis* subsp. *entomocidus. Applied and Environmental Microbiology 61:* 2402-2407.

Sick, A. J., G. E. Schwab, and J. M. Payne (January 1994). *U.S. patent 05281530.*

Sims, S., J. C. Pershing, and B. J. Reich (1996). Field evaluation of transgenic corn containing *Bt* Berliner insecticidal protein gene against *Helicoverpa zea* (Lepidoptera: Noctuidae). *Journal of Entomological Science 31:* 340-346.

Smedley, D. P., and D. J. Ellar (1996). Mutagenesis of three surface-exposed loops of a *Bacillus thuringiensis* insecticidal toxin reveals residues important for toxicity, receptor recognition and possibly membrane insertion. *Microbiology 142:* 1617-1624.

Smith, G. P., and D. J. Ellar (1994). Mutagenesis of two surface-exposed loops of the *Bacillus thuringiensis* CryIC δ-endotoxin affects insecticidal specificity. *Biochemical Journal 302:* 611-616.

Stewart, C. N., M. J. Adang, J. N. All, H. R. Boerma, G. Cardineau, D. Tucker, and W. A. Parrott (1996a). Genetic transformation, recovery and characterization of fertile soybean transgenic for a synthetic *Bacillus thuringiensis cryIAc* gene. *Plant Physiology 112:* 121-129.

Stewart, C. N., M. J. Adang, J. N. All, P. L. Raymer, S. Ramachandran, and W. A. Parrott, (1996b). Insect control and dosage effects in transgenic canola containing a synthetic *Bacillus thuringiensis cryIAc* gene. *Plant Physiology 112:* 115-120.

Strizhov, N., M. Keller, J. Mathur, Z. Koncz-Kalman, D. Bosch, E. Prudovsky, J. Schell, B. Sneh, C. Koncz, and A. Zilberstein (1996). A synthetic *cryIC* gene, encoding a *Bacillus thuringiensis* δ-endotoxin, confers *Spodoptera* resistance in alfalfa and tobacco. *Proceedings of the National Academy of Sciences of the USA 13:* 15012-15017.

Tabashnik, B. E. (1994). Evolution of resistance to *Bacillus thuringiensis*. *Annual Review of Entomology 39:* 47-79.

Teng, P. S., and I. M. Revilla (1996). Technical issues in using crop loss for research prioritization. In *Rice research in Asia: Progress and priorities,* R. E. Evenson, R. W. Herdt, and M. Hossain (Eds.), pp. 261-275. Wallingford, UK: CAB International, in association with International Rice Research Institute.

Tu, J., K. Datta, M. F. Alam, Y. Fan, G. S. Khush, and S. K. Datta (1998). Expression and function of a hybrid *Bt* toxin gene in transgenic rice conferring resistance to insect pest. *Plant Biotechnology 15:* 195-203.

Tu, J., K. Datta, N. Oliva, G. Zhang, C. Xu, G. S. Khush, Q. Zhang, and S. K. Datta (2003). Site-independently integrated transgenes in the elite restorer rice line Minghui 63 allow removal of a selectable marker from the gene of interest by self-segregation. *Plant Biotechnology Journal 1:* 155-165.

Tu, J., G. Zhang, K. Datta, C. Xu, Y. He, Q. Zhang, G. S. Khush, and S. K. Datta (2000). Field performance of transgenic elite commercial hybrid rice expressing *Bacillus thuringiensis* δ-endotoxin. *Nature Biotechnology 18:* 1101-1104.

Vadlamudi, R. K., T. H. Ji, and L. A. Bulla, Jr. (1993). A specific binding protein from *Manduca sexta* for the insecticidal toxin of *Bacillus thuringiensis* subsp. *berliner. Journal of Biological Chemistry 268:* 12334-12340.

Vadlamudi, R. K., E. Weber, I. Ji, T. H. Ji, and L. A. Bulla, Jr. (1995). Cloning and expression of a receptor for an insecticidal toxin of *Bacillus thuringiensis*. *Journal of Biological Chemistry 270:* 5490-5494.

Vaeck, M., A. Reynaerts, H. Hofte, S. Jansens, M. De Beuckeleer, C. Dean, M. Zabeau, M. van Montagu, and J. Leemans (1987). Transgenic plants protected from insect attack. *Nature 327:* 33-37.

Valaitis, A. P., A. Mazza, R. Brousseau, and L. Masson (1997). Interaction analyses of *Bacillus thuringiensis* Cry1A toxins with two aminopeptidases from gypsy moth midgut brush border membranes. *Insect Biochemistry and Molecular Biology 27:* 529-539.

van der Goot, F. G., J. M. Gonzalez-Manas, J. H. Lakey, and F. Pattus (1991). A "molten-globule" membrane-insertion intermediate of the pore-forming domain of colicin A. *Nature 354:* 408-410.

van Emden, H. F. (1987). Cultural methods: The plant. In *Integrated pest management*, A. J. C. Burn, T. J. Coaker, and P. C. Jepson (Eds.), pp. 27-68. London: Academic Press.

van Rie, J., S. Jansens, H. Hofte, D. Degheele, and H. van Mellaert (1989). Specificity of *Bacillus thuringiensis* δ-endotoxins: Importance of specific receptors on the brush border membrane of the mid-gut of target insects. *European Journal of Biochemistry 186:* 239-247.

von Tersch, M. A., S. L. Slatin, C. A. Kulesza, and L. H. English (1994). Membrane-permeabilizing activities of *Bacillus thuringiensis* coleopteran-active toxin CryIIIB2 and CryIIIB domain I peptide. *Applied and Environmental Microbiology 60:* 3711-3717.

Ward, E. S., and D. J. Ellar (1987). Nucleotide sequence of a *Bacillus thuringiensis* var. *israelensis* gene encoding a 130 kDa delta-endotoxin. *Nucleic Acids Research 15:* 7195.

Warren, G. W., N. B. Carozzi, N. Desai, and M. G. Koziel (1992). Field evaluation of transgenic tobacco containing a *Bacillus thuringiensis* insecticidal protein gene. *Journal of Economic Entomology 5:* 1651-1659.

Widner, W. R., and H. R. Whiteley (1989). Two highly related insecticidal crystal proteins of *Bacillus thuringiensis* subsp. *kurstaki* possess different host range specificities. *Journal of Bacteriology 1771:* 965-974.

Willams, S. L. Fredrich, S. Dincher, N. Carozzi, H. Kessmann, E. Ward, and J. Ryals (1992). Chemical regulation of *Bacillus thuringiensis* δ-endototoxin expression in transgenic plants. *Bio/Technology 10:* 540-543.

Wilson, F. D., H. M. Flint, W. R. Deaton, D. A. Fischhoff, F. J. Perlak, T. A. Armstrong, R. L. Fuchs, S. A. Berberich, N. J. Parks, and B. R. Stapp (1992). Resistance of cotton lines containing a *Bacillus thuringiensis* toxin to pink bollworm (Lepidoptera: Gelechiidae) and other insects. *Journal of Economic Entomology 4:* 1516-1521.

Wolfersberger, M. G. (1989). Neither barium nor calcium prevents the inhibition by *Bacillus thuringiensis* δ-endotoxin of sodium or potassium gradient dependent amino acid accumulation by tobacco hornworm midgut brush border membrane. *Archives of Insect Biochemistry and Biophysics 12:* 267-277.

Wu, C., Y. Fan, Q. Zhang, N. Oliva, and S. K. Datta (1997). Transgenic fertile *japonica* rice plants expressing modified *cryIA(b)* gene resistant to yellow stem borer. *Plant Cell Reports 17:* 129-132.

Wu, D., and A. I. Aronson (1992). Localized mutagenesis defines regions of the *Bacillus thuringiensis* δ-endotoxin involved in toxicity and specificity. *Journal of Biological Chemistry 267:* 2311-2317.

Wu, G., G. Y. Ye, H. R. Cui, Q. Y. Shu, and Y. W. Xia (2000). Gus reporter gene assisted selection for striped stem borer resistant transgenic rice carrying a *cry1Ab* gene from Bt. *Chinese Rice Research Newsletter 8:* 5-6.

Wu, S. J., and D. H. Dean (1996). Functional significance of loops in the receptor binding domain of *Bacillus thuringiensis* CryIIIA δ-endotoxin. *Journal of Molecular Biology 255:* 628-640.

Wünn, J., A. Kloti, P. K. Burkhardt, G. C. Ghosh Biswas, K. Launis, V. Iglesia, and I. Potrykus (1996). Transgenic *indica* rice breeding in IR58 expressing a synthetic *cryIA(b)* gene from *Bacillus thuringiensis* provides effective insect pest control. *Bio/Technology 14:* 171-176.

Yaoi, K., T. Kadotani, H. Kuwana, A. Shinkawa, T. Takahashi, H. Iwahana, and R. Isato (1997). Aminopeptidase N from *Bombyx mori* as a candidate for the receptor of *Bacillus thuringiensis* Cry1Aa toxin. *European Journal of Biochemistry 246:* 652-657.

Ye, G. Y., Q. Y. Shu, H. W. Yao, H. R. Cui, X. Y. Cheng, C. Hu, Y. W. Xia, M. W. Gao, and I. Altosaar (2001a). Field evaluation of resistance of transgenic rice containing a synthetic *cry1Ab* gene from *Bacillus thuringiensis* Berliner to two stem borers. *Journal of Economic Entomology 94:* 271-276.

Ye, G. Y., J. Tu, C. Hu, K. Datta, and S. K. Datta (2001b). Transgenic IR72 with fused Bt gene *cry1Ab/cry1Ac* from *Bacillus thuringiensis* is resistant against four Lepidopteran species under field conditions. *Plant Biotechnology 18:* 125-133.

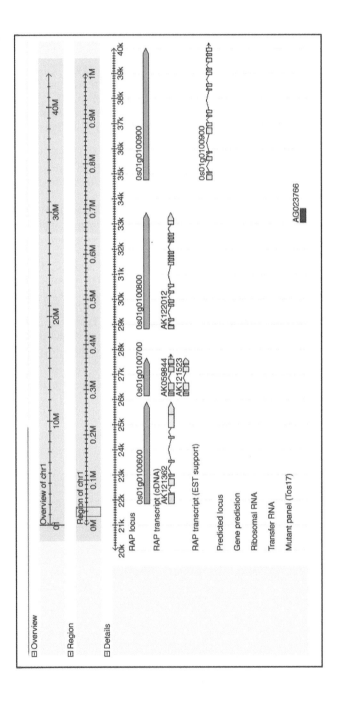

PLATE 1.3. A graphical view of manually curated annotation of the rice genome. Each RAP locus represents gene models manually curated gene using information from rice full-length cDNA sequences and ESTs from various cereal species.

PLATE 6.2. Development and field evaluation of transgenic *Bt* rice. (a) Dead heart symptom; (b) production of putative transgenic rice in greenhouse with *Bt* gene and cut-stem bioassay showing larval mortality (Datta et al., 1998); (c) Southern blot showing integration of 1.8-kb cry gene in the genome of rice cv. *Minghui* 63 (Tu et al., 2003); (d) expression of Cry protein in transgenics of *indica* rice cv. *Tulasi;* (e) resistance reaction of transgenic IR72 *Bt* rice (T) against yellow stem borer; and (f) leaffolder infestation under natural field conditions at Wuhan, China (Ye et al., 2001).

PLATE 7.3. Transgenic rice leaves (cv. IR72 with Xa 21) showing resistance to bacterial blight (PX099) versus control plant leaves with disease symptoms Source: Tu et al., 1998.

PLATE 7.2. Rice plant showing the symptom and damage due to sheath blight (left) and transgenic plant showing enhanced resistance to the disease.

PLATE 7.4. Transgenic rice plants (left) and control rice plants (right) inoculated with PX099 under field condition *Source:* Tu et al., 2000.

PLATE 7.5. Maintainer line for hybrid rice showing the damage caused by bacterial blight (left) and transgenic plants with *Xa21* gene showing resistance to bacterial blight.

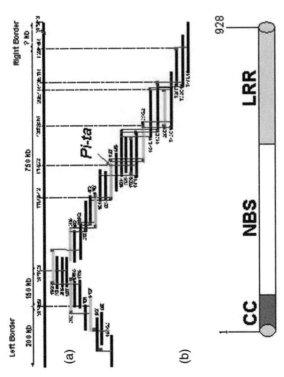

(a)

(b)

PLATE 8.2. (a) The *Pi-ta* gene was identified from the BAC clone 142 that is in the middle of one megabase contig (Adapted from Bryan et al., 2000, *Plant Cell* 12:2033-2045). (b) Predicted structure of the Pi-ta protein. CC indicates the putative coiled coil region at the amino terminus, NBS indicated putative nucleotide binding sites in the middle region, and LRR indicates leucine rich repeats (The *Pi-ta* degenerate LRR is also called LRD).

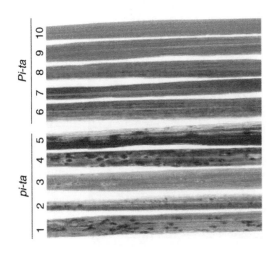

PLATE 8.1. The *Pi-ta* blast resistance gene in rice plays a central role for rice protection. Blast is a serious disease in rice growing areas worldwide. The image shows disease lesions on leaves of C101A51 (lane 1), M202 (lane 2), Nipponbare (lane 3), Wells (lane 4) and breeding line 1153 (lane 5) that do not contain *Pi-ta*, respectively, whereas no disease lesions were evident on Katy (lane 6), Drew (lane 7), Kaybonnet (lane 8) and breeding lines 602 (lane 9) and 605 (lane 10) that contain *Pi-ta*.

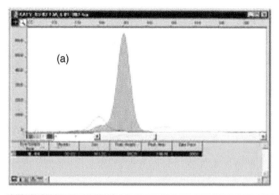

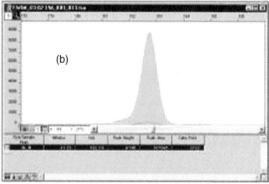

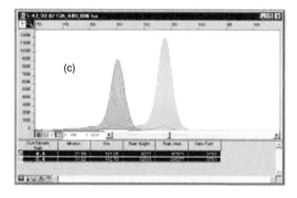

PLATE 8.4. Different locations of peak (visualized using a standard sequencing and genotyping instrument) indicate different alleles were amplified. (a) homozygous *Pi-ta/Pi-ta,* (b) homozygous *pi-ta/pi-ta,* and (c) heterozygous *Pi-ta/pita.* *Source:* M. Redus and Y. Jia, unpublished data.

```
0-137    MLFYSL-FFFHTVAI3AFTNIGTFSHPVYDYNP IPNHIHGDLKRRAYIER 50
0-137    MLFYSLLFFFHTVAISAFTNIGTFSHPVYDYNP IPNHIHGDLKRRAYIER 50


         51
0-137    YSQCSDSQASEIRAALKRCAELASWGYHAUKNDNRLFRL IFKTDSTD IQN 100
IB-49    YSQCSDSQASEIRAALKRCAELASWGYHAUKSDNRLFKL IFKTDSTD IQN 100


         101
0-137    WUQKNFNEIYKECNRDADEISLTCHDKNUYTCUREGUHNL AYAL INEKEI 150
IB-49    WUQNNFNEIYKECNRDADEISLTCHDKNUYTCUREEUHNL AYAL INEKEI 150


         151
0-137    UICPPFFNNPUNSREITAGNQDTUILHEMUH I ILSEWKDYGYEWDGIHKL 200
IB-49    UICPPFFNNPUNSREITAGNQDTILLHEMUH I ILSEWKDYGYEWDGIHKL 200


         201
0-137    DSTES OLM[DSUAOFAQCARULUC 224
IB-49    DSTES IKNPDSYAIFAQCARYKYC 224
```

PLATE 8.6. Amino acid alignment of AVR-Pita proteins from IB-49 and 0-137 (Amino acid sequences of all other U.S. races dispersed in four lineages are indentical).

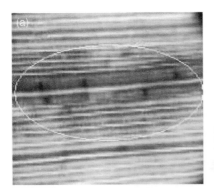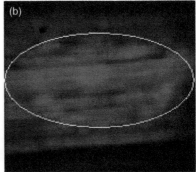

PLATE 8.9. Inoculated area was shown in the oval. Under bright light (a), and the presence of autoflourescence in Katy 24 hours after inoculation with *M. grisea* IC-17 (b). Ten microliters of spore suspension containing 100 spores were spot inoculated and pictures were taken 24 hours after inoculation. *Source:* Y. Jia, unpublished data.

PLATE 8.10. Localized detached leaf inoculation. Five microliters of spore suspension was inoculated on detached rice leaves (left). Resistance: 0-2, Susceptible: 3. Adapted from Jia et al., 2003, *Plant Disease* 87: 129-133.

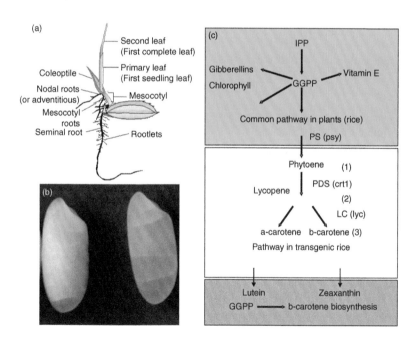

PLATE 11.4. (a) A rice seedling; (b) β-carotene enriched seed, left, control seed, right; and (c) schematic pathway of β-carotene biosynthesis in plant. *Note:* Green tissues contain carotenoids including β-carotene but absent in endosperm tissue.

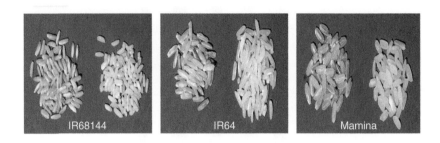

PLATE 11.9. Brown and polished rice of different cultivars.

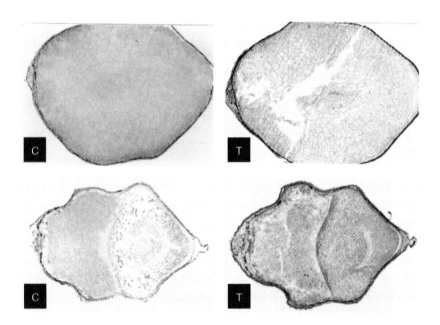

PLATE 11.11. Iron detection in nontransgenic control (C) and transgenic rice grains (T). The accumulation of iron in control material is restricted to the aleurone layer, while iron is present in the entire grain including the endosperm in transgenic seeds. *Source:* Adapted from Krishnan et al., 2003.

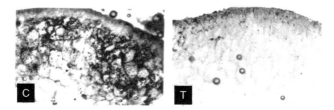

PLATE 11.12. Transverse section of mature polished transgenic rice grain (T) showed blue color immediately after the reaction with Carr-Price reagent indicating the accumulation of β-carotene in endosperm whereas nontransgenic control (C) did not show any color reaction.

PLATE 11.13. Transgenic Golden Rice developed by biolistic (a-d) and *Agrobacterium* transformation (e-f) (β-carotene enriched IR64 and BR29 and high iron grown in greenhouse (pre-field condition) showing good agronomic performance with transgene expression: (a) transgenic BR29 plant (with gene for β-carotene; (b) expression of β-carotene in BR29 seeds (right sample); (c) transgenic IR64 plants in the greenhouse; (d) expression of β-carotene in IR64 seeds; (e) BR29 control seeds and segregating BR29 with simple genetic integration and high β-carotene expression in the seeds of the transformed line.

Lysozyme (Line 159-53)   Lactoferrin (Line 164-12)

PLATE 12.8. Field performance of advanced generation (R₅) maturing transgenic plants producing either rhLYS or rhLF.

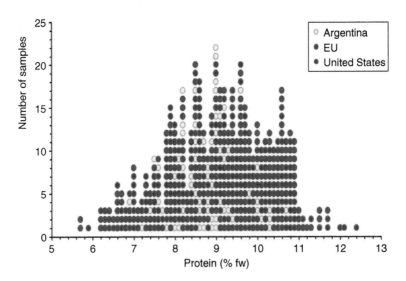

PLATE 15.2. Distribution of protein values in the ILSI database. The numbers of samples having the measured fresh weight (fw) of protein are plotted for maize samples taken over a three-year period from multiple test fields on three continents. *Source:* Adapted from Ridley et al., 2004.

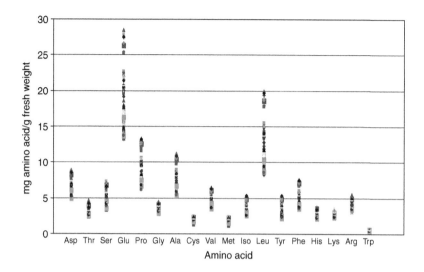

PLATE 15.3. Natural variability in amino acid composition in reference maize hybrids. Seven varieties of maize were grown at six locations during one season. *Source:* Adapted from Ridley et al., 2004.

Chapter 7

Biotechnological Approaches to Disease Resistance in Rice

K. Datta
S. K. Datta

INTRODUCTION

Significant yield losses caused by pathogen attack occur in most cereal crop plants. Global loss due to pathogens is estimated to be 12% of potential crop production (Grover and Pental, 2003). Crop protection from diseases is a major challenge for sustainable agriculture. The use of agrochemicals against diseases has led to serious environmental damage, especially when chemicals are applied repeatedly in the soil, such as in the control of soil-borne fungal pathogens, and this is also a concern for human health (Pingali, 1995). Despite the use of agronomic practices for crop protection, such as integrated pest management and crop rotation, many pathogens remain serious threats to crops. The breeding of a resistant crop depends on the availability of resistance genes in the gene pool, which often shows a limitation in durability. There is a clear need, therefore, for alternatives such as biotechnology. An understanding of the molecular basis of host-pathogen interactions is essential for the effective use of durable genetic resistance to control plant diseases.

Progress has been made in the phenotyping of the plant genome and the isolation and characterization of genes for disease resistance. Microarray biochip technology will accelerate the process of understanding functional genomics for the cutting-edge era of biotechnology-based disease control. The most significant development in crop improvement for disease resistance is the use of the techniques of gene cloning and genetic transformation.

Genetic transformation technology has allowed the genetic modification of almost all important food crops such as rice, wheat, maize, and so forth. This transgenic built-in resistance approach offers a chemical-free and environment-friendly solution for controlling pathogens. The recent rice genome sequencing project developed independently by Monsanto, Syngenta, the Beijing Genomics Institute (BGI), and the International Rice Genome Sequencing Project (IRGSP) will facilitate gene discovery for crop improvement.

Gene transfer technology can be used to enhance plant responses against a pathogen, either by overexpressing existing defense mechanisms (Rommens and Kishore, 2000) or by using a single dominant resistance gene not normally present in the susceptible plant (Keen, 1999). Many resistance genes whose products are involved in recognizing invading pathogens have been identified (Takken and Joosten, 2000). Many of the antimicrobial compounds that are produced by plants for combating infection have been identified (Does and Cornelissen, 1998; Morrissey and Osbourn, 1999; Osbourn, 1996). Several signaling pathways related to pathogen infections have been identified during the last decade. This knowledge has provided several strategies for use in developing transgenic plants resistant to pathogens.

DEFENSE MECHANISMS IN PLANTS

Resistance Genes

Plants have their own network for defense against pathogens, which includes several proteins or other organic compounds produced prior to or during infection. The induced defenses, referred to as *active defense* mechanisms, can be induced by all kinds of pathogens, including fungi, bacteria, and viruses (Jackson and Taylor, 1996). Induced active defense mechanisms are effective against a broad spectrum of diseases. Three different kinds of active defense responses can be identified in responding tissues of resistant plants: primary responses, secondary responses, and systemically acquired responses. *Primary* responses, which involve the recognition of a specific signal molecule, are localized in the cells that are in contact with the pathogen, and the outcome is frequently programmed cell death by hypersensitive reaction (HR). *Secondary* responses are induced into the adjacent cells surrounding the initial infected cell in response to signal molecules or elicitors produced by the primary interaction. *Systemic acquired resistance* (SAR) is hormonally induced throughout the plant and represents a gain in the level of resistance once the plant has been exposed to a given stimulus. Plants are capable of recognizing and responding to the stimuli produced by the pathogen during the early stages of infection. Both plants and pathogens have certain sets of resistance (*R*) and avirulence (*Avr*) genes that determine the outcome of plant-pathogen interactions. These *R* genes represent upstream defense responses. A semi-dominant *R* gene encodes proteins that can recognize a corresponding specific *Avr* gene product, or compatibility factor, of specific races of the pathogen, which is likely to be involved in pathogenicity on the host. Following pathogen recognition, the R proteins activate signaling cascades that coordinate the initial plant defense response. Molecular genetic approaches to the study of plant-pathogen interactions have resulted in the accumulation of information concerning the mechanisms involved in pathogen virulence and resistance in plants. According to the gene-for-gene hypothesis, the *Avr* gene-dependent elicitors bind directly to the *R* gene product, which stimulates the downstream signal transduction pathways in the cell to initiate various defense responses. In most cases, the receptor domain is formed by a leucine-rich repeat (LRR) region of the resistance gene product. Many disease resistance genes have been cloned from different plant sources, for example, *Pto* from tomato (Martin et al., 1993), *Xa21* from wild rice (Song et al., 1995), *Cf-9* from tomato (Jones et al., 1994), *Arabidopsis RPS2* (Bent et al., 1994), and so forth. The tomato *Pto* gene confers resistance to races of *Pseudomonas syringae* pv. *tomato* that carry the *Pto* avirulence gene. This was the first isolated race-specific *R* gene. Resistance genes are commonly located in clusters of closely related genes. In rice, the *Xa21* resistance gene is part of a clustered seven-gene family (Song et al., 1997).

Pathogenesis-Related Proteins

Pathogenesis-related (PR) proteins represent downstream defense responses and are induced in the host plant upon pathogen invasion to prevent or limit the spread of the pathogen. PR proteins are induced not only by biotic stress (i.e., pathogens) but also by other stresses such as wounding, ethylene, UV light, and so on. They can be induced by natural active resistance responses, such as a hypersensitive reaction or SAR. PR proteins were first observed in hypersensitivity reaction in tobacco induced by tobacco mosaic virus. PR proteins have been classified into 12 major groups based on their primary structure and serological, enzymatic, and biological activities.

The functions of most PR proteins are still not clearly known. However, the PR-2 family represents β-1,3 endoglucanases (Kauffmann et al., 1987), and PR-3, PR-4, PR-8, and PR-11 families consist of chitinases with (PR-8) or without lysozyme activity, which suggests that these PR proteins can be directed against the cell walls of fungi and bacteria (Neuhaus, 1999). Overproduction of chitinases in crop plants in some cases increases host resistance to the pathogen. These proteins are becoming very important because of their potential in defense reactions against various pathogens. The first report on developing fungus-resistant transgenic tobacco and *Brassica napus* by constitutive expression of the bean chitinase gene came in 1991 (Broglie et al., 1991).

PR-5 proteins (thaumatin-like; osmotin, NP24) have a sequence similarity to the sweet-tasting protein thaumatin from *Thommatococus danielli*. PR-5 proteins can be induced not only in response to microbial infection, but also by the hormones ethylene and ABA, signal molecules salicylic acid and jasmonic acid, and osmotic stresses. The functional mechanism is still not clear. It may be that a fungal infection causes lysis of the pathogen by permeabilizing the fungal cell wall (Abad et al., 1996). The PR-6 family proteins are proteinase inhibitors. Members of the serine (trypsin, chymotrypsin, or subtilisin) and cysteine proteinase (cystatins) inhibitor families show a wide diversity in structure and regulation. Two other classes of proteinase inhibitors, aspartyl and metalloproteinase, have not been studied extensively.

Ribosome-inactivating proteins (RIPs) are a group of N-glycolytic enzymes that remove a specific adenine residue from the very conserved sequence of 28S rRNA. Because of this modification, the ribosome becomes unable to bind the elongation factor and consequently blocks the translation. Plant RIPs may inactivate foreign ribosomes of distantly related species. RIPs may function in defense against a pathogen or in the mediation of inducible or developmentally programmed cell death (Greenberg, 1997). Evidence for either role is still controversial. Single-chain RIPs, such as pokeweed antiviral protein (PAP), have been shown to possess antiviral activities (Lodge et al., 1993). Transgenic tobacco plants expressing barley RIP 30 under the control of the wound-inducible promoter *win1* of potato showed accumulation of RIP mRNA and protein upon infection by *Rhizoctonia solani* and showed increased tolerance against the pathogen (Logemann et al., 1991). Combinatorial expression of barley seed RIP 30 with chitinase and RIP 30 with glucanase in tobacco showed a significantly higher level of protection against fungal infection than transgenic tobacco with barley RIP alone. However, a better understanding of RIPs is necessary to optimize their use as defensive transgenes.

Signal Transduction Pathway

The interactions of the elicitor protein from the pathogen with the receptive protein from the plant activate the signal transduction pathway leading to programmed cell death and an oxidative burst. The complex regulatory mechanism of *R* gene and PR gene expression due to pathogen infection is still not clearly understood. PR genes that play an important role in plant defense can be induced not only by pathogen infection but also by phytohormones and other kinds of stresses. Upon infection, plants exhibit increased production of reactive oxygen species, ethylene, jasmonates, and salicylic acid (Hammond-Kosack et al., 1996; Yang et al., 1997). These molecules can serve as secondary signals to stimulate plant defenses. Nitric oxide is an important secondary signal molecule for plant defense (Dangl, 1998; Delledonne et al., 1998; Durner et al., 1998). These signals may be inducers of PR gene expression. Salicylic acid (SA) induces PR gene expression during SAR, whereas jasmonates and ethylene stimulate proteinase inhibitors, thionin, and defensin. Cross-interactions between these secondary messengers are also common in signaling pathways (Figure 7.1). The use of mutants with defective secondary signaling may help in the study of the role of secondary messengers in plant defense against pathogens. To obtain broad-spectrum plant disease resistance, one effective strategy has been to exploit the SAR pathway.

The *NPR1* gene encodes a protein containing ankyrin repeats, which are involved in protein-protein interaction. *NPR1* functions as a positive regulator of PR gene expression in the signal transduction pathway. The *npr1* mutant showed inability for PR gene induction, whereas overexpression of *NPR1* has enhanced pathogen-induced PR gene expression in *Arabidopsis* (Cao et al., 1998). The *NPR1* gene induces PR1, PR2, and PR5 gene expression.

DISEASE RESISTANCE IN RICE

Rice is the world's most important food crop, feeding one third of the world population. Every year, there is a significant yield loss because of attacks by different pathogens such as fungi, bacteria, and viruses. Rice disease can cause yield loss as high as 50% in devastated areas. Damages caused by pathogens in rice required the development of strategies to control and to manage crop losses from epidemics. The application of host resistance may be the only reliable method for disease management. Genetic transformation and, to some extent, the marker-assisted breeding approach

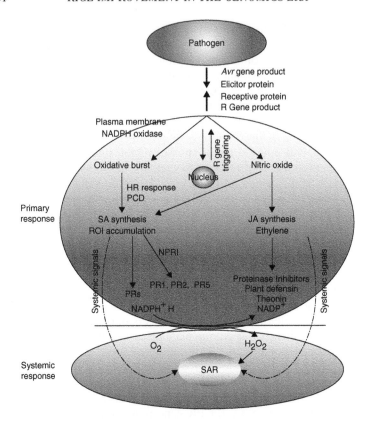

FIGURE 7.1. Diagrammatic presentation of plant-pathogen interaction for an active defense mechanism. R = resistance, *Avr* = avirulence, HR = hypersensitive reaction, PCD = programmed cell death, SA = salicylic acid, JA = jasmonic acid, PRs = pathogenesis-related proteins, SAR = systemic acquired resistance, ROI = reactive oxygen intermediates.

can be effectively used to complement conventional breeding for developing built-in resistance in rice cultivars.

Fungal Resistance in Rice

The major fungal pathogens that attack rice plants are *Rhizoctonia solani*, which is responsible for rice sheath blight disease, and *Pyricularia oryzae*, also known as *Magnaporthe grisea,* which causes rice blast disease.

Sheath Blight Disease Resistance

Sheath blight disease is one of the most devastating diseases of rice, prevalent in all tropical rice-growing countries. Genetic sources for a high level of resistance for this disease have not been identified in cultivated rice or its wild relatives. A large number of PR protein genes have been reported (Datta and Muthukrishnan, 1999) along with a few resistance (*R*) genes (Cao et al., 1998; Hammond-Kosack and Jones, 1997; Richter and Ronald, 2000). The development of genetic engineering techniques for rice opens up the possibility of introducing these genes to combat fungal diseases.

A rice class 1 chitinase gene (PR-3), *Chi11*, under the control of the cauliflower mosaic virus (CaMV) 35S promoter, was introduced to *indica* rice cultivars *Chinsurah Boro II, Vaidehi,* and *Tulsi*. The transgenic plants showed enhanced resistance against sheath blight disease when they were challenged with the pathogen *R. solani* (Figure 7.2). The degree of resistance correlated with the level of chitinase expression (Baisakh et al., 2001; Lin et al., 1995). A basic chitinase, RC24, was isolated from a rice genomic library. The RC24 promoter contained several putative stress-related elements, which indicated its responsiveness to fungal elicitor and wounding stimuli. The expression and function of this gene in transgenic rice plants have been reported (Xu et al., 1996). Another pathogen-inducible chitinase

FIGURE 7.2. Rice plant showing the symptom and damage due to sheath blight (left) and transgenic plant showing enhanced resistance to the disease. See corresponding Plate 7.2 in the central gallery.

cDNA, *RC7,* encoding a class I chitinase, was isolated from *R. solani*-infected rice cultivar IR58. The deduced amino acid sequence of this gene is significantly different from that of the other rice chitinases. Constitutive expression of the *RC7* chitinase gene in transgenic plants (IR72 and IR64) produced an enhanced resistance to sheath blight (Datta et al., 2001). The protective effect of the chitinase transgene on the transgenic homozygous lines was shown to be about 50% in both lesion number and size, although variation exists.

To achieve sheath blight disease resistance, transformation with a thaumatin-like protein (*tlp*) gene, a member of the PR-5 group of PR protein genes, under the control of the CaMV 35S promoter, has been introduced in different *indica* rice cultivars. Accumulation of a varied level of 23 kDa TLP protein has been noticed in individual transformants and was correlated with the bioassay results. Several transgenic plant lines showed limited infection compared with the control plants (Datta et al., 1999). A multiple-gene strategy to introduce two or three different antifungal disease resistance genes can be deployed to achieve a higher degree and longer duration of resistance.

Rice Blast Disease Resistance

Rice blast, caused by *Magnaporthe grisea* (Hebert) Barr (synonym: *Pyricularia oryzae* Sacc.), a heterothallic ascomycete fungus, is responsible for crop yield losses worldwide. The development of resistant cultivars is difficult, as the pathogen can evolve rapidly to overcome resistance genes.

DNA markers that are closely associated with major blast resistant genes have been developed (e.g., *Piz-5*; Mackill and Bonman, 1992). Other genetic markers linked to blast resistance genes, such as *Pi-b*, *Pi-K^h*, and *Pi-ta^2*, are located on chromosomes 2, 11, and 12, respectively. The presence of *Pi-ta^2* and *Pi-K^h*, which is masked by the *Pi-ta^2* resistance gene in cultivars *Madison* and *Kaybonnet*, indicates a multiple genetic mechanism for recognizing and preventing rice blast disease. These markers could be useful to facilitate the stacking of blast resistance genes in combination for broad-spectrum resistance. DNA microsatellite markers (RM 138, 166, and 208) have shown successful transfer of the *Pi-b* disease resistance gene from cultivar *Teqing* into U.S. cultivar *Saber.* Pyramiding of three major *R* genes, *Pi-1*, *Pi-2*, and *Pi-9*, has been done by fine mapping and DNA marker-assisted breeding (Hittalmani et al., 2000).

Rice chitinase genes *Cht-2* and *Cht-3,* under the control of the CaMV 35S promoter, have been introduced to a *japonica* rice cultivar (Nishizawa et al., 1999). Constitutive expression of both genes shows enhanced resistance to two races (007.0 and 333) of *M. grisea.* *Cht-2* expression was targeted extracellularly, whereas the *Cht-3* product accumulated intracellularly. Similarly, a stilbene synthase gene from grapevine under its own promoter has been introduced into a rice cultivar (Stark-Lorenzen et al., 1997). The accumulation of stilbene synthase mRNA was observed in response to inoculation with *Pyricularia oryzae,* and the transgenic plants showed resistance to that fungus. A *YK1* gene, which is similar to the *HM1* gene of maize, cloned from rice, showed increased tolerance of rice blast (Uchimiya et al., 2002).

A gene encoding trichosanthin (TCS), a type-I ribosome-inactivating protein, was isolated from tubers of *Trichosanthes kirilowii.* TCS has been previously reported to have antifungal and antiviral activity (Hu et al., 1999; Zhou et al., 1994). Transgenic rice plants expressing the *tcs* gene have shown significantly increased resistance when they were challenged with rice blast fungus (Yuan et al., 2002).

Antifungal protein (AFP) from *Aspergillus giganteus,* encoded by the *afp* gene, has been reported to possess antifungal activity against various fungal pathogens. Transgenic rice lines (Mediterranean elite *japonica* cultivar *Senia*) were obtained either with the *afp* cDNA sequence from *Aspergillus* or with a chemically synthesized, codon-optimized *afp* gene. In both cases, the DNA encoding signal sequence from the tobacco *AP24* gene was fused N-terminally to the coding sequence of the mature AFP protein. Constitutive expression of the *afp* gene in rice exhibited an enhanced resistance against *M. grisea* in detached leaf infection assays when T2 homozygous *afp* lines were challenged with the fungal infection. The inhibitory activity of the protein from leaf extract of *afp* plants on in vitro growth of *M. grisea* also indicated the biologically active nature of AFP protein produced by these transgenic rice plants (Coca et al., 2004).

Bacterial Disease Resistance in Rice

The major bacterial disease attacking the rice plant is bacterial blight (BB), caused by *Xanthomonas oryzae* (*Xoo*). So far, 26 resistance genes (*Xa*) have been identified from different germplasms of rice and its wild relatives. The most widely used gene in breeding programs is *Xa4*, which has been incorporated in many modern semidwarf rice varieties. The use of molecular markers in breeding for BB resistance resulted from mapping and tagging some dominant and recessive genes. Breeding lines pyramided with different BB resistance genes, such as *Xa4, Xa5, Xa13, Xa7,* and *Xa21,* show a broader spectrum and higher level of resistance than lines with a single gene (Huang et al., 1997).

The cloning and transfer of *R* genes to modern rice varieties eliminates the problem of retaining unwanted, genetically linked traits associated with classical breeding. *Xa1,* which confers resistance to Japanese *Xoo* race 1, has been cloned, but it has not been used much, as it is not effective in many Asian countries. *Xa21* was transferred through conventional methods from the wild species *Oryza longistaminata* to cultivated IR24 (Ikeda et al., 1990; Khush et al., 1990), resulting in a line called IRBB21. *Xa21* confers resistance to all known races of *Xoo* in India and the Philippines. This gene was cloned by the positional cloning approach (Song et al., 1995) and is the only known resistance gene that encodes three structural features: an extracellular domain with leucine-rich repeats, a transmembrane domain, and a cytoplasmic serine/threonine kinase domain.

Xa21 has been introduced into susceptible *japonica* rice T309, which subsequently has shown resistance to BB (Wang et al., 1996). Transgenic elite *indica* rice lines IR64, IR72, and *Minghui 63,* a restorer line for hybrid rice with the *Xa21* gene, showed resistance to the BB pathogen (Tu et al., 1998; Zhang et al., 1998). Transgenic IR72 plants were challenged with two prevalent races (4 and 6) of *X. oryzae.* The level of resistance to race 4 of *Xoo* was higher because of the pyramiding of *Xa21* in addition to the *Xa4* already present in IR72 (Figure 7.3) (Tu et al., 1998). Transgenic plants containing *Xa21* in the IR72 background have been released for testing in field conditions in Wuhan (China), by the Directorate of Rice Research (Hyderabad, India), and in the Philippines. Different isolates of *X. oryzae* from China, Japan, and the Philippines were used for testing. The *Xa21* transgene showed an excellent performance in disease resistance because of the introduced bacterial blight resistance genes (Figure 7.4) (Tu et al., 2000). Field trials of the transgenic IR72 with the *Xa21* gene carried out in the 2002 wet season in the Philippines also showed excellent results.

Genetic engineering can help to develop transgenic plants in less than two years to minimize the effects of a breakdown in resistance in the host plant. The strategic deployment of transgenic rice with gene pyramiding with available resistance genes from different sources may provide desirable bacterial disease resistance in rice breeding. Several rice cultivars have been transformed at IRRI with the *Xa21* gene and are now under evaluation at different stages of development (Figure 7.5).

FIGURE 7.3. Transgenic rice leaves (cv. IR72 with *X*a 21) showing resistance to bacterial blight (PX099) versus control plant leaves with disease symptoms *Source:* Tu et al., 1998. See corresponding Plate 7.3 in the central gallery.

FIGURE 7.4. Transgenic rice plants (left) and control rice plants (right) inoculated with PX099 under field condition. *Source:* Tu et al., 2000. See corresponding Plate 7.4 in the central gallery.

Virus Disease Resistance in Rice

Virus diseases can severely constrain rice productivity. Rice tungro virus disease and rice yellow mottle virus disease are the two major virus diseases of rice in Asia and Africa, respectively. Advances in understanding plant-virus interactions in pathogenesis and resistance along with advances in cloning and sequence analysis of the genomic components of rice viruses and genetic transformation technology have led to new approaches for controlling rice virus diseases. Two approaches can be used to develop virus resistance by genetic engineering. The resistance genes used may be from the virus pathogen—that is, *pathogen-derived resistance* (PDR)—or from other sources based on host resistance genes on other nonpathogenic sources. In PDR, a viral gene—in part or entirely—is inserted into the plant and interferes in the life cycle of the virus. The PDR strategy may include the use

FIGURE 7.5. Maintainer line for hybrid rice showing the damage caused by bacterial blight (left) and transgenic plants with *Xa21* gene showing resistance to bacterial blight. See corresponding Plate 7.5 in the central gallery.

of a coat protein gene from the virus (Powell-Abel et al., 1986), the use of replicase (Golemboski et al., 1990), and a dysfunctional form of the movement protein, which interferes with cell-to-cell movement of the virus (Malyshenko et al., 1993). The nonpathogen-derived resistance strategies include posttranscriptional gene silencing, which is a RNA degradation mechanism; the use of a plant disease resistance gene; ribosome-inactivating proteins, plant proteinase inhibitors, and so forth.

Rice Tungro Virus

Rice tungro disease can cause significant yield losses (U.S. $342.7 million per year in Southeast Asia). It is caused by *rice tungro bacilliform virus* (RTBV), which is responsible for symptom development, and *rice tungro spherical virus* (RTSV), which is responsible for the transmission of both viruses by green leafhoppers, the natural vector of the virus. The coat protein genes *CP1*, *CP2*, and *CP3* of RTSV were introduced singly or in combination into TN1 and *Taipei 309* rice cultivars by particle bombardment. The transgenic plants and their progeny that contained the target gene accumulated transcripts of the chimeric *CP* genes by RNA blot analysis and were subjected to virus inoculation via leafhoppers. From 17-73% of the seedlings escaped infection, and a significant delay in virus replication under greenhouse conditions was noted in the infected plants containing the target *CP* genes (Sivamani et al., 1999).

The RTSV replicase gene in the sense or antisense orientation was introduced in a *japonica* rice cultivar. Plants expressing the gene in an antisense orientation exhibited moderate (60%) resistance to RTSV, whereas plants expressing the full-length or truncated replicase gene in a sense orientation were 100% resistant to different geographically distinct RTSV isolates. The replicase-mediated resistance approach could be used to combat the spread of tungro disease (Huet et al., 1999).

Rice Yellow Mottle Virus

Rice yellow mottle virus (RYMV) is a major limiting factor in African rice production. The estimated yield loss from RYMV infection in susceptible lowland cultivars may be up to 97%, and as high as 54% in tolerant upland cultivars. Natural resistance to RYMV exists in genetically distant African land races, but it has not been possible so far to introduce this resistance into cultivated rice. Replicase genes encoding the RNA-dependent RNA polymerase of RYMV were introduced with the CaMV 35S promoter into widely grown lowland African cultivars (ITA 212, *Bouake 189*, BG 90-2) that are highly susceptible to RYMV disease. Seventeen independent transgenic lines from the three susceptible cultivars revealed variable levels of resistance to RYMV. One selected resistant line (*Bouake 189*) has been tested against five characterized RYMV isolates collected from different areas of Africa. Resistance to all isolates was observed in a preliminary assay and was stable over at least three generations. This RNA-based resistance was manifested either as a delay in symptom development or as a complete suppression of virus multiplication. Using this RNA-based mechanism, functional viral proteins will not be produced (Pinto et al., 1999).

Broad-Spectrum Multiple Disease Resistance in Rice

Regulatory Gene

The recently cloned *Arabidopsis thaliana NPR1* gene has been shown to be a key regulator of the acquired resistance mechanism. When *NPR1* gene expression reaches elevated levels, NPR1 protein is activated, which induces the expression of downstream pathogenesis-related genes. *NPR1* also participates in salicylic acid-independent induced systemic resistance (ISR; Pieterse et al., 1998). The protection provided by a single PR gene is narrow, and the amount of resistance is often less significant. The *NPR1*-regulated SAR can lead to the expression of a battery of PR genes instead of a single gene. Overexpression of *NPR1* in *Arabidopsis* leads to enhanced resistance to both bacterial and oomycete pathogens in a dose-dependent manner (Cao et al., 1998). To obtain broad-spectrum disease resistance against fungi and bacteria in rice, work has been ongoing to enhance immunity by overexpressing *NPR1*, the key regulator of the SAR signaling pathway. Transgenic rice plants *(Taipei 309)* overexpressing the *Arabidopsis NPR1* gene displayed enhanced resistance to *Xoo,* which suggests the presence of a rice (monocotyl) disease resistance pathway similar to the *Arabidopsis* (dicotyl) *NPR1*-mediated signal transduction pathway (Chern et al., 2001).

Antimicrobial Peptides

Small antimicrobial peptides, produced by most multicellular organisms in response to pathogen attack, play an important role in the natural defense mechanism of plants against pathogen infection. These peptides are small, cationic, and have open chain forms. Most of these peptides contain cysteine residues, which form disulfide bonds, creating a rigid and compact structure. These peptides have a net positive charge and can recognize a broad range of microbes, which are generally composed of negatively charged phospholipids in the outer leaflet of the bilayer. Various types of antimicrobial peptides have been identified in plants, including thionin (Bohlmann, 1994), zeamatin (Malehorn et al., 1994), wheat puroindoline (Krishnamurthy et al., 2001), and plant defensins (PDFs) from radish (Broekaert et al., 1995).

Transgenic rice expressing oat cell wall-bound thionin was effective in the control of bacterial diseases like bacterial leaf blight and the disease caused by *Burkholderia plantarii* (Iwai et al., 2002). Expression of wheat puroindolines in transgenic rice showed enhanced resistance to rice blast and sheath blight. Constitutive expression of plant defensin genes from *Brassica oleracea* and *B. campestris* has successfully conferred highly enhanced resistance to rice blast and bacterial leaf blight (Kawata et al., 2003). When the defensin genes of *B. oleracea* and *B. campestris* were modified to substitute a single amino acid at each position and individually introduced to rice, some modified genes produced much higher resistance than the wild type defensin gene. These results suggest that plant defensin conferred resistance to both rice blast and leaf blight and that modification of the wild type defensin gene may lead to an increased level of broad-spectrum resistance. Increasing the antimicrobial activity through the modification of signal peptides of plant defensin from Brassicaceae also may contribute to the improvement of broad disease resistance (Sharma et al., 2000).

GTP-Binding Protein

The involvement of GTP-binding proteins (G proteins) in signal transduction pathways may play an important role in disease resistance. Many studies using inhibitors of heterotrimeric G proteins in several plant species have suggested the involvement of G protein in defense signaling in addition to various other functions (Beffa et al., 1995; Gelli et al., 1997). One possible mechanism of heterotrimeric G protein in defense signaling is an increase in cytosolic Ca^{2+} in the cell, which leads to the activation of downstream signaling pathways. Ca^{2+} activates Ca^{2+}-dependent protein kinase and calmodulins, both of which play an important role in defense signaling. The plant NADPH oxidase, which is directly involved in disease signaling, also seems to be activated by Ca^{2+}.

The small GTPase OsRac1, the Rac homolog of rice, is a regulator of the production of reactive oxygen intermediates, which are associated with hypersensitive reaction (HR) and consequent programmed cell death. The constitutive expression of active OsRac1 in transgenic rice cell cultures has been shown to cause HR-like responses against a virulent race of the rice blast fungus, resistance against a virulent race of bacterial blight, and enhanced production of a phytoalexin. Constitutive expression of OsRac1 in transgenic rice cell cultures induces H_2O_2 production and defense gene activation, even in the absence of pathogens or elicitors, which suggests a general role of OsRac1 in disease resistance in rice (Ono et al., 2001).

The involvement of two different GTPases—the heterotrimeric G protein and the small GTPase OsRac1—in defense signaling has been studied in rice. The role of the heterotrimeric G protein has been studied using rice dwarf 1 (*d1*) mutants lacking a single-copy Gα gene. The *d1* mutants exhibited a reduction of hypersensitive reaction to an avirulent race of rice blast in rice-blast interaction. H_2O_2 production and PR gene expression induced by sphingolipid elicitor were suppressed in *d1* cell cultures. Heterotrimeric G protein (Gα) may function upstream of the GTPase OsRac1 in early stages of defense signaling in rice (Suharsono et al., 2002).

Gene Pyramiding

A combination of genes with different modes of action may provide a multi-method defense for crop protection. Approaches can involve conventional sexual crossing, retransformation, cotransformation, and the use of linked transgenes. A recent report (Narayanan et al., 2004) showed the pyramiding of two major genes into *indica* rice cultivar CO39 using molecular breeding through marker-aided selection (MAS) and transformation. A resistant CO39 near-isogenic line (NIL) carrying a major blast resistance gene (*Piz-5*) was obtained through phenotyping assays and MAS. To develop multiple resistance for blast and blight in this cultivar, this blast-resistant NIL of CO39 was transformed with the cloned bacterial blight resistance gene *Xa21*, which is known to confer resistance to most races of *Xanthomonas oryzae* (Narayanan et al., 2004). Similar work is in progress with another elite *indica* rice cultivar, IR50. An IR50 introgressed line with the *Piz-5* gene was obtained by conventional breeding using a resistant CO39 NIL carrying the *Piz-5* gene as the donor parent, and this introgressed IR50 line was used in transformation with the *Xa21* gene for resistance to BB. The resulting transgenic IR50 line exhibited an increased level of resistance to the blast fungus (*Magnaporthe grisea*) and to bacterial blight (*Xanthomonas oryzae*) (Narayanan et al., 2002).

Pyramiding transgenes for multiple resistance against bacterial blight, yellow stem borer, and sheath blight by the combination of three genes with three different modes of action by sexual crossing of two independent transgenic parental lines has been reported (Datta et al., 2002). Two transgenic homozygous lines of IR72—one carrying the *Xa21* gene (BB resistance) and another carrying the *cry* gene (resistance to stem borer) and *RC7* chitinase gene (resistance to sheath blight)—were used as parents in reciprocal crossing. The transgenes served as markers for identification and molecular selection to achieve stable homozygous lines. The identified pyramided line showed resistance against *X. oryzae* and *Scirpophaga incertulas* (yellow stem borer) and a high tolerance of *R. solani*, the fungal pathogen of sheath blight (Datta et al., 2002). The status of built-in disease resistance development in rice is reported in Table 7.1.

Genomic Approach to Rice Disease Resistance

The genomic approach has potential for the understanding of plant disease resistance genes and the genes that they regulate. Structural genomics investigates the genome structure at the sequence level. The complete genomic sequences of a variety of microorganisms and of some model

TABLE 7.1. Significant biotechnological developments in disease resistance in rice.

Disease	Pathogen type	Genes/transgenes employed	Method employed	References
Sheath blight	*Rhizoctonia solani* (fungus)	Rice *Chi11*	Genetic engineering	Lin et al., 1995; Baisakh et al., 2001
		Rice *Rc24*	Genetic engineering	Xu et al., 1996
		Rice *TLP*	Genetic engineering	Datta et al., 1999
		Rice *RC7*-Chitinase	Genetic engineering	Datta et al., 2001
		Arabidopsis NPR1	Genetic engineering	Baisakh et al. (unpublished)
Blast	*Magnaporthe grisea* (fungus)	Stilbene synthase	Genetic engineering	Stark-Lorenzen et al., 1997
		Rice *Cht2*, *Cht3*	Genetic engineering	Nishizawa et al., 1999
		Rice *Pi-1*, *Pi-2*, *Pi-9*	Fine mapping and MAS	Hittalmani et al., 2000
		Trichosanthin (*tcs*) gene	Genetic engineering	Yuan et al., 2002
		Rice *YK1*	Genetic engineering	Uchimiya et al., 2002
Bacterial blight	*Xanthomonas oryzae* (bacterium)	*Xa21*	Conventional breeding	Khush et al., 1990
			Genetic engineering	Song et al., 1995; Tu et al., 1998
		Xa4, *Xa5*, *Xa13*, *Xa7*, *Xa21*	Mapping and tagging	Huang et al., 1997
		Arabidopsis NPR1	Genetic engineering	Chern et al., 2001
Tungro disease	RTBV and RTSV (virus)	*CP2*, *CP3*	Genetic engineering	Sivamani et al., 1999
		RTSV replicase	Genetic engineering	Huet et al., 1999
Yellow mottle virus disease	RYMV	RNA-dependent RNA polymerase	Genetic engineering	Pinto et al., 1999

TABLE 7.1 *(continued)*

Disease	Pathogen type	Genes/transgenes employed	Method employed	References
Blast + bacterial blight	*M. grisea* + *X. oryzae*	*Piz-5* + *Xa21*	MAS + genetic engineering	Narayanan et al., 2002, 2004
Bacterial blight + sheath blight	*X. oryzae* + *R. solani*	*Xa21* + *RC7-Chi*	Genetic engineering + conventional breeding	Datta et al., 2002

Note: MAS = marker-aided selection; RTBV = rice tungro bacilliform virus; RTSV = rice tungro spherical virus; RYMV = rice yellow mottle virus.

plants, such as *Arabidopsis* and rice, are being determined. Sequencing of resistance gene clusters may help in understanding and cloning the resistance gene. The majority of disease resistance genes cloned from different species encode a predicted protein with nucleotide binding sites (NBS) carrying leucine-rich repeats (LRR) at the N-terminal end. Currently, available BAC end sequences, representing approximately 5% of the rice genome, suggest the existence of 750-1500 NBS-encoding genes in rice (Meyers et al., 1999).

Comparative genomics investigates the molecular basis of differences between organisms. Macrosynteny based on linkage analysis is well documented in monocotyledoneous species (Gale and Devos, 1998). As the sequencing of the genome of rice and several other plant species nears completion, once the extent of synteny has been established, it may be possible to predict the position of the genes in each part of the genome. High-throughput genotyping will facilitate the accurate mapping of quantitative trait loci (QTLs) for disease resistance. QTL mapping with genomic sequence data and information on allelic differences may help in candidate gene approaches to clone the QTLs for disease resistance. Comparative genomics is helpful for understanding the molecular basis of specificity. From a few reports, it has been suggested that the LRR region of the NBS-encoding gene is not the only determinant of specificity (Ellis et al., 1999).

Functional genomics focuses on the analysis of the function of genes. It involves high-throughput technology, such as microarray for mRNA expression, insertion or deletion mutagenesis, entrapment, or activation of genes. The global analysis of gene expression, together with prediction from DNA sequence data to understand gene and protein function, is still in its infancy.

Rice, with its small genome size (approximately 450 Mbp), has become a model plant because of its well-developed genetic and molecular information (Shimamoto, 1995). A large effort is under way to determine the function of its disease resistance genes.

CONCLUSIONS

Attempts have been made to engineer durable resistance in important crop plants ever since the initial discovery of genes involved in plant disease resistance. Based on our knowledge of molecular events occurring during plant-pathogen interactions, several options have opened up for developing plant varieties with resistance to pathogens. For long-term resistance, and specifically for fungal resistance, the broad range of resistance in transgenic plants could be increased by a combination of new genes.

R genes have been used by plant breeders against various diseases. But when deployed in monoculture, resistance frequently breaks down, as races of the pathogen can overcome the *R* gene through recessive mutations in the corresponding *Avr* gene. A cloned *R* gene can now be used to improve the efficiency of plant breeding strategies for accelerating the introgression of useful *R* genes from related species by marker-assisted breeding and transformation (Crute and Pink, 1996; Michelmore, 1995). Plant biotechnologists can attempt to manipulate both *Avr* and *R* gene sequences to obtain durable broad-spectrum disease control (Hammond-Kosack et al., 1996; Staskawicz et al., 1995). Rice, with its whole genome sequenced, will be of tremendous use in decoding the function of various defense genes and pathways and in the isolation of more *R* genes.

Various defense mechanisms can be activated by virulent pathogens that are not recognized by *R* genes, which suggests the possible existence of other pathogen surveillance mechanisms that attenuate disease severity (Delaney, 1997). Understanding of these mechanisms may provide us with options for developing disease resistance in crop plants. The combinatorial deployment of these strategies could also be exploited to attain effective, durable resistance.

Genetic manipulation of regulatory mechanisms and the signal transduction pathways controlling the induction of multiple defense responses might be another efficient approach for obtaining plant disease resistance. Precise knowledge of the regulation of the signaling process involved and the subsequent downstream metabolic pathways that get triggered is required to develop plant defense using regulatory mechanisms. Manipulation of single regulatory genes, such as transcription factors, can be used to alter the activity of the set of downstream genes influenced by the transcription factor. The signal transduction pathways are conserved in broad families of plants, which enables the use of new resistance genes identified in unrelated plant species to create transgenic plants with novel disease resistance.

Significant progress has been made in rice biotechnology for a better understanding of the biology of rice at the molecular level and its interactions with various pathogens. Rice breeding has benefited from marker-assisted selection and genetic engineering to widen the gene pool for improving plant protection against diseases and increasing yield. It is still a significant challenge to be able to engineer rice for complete protection against fungal pathogens, particularly sheath blight and blast.

"The challenge of feeding an additional 3 billion human beings, 95 percent of them in the poor developing countries, on the same amount of land and water currently available requires a dramatic transformation of rural economies and intensified agriculture. All possible tools that can help promote sustainable agriculture for food security must be marshaled, and biotechnology, safely deployed, could be a tremendous help in that fight," says Ismail Serageldin, the former chairman of the CGIAR and World Bank vice president for environmentally and socially sustainable development (Serageldin, 1997). No commercial transgenic product with enhanced disease resistance is currently available. With the maturity of our knowledge of pathogenesis and defense mechanisms, together with genomics-based strategies for gene discovery and the application of advanced genetic engineering technology in the coming years, we hope to see the first genetically modified disease-resistant crop appear on the market.

REFERENCES

Abad, L. R., M. P. D' Urzo, D. Liu, M. L. Narasimhan, M. Reuveni, J. K. Zhu, X. Niu, N. K. Singh, P. M. Hasegawa, and R. A. Bressan (1996). Antifungal activity of tobacco osmotin has specificity and involves plasma membrane permeabilization. *Plant Science 118:* 11-23.

Baisakh, N., K. Datta, N. Oliva, I. Ona, G. J. N. Rao, T. W. Mew, and S. K. Datta (2001). Rapid development of homozygous transgenic rice using anther culture harboring rice *chitinase* gene for enhanced sheath blight resistance. *Plant Biotechnology 18:* 101-108.

Beffa. R., M. Szell, P. Meuwly, A. Pay, R. Vogeli-Lange, J. P. Metraux, G. Neuhaus, F. Meins, Jr., and F. Nagy (1995). Cholera toxin elevates pathogen resistance and induces pathogenesis-related gene expression in tobacco. *EMBO Journal 14:* 5753-5761.

Bent, A. F., B. N. Kunkel, D. Dahlbeck, K. L. Brown, R. Schmidt, J. Giraudat, J. Leung, and B. J. Staskawicz (1994). RPS2 of *Arabidopsis thaliana:* A leucine-rich repeat class of plant disease resistance genes. *Science 265:* 1856-1860.

Bohlmann, H. (1994). The role of thionins in plant protection. *Critical Reviews in Plant Sciences 13:* 1-16.

Broekaert, W. F., F. R. G. Terras, B. P. A. Cammue, and R. W. Osborn (1995). Plant defensins: Novel antimicrobial peptides as components of the host defense system. *Plant Physiology 108:* 1353-1358.

Broglie, K., I. Chet, M. Holliday, R. Cressman, P. Briddle, S. Knowlton, C. J. Mauvais, and R. Broglie (1991). Transgenic plants with enhanced resistance to the fungal pathogen *Rhizoctonia solani. Science 254:* 1194-1197.

Cao, H., X. Li, and X. Dong (1998). Generation of broad-spectrum disease resistance by overexpression of an essential regulatory gene in systemic acquired resistance. *Proceedings of the National Academy of Sciences of the USA 95:* 6531-6536.

Chern, M. S., H. A. Fitzgerald, R. C. Yadav, P. E. Canlas, X. Dong, and P. C. Ronald (2001). Evidence for a disease resistance pathway in rice similar to the NPR1 mediated signaling pathway in *Arabidopsis. The Plant Journal 27:* 101-113.

Coca, M., C. Bortolotti, M. Rufat, G. Penas, R. Eritja, D. Tharreau, D. A. M. Pozo, J. Messeguer, and B. S. Segundo (2004). Transgenic rice plants expressing the antifungal AFP protein from *Aspergillus giganteus* show enhanced resistance to the rice blast fungus *Magnaporthe grisea. Plant Molecular Biology 54:* 245-259.

Crute, I. R., and D. A. C. Pink (1996). The genetics and utilization of pathogen resistance in plants. *Plant Cell 8:* 1747-1755.

Dangl, J. L. (1998). Plants just say NO to pathogens. *Nature 394:* 525-527.

Datta, K., N. Baisakh, K. M. Thet, J. Tu, and S. K. Datta (2002). Pyramiding transgenes for multiple resistance in rice against bacterial blight, yellow stem borer and sheath blight. *Theoretical and Applied Genetics 106:* 1-8.

Datta, K., J. Tu, N. Oliva, I. Ona, R. Velazhahan, T. S. Mew, S. Muthukrishnan, and S. K. Datta (2001). Enhanced resistance to sheath blight by constitutive expression of infection-related rice chitinase in transgenic elite *indica* rice cultivars. *Plant Science 160:* 405-414.

Datta, K., R. Velazhahan, N. Oliva, T. Mew, G. S. Khush, S. Muthukrishnan, and S. K. Datta (1999). Overexpression of cloned rice thaumatin-like protein (PR-5) gene in transgenic rice plants enhances environmental friendly resistance to *Rhizoctonia solani* causing sheath blight disease. *Theoretical and Applied Genetics 98:* 1138-1145.

Datta, S. K., and S. Muthukrishnan (Eds.). (1999). *Pathogenesis-related proteins in plants.* Boca Raton, FL: CRC Press.

Delaney, T. P. (1997). Genetic dissection of acquired resistance to disease. *Plant Physiology 113:* 5-12.

Delledonne, M., Y. Xia, R. A. Dixon, and C. Lamb (1998). Nitric oxide functions as a signal in plant disease resistance. *Nature 394:* 585-588.

Does, M. P., and B. J. C. Cornelissen (1998). Emerging strategies to control fungal diseases using transgenic plants. In *Crop productivity and sustainability: Shaping the future,* V. L. Chopra, R. B. Singh, and A. Verma (Eds.), pp. 233-244. New Delhi: Oxford/IBH.

Durner. J., D. Wendehenne, and D. F. Klessig (1998). Defense gene induction in tobacco by nitric oxide, cyclic GMP and cyclic ADP ribose. *Proceedings of the National Academy of Sciences of the USA 95:* 10328-10333.

Ellis, J. G., G. J. Lawrence, J. E. Luch, and P. N. Dodds (1999). Identification of regions in alleles of the flax rust resistance gene *L* that determine differences in gene-for-gene specificity. *Plant Cell 11:* 495-506.

Gale, M., and K. Devos (1998). Plant comparative genetics after 10 years. *Science 282:* 656-658.

Gelli, A., V. J. Higgins, and E. Blumwald (1997). Activation of plant plasma membrane Ca^{2+}-permeable channels by race-specific fungal elicitors. *Plant Physiology 113:* 269-279.

Golemboski, D. B., G. P. Lomonossoff, and M. Zaitlin (1990). Plants transformed with a tobacco mosaic virus non-structural gene sequence are resistant to the virus. *Proceedings of the National Academy of Sciences of the USA 87:* 6311-6315.

Greenberg, J. T. (1997). Programmed cell death in plant-microbe interactions. *Annual Review of Plant Physiology and Plant Molecular Biology 48:* 525-545.

Grover, A., and D. Pental (2003). Breeding objectives and requirements for producing transgenics for major field crops of India. *Current Science 84:* 310-320.

Hammond-Kosack, K. E., and J. D. G. Jones (1997). Plant disease resistance genes. *Annual Review of Plant Physiology and Plant Molecular Biology 48:* 575-607.

Hammond-Kosack, K. E., D. A. Jones, and J. D. G. Jones (1996). Ensnaring microbes: The components of plant disease resistance. *The New Phytologist 133:* 11-24.

Hittalmani, S., A. Parco, T. W. Mew, R. S. Zeigler, and N. Huang (2000). Fine mapping and DNA marker-assisted pyramiding of the three major genes for blast resistance in rice. *Theoretical and Applied Genetics 100:* 1121-1128.

Hu, P., C. C. An, V. Li, and Z. L. Chen (1999). Prokaryotic expressed trichosanthin and two other proteins have antifungal activity in vitro. *Acta Microbiologica Sinica 39:* 234-240.

Huang, N., E. R. Angeles, J. Domingo, G. Magpantay, S. Singh, G. Zhang, N. Kumaradivel, J. Bennett, and G. S. Khush (1997). Pyramiding of bacterial blight resistance genes in rice: Marker-assisted selection using RFLP and PCR. *Theoretical and Applied Genetics 95:* 313-320.

Huet, H., S. Mahendra, J. Wang, E. Sivamani, C. A. Ong, L. Chen, A. De Kochko, R. N. Beachy, and C. Fauquet (1999). Near immunity to rice tungro spherical virus achieved in rice by a replicase-mediated resistance strategy. *Phytopathology 89:* 1022-1027.

Ikeda, R., G. S. Khush, and R. E. Tabien (1990). A new resistance gene to bacterial blight derived from *O. longistaminata. Japanese Journal of Breeding 40:* 280-281.

Iwai, T., H. Kaku, R. Honkura, S. Nakamura, H. Ochiai, T. Sasaki, and Y. Ohashi (2002). Enhanced resistance to seed-transmitted bacterial disease in transgenic rice plants over producing an oat cell-wall-bound thionin. *Molecular Plant-Microbe Interactions 15:* 515-521.

Jackson, A. O., and C. B. Taylor (1996). Plant-microbe interactions: Life and death at the interface. *Plant Cell 8:* 1651-1658.

Jones, D. A., C. A. Thomas, K. E. Hammond-Kosack, P. J. Balint-Kurti, and J. D. G. Jones (1994). Isolation of the tomato *Cf-9* gene for resistance to *Cladosporium fulvum* by transposon tagging. *Science 266:* 789-793.

Kauffmann, S., M. Legrand, P. Geoffroy, and B. Fritig (1987). Biological function of "pathogenesis-related" proteins: Four PR proteins of tobacco have 1,3-β-glucanase activity. *EMBO Journal 6:* 3209-3212.

Kawata, M., T. Nakajima, T. Yamamoto, K. Mori, T. Oikawa, F. Fukumoto, and S. Kuroda (2003). Genetic engineering for disease resistance in rice (*Oryza sativa* L.) using antimicrobial peptides. *Japan Agricultural Research Quarterly 37:* 71-76.

Keen, N. T. (1999). Plant disease resistance: Progress in basic understanding and practical application. *Advances in Botanical Research 30:* 292-328.

Khush, G. S., E. Bacalanco, and T. Ogawa (1990). A new gene for resistance to bacterial blight from *O. longistaminata. Rice Genetics Newsletter 7:* 121-122.

Krishnamurthy, K., C. Balconi, J. E. Sherwood, and M. Giroux (2001). Wheat puroindolines enhance fungal disease resistance in transgenic rice. *Molecular Plant-Microbe Interactions 14:* 1255-1260.

Lin, W., C. S. Anuratha, K. Datta, I. Potrykus, S. Muthukrishnan, and S. K. Datta (1995). Genetic engineering of rice for resistance to sheath blight. *Bio/Technology 13:* 686-691.

Lodge, J. K., W. K. Kaniewski, and N. E. Tumer (1993). Broad-spectrum virus resistance in transgenic plants expressing pokeweed antiviral protein. *Proceedings of the National Academy of Sciences of the USA 90:* 7089-7093.

Logemann, J., G. Jach, H. Tommerup, J. Mundy, and J. Schell (1991). Expression of a barley ribosome-inactivating protein leads to increased fungal protection in transgenic tobacco plants. *Bio/Technology 10:* 305-308.

Mackill, D. J., and J. M. Bonman (1992). Inheritance of blast resistance in near isogenic lines of rice. *Phytopathology 82:* 746-749.

Malehorn, D. E., J. R. Borgmeyer, C. E. Smith, and D. M. Shah (1994). Characterization and expression of an antifungal zeamatin-like protein *(Z1p)* gene from *Zea mays. Plant Physiology 106:* 1471-1481.

Malyshenko, S. L., O. A. Kondakova, J. U. V. Nazarova, I. B. Kaplan, M. E. Taliansky, and J. G. Atabekov (1993). Reduction of tobacco mosaic virus accumulation in transgenic plants producing non-functional viral transport proteins. *Journal of General Virology 74:* 1149-1156.

Martin, G. B., S. H. Brommonschenkel, J. Chunwongse, A. Frary, and M. W. Ganal (1993). Map-based cloning of a protein kinase gene conferring disease resistance in tomato. *Science 262:* 1432-1436.

Meyers, B. C., A. W. Dickerman, R. W. Michelmore, S. Sivaramakrishnan, B. W. Sobral, and N. D. Young (1999). Plant disease resistance genes encode members of an ancient and diverse protein family within the nucleotide-binding superfamily. *The Plant Journal 20:* 317-332.

Michelmore, R. (1995). Molecular approaches to manipulation of disease resistance genes. *Annual Review of Phytopathology 15:* 393-427.

Morrissey, J. P., and A. D. Osbourn (1999). Fungal resistance to plant antibiotics as a mechanism of pathogenesis. *Microbiology and Molecular Biology Reviews 63:* 708-724.

Narayanan, N. N., N. Baisakh, N. P. Oliva, C. M. Vera Cruz, S. Gnanamanickam, K. Datta, and S. K. Datta (2004). Molecular breeding: Marker-assisted selection combined with biolistic transformation for blast and bacterial blight resistance in *indica* rice (cv. CO39). *Molecular Breeding 14:* 61-71.

Narayanan, N. N., N. Baisakh, C. M. Vera Cruz, S. S. Gnanamanickam, K. Datta, and S. K. Datta (2002). Molecular breeding for the development of blast and bacterial blight resistance in rice cv. IR50. *Crop Science 42:* 2072-2079.

Neuhaus, J. M. (1999). Plant chitinases (PR-3, PR-4, PR-8, PR-11). In *Pathogenesis-related proteins in plants,* S. K. Datta and S. Muthukrishnan (Eds.), pp. 77-105. Boca Raton, FL: CRC Press.

Nishizawa, Y., Z. Nishio, K. Nakazono, M. Soma, E. Nakajima, M. Ugaki, and T. Hibi (1999). Enhanced resistance to blast *(Magnaporthe grisea)* in transgenic *japonica* rice by constitutive expression of rice chitinase. *Theoretical and Applied Genetics 99:* 383-390.

Ono, E., H. L. Wong, T. Kawasaki, M. Hasegawa, O. Kodama, and K. Shimamoto (2001). Essential role of the small GTPase Rac in disease resistance of rice. *Proceedings of the National Academy of Sciences of the USA 98:* 759-764.

Osbourn, A. D. (1996). Preformed antimicrobial compounds and plant defense against fungal attack. *Plant Cell 8:* 1821-1831.

Pieterse, C. M. J., S. C. M. Van Wees, J. A. Van Pelt, M. Knoester, R. Laan, H. Gerrits, P. J. Weisbeek, and L. C. Van Loon (1998). A novel signaling pathway controlling induced systemic resistance in *Arabidopsis. Plant Cell 10:* 1571-1580.

Pingali, P. L. (1995). The impact of pesticides on farmer health: A medical and economic analysis in the Philippines. In *Impact of pesticides on farmer health and the rice environment,* P. L. Pingali and P. A. Roger (Eds.), pp. 343-360. Kluwer Academic Publishers, Norwell, MA.

Pinto, Y. M., R. A. Kok, and D. C. Baulcombe (1999). Resistance to rice yellow mottle virus (RYMV) in cultivated African rice varieties containing RYMV transgenes. *Nature Biotechnology 17:* 702-707.

Powell-Abel, P., R. S. Nelson, B. De, N. Hoffman, S. G. Rogers, R. T. Fraley, and R. N. Beachy (1986). Delay of disease development in transgenic plants that express the tobacco mosaic virus coat protein gene. *Science 232:* 738-743.

Richter, T. E., and P. C. Ronald (2000). The evolution of disease resistance genes. *Plant Molecular Biology 42:* 195-204.

Rommens, C. A., and G. M. Kishore (2000). Exploiting the full potential of disease resistance genes for agricultural use. *Current Opinion in Biotechnology 11:* 120-125.

Serageldin, I. (1997). *Bioengineering of crops could help feed the world. Crop increases of 10-15 percent possible,* pp. 1-7. CGIAR Press Release. Washington, DC: CGIAR.

Sharma, A., R. Sharma, M. Imamura, M. Yamakaua, and H. Machii (2000). Transgenic expression of cecropin B, an antibacterial peptide from *Bombyx mori* confers enhanced resistance to bacterial leaf blight in rice. *FEBS Letters 484:* 7-11.

Shimamoto, K. (1995). The molecular biology of rice. *Science 270:* 1772-1773.

Sivamani, E., H. Huet, P. Shen, C. A. Ong, A. de Kochko, C. Fauquet, and R. N. Beachy (1999). Rice plants (*Oryza sativa* L.) containing rice tungro spherical virus (RTSV) coat protein transgenes are resistant to virus infection. *Molecular Breeding 5:* 177-185.

Song, W. Y., L. Y. Pi, G. L. Wang, J. Gardner, and T. Holsten (1997). Evolution of the rice *Xa21* disease resistance gene family. *Plant Cell 9:* 1279-1287.

Song, W. Y., G. L. Wang, L. L. Chen, H. S. Kim, L. Y. Pi, T. Hosten, J. Gardner, B. Wang, W. X. Zhai, L. H. Zhu, et al. (1995). A receptor kinase-like protein encoded by the rice disease resistance gene, *Xa21*. *Science 270:* 1804-1806.

Stark-Lorenzen, P., B. Nelke, G. Hanssler, H. P. Muhlbach, and J. E. Tomzik (1997). Transfer of a grapevine stilbene synthase gene to rice (*Oryza sativa* L.). *Plant Cell Reports 16:* 668-673.

Staskawicz, B. J., F. M. Ausubel, B. J. Baker, J. G. Ellis, and J. D. G. Jones (1995). Molecular genetics of plant disease resistance. *Science 268:* 661-667.

Suharsono, U., Y. Fujisawa, T. Kawasaki, Y. Iwasaki, H. Satoh, and K. Shimamoto (2002). The heterotrimeric G protein α subunit acts upstream of the small GTPase Rac in disease resistance of rice. *Proceedings of the National Academy of Sciences of the USA 99:* 13307-13312.

Takken, F. L. W., and M. H. A. J. Joosten (2000). Plant resistance genes: Their structure, function and evolution. *European Journal of Plant Pathology 106:* 699-713.

Tu, J., K. Datta, G. S. Khush, Q. Zhang, and S. K. Datta (2000). Field performance of *Xa21* transgenic *indica* rice (*Oryza sativa* L.), IR72. *Theoretical and Applied Genetics 101:* 15-20.

Tu, J., I. Ona, Q. Zhang, T. W. Mew, G. S. Khush, and S. K. Datta (1998). Transgenic rice variety IR72 with *Xa21* is resistant to bacterial blight. *Theoretical and Applied Genetics 97:* 31-36.

Uchimiya, H., S. Fujii, J. Huang, T. Fushimi, M. Nishioka, K. M. Kim, M. K. Yamada, T. Kurusu, K. Kuchitsu, M. Tagawa, et al. (2002). Transgenic rice plants conferring increased tolerance to rice blast and multiple environmental stresses. *Molecular Breeding 9:* 25-31.

Wang, G. L., W. Y. Song, D. L. Ruan, S. Sideris, and P. Ronald (1996). The cloned gene, *Xa21*, confers resistance to multiple *Xanthomonas oryzae* pv. *oryzae* isolates in transgenic plants. *Molecular Plant-Microbe Interactions 9:* 850-855.

Xu, Y., Q. Zhu, W. Panbangred, K. Shirasu, and C. Lamb (1996). Regulation, expression and function of a new basic chitinase gene in rice (*Oryza sativa* L.). *Plant Molecular Biology 30:* 387-401.

Yang, Y., J. Shah, and D. F. Klessig (1997). Signal perception and transduction in plant defense responses. *Genes and Development 11:* 1621-1639.

Yuan, H., X. Ming, L. Wang, P. Hu, C. An, and Z. Chen (2002). Expression of a gene encoding trichosanthin in transgenic rice plants enhances resistance to fungus blast disease. *Plant Cell Reports 20:* 992-998.

Zhang, S. P, W. Y. Song, L. L. Chen, D. L. Ruan, N. Taylor, P. Ronald, R. Beachy, and C. Fauquet (1998). Transgenic elite *indica* rice varieties, resistant to *Xanthomonas oryzae* pv. *oryzae*. *Molecular Breeding 4:* 551-558.

Zhou, P, Y. X. Zhu, W. H. Zheng, H. Li, and Z. L. Chen (1994). Prevention of disease development in transgenic plants expressing the trichosanthin gene. *Asia Pacific Journal of Molecular Biology 2:* 200-204.

Chapter 8

Molecular Aspects of Rice Blast Disease Resistance: Insights from Structural and Functional Analysis of the *Pi-ta* and *AVR-Pita* Gene Pair

Y. Jia
B. Valent

INTRODUCTION

Rice blast is one of the most devastating diseases in most rice-growing areas worldwide (Zeigler et al., 1994). Crop reduction due to blast and the cost for blast control are still substantial. Although the control of blast disease with fungicides has had reasonable success, there is continued pressure to lower the cost of inputs for rice farmers and to reduce the use of chemicals. The use of resistant cultivars containing major resistance (*R*) genes plays a central role in crop protection. Understanding how *R* genes work will lead to the development of molecular tools to control disease.

The authors thank the Arkansas Rice Research and Promotion Board for financial support for portions of this research and Melissa H. Jia for proofreading. For critical comments, the authors thank Roger Thilmony and Benildo G. de los Reyes. Research was conducted in Dr. Jia's lab in the Molecular Plant Pathology Program of USDA-ARS Dale Bumpers National Rice Research Center. Research conducted in Dr. Valent's lab was supported by the DuPont Company, USDA-NRI, and the Kansas State University Agricultural Experiment Station.

Current advancement of rice genomics has facilitated the identification and tagging of *R* genes for blast disease control (Gowda et al., 2003). Blast *R* genes follow a classical gene-for-gene system (Flor, 1971). Each major *Pi* gene in rice is effective in preventing infection by specific races of the rice blast fungus, *Magnaporthe grisea* (Hebert) Barr (anamorph: *Pyricularia grisea;* Rossman et al., 1990), containing the corresponding avirulence (*AVR*) gene (Flor, 1971; Silué et al., 1992). Molecular characterization of blast *R* genes has allowed us to obtain a better understanding of the molecular mechanisms of *R* gene-mediated disease resistance. *Pi-b* (Wang et al., 1999) and *Pi-ta* (Bryan et al., 2000) were the first two blast *R* genes that were molecularly characterized. *Pi-ta* will be described in more detail in this chapter. *Pi-b* encodes a predicted cytoplasmic protein containing a centrally localized nucleotide binding site (NBS) and a leucine-rich repeat (LRR) domain at the carboxyl terminus. Similar to most characterized *R* genes, *Pi-b* is a member of a small gene family. In contrast to most *R* genes, however, the expression of *Pi-b* is induced by altered environmental conditions such as altered temperatures and darkness (Wang et al., 1999, 2001).

The *gene-for-gene* theory suggests that resistance occurs when products of *R* genes interact with the products of their cognate *AVR* genes. The ligand-receptor model provides the simplest explanation for the gene-for-gene relationship. In this model, the *R* gene encodes a receptor that might directly detect the ligand—the pathogen molecule encoded by the *AVR* gene.

Short-term benefits of cloning an *R* gene would result from developing DNA markers for blast disease control by marker-assisted selection (MAS; Jia, 2003) or from incorporating this cloned *R* gene into various elite rice cultivars by genetic transformation. MAS is a selection method based on DNA sequence that is closely linked to a given trait, and it is particularly effective for disease resistance because resistance is often controlled by single genes (Young, 1996). Use of MAS can avoid the following obvious problems in developing rice blast control:

1. Quarantine restrictions prevent the exchange of pathogen isolates between different rice production areas and thus prevent the verification of a particular *R* gene.
2. *R* gene identification is complicated by overlapping resistance genes because resistance is triggered by any single pair of the *R* gene in rice and its cognate *AVR* gene in *M. grisea.*
3. There often exist multiple *AVR* genes in a single field isolate of the pathogen, any one of which may be sufficient to trigger resistance in the presence of its corresponding *R* gene.

Similar to the short-term benefit of cloning *R* genes, the cloning of *AVR* genes allows the analysis of the *AVR* gene structure in field rice blast populations and, more important, allows the prediction of the stability of resistance in the rice cultivars that are currently grown. Early work on the population biology of *M. grisea* was focused on general characterization using the dispersed, middle-repetitive DNA sequence MGR586. Fingerprinting using MGR586 does not necessarily reveal a specific pathotype of the rice blast fungus, although lineages often include a predictable subset of pathotypes (Zeigler et al., 1995).

In this chapter, we describe progress on the molecular characterization of the *Pi-ta-AVR-Pita* gene pair from the rice blast system, and describe how the *Pi-ta* gene may function on a molecular level to detect the pathogen signal molecule. New knowledge derived from our research can benefit rice cultivar improvement. *Pi-ta* encodes a predicted cytoplasmic NBS-LRR protein with a degenerate LRR domain, termed leucine-rich domain (LRD; Bryan et al., 2000). *Pi-ta* prevents the infection of *M. grisea* expressing *AVR-Pita*. *AVR-Pita*, formerly referred to as *AVR2-YAMO*, after the Japanese *Pi-ta* differential cultivar *Ya shiro-mochi*, appears to encode a metalloprotease (Orbach et al., 2000). The Pi-ta protein has been shown to interact physically with the AVR-Pita protein in triggering resistance (Jia et al., 2000). Sequence analysis of rice germplasm identified one *Pi-ta* allele conferring resistance and three *pi-ta* alleles conferring susceptibility (Jia et al., 2003a). In the United States, *Pi-ta* has been successfully introduced into diverse *japonica* rice cultivars by classical plant breeding. *Pi-ta* is still effective in preventing infection by the most prevalent rice blast field races IB-49 and IC-17 since its release in the 1980s (Jia et al., 2004). Both dominant and codominant markers for *Pi-ta* have been developed from the *Pi-ta* sequence to accelerate its incorporation into advanced rice breeding lines in the United States (Jia et al., 2002). The resistant *Pi-ta* allele was found in all *Pi-ta-* and *Pi-ta²*-containing rice cultivars, and these cultivars are resistant to both IB-49 and IC-17. The relationship between *Pi-ta* and *Pi-ta²* will be discussed. Sequence analysis of *AVR-Pita* alleles from IB-49 and IC-17 has identified new predicted functional *AVR-Pita* genes (Singh and Jia, unpublished data). Current progress in investigating molecular signaling in the *Pi-ta* pathway will be described.

RICE Pi-Ta *GENE*

The *Pi-ta* blast resistance gene is a component of blast control in a number of rice-growing regions worldwide (Figure 8.1). The common donor for

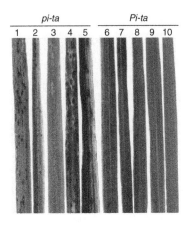

FIGURE 8.1. The *Pi-ta* blast resistance gene in rice plays a central role for rice protection. Blast is a serious disease in rice-growing areas worldwide. The image shows disease lesions on leaves of C101A51 (lane 1), M202 (lane 2), Nipponbare (lane 3), Wells (lane 4) and breeding line 1153 (lane 5) that do not contain *Pi-ta*, respectively, whereas no disease lesions were evident on Katy (lane 6), Drew (lane 7), Kaybonnet (lane 8) and breeding lines 602 (lane 9) and 605 (lane 10) that contain *Pi-ta*. See corresponding Plate 8.1 in the central gallery.

the *Pi-ta* and *Pi-ta*^{2} genes in Japan was a land race cultivar, *Tadukan*, from the Philippines (Shigemura and Kitamura, 1954), although some questions remain about the origin of *Pi-ta* in the differential variety *Yashiro-mochi* (Rybka et al., 1997). In the United States, another land race cultivar, *Tetep* from Vietnam, was used as the donor for *Pi-ta* (Moldenhauer et al., 1990). The U.S. cultivar *Katy* was first reported to contain *Pi-ta* (Bryan et al., 2000; Jia et al., 2003a; Moldenhauer et al., 1990), and subsequently *Katy* was used as the *Pi-ta* donor for additional *japonica* cultivars, *Drew, Kaybonnet,* and *Aherent* (Jia, unpublished data; Jia et al., 2003a; Moldenhauer et al., 1990, 1998). *Pi-ta* was first thought to be located on chromosome 9 but was later localized on chromosome 12 (Ise, 1993). The *Pi-ta* gene was suggested to be the same as *Pi-4(t)*, which was positioned 15.3 cM from RG869 (Inukai et al., 1994), but pathotypes that differentiate these two genes have been identified (B. Valent and M. Levy, unpublished data). An *R* gene in the Japanese cultivar *Yashiro-mochi,* originally named *Pi-62,* conditioned resistance to the Chinese *M. grisea* field isolate O-137 containing the *AVR2-YAMO* gene (Wu et al., 1996). *Pi-62* was found to be closely linked to RG869 near the centromere of chromosome 12 (Wu et al., 1996), in the general vicinity of *Pi-ta*. The *Pi-ta* differential rice variety K1 (Kiyosawa, 1984) showed an identical infection spectrum to that of *Yashiro-mochi* in inoculations with strain O-137. Japanese avirulent pathogen strains Ken54-20, Ina72, and Ken54-04, and virulent strains Ina168, Ken53-33, and P-2b, were originally used to identify the *Pi-ta* gene (Kiyosawa, 1967, 1971). Infection assays using these Japanese strains produced avirulence patterns consistent with *Pi-ta* in members of the rice mapping population that identified *Pi-62* (Wu et al., 1996). Thus, we renamed *Pi-62* in *Yashiro-mochi* as *Pi-ta* and its corresponding *AVR2-YAMO* gene as *AVR-Pita* (Bryan et al., 2000; Orbach et al., 2000).

Pi-ta *is a Putative Cytoplasmic Receptor*

We initiated the molecular cloning of the *Pi-ta* gene in the 1990s, and this work was completed in 2000 (Bryan et al., 2000). Bulked segregant analysis was initially used to identify DNA markers linked to *Pi-ta* (Wu et al., 1996). *Pi-ta* was mapped to chromosome 12 between flanking random amplified polymorphic DNA (RAPD) markers SP4B9 and SP9F3. One recombination event between *Pi-ta* and SP4B9 and one recombination event between *Pi-ta* and SP9F3 was identified in a population of 990 segregating progeny. Co-segregating restriction fragment length polymorphism (RFLP) marker p7C3 was identified and used for chromosome walking (Bryan et al., 2000; Wu et al., 1996). An overlapping contig of rice bacterial artificial chromosome (BAC) clones spanning 850 kbp at the *Pi-ta* locus was constructed (Figure 8.2(a)). The left border was closed when SP4B9 was identified in the BAC contig. SP9F3 was not identified from any BAC clones in the contig. While continuing to expand the contig to the right, a shotgun sequencing strategy was undertaken to identify *Pi-ta* candidate genes. Overlapping BAC clones in the contig were sheared to produce 1-3-kbp DNA fragments for subcloning and sequencing. Sequence data representing 99% coverage of the BAC contig region identified a *Pi-ta* candidate on BAC 142. This clone contains an open reading frame encoding a protein sharing homology to the nucleotide binding site (NBS) region of *R* genes, and this candidate *R* gene was subsequently introduced into the susceptible rice cultivar *Nipponbare* using an He-gene gun. The resulting transgenic plants were resistant to *M. grisea* strain O-137 expressing *AVR-Pita*, confirming that this gene is *Pi-ta* (Bryan et al., 2000).

Pi-ta is a single-copy gene that differs from *Pi-b* and most other characterized *R* genes because no highly related homologous family members were found in the rice genome. The expression of *Pi-ta* was undetectable using rice total RNA and was detected only when a large quantity of poly-

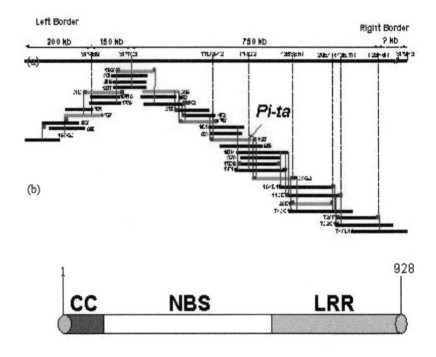

FIGURE 8.2. (a) The *Pi-ta* gene was identified from the BAC clone 142 that is in the middle of one megabase contig (Adapted from Bryan et al., 2000, *Plant Cell* 12:2033-2045). (b) Predicted structure of the Pi-ta protein. CC indicates the putative coiled coil region at the amino terminus, NBS indicated putative nucleotide binding sites in the middle region, and LRR indicates leucine rich repeats (The *Pi-ta* degenerate LRR is also called LRD). See corresponding Plate 8.2 in the central gallery.

A$^+$ mRNA was used in northern blots. Further analysis using northern blots revealed that transcripts of *Pi-ta* are present at extremely low levels in both healthy resistant and susceptible rice plants. Induction of *Pi-ta* by virulent or avirulent *M. grisea* races was not observed. *Pi-ta* encodes a predicted cytoplasmic protein (Figure 8.2(b)) with an NBS region characteristic of the largest class of plant *R* genes (Bent, 1996; Hammond-Kosack and Jones 1997). *Pi-ta* lacks both a leucine zipper motif and the Toll/ Interleukin-1 receptor homology reported as key features of dicotyledoneous *R* genes. The C-terminal portion of the Pi-ta protein contains a leucine-rich domain (LRD), which is degenerate compared to LRRs in other *R* gene products. *Pi-ta* belongs to the CC-NBS-LRR class of *R* genes (Bryan et al., 2000;

Dangl and Jones, 2001). Homology searching using DNA sequences from BAC clones spanning the *Pi-ta* region identified no other obvious candidate *R* genes. However, a relatively high abundance of repetitive DNA sequences, including the centromere-specific sequence RCE1, was present within the *Pi-ta* region (Dong et al., 1998; Singh et al., 1996). This result is consistent with a previous report that the region near *Pi-ta* is high in repetitive DNA sequences (Nakamura et al., 1997) and also confirmed that *Pi-ta* is located near a centromere. The right border of the contig was not closed in our study, presumably due to repressed recombination at the *Pi-ta* locus. The ratio of genetic and physical distance in this region is > 1200kbp/cM, which is much larger than the average of 244 kbp/cM in rice. Such a skewed ratio of physical to genetic distance has also been observed in the centromeric region of rice chromosomes 1, 4, and 10 (Chen et al., 2002). Completion of the rice genome sequence will allow the localization of the marker SP9F3 for an accurate measurement of the ratio of physical to genetic distance at the centromere of chromosome 12.

Natural Variation at the Pi-ta *Locus*

One way to determine the critical amino acid residues for the function of the Pi-ta protein is to analyze the structure of the products of natural alleles of *Pi-ta* in resistant and susceptible rice cultivars (Bryan et al., 2000; Jia et al., 2003a). *Pi-ta* was introduced into diverse cultivars worldwide, and thus representative cultivars from different rice production areas were used for this analysis. These are *Katy, Kaybonnet,* and *Drew* from the United States; 93-11 from China; K1, *Nipponbare, Tsuyuake,* and *Yashiro-mochi* from Japan; *Sariceltik* from Turkey; *Tetep* from Vietnam; *Cica 9* and *El Paso 144* from Latin America; and IR64 and *Tadukan* from the Philippines. To date, a total of four *Pi-ta* alleles have been identified (Table 8.1). The identical *Pi-ta* resistance allele was found in *Yashiro-mochi, K1, Tetep, Tadukan, Katy,* and *Drew.* Three *pi-ta* alleles, represented by C101A51, *El Paso 144*, and *Tsuyuake,* respectively, confer susceptibility to *AVR-Pita*-expressing fungus (Jia et al., 2003a). All four alleles encode a same-length CC-NBS-LRR protein. Overall, five amino acid differences were identified between the product of the *indica* resistant *Pi-ta* allele and the product of the *japonica* susceptible allele, and the *indica pi-ta* alleles contained subsets of these 5 differences. One amino acid difference was perfectly correlated with resistance versus susceptibility. If a *Pi-ta* haplotype encodes alanine at position 918, the cultivar is resistant, and if a *pi-ta* haplotype encodes serine at position 918, the cultivar is susceptible. The amino acid 918 *functional polymorphism* resides in the LRR region. In contrast to most other R proteins,

TABLE 8.1. Naturally occurring *Pi-ta* haplotypes.

		Nucleotide position[a]						
Haplotype	Cultivar	17	384	444	474	527	2388	2752
Pi-ta	Yashiro-mochi[b]	T	G	G	C	A	G	G
pi-ta	C101A51	T	C	G	C	A	G	T
pi-ta	El Paso 144[c]	G	C	G	C	A	G	T
pi-ta	Tsuyuake[d]	G	C	C	G	T	A	T

Source: Adapted from Jia et al. (2003b).

[a]Nucleotide position in the cDNA coding sequence. Polymorphisms at nucleotides 17, 444, 474, 527, and 2752 result in amino acid substitutions.

[b]Rice cultivars K1, *Reiho, Tetep, Tadukan, Katy,* and *Drew* contain this allele.

[c]*indica* rice cultivar *Cica 9* also contains this allele. The sequenced Chinese *indica* cultivar 93-11 contains this allele and is therefore predicted to be susceptible.

[d]*japonica* rice cultivars *Nipponbare* and *Sariceltik* contain this allele.

the Pi-ta LRR does not appear to be undergoing diversified selection. The functional importance of amino acid 918 is also supported by molecular interactions of Pi-ta and AVR-Pita that will be described later in this chapter. Further sequence analysis of the *Pi-ta* alleles from wild relatives of rice should provide a better understanding of the molecular evolution of the *Pi-ta* gene.

Pi-ta *Is Effective in Preventing Blast Disease in the USA*

In U.S. rice fields, there are relatively few blast pathogen races, dispersed among a few lineages (Correll et al., 2000). Surveys have indicated that rice blast races IB-49 and IC-17 are the most common in the southern United States (Xia et al., 2000). We determined that *Pi-ta* is effective in preventing the infection of IB-49 and IC-17 (Jia et al., 2004). The presence of *Pi-ta* in tested rice cultivars was correlated with resistance to these two major pathotypes. We investigated the inheritance of resistance to IC-17 using a marker for the resistant *Pi-ta* allele in an F_2 population of 1,345 progeny from a cross of cultivar *Katy* with experimental line RU9101001, possessing and lacking *Pi-ta,* respectively. The results showed that a single dominant gene that cosegregates with *Pi-ta* controlled resistance to IC-17, suggesting that *Pi-ta* is responsible for resistance to IC-17. A second F_2 population of 377 progeny from a reciprocal cross between *Katy* and RU9101001 was used to verify the conclusion that resistance to IC-17 was conditioned by *Pi-ta*. In this cross, individuals resistant to IC-17 were also resistant to IB-49, as determined by a detached leaf method (Jia et al., 2004). Furthermore, the presence of *Pi-ta* and resistance to IB-49 was also correlated in additional crosses between *Kaybonnet* and M-204, which possess and lack *Pi-ta*, respectively. We suggest that *Pi-ta* is responsible for resistance to IB-49 and IC-17. Because *Pi-ta* is effective in preventing infection of *M. grisea* containing *AVR-Pita*, we predicted that both IB-49 and IC-17 races would contain *AVR-Pita* genes. Subsequent sequence analysis of the *AVR-Pita* alleles from both IB-49 and IC-17 will be described later in this chapter.

Dominant and Codominant Markers for Molecular Breeding

Incorporation of *Pi-ta* into improved rice germplasm has been traditionally accomplished using linked DNA markers and pathogenicity screening. A short-term benefit of cloning an *R* gene is the chance to develop DNA markers corresponding to functional polymorphisms within the *R* gene itself; this type of marker is thus a so-called *perfect marker*. We developed both dominant and codominant DNA markers from *Pi-ta* to accelerate its incorporation by MAS (Jia et al., 2002; M. Redus and Y. Jia, unpublished data). Three pairs of DNA primers specific to the *Pi-ta* allele were designed to amplify portions of *Pi-ta* by polymerase chain reaction (PCR) (Table 8.2). PCR products amplified by these *Pi-ta* specific primers were cloned and sequenced. Sequence analysis confirmed the presence of the targeted portions of the dominant *indica Pi-ta* allele. These *Pi-ta* primers were used to examine the existence of *Pi-ta* alleles in advanced Arkansas rice breeding lines and in a larger set of U.S. rice germplasm. The *Pi-ta*-containing rice lines, as determined by PCR analysis, were resistant to both IB-49 and IC-17 in standard pathogenicity assays. In contrast, rice lines lacking the resistant *Pi-ta* allele were susceptible to both IB-49 and IC-17. The presence of *Pi-ta* markers was correlated with *Pi-ta* resistance. These DNA markers are ideal for selection of progeny from crosses of *Pi-ta*-containing parents.

For breeding purposes, massive numbers of progeny are needed for screening, and automated genotyping machines are often used for this type of analysis. However, these markers for *Pi-ta* are not suitable for fluorescence-based fragment analysis. We compared the *Pi-ta* allele from resistant and susceptible rice cultivars and identified a single-nucleotide length polymor-

TABLE 8.2. Sequences of 19- or 20-mer oligonucleotide primers for *Pi-ta*-specific dominant markers.

Locus[a]	Primer	Sequence (5'-3') (number of bp)	Location of *Pi-ta*[b]
Pi-ta$_{403}$	YL100	CAATGCCGAGTGTGCAAAGG$_{(20)}$	6257-6276
	YL102	TCAGGTTGAAGATGCATAGC$_{(20)}$	6659-6640
Pi-ta$_{440}$	YL153	CAACAATTTAATCATACACG$_{(20)}$	2021-2040
	YL154	ATGACACCCTGCGATGCAA$_{(19)}$	2460-2442
Pi-ta$_{1042}$	YL155	AGCAGGTTATAAGCTAGGCC$_{(20)}$	4409-4428
	YL87	CTACCAACAAGTTCATCAAA$_{(20)}$	5450-5431

Source: Adapted from Jia et al. (2002).

[a]The subscript number refers to the size in bp of the amplified dominant product.

[b]The location of the *Pi-ta* gene is based on GenBank accession no. AF207842.

phism (SNLP) in the intron region of the *Pi-ta* gene. Figure 8.3 shows the location and primers for this SNLP, and Figure 8.4 shows how the data were analyzed using an automated sequencing/genotyping instrument (M. Redus and Y. Jia, unpublished data). The DNA primer specific to the resistant *Pi-ta* allele was labeled with blue dye as a forward primer, the DNA primer specific to the susceptible *pi-ta* allele was labeled with green dye as another forward primer, and the DNA primer shared between both *Pi-ta* and *pi-ta* alleles was unlabeled as the reverse primer. PCR using these three primers produced a 181-bp blue peak in homozygous resistant, a green peak of 182-183 bp in homozygous susceptible, and both peaks in heterozygous plants. The utility of this SNLP marker in an automated genotyping system was verified using a segregating F_2 population and pathogenicity testing. Fourteen rice cultivars were confirmed to possess *Pi-ta* in a rapid survey of 137 U.S. rice cultivars using a *Pi-ta* gene marker (Table 8.3). These *Pi-ta*-containing rice lines are also resistant to IB-49 and IC-17. Thus, these *Pi-ta* gene markers provide a basis for stacking *Pi-ta* with other blast *R* genes into improved germplasm.

Transgenic Strategies Using the **Pi-ta** Gene

As discussed in the previous section, the natural variation in the *Pi-ta* gene can be used to develop allele-specific markers to accelerate conventional crop improvement. Another attractive feature of *Pi-ta* cloning is our ability to transfer *Pi-ta* into diverse rice cultivars by genetic transformation. A trans-

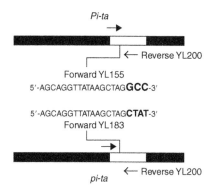

FIGURE 8.3. Schematic location of SNLP of the *Pi-ta* gene and sequences of gene-specific primers for the resistant and susceptible *pi-ta* alleles are shown. *Source:* M. Redus and Y. Jia, Unpublished data.

genic approach is the fastest method to introduce *Pi-ta* and is particularly attractive for *Pi-ta* due to its centromeric location. Skewed physical to genetic distance at the *Pi-ta* locus will exacerbate linkage drag during introgression by molecular breeding. Concerns for the safety of genetically modified crops and intellectual property issues currently limit transgenic strategies. Specific concerns need to be resolved before this potentially valuable approach can be widely accepted.

Relationship Between Pi-ta *and* Pi-ta^2

As determined by different rice blast isolates, *Pi-ta*^{2} is another *R* gene that cosegregates with *Pi-ta* (Kiyosawa, 1971; Rybka et al., 1997). It has not been confirmed whether the *Pi-ta*^{2} gene is an allele of *Pi-ta* or another *R* gene located within that region (Kiyosawa, 1971; Rybka et al., 1997; Valent et al., 2001). By sequencing, we identified the resistant *Pi-ta* allele in *Pi-ta*^{2}-containing cultivars *Tetep, Tadukan, Reiho,* and *Katy* (Bryan et al., 2000; Jia et al., 2003a). We then used two pairs of well-characterized strains of the rice blast pathogen to address the relationship of *Pi-ta* and *Pi-ta*^{2}. The strain O-137, which contains *AVR-Pita,* is avirulent on both *Pi-ta* and *Pi-ta*^{2} rice cultivars, whereas the spontaneous O-137 mutant CP3337, lacking *AVR-Pita,* has gained the ability to infect both genotypes of host. Thus, *AVR-Pita* in strain O-137 is responsible for triggering resistance in *Pi-ta*^{2} plants as well as in *Pi-ta* plants, and O-137 lacks other *AVR* genes recognized by those cultivars.

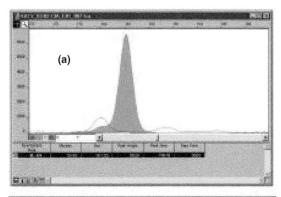

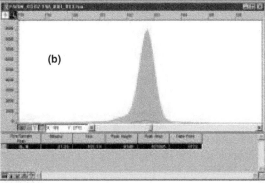

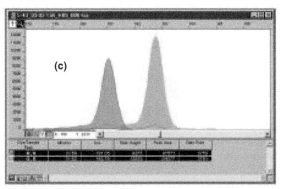

FIGURE 8.4. Different locations of peak (visualized using a standard sequencing and genotyping instrument) indicate different alleles were amplified. (a) homozygous *Pi-ta/Pi-ta,* (b) homozygous *pi-ta/pi-ta,* and (c) heterozygous *Pi-ta/pita. Source:* M. Redus and Y. Jia, unpublished data. See corresponding Plate 8.4 in the central gallery.

TABLE 8.3. A survey of the U.S. germplasm for the *Pi-ta* gene.

Pi-ta	*pi-ta*
Ahrent, Drew, Katy, Gui Chow, Dom Siah, IR36, IR64, *Madison, Kaybonnet, Kho Dauk Mali,* PI-5, *Te-Qing, Tetep, Tadukan*	A201, A301, *Amber, Arborio, Arkrose,* Azucena, B8462T3-710, *Baldo, Basmati, Bluebonnet 50, Bluebonnet, Bengal, Bellemont, Belle-Patna, Bluebelle, Bluerose, Bonnet 73, Bolivar, Brazos, Cadet, Calrose, Calikikari, Caloro, Carolina gold,* Co39, *Cocodrie,* Ci5309, Ci9122, Ci9515, Ci9545, Ci9881, Ci9902, *Colusa, Century-Patina 231, Cypress, Dawn, Delitus, Dellmati, Dellmont, Dixiebelle, Doon-Gara,* DM-2, *Dragon eyeball, Fortuna, Gulfmont, Gulfrose, Hill* long grain, *Hill* selection long grain, IR8, IRGA409, IRGA410, IRGA411, IRGA 415, *Jacinto, Jasmine 85, Jefferson, Koshihikari,* L201, L202, L205, *Labelle, Lacrosse, Lady Wright, LaGrue, Leah, Lebonnet, Lemont, Lafitte,* LSBR-33, M201, M202, M204, M401, *Magnolia, Mars, Maybelle, Mercury, Myling, Newbonnet, Newrex, Ngakyauk, Nipponbare, Nira, Norin-29, Nortai, Northrose, Nova, Nova66, Orion, Panda, Pecos, Pelde,* Pi331581, Pi408449, *Priscilla, Puntal, Reimi, Rexark, Rexmont, Rexoro, Rico 1, Rojele, Rosemont, Saber, Sadri, Saturn, Shomed,* Short-stemmed *Starbonnet, Somewake, Starbonnet,* STG533187, STG59D1350, STG697717, *Tainun Iku 487, Tebonnet, Toro2,* TP49, *Vegold, Vista, Wells, Zenith,* Zhe733

Source: Z. Wang and Y. Jia, unpublished data.

The presence of the *Pi-ta* gene was determined by a dominant marker for the resistant *Pi-ta* gene (Jia et al., 2002), and the presence of the *pi-ta* gene was determined by a dominant marker for the susceptible *pi-ta* gene (Jia et al., 2004).

    Another Chinese field isolate, O-135, which lacks sequences that hybridize to *AVR-Pita*, failed to trigger *Pi-ta*-mediated resistance in *Yashiro-mochi* and K1. However, O-135 is avirulent on *Pi-ta²* cultivars, and a spontaneous mutant of O-135, named CP753, has gained virulence toward these cultivars. Thus, the O-135/CP753 isogenic blast strains differ in a novel *AVR* gene, corresponding to a second *R* gene that correlates with *Pi-ta²* specificity in *Reiho, Katy,* and *Drew* (Jia et al., 2003a, B. Valent, unpublished data). Taken together, these results suggest that the broader spectrum *Pi-ta²* gene is a combination of *at least* two genes: *Pi-ta* and the second *R* gene, which were not separated in a total of at least 2,000 F_2 segregating progeny involving *Pi-ta* and *Pi-ta²* cultivars (Wang and Jia, unpublished data). This is interesting in light of reports that *Pi-ta* is required for the function of *Pi-ta²* (Kiyosawa, 1967, 1971; Rybka et al., 1997; Silué et al., 1993).

Southern blot analysis has indicated that *Pi-ta* is a single-copy gene in rice, and no obvious candidate for *Pi-ta²* was identified from sequencing the *Pi-ta* contig. Genetic distance near the centromere provides misleading estimates of physical distance (Chen et al., 2002), and novel strategies may be required for cloning *Pi-ta²* (Wang et al., 2002). Regardless of where *Pi-ta²* is located, the *Pi-ta* and *Pi-ta²* genes tend to act together for resistance. Other *R* genes, including *Pi-4ᵃ(t)*, *Pi-4ᵇ(t)*, *Pi-6(t)*, and *Pi-12(t)*, also map to the centromeric region of chromosome 12 (Inukai et al., 1994; Yu et al., 1996; Zheng et al., 1996). The cultivar *Katy* was previously reported to contain a tightly linked cluster of at least seven *R* genes that mapped in the same region as *Pi-ta* and *Pi-ta²* (Chao et al., 1999; Moldenhauer et al., 1992). The availability of the rice genome sequence will facilitate the analysis of other *R* genes in this complex chromosomal region.

MAGNAPORTHE GRISEA AVR-Pita *GENE*

AVR-Pita belongs to the first *AVR* gene-*R* gene pair to be cloned from the rice blast system. This fungal *AVR* gene was originally called *AVR2-YAMO*, because it conferred avirulence toward cultivar *Yashiro-mochi* (Valent, 1997). The gene was identified by genetic crosses in the 1980s, but cloning it required first building an RFLP map of the fungus (Sweigard et al., 1993). *AVR-Pita* mapped near a telomere of chromosome 3 in the integrated *M. grisea* map (Nitta et al., 1997). This *AVR* gene was determined to be directly adjacent to its telomere by analysis of spontaneous virulent mutants that simultaneously lost AVR function and showed size polymorphisms in their telomeric end fragments (Orbach et al., 2000). A challenge for the cloning of *AVR-Pita* was its telomeric location, as it was too near the chromosome end to be included in standard genomic libraries. Moreover, sequences directly adjacent to the telomeric repeats were required for *AVR-Pita* function, but the telomeric repeat sequence itself complicated functional analysis by fungal transformation and complementation (Orbach et al., 2000).

AVR-Pita *is a Putative Zinc Metalloprotease*

The gene-for-gene theory predicts that the efficacy of an *R* gene depends on the biological function of and potential for mutation in its corresponding *AVR* gene. *AVR-Pita* encodes a 223-amino acid protein that possesses features characteristic of fungal secreted metalloproteases (Orbach et al., 2000). Homology between the AVR-Pita protein and the metalloprotease NpII from *Aspergillus oryzae* is confined to the C-terminal 176 amino acids, corresponding to the mature region of the NPII pre-pro-enzyme (Tatsumi et al., 1991). Amino acids 173-182 of AVR-Pita correspond to a motif found in neutral zinc metalloproteases (Figure 8.5). We predicted that AVR-Pita may contain a pre-pro region that is removed to produce a mature AVR-Pita protease. Mutational analysis of *AVR-Pita* indicates that maintenance of the intact protease motif is important for avirulence function (G. T. Bryan and B. Valent, unpublished results).

AVR-Pita *Acts Inside the Host Cell*

We investigated the potential site of action of *AVR-Pita* relative to its host cell. Various forms of recombinant AVR-Pita protein were expressed in *E. coli,* purified using affinity columns, and introduced into intercellular spaces of rice by vacuum infiltration or other apoplastic application methods. *Pi-ta*-containing leaf tissue failed to respond with a hypersensitive response (HR), suggesting that the presence of AVR-Pita outside its host cell was not sufficient to induce the *Pi-ta*-mediated HR (G. T. Bryan and B. Valent, unpublished results).

We used a biolistic transient expression assay to investigate if *AVR-Pita* can trigger *Pi-ta*-mediated resistance when expressed inside rice cells (Jia et al., 2000). For this assay, a GUS reporter gene (*E. coli uidA* linked to the constitutive 35S promoter from cauliflower mosaic virus) was used to transform approximately 1-week-old intact rice seedlings by biolistic bombardment. Bombarded cells expressed the β-glucuronidase (GUS) enzyme, which is detected histochemically. GUS activity is detected in live plant

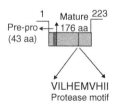

FIGURE 8.5. Predicted structure of the AVR-Pita protein including the protease motif. The pre-proamino region contains 43 amino acids and the putative mature protein contains 176 amino acids.

cells, but not in cells that have undergone hypersensitive cell death. The full-length *AVR-Pita* gene and *AVR-Pita$_{176}$* encoding the predicted mature protease (Figure 8.5) were engineered for direct expression in rice cells. Each of these AVR-Pita constructs was co-bombarded along with the GUS reporter gene in the transient assay. We observed that GUS activity was inhibited whenever *AVR-Pita$_{176}$* was expressed in rice cells containing *Pi-ta*. This inhibition of GUS activity was observed regardless of whether the experiment was done with rice cells expressing the native *Pi-ta* gene, or whether a *Pi-ta* transgene was transiently expressed along with the GUS reporter and *AVR-Pita$_{176}$* genes in cultivars lacking *Pi-ta*. These results suggest that AVR-Pita recognition occurs inside rice cells and that the predicted mature AVR-Pita$_{176}$ protein is the active form in this assay. HR was not observed when virulent forms of *avr-pita$_{176}$* were used in the transient assay (Bryan et al., 2000; Jia et al., 2000). These findings suggest that AVR-Pita protein is secreted into host cells during or after penetration where it functions to induce *Pi-ta*-mediated HR.

AVR-Pita *Protein Binds to the* Pi-ta *Protein*

The finding that *AVR-Pita* induces *Pi-ta*-mediated HR when expressed inside rice cells suggested that AVR-Pita may bind to the Pi-ta protein. We tested the hypothesis that the Pi-ta protein might be a receptor for the AVR-Pita elicitor using the yeast two-hybrid system and in vitro binding assays (Jia et al., 2000).

The full-length and mature AVR-Pita$_{176}$ proteins were tested for interaction with the full-length Pi-ta protein or with a Pi-ta LRD polypeptide in the yeast two-hybrid system. When the AVR-Pita$_{176}$ fusion protein together with the Pi-ta LRD fusion protein were expressed together in yeast, they interacted to produce a functional transcription factor that activated the expression of two reporter genes. This suggested that the AVR-Pita$_{176}$ protein binds directly to the LRD region from Pi-ta. Interaction was not detected with the susceptible LRD region with serine instead of alanine at amino acid position 918 (LRD^{A918S}) or with virulent avr-pita$_{176}$ proteins. Interaction specificity in the two-hybrid system is in agreement with the functional polymorphism of the Pi-ta protein.

We next verified the physical interaction between the Pi-ta LRD and AVR-Pita$_{176}$ on a membrane using an in vitro technique called a *Far-Western*. Recombinant AVR-Pita protein was separated using SDS-PAGE gel electrophoresis, immobilized on a membrane, and allowed to refold in the presence of ZnCl$_2$ at 4°C. Membranes containing refolded AVR-Pita proteins were incubated with crude extracts of *E. coli* total proteins containing soluble S-tagged Pi-ta LRD polypeptides and rinsed extensively. The Pi-ta LRD polypeptides that bound to AVR-Pita polypeptides on the membrane were detected with the S-antibody (Novagen). The Pi-ta LRD polypeptide bonded specifically to AVR-Pita$_{176}$ and not to longer forms of the AVR protein. The pi-ta LRD^{A918S} polypeptide did not bind to AVR-Pita$_{176}$ under the same conditions. We further showed that the Pi-ta protein could bind to the AVR-Pita$_{176}$ polypeptide, although in this case the specificity of binding was not maintained. Based on the data from transient assays, the yeast two-hybrid interaction, and the in vitro binding assays, we suggest that the AVR-Pita$_{176}$ protein binds to the Pi-ta protein within the invaded plant host cell. Consistent with the prediction obtained from allelic variation of the *Pi-ta* gene discussed earlier, the amino acid at position 918 in the LRD region is a critical residue in determining the function of the Pi-ta protein and its ability to interact with the elicitor AVR-Pita$_{176}$.

Allelic Variation of AVR-Pita Alleles

Our analysis indicates that *Pi-ta* is effective for rice blast control in the United States (Jia et al., 2004). We predicted that the common U.S. races IB-49 and IC-17 contained functional *AVR-Pita* genes. We sequenced an allele of *AVR-Pita* from IB-49 isolate ZN61 and from IC-17 isolate ZN57 (P. Singh and Y. Jia, unpublished data; Correll et al., 2000). *AVR-Pita* gene-specific DNA primers were designed to amplify the orthologous *AVR-Pita* alleles. DNA fragments of 1,086 bp containing the *AVR-Pita* coding region were amplified by PCR from fungal genomic DNAs using these primers, and the PCR products were cloned and sequenced (P. Singh and Y. Jia, unpublished data). DNA sequences of the *AVR-Pita* genes from IB-49 and IC-17 were identical. Each encodes an open reading frame interrupted by three introns, as in *AVR-Pita* derived from O-137. Alignment of amino acid sequences encoded by the *AVR-Pita* alleles from IB-49 and IC-17 with the sequence encoded by the O-137 *AVR-Pita* identified an insertion at position 7 and amino acid substitutions at positions 82, 88, 104, 136, and 174 (Figure 8.6) (Orbach et al., 2000; P. Singh and Y. Jia, unpublished data). Isolate ZN61 for IB-49 belongs to the lineage group C, and isolate ZN57 for IC-17 belongs to lineage group A, based on DNA fingerprinting using MGR586 (Correll et al., 2000). Rice blast populations are commonly composed of dozens of lineages, with each lineage derived through clonal propagation from a common ancestor in major rice-growing areas (Levy et al., 1991, 1993; Xia et al., 1993; Zeigler et al., 1994). Allelic diversity in *AVR-Pita* alleles has been documented from Columbian lineages (Montenegro-Chamorro 1997; Valent et al., 2001). In contrast, all orthologous

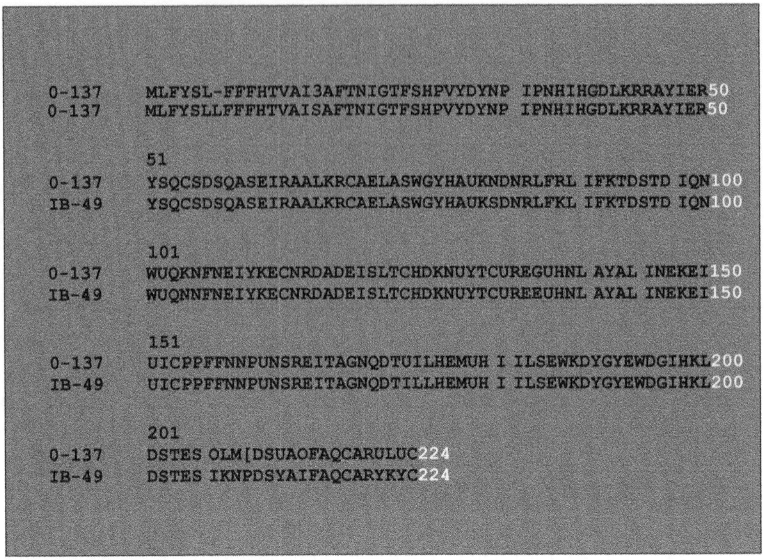

```
0-137    MLFYSL-FFFHTVAI3AFTNIGTFSHPVYDYNP IPNHIHGDLKRRAYIEF50
0-137    MLFYSLLFFFHTVAISAFTNIGTFSHPVYDYNP IPNHIHGDLKRRAYIEF50


         51
0-137    YSQCSDSQASEIRAALKRCAELASWGYHAUKNDNRLFRL IFKTDSTD IQN100
IB-49    YSQCSDSQASEIRAALKRCAELASWGYHAUKSDNRLFKL IFKTDSTD IQN100


         101
0-137    WUQKNFNEIYKECNRDADEISLTCHDKNUYTCUREGUHNL AYAL INEKEI150
IB-49    WUQNNFNEIYKECNRDADEISLTCHDKNUYTCUREEUHNL AYAL INEKEI150


         151
0-137    UICPPFFNNPUNSREITAGNQDTUILHEMUH I ILSEWKDYGYEWDGIHKL200
IB-49    UICPPFFNNPUNSREITAGNQDTILLHEMUH I ILSEWKDYGYEWDGIHKL200


         201
0-137    DSTES OLM[DSUAOFAQCARULUC224
IB-49    DSTES IKNPDSYAIFAQCARYKYC224
```

FIGURE 8.6. Amino acid alignment of AVR-Pita proteins from IB-49 and 0-137 (Amino acid sequences of all other U.S. races dispersed in four lineages are indentical). See corresponding Plate 8.6 in the central gallery.

AVR-Pita alleles from 14 U.S. isolates, including IB-49 (ZN61) and IC-17 (ZN57) are identical (P. Singh and Y. Jia, unpublished data). It is of interest to analyze *AVR-Pita* structures from naturally occurring "race-shift" isolates that gain virulence toward rice with *Pi-ta*.

CURRENT CHALLENGES IN RICE BLAST RESEARCH

AVR-Pita's Dual Role in Rice Blast Biology

The products of *AVR* genes like *AVR-Pita* are produced by the pathogen for its own purposes, and they just happen to be "recognized" in the presence of a particular *R* gene. It is this other role of AVR genes—most likely in promoting pathogenicity—that determines the durability of their corresponding *R* genes. *AVR* genes might be lost completely from particular races if this loss does not result in a fitness penalty to the pathogen, resulting in the "breakdown" of its *R* gene. Or mutations might eliminate the recognition function in triggering resistance but not affect the biological role for the pathogen. Alternatively, essential genes without potential to lose recognition and retain their essential function should correspond to durable *R* genes. Molecular understanding of both roles of *AVR* genes will contribute to future disease control strategies.

Rice blast disease has become a powerful model system for understanding how a fungus penetrates into plant tissue using a highly elaborate appressorium, and many genes involved in appressorium formation and penetration have been characterized (Talbot, 2003). Still, little is known about the genes required for growth in the plants after penetration. Successful infection by the hemibiotrophic rice blast fungus (a compatible interaction) requires an initial biotrophic interaction in which the pathogen produces bulbous invasive hyphae inside living rice cells (Heath et al., 1992; Koga, 2001). Once established inside the plant, the fungus switches to necrotrophic growth, killing plant cells and ramifying throughout the tissue. Figure 8.7 shows fungal invasive hyphae inside rice epidermal cells 48 hours after in-

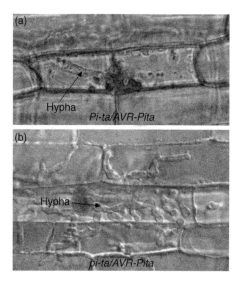

FIGURE 8.7. Light micrographs of *M. grisea* in the rice leaf sheath tissue 48 hours after inoculation. (a) *Pi-ta* containing rice cultivar inoculated with *M. grisea* with *AVR-Pita;* invaded epidermal cell shows HR and hypha fails to grow. (b) *pi-ta* containing rice cultivar inoculated with *M. grisea* containing *AVT-Pita;* invasive hyphae fill the first invaded epidermal cells by 36 hours and then spread in several directions at once into neighboring cells. *Source:* P. Kankanala and B. Valent, unpublished data.

oculation with *M. grisea* with *AVR-Pita*. Contrasting outcomes are shown depending on the presence or absence of *Pi-ta*. The evidence that *AVR-Pita* acts inside host cells raises the question of when and how the mature AVR-Pita protein is delivered inside the host cell. The decision between a compatible or *AVR-Pita/Pi-ta*-induced incompatible interaction appears to occur soon after penetration (24-36 hours post inoculation), suggesting that *AVR-Pita* is delivered into the plant cell at this time. The cellular mechanisms by which invasive hyphae grow within apparently healthy plant cells are poorly understood, although it is clear that plant membrane integrity must be maintained in the early biotrophic phase for the cell to be alive. It is not known if the fungal invasive hypha invaginates or breaches the plant plasma membrane as it grows within the plant cell (Heath et al., 1992; Koga, 2001). The mechanism of delivery of *AVR-Pita* into the plant cytoplasm, where it can interact with a cytoplasmic *R* gene, depends on the nature of this fungus-plant cell interface.

Another important question concerns the timing of the host plant's initial perception of the fungus. We are beginning to address this question by determining the timing of defense gene expression at different intervals after inoculation with *M. grisea* containing *AVR-Pita*. Although the cause-effect relationship of these defense genes in the *Pi-ta*-mediated defense pathway is unclear, the accumulation of defense gene transcripts suggests that the plant has recognized and responded to the pathogen. As shown in Figure 8.8, transcripts of phenylalanine ammonia lyase (*PAL*) and β-glucanase were observed perhaps by six hours and certainly by 16 hours after inoculation. PR-1 accumulation was obvious 24 hours after inoculation. Using ultraviolet light, the accumulation of autofluorescent material is visible at the macroscopic level by 24 hours after inoculation in the incompatible interaction, whereas autofluorescent material is not visible in the compatible interaction (Figure 8.9). The presence of autofluorescence is an indication of the rapid cell death in the hypersensitive reaction. In general, *M. grisea* conidia form appressoria and begin penetration by 24 hours after inoculation (Y. Jia and B. Valent, unpublished data). Thus, the timing of differential defense gene expression suggests that the plant may perceive the fungus before penetration.

Localized Blast Inoculation Method

Intracellular invasive hyphae are morphologically distinct from the filamentous hyphae produced on agar media and from the intercellular hyphae produced in the necrotrophic stages of lesion formation. Future research focus will be on gene expression in the invasive hyphae in order to identify

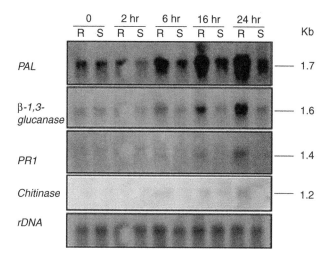

FIGURE 8.8. Analysis of rice pathogenesis-related gene *(PR)* expression in compatible and incompatible interactions. R indicates resistance and S indicates susceptible. Leaves from two-week old plants of both YT14 (*Pi-ta*) and YT16 (*pi-ta*) were inoculated with 2.5 × 10^6 spore/ml conidial suspensions of *M. grisea* expressing *AVR-Pi-ta*. Leaf tissues were harvested at the indicated time after inoculation. Total RNA was extracted, blotted onto hybond N, and hybridized with radiolabeled probes. The same blot was hybridized with genes encoding PAL, β-1.3-glucanase, PR-1, chitinase and rDNA. Between each probing, the blot was stripped and the removal of the previous probe was verified. The experiments were repeated. The sizes of the transcripts (Kb) are indicated. *Source:* Y. Jia and B. Valent, unpublished data.

fungal molecules critical to establishing successful infection. Inherent non-uniform dispersal of developing fungal infection sites prevents the prediction of the locations of these sites in inoculated rice tissue in the first four to five days, before macroscopic symptoms develop. This disadvantage makes detecting differentially expressed genes difficult, because mRNA associated with the developing lesions (particularly biotrophic fungal invasive hyphae) will be in low abundance relative to healthy plant RNA at early infection stages. A novel spot inoculation method resulting in dense infection sites within a highly contained area will facilitate studies of molecular interactions between rice and *M. grisea* (Figure 8.10) (Jia et al., 2003b). For the development of this assay, serially diluted Tween-20 was added to *M. grisea* conidial suspensions in 0.25% (w/v) gelatin to promote adherence of conidia to detached rice leaves in petri dishes. Standard pathogenicity assays indicated no

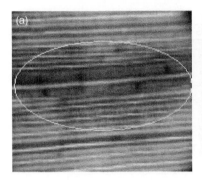
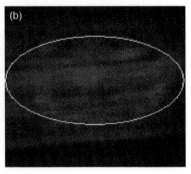

FIGURE 8.9. Inoculated area was shown in the oval. Under bright light (a), and the presence of autoflourescence in Katy 24 hours after inoculation with *M. grisea* IC-17 (b). Ten microliters of spore suspension containing 100 spores were spot inoculated and pictures were taken 24 hours after inoculation. *Source:* Y. Jia, unpublished data. See corresponding Plate 8.9 in the central gallery.

0   1   2   3

FIGURE 8.10. Localized detached leaf inoculation. Five microliters of spore suspension was inoculated on detached rice leaves (left). Resistance: 0-2, Susceptible: 3. Adapted from Jia et al., 2003, *Plant Disease* 87: 129-133. See corresponding Plate 8.9 in the central gallery.

deleterious effects of Tween-20 on blast development, and 0.02% (v/v) Tween-20 was necessary for adherence of spore suspensions to the detached leaves. Disease reactions of the host were determined 6-10 days after inoculation. The spot inoculation method was evaluated using three well-characterized races of *M. grisea* and validated with standard pathoge-

nicity assays in the greenhouse. Disease reactions of rice plants to four predominant races of *M. grisea* were tested concurrently using spot inoculation and standard pathogenicity assays. This assay is complementary to
standard pathogenicity assays that emphasize overall disease reactions. It is
also potentially useful for those laboratories that handle exotic isolates and
for those that evaluate pathogenicity of different *M. grisea* isolates in the
same plant.

A Model of Pi-ta-Mediated Resistance

The current model is that a mature AVR-Pita protease triggers *Pi-ta*-
mediated resistance by binding directly to the cytoplasmic Pi-ta receptor
protein (Figure 8.11). In this model, the AVR-Pita protein is processed and
transferred into plant cells by unknown mechanisms during penetration.
Binding of the elicitor by the Pi-ta protein activates defense pathways and
leads to effective resistance. It is unknown if there are other proteins (e.g.,
kinases) involved in the initiation of the defense response and whether or
not the mature AVR-Pita$_{176}$ is an active protease that cleaves Pi-ta or other
host proteins (Figure 8.11).

The NBS-LRR type of *R* gene represents a majority among the 48 *R* genes
cloned from diverse plant species (Bent, 1996; Hammond-Kosack and Jones,
1997; Martin et al., 2003; Wang, 2003). *Pi-ta* is the first NBS-LRR type *R*
gene to support the ligand-receptor model of the gene-for-gene theory. Ini-

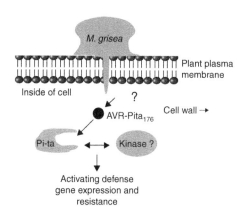

FIGURE 8.11. A model of the *Pi-ta* gene mediated resistance. The pi-ta protein
acts as a putative cytoplasmic receptor that binds to the elicitor AVR-Pita$_{176}$ acti-
vating defense genes and leading to resistance.

tial support of the ligand-receptor model came from the study of molecular interactions of a tomato R gene with an AVR gene of *Pseudomonas syringae* pv. *tomato*. The tomato *Pto* gene encodes an active protein kinase that binds to Avr-Pto, an effector of *Pseudomonas syringae* pv. *tomato,* triggering the *Pto*-mediated defense response (Scofield et al., 1996; Tang et al., 1996). In the tomato system, another gene, *Prf,* encoding a predicted NBS-LRR protein is required in the *Pto* pathway (Salmeron et al., 1996). Despite the difficulty of demonstrating the role of *Prf* in the *Pto/AVR-Pto* complex, comprehensive expression profiling of the *Pto*-mediated defense response indicates that *Prf* can also function as an independent host recognition determinant of bacterial infection (Mysore et al., 2002). In *Arabidopsis,* the resistance gene *PRs1-R* is an NBS-LRR type R gene, and the product of *PRs1-R* was shown to interact with the product of avirulence gene *PopP1* in the yeast split-ubiquitin two-hybrid system (Deslandes et al., 2003). The technical difficulty of demonstrating the physical interaction of R and AVR genes in other systems has led to a *guard hypothesis,* in which R gene products are proposed to "guard" virulent targets (Van der Biezen and Jones, 1998; for comments, see review by Martin et al., 2003).

FUTURE PERSPECTIVES

In the interaction of *Pi-ta* and *AVR-Pita,* the most important challenges lie ahead: in defining the complete signal transduction pathway downstream of the Pi-ta protein, in the biochemical demonstration of AVR-Pita protease activity, in defining AVR-Pita's natural substrate and its role in establishing disease, and in determining when and how the AVR-Pita protein enters the plant cell during infection. Recent efforts have focused on developing robust blast infection assays in vitro, which would allow the detection of genes that are involved in the earliest stages of the rice-blast interaction. Techniques such as DNA microarray (Schena et al., 1995), SAGE (Matsumura et al., 1999), and analysis of genes isolated from subtracted cDNA libraries (Xiong et al., 2001) will facilitate understanding of differentially expressed genes. An improved yeast two-hybrid system is being employed to identify additional interacting proteins, and mutant identification in the *Pi-ta* pathway using genetic screening of mutagenized populations has been undertaken. A total of 20,000 putative M_2 mutant lines of *Pi-ta-* and *Pi-ta²*-containing cultivar *Katy* induced by fast neutrons and ethyl methane sulfonate (EMS) were screened, and 44 putative blast-susceptible mutants were recovered from the fast neutron treatment. The verification of blast susceptibility in M_3 is in progress. Confirmed mutants will be useful for identifying additional components in the *Pi-ta* pathway and understanding structure-function relationships of *Pi-ta*. In the long term, understanding the molecular basis for *AVR* gene triggering of resistance will lead to the development of novel strategies for broad-spectrum, durable disease control using the native plant defense systems.

In the short term, a study of the structure of the *AVR-Pita* alleles in the rice blast field populations should allow a better understanding of the stability of resistance conferred by *Pi-ta* in the rice cultivars that are currently grown. In the southern United States, *Pi-ta* has been effective since its release in the 1980s. The relative longevity of *Pi-ta* provides clues to investigate the role of *AVR-Pita* in the pathogenicity process. Analysis of rice blast populations worldwide certainly will enhance our understanding of the evolution of the pathotypes of the rice blast fungus in rice-growing areas. Furthermore, continued cloning and characterization of other matched *R-AVR* gene pairs from the rice blast system (Farman et al., 2002) should help to understand how blast *R* genes work. Such efforts will help to understand sophisticated molecular interaction mechanisms between the rice blast fungus and plant cells. The resulting knowledge will contribute to the development of improved resistant cultivars by both conventional and novel strategies.

REFERENCES

Bent, A. F. (1996). Plant disease resistance genes: Function meets structure. *Plant Cell 8:* 1757-1771.

Bryan, G. T., K. S. Wu, L. Farrall, Y. Jia, H. P. Hershey, S. A. McAdams, K. N. Faulk, G. K. Donaldson, R. Tarchini, and B. Valent (2000). A single amino acid difference distinguishes resistant and susceptible alleles of the rice blast resistance gene *Pi-ta*. *Plant Cell 12:* 2033-2045.

Chao, C. T., K. A. K. Moldenhauer, and A. H. Ellingboe (1999). Genetic analysis of resistance/susceptibility in individual F_3 families of rice against strains of *Magnaporthe grisea* containing different genes for avirulence. *Euphytica 109:* 183-190.

Chen, M., G. Presting, W. B. Barbazuk, J. L. Goicoechea, B. Blackmon, G. Fang, H. Kim, D. Frisch, Y. Yu, S. Sun, et al. (2002). An integrated physical and genetic map of the rice genome. *Plant Cell 14:* 537-545.

Correll, J. C., T. L. Harp, J. C. Guerber, R. S. Zeigler, B. Liu, R. D. Cartwright, and F. N. Lee (2000). Characterization of *Pyricularia grisea* in the United States using independent genetic and molecular markers. *Phytopathology 90:* 1396-1404.

Dangl, J. L., and J. D. G. Jones (2001). Plant pathogens and integrated defense response to infection. *Nature 411:* 826-833.

Deslandes, L., J. Olivierier, N. Peeters, D. X. Feng, M. Khounlotham, C. Boucher, I. Somssich, S. Genin, and Y. Marco (2003). Physical interaction between RRS1-

R, a protein conferring resistance to bacterial wilt, and PopP2, a type III effector targeted to the plant nucleus. *Proceedings of the National Academy of Sciences of the USA 100:* 8024-8029.

Dong, F., J. T. Miller, S. A. Jackson, G. L. Wang, P. C. Ronald, and J. Jiang (1998). Rice *(Oryza sativa)* centromeric regions consist of complex DNA. *Proceedings of the National Academy of Sciences of the USA 95:* 8135-8140.

Farman, M. L., Y. Eto, Y. Nakao, Y. Tosa, H. Nakayashiki, S. Mayama, and S. A. Leong (2002). Analysis of the structure of the *Avr1-CO39* avirulence locus in virulent rice-infecting isolates of *Magnaporthe grisea*. *Molecular Plant-Microbe Interactions 15:* 6-16.

Flor, H. H. (1971). Current status of the gene-for-gene concept. *Annual Review of Phytopathology 9:* 275-296.

Gowda, M., R. C. Venu, K. Roopalakshmi, M. V. Sreerekha, and R. S. Kulkarni (2003). Advances in rice breeding, genetics and genomics. *Molecular Breeding 11:* 337-352.

Hammond-Kosack, K. E., and J. D. G. Jones (1997). Plant disease resistance genes. *Annual Review of Plant Physiology and Plant Molecular Biology 48:* 575-607.

Heath, M. C., R. J. Howard, B. Valent, and F. G. Chumley (1992). Ultrastructural interactions of one strain of *Magnaporthe grisea* with goosegrass and weeping lovegrass. *Canadian Journal of Botany 70:* 779-787.

Inukai, T., R. J. Nelson, R. S. Zeigler, S. Sarkarung, D. J. Mackill, J. M. Bonman, I. Takamure, and T. Kinoshita (1994). Allelism of blast resistance genes in near-isogenic lines of rice. *Phytopathology 84:* 1278-1283.

Ise, K. (1993). A close linkage between blast gene Pi-ta^2 and a marker on chromosome 12 in *japonica* rice. *International Rice Research Institute Notes 18:* 14.

Jia, Y. (2003). Marker assisted selection for the control of rice blast disease. *Pesticide Outlook 14:* 150-152.

Jia, Y., G. T. Bryan, L. Farrall, and B. Valent (2003a). Natural variation at the *Pi-ta* rice blast resistance locus. *Phytopathology 93:* 1452-1459.

Jia, Y., S. A. McAdams, G. T. Bryan, H. P. Hershey, and B. Valent (2000). Direct interaction of resistance gene and avirulence gene products confers rice blast resistance. *EMBO Journal 19:* 4004-4014.

Jia, Y., B. Valent, and F. N. Lee (2003b). Determination of host responses to *Magnaporthe grisea* on detached rice leaves using a spot inoculation method. *Plant Disease 87:* 129-133.

Jia, Y., Z. Wang, R. G. Fjellstrom, K. A. K. Moldenhauer, M. A. Azam, J. Correll, F. N. Lee, Y. Xia, and J. N. Rutger (2004). Rice *Pi-ta* gene confers resistance to the major pathotypes of the rice blast fungus in the United States. *Phytopathology 94:* 296-301.

Jia, Y., Z. Wang, and P. Singh (2002). Development of dominant rice blast resistance *Pi-ta* gene markers. *Crop Science 42:* 2145-2149.

Kiyosawa, S. (1967). Inheritance of resistance of the rice variety Pi No. 4 to blast. *Japanese Journal of Breeding 17:* 165-172.

Kiyosawa, S. (1971). Genetic approach to the biochemical nature of plant disease resistance. *Japan Agricultural Research Quarterly 6:* 73-80.

Kiyosawa, S. (1984). Establishment of differential varieties for pathogenicity test of ice blast fungus. *Rice Genetics Newsletter 1:* 95-97.

Koga, H. (2001). Cytological aspects of infection by the rice blast fungus *Pyricularia oryzae*. In *Major fungal diseases of rice: Recent advances*, S. Sreenivasaprasad and R. Johnson (Eds.), pp. 87-110. Dordrecht, Netherlands: Kluwer.

Levy, M., F. J. Correa-Victoria, R. S. Zeigler, S. Xu, and J. E. Hamer (1993). Genetic diversity of the rice blast fungus in a disease nursery in Colombia. *Phytopathology 83:* 1427-1433.

Levy, M., J. Romao, M. A. Marchetti, and J. E. Hamer (1991). DNA fingerprinting with a dispersed repeated sequence resolves pathotype diversity in the rice blast fungus. *Plant Cell 3:* 95-102.

Martin, G. B., A. J. Bogdanove, and G. Sessa (2003). Understanding functions of plant disease resistance proteins. *Annual Review of Plant Biology 54:* 1-5.

Matsumura, H., S. Nirasawa, and R. Teruchi (1999). Transcript profiling in rice *(Oryza sativa)* using serial analysis of gene expression. *The Plant Journal 20:* 719-726.

Moldenhauer, K. A. K., A. O. Bastawisi, and F. N. Lee (1992). Inheritance of resistance in rice to races IB-49 and IC-17 of *Pyricularia grisea* rice blast. *Crop Science 32:* 584-588.

Moldenhauer, K. A. K., K. A. Gravois, F. N. Lee, R. J. Norman, J. L. Bernhardt, B. R. Well, R. H. Dilday, M. M. Blocker, P. C. Rohman, and T. A. McMinn (1998). Registration of 'Drew' rice. *Crop Science 38:* 896-897.

Moldenhauer, K. A. K., F. N. Lee, R. J. Norman, R. S. Helms, R. H. Well, R. H. Dilday, P. C. Rohman, and M. A. Marchetti (1990). Registration of 'Katy' rice. *Crop Science 30:* 747-748.

Montenegro-Chamorro, M. V. (1997). *Allellic diversity of an avirulence gene in Colombian field isolates of the rice blast fungus*. MS thesis, Purdue University, West Lafayette, IN.

Mysore, K. S., O. R. Crasta, R. P. Tuori, O. Folkerts, P. B. Swisky, and G. B. Martin (2002). Comprehensive transcript profiling of *Pto*-and *Prf*-mediated host defense response to infection by *Pseudomonas syrainae* pv. *tomato. The Plant Journal 32:* 299-315.

Nakamura, S., S. Asakawa, N. Ohmido, K. Fukui, N. Shimizu, and S. Kawasaki (1997). Construction of an 800-kb contig in the near-centromeric region of the rice blast resistance gene *Pi-ta²* using a highly representative rice BAC library. *Molecular and General Genetics 254:* 611-620.

Nitta, N., M. L. Farman, and S. A. Leong (1997). Genome organization of *Magnaporthe grisea*: Integration of genetic maps, clustering of transposable elements and identification of genome duplications and rearrangements. *Theoretical and Applied Genetics 95:* 20-32.

Orbach, M. J., L. Farrall, J. A. Sweigard, F. G. Chumley, and B. Valent (2000). The fungal avirulence gene *AVR-Pita* determines efficacy for the rice blast resistance gene *Pi-ta. Plant Cell 12:* 2019-2032.

Rossman, A. Y., R. J. Howard, and B. Valent (1990). *Pyricularia grisea,* the correct name for the rice blast fungus. *Mycologia 82:* 509-512.

Rybka, K., M. Miyamoto, I. Ando, A. Saito, and S. Kawasaki (1997). High resolution mapping of the *indica*-derived rice blast resistance genes. II. *Pi-ta²* and *Pi-ta* and a consideration of their origin. *Molecular Plant-Microbe Interactions 10:* 517-524.

Salmeron, J. M., G. E. D. Olyroyd, C. M. T. Rommens, S. R. Scofield, H. S. Kim, D. T. Lavelle, D. Dahlbeck, and B. J. Staskawicz (1996). Tomato *Prf* is a member of the leucine-rich repeat class of plant disease resistance genes and lies embedded within the *Pto* kinase gene cluster. *Cell 86:* 123-133.

Schena, M. D., D. Shalon, R. W. Davis, and P. O. Brown (1995). Quantitative monitoring of gene expression patterns with a complementary DNA microarray. *Science 270:* 467-470.

Scofield, S. R., C. M. Tobias, J. P. Rathjen, J. H. Chang, D. T. Lavelle, R. W. Michelmore, and B. J. Staskawicz (1996). Molecular basis of gene-for-gene specificity in bacterial speck disease of tomato. *Science 268:* 661-667.

Shigemura, S., and E. Kitamura (1954). Breeding of blast resistant cultivars with crossing of *japonica* and *indica* rices. *Journal of Agricultural Science Tokyo Nogyo Daigaku 9:* 321-323 (in Japanese).

Silué, D., J. L. Notteghem, and D. Tharreau (1992). Evidence for a gene-for-gene relationship in the *Oryza sativa-Magnaporthe grisea* pathosystem. *Phytopathology 82:* 577-580.

Silué D., D. Tharreau, and J. L. Notteghem (1993). Genetic control of *Magnaporthe grisea* avirulence to rice cultivar Pi-No4 and K1. *Abstracts, 6th International Congress of Plant Pathology,* Montreal, Canada. p. 159.

Singh, K., T. Ishii, A. Parco, N. Huang, D. S. Brar, and G. S. Khush (1996). Centromere mapping and orientation of the molecular linkage map of rice (*Oryza sativa* L.). *Proceedings of the National Academy of Sciences of the USA 93:* 6163-6168.

Sweigard, J. A., B. Valent, M. J. Orbach, A. M. Walter, A. Rafalski, and F. G. Chumley (1993). Genetic map of the rice blast fungus *Magnaporthe grisea.* In *Genetic maps,* S. J. O'Brien (Ed.), pp. 3.112-3.117. Plainview, NY: Cold Spring Harbor Laboratory Press.

Talbot, N. J. (2003). On the trail of a cereal killer: Exploring the biology of *Magnaporthe grisea. Annual Review of Microbiology 57:* 177-202.

Tang, X., R. Frederick, J. Zhou, D. A. Halterman, Y. Jia, and G. B. Martin (1996). Initiation of plant disease resistance by physical interaction of AvrPto and Pto kinase. *Science 274:* 2060-2063.

Tatsumi, H., S. Murakami, R. F. Tsuji, Y. Ishida, K. Murakami, A. Masaki, H. Kawabe, H. Arimura, E. Nakano, and H. Motai (1991). Cloning and expression in yeast of a cDNA clone encoding *Aspergillus oryzae* neutral protease II, a unique metalloprotease. *Molecular and General Genetics 228:* 97-103.

Valent, B. (1997). The rice blast fungus, *Magnaporthe grisea.* Plant relationships. In *The mycota V Part B,* G. C. Carroll and P. Tudzynoski (Eds.), pp. 37-54. Berlin: Springer-Verlag.

Valent, B., G. T. Bryan, Y. Jia, L. Farrall, S. A. McAdams, K. N. Faulk, and M. Levy (2001). Enhancing deployment of genes for blast resistance: Opportunities from cloning a resistance gene/avirulence gene pair. In *Rice genetics IV,* G. S. Khush,

D. S. Brar, and B. Hardy (Eds.), pp. 309-322. Los Baños, Philippines: International Rice Research Institute.

Van der Biezen, E. A., and J. D. G. Jones (1998). The NB-ARC domain: A novel signaling motif shared by plant disease resistance gene products and regulators of cell death in animals. *Current Biology 8:* R226-227.

Wang, C., K. Hirano, and S. Kawasaki (2002). Cloning of Pi-ta[2] in the centromeric region of Chromosome 12 with HEGS: High efficiency genome scanning. In *3rd International Rice Blast Conference,* Tsukuba, Japan, p. 25.

Wang, Z. (2003). *Development of molecular markers of the rice blast resistance gene Pi-ta and its application.* PhD dissertation. Institute of Nuclear Agricultural Sciences, Zhejiang University, China. (in Chinese with English abstract).

Wang, Z. X., U. Yamanouchi, Y. Katayose, T. Sasaki, and M. Yano (2001). Expression of the *Pib* rice-blast-resistance gene family is upregulated by environmental conditions favouring infection and by chemical signals that trigger secondary plant defenses. *Plant Molecular Biology 47:* 653-661.

Wang Z. X., M. Yano, U. Yamanouchi, M. Iwamoto, L. Monna, H. Hayasaka, Y. Katayose, and T. Sasaki (1999). The *Pib* gene for rice blast resistance belongs to the nucleotide binding and leucine-rich repeat class of plant disease resistance genes. *The Plant Journal 19:* 55-64.

Wu, K. S., C. Martínez, Z. Lentini, J. Tohme, F. G. Chumley, P. A. Scolnik, and B. Valent (1996). Cloning a blast resistance gene by chromosome walking. In *Rice genetics III,* G. S. Khush (Ed.), pp. 669-674. Proceedings of the Third International Rice Genetics Symposium, October 16-20, 1995, Manila, Philippines. Los Baños, Philippines: International Rice Research Institute.

Xia, J. Q., J. C. Correll, F. N. Lee, M. A. Marchetti, and D. D. Rhoades (1993). DNA fingerprinting to examine microgeographic variation in the *Magnaporthe grisea (Pyricularia grisea)* population in two rice fields in Arkansas. *Phytopathology 83:* 1029-1035.

Xia, J. Q., J. C. Correll, F. N. Lee, W. J. Ross, and D. D. Rhoades (2000). Regional population diversity of *Pyricularia grisea* in Arkansas and the influence of host selection. *Plant Disease 84:* 877-884.

Xiong, L., M. Qi, M. Lee, and Y. Yang (2001). Identification of rice defense-related genes by subtractive cloning and differential screening. *Molecular Plant-Microbe Interactions 14:* 685-692.

Young, N. D. (1996). QTL mapping and quantitative disease resistance in plants. *Annual Review of Phytopathology 34:* 479-501.

Yu, Z. H., D. J. Mackill, J. M. Bonman, S. R. McCouch, E. Guiderdoni, J. L. Notteghem, and S. D. Tanksley (1996). Molecular mapping of genes for resistance to rice blast (*Pyricularia grisea* Sacc.). *Theoretical and Applied Genetics 93:* 859-863.

Zeigler, R. S., L. X. Cuoc, R. P. Scott, M. A. Bernardo, D. H. Chen, B. Valent, and R. J. Nelson (1995). The relationship between lineage and virulence in *Pyricularia grisea* in the Philippines. *Phytopathology 85:* 443-451.

Zeigler, R. S., S. A. Leong, and P. S. Teng (Eds.). (1994). *Rice blast disease.* Wallingford, UK: CAB International and International Rice Research Institute.

Zheng, K. L., J. Y. Zhuang, J. Lu, H. R. Qian, and H. X. Lin (1996). Identification of DNA markers tightly linked to blast resistance genes in rice. In *Rice genetics III*, G. S. Khush (Ed.), pp. 565-569. Proceedings of the Third International Rice Genetics Symposium, October 16-20, 1995, Manila, Philippines. Los Baños, Philippines: International Rice Research Institute.

Chapter 9

Abiotic Stress Tolerance in Rice

A. Grover
A. Chandramouli
S. Agarwal
S. Katiyar-Agarwal
M. Agarwal
C. Sahi

INTRODUCTION

Rice is the most important food crop in the world, providing 80% of the diet for half the world's population. The generation of high-yield varieties and improved farming methods during the "green revolution" provided great impetus to rice production in the 1970s and 1980s. However, rice production has been declining for the past decade due to a combination of factors. The sensitivity of rice plants toward abiotic stresses is one of the important reasons for the diminishing yield.

WORLDWIDE IMPACT OF ABIOTIC STRESSES IN RICE CULTIVATION

Rice is grown in a wide range of ecological environments at varying altitudes, in diverse climates, and on different soil types. Asia is the home of

The authors are grateful to the Department of Science and Technology (DST) and the Department of Biotechnology (DBT), Government of India, and the National Agricultural Technology Project (NATP), Indian Council of Agricultural Research (ICAR), Government of India, for financial support to the group. S. Agarwal, S. Katiyar-Agarwal, M. Agarwal, and C. Sahi are thankful to Council of Scientific and Industrial Research (CSIR), Government of India, for their fellowship awards.

rice cultivation, contributing 91% of the total world production. China and India are the major rice-producing countries in Asia. In southeastern China, high temperature and adequate rainfall make an ideal environment for rice growth, but northern regions of China are not highly conducive to cultivation of this crop owing to low temperature, short growth period, and scanty rainfall. Serious water losses combined with soil erosion, expanding desertification, salinization, droughts, floods in lowland areas, waterlogging, and severe cold snaps are the major constraints toward higher productivity of rice in this country. Rice is cultivated as a "kharif" crop in most parts of northern and southern India. The sustainable rice production in India suffers due to excess salinity (which most severely affects coastal areas), drought (which affects both lowland and upland rice ecosystems), and flood conditions (particularly affecting rain-fed areas of the eastern states, such as Orissa and West Bengal). The rise in atmospheric temperature due to global warming is also emerging as a major threat to rice cultivation. Furthermore, the acidic soils of southern and eastern India and semi-arid zones, especially in northern states, curtail the optimum production of rice.

Other countries in Asia that make sizeable contributions to rice production include Indonesia (contributing 8.45% of total world production), Bangladesh (6.6%), Vietnam (5.37%), Thailand (4.27%), Myanmar (3.47%), the Philippines (2.19%), Japan (1.91%), South and North Korea (1.58%), Pakistan (1.13%), Nepal (0.71%), Cambodia (0.69%), Sri Lanka (0.48%), Laos (0.37%), Malaysia (0.37%), and Bhutan (0.008%). Rice production in Indonesia is limited by prolonged drought seasons, and about 10% of the land is prone to frequent floods. Bangladesh mostly has a low, flat, and fertile land, as the alluvial soil is being continuously enriched by heavy silt deposition by rivers. Global warming and the consequent water level rise may result in the loss of arable land to a notable extent. Droughts and flooding are frequent, and high salinity levels affect soils in the coastal areas of Bangladesh. Moreover, low temperatures during early phases of *Boro* rice cultivation affect net production. In Cambodia and Vietnam, erratic rainfall, drought, floods, and poor soil quality take a heavy toll on the rice crop. In Thailand, inland saline soils in the rain-fed lowlands of northeastern regions, acid sulfate soils in the central plains, and acidic saline soils in southern coastal areas are the major abiotic constraints. Low temperatures in northern hilly areas and dry weather in the north and northeast also bring down the production markedly. In Myanmar, problems encountered in deepwater rice lands are unfavorable irrigation conditions, physical soil problems, and excessive water. Rice cultivation in the Philippines is often endangered by typhoons and drought stress. In Japan, drought, low temperature (particularly in northern Japan), and heavy rainfall, resulting in flooding, severely damage rice plants. The rice crop also endures losses due to typhoons, particularly in southern Japan. In Korea, uneven rainfall distribution during the months of April-June and heavy rains during July-October pose a major environmental threat to rice cultivation. Low soil fertility, occasional salinity problems, and high acidity (pH 4-5) in mountain areas also aggravate the problem. Flooding due to heavy rainfall is another major environmental constraint that results in serious losses. The major problems in rice cultivation in Pakistan seem to be waterlogging, salinity, and inadequate supply of irrigation water. In Nepal, rainfall and soil and topographic variations produce a range of field water regimes, which cause major differences in rice production potential. Prolonged drought, cold injury, strong winds, and hailstorms are the major environmental stresses that limit net rice production at higher altitudes. In Sri Lanka, rice lands are located mainly in flood plains and valley bottoms, where soils are clay type with high water tables, which are unfit for cultivation. Seasonal flooding is a regular occurrence in the Sri Lankan coastal plains, which are peaty and acidic and not very suitable for the growth of rice. The major constraint to rice production in Laos is the dependence on a rainfed system of cultivation. In addition to the problem of periodic droughts, floods also affect large areas of rice production along the tributaries of the Mekong River. Periodic drought, irregular and erratic rainfall regimes, and seasonal fluctuations of monsoon floods are threats to the rice crop in Malaysia. Rice cropping in Bhutan is negatively affected by temperature, rainfall, and a shortage of irrigation water.

Africa (2.86%) and Australia (0.2%) make a small but definite input to the world rice bowl. On the African continent, rice is basically cultivated as a summer crop. In Egypt, soil salinity and high alkalinity affects about 30% of the potential growth area. Low temperature regimes, dependence on variable rains, and poor water control management limit the production in the highlands of Madagascar. The major setback in rice production in Australia is salinization of shallow ground water due to rising water tables.

From the aforementioned data, it is amply clear that the situations arising due to drought, flooding, high salinity, soil acidity, and extremes of temperature are of a cumulative nature. As the severity of most abiotic stresses is on the rise due to intense cultivation practices and environmental deterioration caused by the greenhouse effect, the raising of abiotic stress tolerant rice needs proactive consideration. The problem of rice abiotic stress interactions has been highlighted by rice workers in a large number of publications (Bennett, 2001; Datta, 2002; Hossain, 1996; Khush, 1998, 1999; Khush and Toenniessen, 1991; Swaminathan, 1982; Widawsky and O'Toole, 1990).

BASIC TENETS OF ABIOTIC STRESS RESPONSES IN RICE

An introductory account of the biology of rice abiotic stress responses follows. For details, readers may refer to a number of monographs and special papers published by the International Rice Research Institute (IRRI; 2006) of the Philippines.

Drought Stress

Drought occurs frequently in rain-fed uplands and limits rice growth and yield. Water stress during the vegetative stage reduces leaf expansion and CO_2 assimilation, plant height, tiller number, and leaf area. Leaf rolling and leaf tip drying are common symptoms of drought stress. Cell enlargement requires turgor to extend the cell wall and a gradient in water potential to draw water into the enlarging cell. Water stress decreases leaf area, which in turn reduces the intercepted solar radiation. Decrease in leaf water potential closes stomata and decreases transpiration, which in turn increases leaf temperature. There is a marked genotypic variation in rooting pattern in rice in response to water stress. Rice is most susceptible to water stress during the period from about ten days before flowering to the end of flowering (Yoshida et al., 1981). Water stress during this stage inhibits panicle exertion and spikelet filling and causes high sterility, leading to decreased yield (Ekanayake et al., 1989). The sterility is irreversible, as water supply at later stages does not restore the fertility. Water stress at or before panicle initiation reduces panicle number the most, whereas stress after panicle initiation reduces the potential spikelet number. Anthesis and ripening stages are highly sensitive to water stress. Water stress during anthesis increases the number of unfilled spikelets. Stress during grain filling decreases translocation of assimilates to the grain, which decreases grain weight and increases empty grains.

Flooding Stress

The adverse effects of submergence on rice are due to mechanical damage, silt deposition on leaves, reduced light, leaching of solutes from plant tissues, increased susceptibility to pests and diseases, and limited gas diffusion (Setter and Ella, 1994). The reduced gas diffusion is the dominant factor that limits the survival of rice plants during submergence. Slow diffusion results in (1) restricted CO_2 influx during photosynthesis and O_2 efflux from leaves during the day, causing enhanced photorespiration; (2) O_2 deficiency in shoots and roots during the night; and (3) increased accumulation of ethylene (C_2H_4) (Setter et al., 1989). In most environments, floodwater O_2 concentration is usually below air saturation, which leads to anoxia (a condition in which there is a total absence of O_2, and fermentation is the only significant source of ATP) or hypoxia (a condition where partial O_2 is present, allowing concurrent fermentation and respiration; Perata and Alpi, 1993). These conditions lead to a reduction in the energy production needed for survival or growth processes. Limited O_2 supply to roots causes enhanced breakdown of carbohydrates and reduced nutrient uptake (Setter et al., 1987). Rice has a built-in ability to transport O_2 efficiently from the aerial parts of the plants to the roots, but the problem gets accentuated when the plant is totally submerged. During waterlogging, stomatal closure and wilting are noted. Leaf growth and stem elongation are severely restricted by root anoxia in the short term. The adaptive significance of this response is to reduce the shoot/root ratio as a final adjustment to an impaired root system. In contrast, flooding promotes coleoptile growth in seedlings and internode elongation in adult plants in deepwater rice (Vergara et al., 1976). Under anoxic conditions, C_2H_4 synthesis is enhanced, rendering internodes more sensitive to the growth-promoting hormone gibberellic acid, which enhances cell elongation in the intercalary meristem (Sauter and Kende, 1992). Short-term flooding causes reorientation of growth processes, involving hypertrophic growth and leaf epinasty, whereas long-term stress promotes senescence and leaf abscission. Flooding stress also causes changes in cell ultrastructure, leading to cell injury and death.

Salinity Stress

Osmotic stress, ionic imbalance/poisoning, and high pH affect rice plants under excess salt conditions. The severity of these effects is dependent on the growth stage of the plant and the magnitude of the stress. The whole-plant effects of salt stress include reduced photosynthesis and photosynthetic translocation, stimulation of photorespiration, degradation of chlorophylls, stomatal closure, and a change in chloroplastic pH. The Fv/Fm ratio has been observed to decrease in response to salt stress (Lutts et al., 1996). An increase in the concentration of Na^+ and Cl^- invariably affects enzyme activities. The grain yield has been found to be inversely proportional to Na^+ and Na^+/K^+ ratio, and directly proportional to K^+ levels. Specific lipid domains may be formed in membranes, resulting in inverted micelles and increased leakiness. A prominent and sharp increase in levels of malondialdehyde during salt stress has been reported (Lutts et al., 1996). Disruption of microbodies, deformation of thylakoid membranes, increase in plast globuli in chloroplasts, separation of thylakoid membranes from appressed regions, appearance of fat droplets in the cytosol, and rearrangement of chromatin threads are some of the notable effects of salt stress. Other cellular effects triggered by excess salinity in rice include cell wall damage, accumulation of electron-dense proteinaceous particles, plasmolysis, cytoplasmic lysis, and damage to endoplasmic reticulum (Pareek et al., 1997). Rice cultivars grown in flood-prone areas are particularly affected by salinity.

Low Temperature Stress

Poor germination, delayed seedling emergence, stunted growth, and leaf discoloration are the notable effects of low temperature on the vegetative growth of rice. Accumulation of toxins such as C_2H_4 and acetaldehyde is also common. Cold stress creates concentrated solutions of solutes, thereby subjecting the plant to water stress. Low temperature stress causes ultra-structural changes, including alterations in the structure of chloroplasts and mitochondria and in membranes associated with organelles and vacuoles (Kratsch and Wise, 2000). Chilling can affect enzymatic functions and solubility of proteins. Low temperature stress affects O_2 evolution, organic acids, sugars, polyphenols, phospholipids, proteins, and ATP (Hirano and Sano, 1998). Other symptoms of low temperature injury include decreased protoplasmic streaming, electrolyte leakage and plasmolysis, increase or decrease in respiration, production of abnormal metabolites due to anaerobic conditions, loss of vigor, splitting, dieback of stem, death, and senescence. During the reproductive stages, degeneration of the panicle tip, incomplete panicle exertion, delayed flowering, failure of dehiscence and fertilization, high spikelet sterility, and irregular maturity are commonly observed in low temperature conditions (Yoshida et al., 1981). Low temperature treatment during the flowering stages of the rice plant causes abnormal digestion of starch in mature pollen grains, which reduces pollen viability. Low temperature conditions also lead to excessive shattering of grains in rice.

High Temperature Stress

High temperature stress is especially harmful at the reproductive stage. Generally, the growth rate of rice is optimal in the temperature range of 23 to 31 ± 2°C. Rice seed germination is drastically reduced at high temperatures. Elevated temperatures tend to reduce the vigor of rice seedlings and cause abnormal branching patterns of roots (Yoshida et al., 1981). The rate of plant development increases with temperature, and as a result, the duration of each developmental phase declines as temperature rises (Slafer and Rawson, 1994). In rice, the number and height of tillers and the duration of tillering is severely reduced in response to high temperature (Yoshida et al., 1981). Rice is most susceptible to heat injury during flowering, as pollen viability is particularly sensitive to heat stress. Even one to two hours of high temperature at anthesis results in high spikelet sterility (Satake and Yoshida, 1978). The duration of grain filling in rice is highly sensitive to elevated temperatures. Reduced grain filling results in a decline in grain mass and eventually grain yield. Besides grain production, the quality of grain is also affected by a rise in temperature. Because rice is consumed as a whole grain product, the physical appearance of the grain holds importance in the market. High temperature affects both the carbohydrate composition and the physical appearance of rice grains. Grain filling during periods of high temperature induces a chalky appearance in kernels, thereby reducing their market value (Tashiro and Wardlaw, 1991). The amount of amylose determines the water absorption and expansion of rice grains during cooking. High temperatures during the postanthesis period decrease the proportion of amylose in rice starch (Asaoka et al., 1984, 1985). It is therefore evident that the overall growth, production, and quality of rice are severely affected by high temperature stress. Increased high temperature tolerance may enable changing the dates of plantation of rice as may be desired to suit the cultivation of preceding or succeeding crops or extending the cultivation of rice to areas that are currently not suitable due to constraints of high temperature sensitivity.

MOLECULAR CHANGES ASSOCIATED WITH ABIOTIC STRESSES

Since the late 1970s, there have been intense efforts in the identification and characterization of stress proteins through comparison of stressed and unstressed (i.e., control) tissues employing one-dimensional and two-dimensional protein gel electrophoresis followed by Coomassie staining, silver staining, autoradiographic detection, and specific detection by western blotting methods. As a result, a large number of proteins associated with high temperature stress (heat shock proteins or HSPs), water and salt stress (osmotin, dehydrins, Rab or responsive to abscisic acid proteins, etc.), low temperature stress (cold-regulated or Cor proteins), and anaerobic stress (anaerobic stress proteins or ANPs) have been documented in a range of plant species. Detailed proteome maps of salt-, water-, low temperature-, high temperature-, and abscisic acid (ABA)-stressed rice tissues have been documented (Pareek et al., 1998a,b, 1999). The proteome maps of flooding-stressed tissues from contrasting rice types have also been reported (Dubey et al., 2003).

A great deal of progress has also been made on the isolation and cloning of genes that encode stress proteins. Such genes have been isolated through screening of gene libraries using specific homologous/heterologous probes, differential screening of cDNA libraries constructed using RNA from stressed tissues, and subtractive hybridization-based approaches where stress-responsive genes are enriched. A large number of stress-associated genes have been either completely or partially sequenced using expressed sequence tags (ESTs), and the expression of the corresponding transcripts has been examined. The induction response of these genes has been unveiled to an appreciable extent through analysis of stress promoters. Specific cis-acting sequences in the promoters of genes that govern stress inducibility have been delineated. Specific heat shock elements (HSEs), anaerobic responsive elements (AREs), and ABA-responsive elements (ABREs), which underline responsivity to heat shock, anaerobic stress, and abscisic acid (ABA), respectively, have been well characterized. Salt-, low temperature-, and ABA-responsive promoter sequences obtained from the *em* gene from wheat (Marcotte et al., 1989), *pin2* gene from potato (Xu et al., 1993), *rab16* (Ono et al., 1996) and *rab21* (Mundy and Chua, 1988; Mundy et al., 1990) genes from rice, and *rab28* gene (Pla et al., 1993) from maize have been tested using rice as a trans-host system. Likewise, an anaerobic stress-responsive promoter sequence obtained from *adh* gene from maize has been expressed in a rice *trans*-host (Kyozuka et al., 1994). In specific cases, genes that encode *trans*-acting factor proteins that interact with cis-acting promoter sequences of the stress genes have been cloned and characterized. Selected examples of stress-associated proteins and genes from rice are shown in Table 9.1.

DEVELOPMENT OF ABIOTIC STRESS-TOLERANT TRANSGENIC RICE

In the early 1990s, low temperature-tolerant transgenic tobacco plants were raised by overexpressing a desaturase gene isolated from *Arabidopsis* or cucurbits, heralding the beginning of a transgenic solution to the problems of abiotic stresses (Murata et al., 1992). During the past ten years of research (1992-2002), there have been nearly 75 reports on the production of abiotic stress-tolerant transgenic plants, providing testimony that this approach has great potential in alleviating abiotic stress-induced injuries (Grover et al., 2003). Nearly 45 different genes have proven useful for combating abiotic stresses. To obtain faster results—which is desirable when a large number of candidate stress genes are to be tested—the raising of abiotic stress-tolerant transgenics in most cases has been carried out using tobacco and *Arabidopsis*. With the increased standardization of transformation

TABLE 9.1. Selected examples of genes and proteins induced in rice by abiotic stresses.

Genes/proteins	Characteristic features	References
Salt and water stress		
rab16A, rab16B, rab16C, rab16D (Rab proteins)	Induced when plants are subjected to water stress, *rab16* mRNA and protein accumulate in rice embryos, leaves, roots, and callus-derived suspension cells upon treatment with NaCl or ABA; *rab 16A* promoter specifically binds nuclear protein factors	Mundy and Chua, 1988; Mundy et al., 1990; Yamaguchi-Shinozaki et al., 1990
SalT	mRNA accumulates rapidly in shoots and roots of mature seedlings treated with ABA, salt, PEG, NaCl, and KCl	Claes et al., 1990
em (early methionine-labelled protein)	Induced by ABA and salt stress, salt interacts synergistically with ABA	Bostock and Quatrano, 1992
Flooding stress		
ss (sucrose synthase)	Enzyme activity increases under anoxia	Ricard et al., 1991
gpi (glucose phosphate isomerase), *pfk* (phosphofruc-tokinase), *ald* (aldolase), *gpd* (glucose phosphate dehydrogenase), *tpi* (triose phosphate isomerase), *pgk* (phosphogluco kinase), *eno* (enolase), *pk* (pyruvate kinase)	Transcript levels increase under anoxia stress; enzyme activity of Gpd and Eno noted to be higher under anoxia	Lal et al., 1998; Minhas and Grover, 1999b; Nozue et al., 1996; Ricard et al., 1989; Umeda and Uchimiya, 1994
pdc (pyruvate decarboxylase), *adh* (alcohol dehydrogenase)	Four cDNA clones of *pdc* (*pdc1, pdc2, pdc3, pdc4*) and two cDNA clones of *adh* (*adh1, adh2*) have been identified in rice; transcript accumulation of both noted under oxygen limiting conditions; Pdc α subunit accumulation seen in early periods of anoxia stress, whereas Pdc β subunit accumulation is seen under longer durations of stress; increase in Adh enzyme activity was also observed under anoxia	Hossain et al., 1994, 1996; Huq et al., 1995; Laszlo and Lawrence, 1983; Minhas and Grover, 1999b; Ricard et al., 1986; Rivoal et al., 1990, 1997; Umeda and Uchimiya, 1994; Xie and Wu, 1989

TABLE 9.1 *(continued)*

Genes/proteins	Characteristic features	References
Osexp1, 2, 4, Osexpb (1-14)	α expansins mediate cell wall extension and are induced by submergence; β expansins are induced by gibberellin and wounding	Cho and Kende, 1997; Lee and Kende, 2001
ACC synthase	Enzyme activity induced by hypoxia	Cohen and Kende, 1987
Cold stress		
95, 75, 25 21 kDa protein	Induced at low temperature	Hahn and Walbot, 1989
rab16A	Induced by low temperature, water stress, and ABA	Hahn and Walbot, 1989
lip5, lip9, lip19 (low temperature-induced proteins)	Induced by low temperature; *lip5* and *lip19* also stimulated by ABA	Aguan et al., 1991
Heat stress		
hsp16.9, hsp16.9A, hsp16.9B, hsp17.4, hsp18	Class I low molecular weight heat inducible HSPs that confer tolerance in *E. coli; Oshsp16.9A* and *Oshsp16.9B* genes share 98.8% homology at the nucleotide level and 99.3% homology at the deduced amino acid level; transcripts corresponding to these genes show differential induction on heat shock; selective transcripts also noted to be induced by different abiotic stresses	Lee et al., 1995; Tzeng et al., 1992; Young et al., 1999; M. Agarwal, C. Sahi and A Grover (unpublished data)
hsp 26	Class I low molecular weight HSP induced independently under heat and oxidative stress	Lee et al., 2000
33 kDa HSP	Synthesis occurs at high temperature	Fourre and Lhoest, 1989
hsp70	DnaK-type molecular chaperone; transcript corresponding to hsp70 induced by heat shock, salinity stress, and ethanol	Borkird et al., 1991; Van Breusegem et al., 1994
pts1 and *pts3*	Encode 16-20 kDa HSPs; also synthesized in response to heavy metal stress	Tseng et al., 1993
hsp82 a, b, c	4-5 copies of *hsp82*-related genes are present in rice;	Van Breusegem et al., 1994

TABLE 9.1 *(continued)*

Genes/proteins	Characteristic features	References
	transcript accumulates in response to heat shock; not induced by osmotic stress, ethanol, ABA, or salicylic acid	
hsp90	Protein induced by heat shock, salinity, water stress, low temperature, and ABA; high uninduced amounts of protein exist in lemma, palea, culm tissues, seeds, and seed-derived callus	Pareek et al., 1995, 1997
hsp100	Protein induced by salinity, water stress, heat stress, low temperature, and ABA; hsp100 cDNA encoding ClpB exhibits high degree of conservation with other ClpB; complements thermotolerance defect of yeast cells; probably a single gene encodes the protein in rice genome; Oshsp100 transcript is tightly heat regulated	Agarwal et al., 2003; Singla et al., 1998; Young et al., 2001

protocols, genes shown to endow stress tolerance to model plant systems are being introduced and tested in crops. The ease with which rice can be genetically transformed has encouraged the introduction of several stress genes in rice (Grover and Minhas, 2000). Table 9.2 provides a compilation of different reports in this respect, and the salient features of this research are detailed in the following sections for different abiotic stresses.

Drought and Salinity Stresses

The *hva1, cor47, p5cs,* and *adc* genes have been employed for the production of drought-tolerant transgenic rice. For tolerance against salt stress, *hva1, codA, gs2, p5cs,* and *Oscdpk7* genes have been introduced in rice. Notably, there is a great deal of cross-protection in salt and drought stress responses. Transgenic *Nipponbare* rice that overexpressed the *hva1* gene showed significantly increased tolerance to water deficit and salinity in the R_1 generation. Transgenic rice plants in this study maintained higher growth rates than nontransformed control plants under stress conditions. The increased tolerance was also reflected by the delayed development of damage

TABLE 9.2. Selected reports on the production of abiotic stress-tolerant transgenic rice.

Gene	Protein	Source	Stress	Construct[a]	References
adc	Arginine decarboxylase	*Avena sativa*	Drought	CaMV35Sp (double enhancer)-*adc*-CaMVpolyA (CaMV35Sp-*hyg-nos*)	Capell et al., 1998
adh1	Alcohol dehydrogenase	*Gossypium hirsutum*	Flooding	ubiquitin1p-*adh1-nos* (CaMV35 Sp-*hyg-tml*)	Rahman et al., 2001
codA	Choline oxidase A	*Arthrobacter globiformis*	Salt and cold	CaMV35Sp-catalase intron-transit peptide-*codA-nos* (CaMV35 Sp-*hyg-nos*)	Sakamoto et al., 1998; Mohanty et al., 2002
cor47	Cold-regulated protein	*Arabidopsis thaliana*	Drought	actin1p-*cor47-pin2* (CaMV35Sp-*bar-nos*)	Cheng and Wu, 1998
gpat	Glycerol-3-phosphate o-acetyltransferase	*A. thaliana*	Low temperature	ubiquitin1p-*gpat-nos* (CaMV35 Sp-*hyg-nos*)	Yokoi et al., 1998
gs2	Glutamine synthetase	*Oryza sativa*	Salt	CaMV35Sp-intron-transit peptide-*gs2-nos* (CaMV35 Sp-*gus-nos*)	Hoshida et al., 2000
hsp101	Heat shock protein	*A. thaliana*	High temperature	ubiquitin1p-*hsp101-nos* (CaMV35 Sp-*hyg-*CaMVpolyA)	Katiyar-Agarwal et al., 2003
hva1	Lea protein	*Hordeum vulgare*	Water deficit and salt	actin1p-*hva1-pin2* (CaMV35 Sp-*bar-nos*)	Xu et al., 1996
Oscdpk7	Calcium-dependent protein kinase	*O. sativa*	Cold and salt/drought	CaMV35Sp-*Oscdpk7-nos* (hygromycin-based selection; construct details not given)	Saijo et al., 2000

TABLE 9.2 *(continued)*

Gene	Protein	Source	Stress	Construct[a]	References
otsA and *otsB*	Trehalose-6-phosphate synthase and trehalose-6-phosphate phosphatase fusion protein (TPSP)	*Escherichia coli*	Salt, drought, and low temperature	ABA-inducible promoter-TPSP-*pin2* (CaMV35Sp-*bar*-nos); *rbcS*-TPSP-*pin2*	Garg et al., 2002
p5cs	Δ^1-pyrroline-5-carboxylate synthetase	*Vigna aconitifolia*	Salt and water deficit	ABA-inducible promoter complex-*p5cs*-*pin2* (CaMV35Sp-*bar*-nos)	Zhu et al., 1998
pdc1	Pyruvate decarboxylase	*O. sativa*	Submergence	actin1-intron-transit peptide-*pdc1*-nos (CaMV35Sp-*hyg*-nos)	Quimio et al., 2000

[a]The constructs shown in brackets were employed for the selection of the transformants. nos: nopaline synthetase gene, p: promoter, pin2: proteinase inhibitor gene, tml: tumor morphology large gene.

symptoms caused by stress and by improved recovery upon the removal of stress conditions (Xu et al., 1996). Transgenic *Pusa Basmati 1* rice plants containing the *hva1* gene have also been raised. Detailed studies are underway to test the stress tolerance of these transgenics (Rohila et al., 2001). Transgenic *Nipponbare* plants overexpressing the *codA* gene in chloroplasts as well as in cytoplasm have been raised (Sakamoto et al., 1998). In this study, treatment with 150 mM NaCl inhibited the growth of both wild type and transgenic plants. However, after removal of the salt stress, transgenic plants began to grow again at the normal rate in a shorter time period than the wild type plants. The chloroplast-expressed *codA* transgenics were more tolerant than the cytoplasm-expressed transgenic plants to photoinhibition under salt stress. The work with *codA* gene has been extended to *indica* rice cultivar *Pusa Basmati 1,* and higher salt tolerance has been marked in its transgenic progeny (Mohanty et al., 2002).

Transgenic rice overexpressing the *cor47* gene in the genetic background of the *Kenfeng* cultivar has been raised. When drought stress was applied for five days and then plants were allowed to recover for two days, the R_1 generation transgenic plants were more tolerant than control plants (Cheng and Wu, 1998). Transgenic R_1 rice overexpressing the *p5cs* gene showed more biomass than the control plants under both salt stress and drought stress conditions (Zhu et al., 1998). Transgenic rice plants overexpressing oat *adc* cDNA were found to contain increased levels of putrescine in calli and regenerated plants. The normal morphogenic development and differentiation of callus lines expressing high levels of *adc* mRNA was found to be blocked by exposure to light. Preliminary experiments showed that chlorophyll loss was negligible in the transgenic rice plants after eight days of drought stress (Capell et al., 1998). However, detailed experiments to test the level of stress tolerance of these plants have not been reported.

Transgenic *Kinuhikari* rice plants overexpressing *gs2* have also been produced. Upon exposing seedlings to 150 mM NaCl, control plants completely lost their photosystem II activity, but transgenic plants retained more than 90% activity after two weeks of salt stress treatment (Hoshida et al., 2000). Overproduction of *gs2* also reduced the increase of ammonium ion (NH_4^+) content at high salinity, presumably due to a decrease in electron transport chain rates. Introduction of *Oscdpk7* into *Nipponbare* rice cells caused no obvious effects on plant growth and development under normal growth conditions. After being subjected to 200 mM NaCl, untransformed plants showed wilting of leaves after three days, whereas transgenic plants exhibited significantly greater salt tolerance (Saijo et al., 2000). Transgenic plants also showed increased drought tolerance. In this study, several stress-responsive genes (i.e., *rab16, salT,* and *wsi18)* were overexpressed in transgenic plants under salt/drought conditions.

Also, transgenic *indica* rice was transformed with genes encoding trehalose 6-phosphate synthase (Tps) and trehalose 6-phosphate phosphatase (Tpp) enzymes from *E. coli.* These transgenics exhibited vigorous shoot and root growth compared to untransformed control plants under salt and drought stress (Garg et al., 2001). *E. coli* Tps and Tpp, encoded by the *otsA* and *otsB* genes, respectively, have been fused and transformed in *Pusa Basmati 1* type rice. Transgenic plants overexpressing the fused gene construct exhibited sustained plant growth, less photo-oxidative damage, and more favorable mineral balance under salt, drought, and low-temperature conditions compared to nontransgenic rice (Garg et al., 2002).

Flooding Stress

The *pdc1* and *adh1* genes have so far been employed for the production of flooding-tolerant transgenic rice. Transgenic *Taipei 309* plants overexpressing rice *pdc1* cDNA driven by the actin1 gene promoter were found to contain high levels of ethanol compared to the untransformed control (Quimio et al., 2000). The *pdc1* transgenics showed higher survival rates than untransformed lines during the recovery phase after 14 days complete submergence. In another study, *Taipei 309* transgenics that overexpressed rice *pdc1* cDNA driven by an anaerobically inducible 6XARE promoter have been raised (Rahman et al., 2001). These plants exhibited a moderate increase in Pdc enzyme activities and ethanol production in shoots under anaerobic conditions compared to untransformed controls. However, it was noted that the survival of the transgenics during the recovery period following stress was lower than that of untransformed plants and of the submergence-intolerant cultivar IR42. The same group also reported the production of transgenic rice underexpressing Adh (using antisense rice *adh1* cDNA) and overexpressing cotton *adh1* cDNA (Rahman et al., 2001). Transgenics underexpressing Adh showed reduced Adh activity and ethanol production, whereas the transgenics overexpressing Adh showed increased Adh activity and slightly higher ethanol production under anoxic conditions. There was no increase in the survival of Adh-overexpressing transgenics under anoxic conditions, whereas antisense Adh lines were found to show reduced survival under anoxic stress as compared to the untransformed controls. Progeny from sexually crossed Pdc- and Adh-overexpressing transgenic plants were found to contain high levels of both Pdc and Adh. These transgenic rice plants showed increased anoxia tolerance and reduced acetaldehyde production. Rice *Pusa Basmati 1* transgenics overexpressing and underexpressing rice *pdc1* have also been raised (Grover et al., 2002). Likewise, *Pusa Basmati 1* transgenics overexpressing cotton *adh1* have also been raised (S. Agarwal, A. Kapoor, S. Katiyar-Agarwal, and A. Grover, unpublished results). Detailed molecular characterization of these rice plants is in progress.

Low Temperature Stress

The *gpat, Oscdpk7, codA,* and *gs2* genes have so far been employed for the production of low temperature-tolerant transgenic rice. Transgenic *gpat*-overexpressing *Yamahoushi* rice plants showed 28% higher levels of unsaturated fatty acids in the phosphatidylglycerol pathway than untransformed controls. The net photosynthetic rate of transformed rice plants was 20% higher than that of the wild type at 17°C (Yokoi et al., 1998). Seedlings of transgenic rice that overexpressed *Oscdpk7* were exposed to 4°C for 24 hours and then allowed to recover. Fv/Fm values recovered to normal in transgenic plants, whereas there was a progressive lowering of these values in nontransgenic plants (Saijo et al., 2000). Transgenic rice harboring *gs2* and *codA* engineered for increased salt tolerance have been found to exhibit tolerance to cold stress as well (Hoshida et al., 2000; Sakamoto and Murata, 1998).

High Temperature Stress

Genetic improvement for high temperature tolerance is the least researched issue in rice abiotic stress biology. Hsp101 is an important component in the high-temperature response of rice (Agarwal et al., 2001, 2002, 2003; Katiyar-Agarwal et al., 2001; Pareek et al., 1995; Singla and Grover, 1993, 1994; Singla et al., 1997). Overexpression of this gene in transgenic *Pusa Basmati 1* rice has recently been achieved (Katiyar-Agarwal et al., 2003). The transgenic plants that overexpress *Athsp101* cDNA appeared normal, showing that there is no ill effect of transgene expression on overall growth and development of transformants. The survival of plants was compared in terms of recovery after exposure to 45, 47, and 50°C heat stress for different durations. The transgenic rice lines showed significant regrowth in the recovery phase, whereas the untransformed plants could not recover (Katiyar-Agarwal et al., 2003). Apart from *hsp101*, several small molecular weight *hsp* cDNAs (such as *hsp18, hsp17.3,* and *hsp16.9*) have also been cloned from rice (M. Agarwal, C. Sahi, and A. Grover, unpublished results). Experiments to unveil the physiological roles of these cDNA in imparting abiotic stress tolerance are in progress.

The foregoing account of abiotic stress-tolerant transgenic plants has specifically dealt with rice. For more details on plant abiotic stress interactions and the possibilities of raising abiotic stress-tolerant transgenic crops (including rice), readers may refer to Agarwal et al. (2002), Bajaj et al. (1999), Bartels (2001), Boamfa et al. (2003), Dhaliwal et al. (1998), Dubey and Grover (2001, 2003), Grover (1999, 2000, 2002), Grover et al. (1993, 1995, 1998a,b, 1999, 2000, 2001a,b, 2003, 2004), Grover and Chandramouli (2002), Grover and Minhas (2000), Grover and Pental (2003), Jackson and Ram (2003), Katiyar-Agarwal et al. (1999, 2001), Khanna, Chopra, and Sinha (1998), Kumria et al. (1999), Minhas and Grover (1999a), Mustroph and Albrecht (2003), Nuccio et al. (1999), Pareek et al. (1995, 1997), Singla et al. (1997), and Zeng et al. (2003).

GENOMICS AND PROTEOMICS OF ABIOTIC STRESSES IN RICE

The precise physiological processes, biochemical enzymes, molecular mechanisms, and proteins and genes that impart stress tolerance are not yet well understood. This poses a major limitation in the production of high-level abiotic stress-tolerant transgenic plants. Although efforts to unveil these details by conventional biochemical and molecular approaches continue, the recent upsurge in gene discovery methods has opened up certain avenues of research that were not available earlier (Chen et al., 2002; Cushman and Bohnert, 2000). The publication and release of complete nucleotide sequence information in public databases of various genomes (e.g., see Arabidopsis Information Resource, 2006; Syngenta, 2007; http://www.irri.org), including rice (Yu et al., 2002), has made a great impact on contemporary biological science (Goff, 1999; Goff et al., 2002; Grover et al., 2003). New techniques such as the use of microarrays, subtractive cloning, and protein microsequencing have been employed for deciphering genes involved in plant responses to environmental stress. Significant work has been carried out on gene discovery related to stress tolerance in *Arabidopsis* (Agarwal et al., 2001; Gong et al., 2001; Seki et al., 2001). Work on gaining insights in the structural and functional organization of novel stress-related genes in rice is also being undertaken. The Rice Full-Length cDNA Consortium (2003) has reported the collection, mapping, and annotation of more than 28,000 cDNA clones from rice. A network of genes, associated with developmental and stress responses in rice, has been built by several molecular approaches (Cooper et al., 2003). This data set suggests that similar genes respond to environmental cues and stresses, and some may also regulate development. Transcript regulation (transcript abundance and expression patterns from 15 min to one week) in response to high salinity has been examined in salt-tolerant *Pokkali* rice using microarrays, including 1,728 cDNAs from libraries of salt-stressed roots (Kawasaki et al., 2001). In this work, approximately 10% of the transcripts were significantly up- or downregulated within one hour of salt stress. These results suggested that salt-tolerant *Pokkali* rice can overcome the stress due to its ability to induce transcripts that stimulate protein synthesis and components of signaling circuits, whereas salt-sensitive IR29 shows a delay in responding by upregulation and fewer responses in total, which may bring about a general decline in transcription and death within 24 hours. The transcript expression profiles in *Pokkali* were categorized into at least ten patterns using cluster analysis, which could be correlated with several biochemical pathways, such as ion and metabolite transport processes, hormonal control, cell structure and organ growth, and cell signaling.

Several cDNAs have been identified from libraries constructed using drought-stressed leaf and root tissues of upland rice cultivar *Nagina 22* (Reddy et al., 2002). The analysis of expressed sequence tags (ESTs) generated from leaf and root cDNA libraries revealed that genes involved in metabolism constitute the most abundant class among ESTs, and those genes related to drought stress response are highly represented in the stressed seedlings. Metallothionein-like genes were the most abundant class in the normalized leaf library. Several ESTs showed significant sequence similarity to genes encoding glyceraldehyde-3-phosphate dehydrogenase, aldolase, Lea proteins, and HSPs that have been shown to be affected by ABA, drought, and other environmental stresses. Several transcription factor genes, such as *dreb1A, myb, myc, apc,* and Zn-finger containing proteins were identified. Several classes of protein kinases, including calcium-dependent protein kinase (Cdpk), and protein phosphatases, homologues of Lea proteins, cytochrome P450 enzymes, catalases, glycine-rich proteins, and peroxidases were reported. In an IRRI study, the overexpression of the *dreb1A* gene driven by stress-inducible promoter rd29A showed enhanced drought tolerance in transgenic *indica* rice (K. Datta et al., unpublished data).

An improved PCR-based technique has been employed for the isolation of differentially induced genes in three contrasting rice types, namely, *Pusa Basmati 1,* CSR 27, and *Pokkali* (Sahi et al., 2003). Four subtracted cDNA libraries were constructed in this study. A database search of 85 sequenced clones showed that 22 clones were homologous to genes that had earlier been implicated in stress response. This category includes genes that encode SalT, glycine-rich RNA-binding protein, ADP-ribosylation factor, NADP-dependent malic enzyme, ubiquitin fusion protein, tumor suppressor wound inducible protein, nucleotide diphosphate kinase (Ndpk), and tumor protein. Nineteen clones were homologous to sequences in the rice genomic database. Some of the annotated clones included H^+-ATPase, peroxidase, and tricopeptide repeat (TPR) containing protein. Some of the clones in this study (such as T4D2-165) although not annotated, were found to be salt inducible. Thirty-four clones were found to be novel with respect to their function. Some of the clones were found to be salt inducible, whereas others showed homology to ESTs from drought- or salt-stressed libraries of rice, *Sorghum, Mesembryanthemum, Medicago sativa,* and so on. Six clones in this study showed no homology to any sequences in the public database. Apart from these, there were a few clones that showed little sequence homology to transcription factors like ethylene responsive element binding protein and zinc-finger binding protein. Other stress-associated genes in this category were those encoding metallothionein, copper chaperone, and alanine aminotransferase.

In maize, the expression of as many as 262 individual proteins was altered on 2D electrophoretograms with changes in O_2 tension regime (Chang et al., 2000). Furthermore, of the 48 protein spots analyzed by MS, 46 were identifiable on the basis of a database search that showed a wide range of functions. Although, as mentioned earlier, proteome maps of stressed rice tissues have been constructed, there is a need to extend the high-throughput proteomics approach for unveiling details on stress-induced proteins in rice.

RESEARCH GAPS AND FUTURE PROSPECTS

Tremendous progress has been made on the molecular biology and bio-technology of rice in the past two decades (Datta, 1999, 2000; Grover and Minhas, 2000; Shimamoto, 1999; Tyagi et al., 1999; Tyagi and Khurana, 2003; Tyagi and Mohanty, 2000). A number of different rice transgenics have been developed for various traits. However, progress on the production of abiotic stress resistance has not been as spectacular as desired and noted for other traits, including resistance to biotic stresses. The main reason behind the slow progress in generating abiotic stress-resistant transgenics is the higher genetic complexity of the underlying responses. The attempts made to generate abiotic stress-tolerant transgenics in spite of this hurdle must therefore be appreciated. There may not be single-gene-based solutions for high-level abiotic stress tolerance, but the single-gene transformations open up avenues for possible pyramiding of various stress related genes at a later stage.

The following points merit attention in order to achieve high-level abiotic stress tolerance in rice plants:

1. Various genes that are involved in abiotic stress responses need to be catalogued, and their hierarchy in expression needs to be worked out. Genomic approaches involving the application of high-throughput sequencing and expression analysis need to be fully exploited. In vitro analysis of the rice genome can provide valuable leads. For instance, whereas only one *hsp101* gene was cloned, in vitro analysis of the *Arabidopsis* genome sequence has revealed that there are seven homologues of this gene that are yet to be cloned (Agarwal et al., 2001). Likewise, only one *hsp101* gene has been cloned as yet in rice (Agarwal et al., 2003), but the high-throughput genome sequence data have shown that there are at least two copies of *hsp101* sequences present on chromosome 2. The understanding of the global expression of different abiotic stress genes is expected to receive a great impetus from microarray-based analyses.

2. Every candidate gene that shows inducibility upon stress treatment may not have a primary role in stress adaptation. There is a need to devise suitable experimental approaches to delineate the genes responsible for

primary effects from those that represent secondary or tertiary effects. Comparative analysis of the molecular mechanisms using contrasting rice cultivars may be extremely useful in distinguishing genes that play a role in stress adaptation from those that are not linked to stress adaptation (Dubey et al., 2003; Sahi et al., 2003). Wild rice cultivars are also important for studying the response of rice to different abiotic stresses and isolating genes related to stress tolerance (Grover and Pental, 1992); due considerations must be given to these rice varieties.

3. Although single-gene transfers are easier to achieve, such transfers may not be efficient for the desired level of abiotic stress tolerance. This handicap must be overcome by judicious choice of the targeted single genes. Possibly, single genes that can affect the cascade of gene expression changes need to be employed for rice transformation leading to enhanced stress tolerance. Physiological, genetic, and biochemical approaches have generated a great deal of information about stress-signalling components (e.g., G proteins; cAMP; Ca^{2+}/CaM kinases and phosphatases; inositol-1,4,5-trisphosphate; and phosphatidic acid) and trans-acting factors, such as the basic leucine zipper motif, Myb, Myc, and Zn-finger transcription factors, including Rd22bp1 (Myc), Atmyb2 (Myb), Dreb1A and Dreb2A (AP2 domain), and Alfin1 in plants. The genes encoding such proteins may prove useful (Jaglo-Ottosen et al., 1998; Kasuga et al., 1999; Kim et al., 2001; Park et al., 2001). For the extensive application of this approach, there is a need to identify, clone, and characterize more stress-responsive signaling components and transcription factors encoding genes. There is limited information on abiotic stress-induced promoters. It is important not only to work out the detailed characterization of these promoters but to clone them in binary cloning vectors, so that they can be routinely used to drive transgene expression in rice. Zhu et al. (1998) employed a stress-induced AIPC-ABA promoter complex to drive the expression of *p5cs* cDNA and noted that stress-inducible transgene expression conferred water stress tolerance to transgenic rice.

4. *Agrobacterium*-mediated rice transformation has emerged as the most favored approach. However, very few binary cloning vectors are available that enable efficient rice transformation for abiotic stress tolerance (Katiyar-Agarwal et al., 2002). There is an urgent need for the construction of improved binary vector systems for rice transformation that take into account different stress-induced promoters, genetic elements for enhanced expression, selection markers that have wider acceptability, and so on. Pyramiding of different genes in the same plant is a powerful approach for achieving better stress tolerance.

However, there are no reports of such studies for abiotic stresses. Vector systems that enable gene pyramiding in rice need to be optimized.
5. The transgenic rice lines engineered for resistance to abiotic stresses have been mostly confined to the laboratory so far, and there are no reports of their use in field trials as yet. The stress response in laboratory conditions cannot be directly correlated with that in field conditions. Further work will warrant rigorous testing of the transgenics by plant physiologists and agronomists.

REFERENCES

Agarwal, M., S. Katiyar-Agarwal, and A. Grover (2002). Plant Hsp100 proteins: Structure, function and regulation. *Plant Science 163:* 397-405.

Agarwal, M., S. Katiyar-Agarwal, C. Sahi, D. R. Gallie, and A. Grover (2001). *Arabidopsis thaliana* Hsp10 proteins: Kith and kin. *Cell Stress and Chaperones 6:* 219-224.

Agarwal, M., C. Sahi, S. Katiyar-Agarwal, S. Agarwal, T. Young, D. R. Gallie, V. M. Sharma, K. Ganesan, and A. Grover (2003). Molecular characterization of rice hsp101: Complementation of yeast hsp104 mutation by disaggregation of protein granules and differential expression in *indica* and *japonica* rice types. *Plant Molecular Biology 51:* 543-553.

Aguan, K., K. Sugawara, and T. Suzuki-Kusano (1991). Isolation of genes for low temperature-induced proteins in rice by a simple subtractive method. *Plant Cell Physiology 32:* 1258-1289.

Arabidopsis Information Resource (2006). Search overview. In *TAIR: The Arabidopsis information resource.* http://www.arabidopsis.org/index.jsp (last accessed February 26, 2007).

Asaoka, M., K. Okuno, K. Hara, and H. Fuwa (1985). Effect of environmental temperature at the milky stage on amylose content and fine structure of amylopectin of waxy and non waxy endosperm starches of rice (*Oryza sativa* L.). *Agricultural Biological Chemistry 49:* 373-379.

Asaoka, M., K. Okuno, Y. Sugimoto, J. Kawakami, and H. Fuwa (1984). Effect of environmental temperature during development of rice plants on some properties of endosperm starch. *Starch/Stärke 36:* 189-193.

Bajaj, S., J. Targolli, L. F. Liu, T. H. D. Ho, and R. Wu (1999). Transgenic approaches to increase dehydration-stress tolerance in plants. *Molecular Breeding 5:* 493-503.

Bartels, D. (2001). Targeting detoxification pathways: An efficient approach to obtain plants with multiple stress tolerance. *Trends in Plant Science 6:* 284-286.

Bennett, J. (2001). Summing-up: Cutting-edge science for rice improvement—breakthroughs and beneficiaries. In *Rice biotechnology: Improving yield, stress tolerance and grain quality,* J. Goode and D. Chadwick (Eds.), pp. 242-251. Proceedings from a symposium held at IRRI, Los Baños, Laguna, Philippines, March 27-29, 2000. Novartis Foundation and John Wiley.

Boamfa, E. I., P. C. Ram, M. B. Jackson, J. Reuss, and F. J. M. Harren (2003). Dynamic aspects of alcoholic fermentation of rice seedlings in response to anaerobiosis and to complete submergence: Relationship to submergence tolerance. *Annals of Botany 91:* 279-290.

Borkird, C., C. Simoens, R. Villarroel, and M. Van Montagu (1991). Gene expression associated with water-stress adaptation of rice cells and identification of two genes as hsp70 and ubiquitin. *Physiologia Plantarum 82:* 449-457.

Bostock, R. M., and R. S. Quatrano (1992). Regulation of *Em* gene expression in rice: Interaction between osmotic stress and ABA. *Plant Physiology 98:* 1356-1363.

Capell, T., C. Escobar, H. Liu, D. Burtin, O. Lepri, and P. Christou (1998). Overexpression of the oat arginine decarboxylase cDNA in transgenic rice (*Oryza sativa* L.) affects normal development patterns in vitro and results in putrescence accumulation in transgenic plants. *Theoretical and Applied Genetics 97:* 246-254.

Chang, W. W. P., L. Huang, M. Shen, C. Webster, A. L. Burlingame, and K. M. Roberts (2000). Patterns of protein synthesis and tolerance of anoxia in root tips of maize seedlings acclimated to low-oxygen environment, and identification of proteins by mass spectrometry. *Plant Physiology 122:* 295-317.

Chen, M., G. Presting, W. B. Barbazuk, J. L. Goicoechea, B. Blackmon, G. Fang, H. Kim, D. Frisch, Y. Yu, S. Sun, et al. (2002). An integrated physical and genetic map of the rice genome. *Plant Cell 14:* 537-545.

Cheng, W., and R. Wu (1998). Production and analysis of *cor47* gene-containing transgenic rice plants. *Rice Genetics Newsletter 15:* 178-180.

Cho, H.-D., and H. Kende (1997). Expression of expansin genes is correlated with growth in deepwater rice. *Plant Cell 9:* 1661-1671.

Claes, B., R. Dekeeyser, R. Villarroel, M. V. Blucke, G. Bauw, M. V. Montagu, and A. Caplan (1990). Characterization of a rice gene showing organ-specific expression in response to salt stress and drought. *Plant Cell 2:* 19-27.

Cohen, E., and H. Kende (1987). *In vivo* 1-aminocyclopropane-1-carboxylase synthase activity in internodes of deepwater rice. *Plant Physiology 84:* 282-286.

Cooper, B., J. D. Clarke, P. Budworth, J. Kreps, D. Hutchison, S. Park, S. Guimil, M. Dunn, P. Luginbuhl, C. Ellero, S. A. Goff, and J. Glazebrook (2003). A network of rice genes associated with stress response and seed development. *Proceedings of the National Academy of Sciences of the USA 100:* 4945-4950.

Cushman, J. C., and H. J. Bohnert (2000). Genomic approaches to plant stress tolerance. *Current Opinion in Plant Biology 3:* 117-124.

Datta, S. K. (1999). Transgenic cereals: *Oryza sativa* (rice). In *Molecular improvement of cereal crops,* I. K. Vasil (Ed.), pp. 149-187. Dordrecht, Netherlands: Kluwer.

Datta, S. K. (2000). Transgenic rice: Development and products for environmentally friendly sustainable agriculture. In *Proceedings of the 12th Toyota conference: Challenge of plant and agriculture sciences to the crisis of biosphere on the earth in the 21st century,* K. Watanabe and A. Komamine (Eds.), pp. 237-246. Georgetown, TX: Landes Bioscience.

Datta, S. K. (2002). Recent development in transgenics for abiotic stress tolerance in rice. In *Genetic engineering of crop plants for abiotic stress, JIRCAS working report,* M. Iwanaga (Ed.), pp. 43-53. Tsukuba, Japan: International Research Center for Agricultural Sciences.

Dhaliwal, H. S., M. Kawai, and H. Uchimiya (1998). Genetic engineering for abiotic stress tolerance in plants. *Plant Biotechnology 15:* 1-10.

Dubey, H., G. Bhatia, S. Pasha, and A. Grover (2003). Proteome maps of flood tolerant FR 13A and flood sensitive IR54 rice types depicting proteins associated with deprivation stress and recovery regimes. *Current Science 84:* 83-89.

Dubey, H., and A. Grover (2001). Current initiatives in proteomics research: The plant perspective. *Current Science 80:* 262-269.

Dubey, H., and A. Grover (2003). Respiratory pathway enzymes are differentially altered in flood tolerant and sensitive rice types during O_2 deprivation stress and post-stress recovery phase. *Plant Science 164:* 815-821.

Ekanayake, I. J., S. K. Datta, and P. L. Stepoukus (1989). Spikelet sterility and flowering response of rice to water stress at anthesis. *Annals of Botany 63:* 257-264.

Fourre, J. L., and J. Lhoest (1989). Protein synthesis and modification by heat in rice cell culture. *Plant Science 61:* 69-74.

Garg, A. K., J. K. Kim, T. G. Owens, A. P. Ranwala, Y. D. Choi, L. V. Kochian, and R. J. Wu (2002). Trehalose accumulation in rice plants confers high tolerance levels to different abiotic stresses. *Proceedings of the National Academy of Sciences of the USA 99:* 15898-15903.

Garg, A. K., J. K. Kim, A. Ranwala, and R. Wu (2001). Accumulation of trehalose in transgenic *indica* rice using bifunctional fusion enzyme of trehalose-6-phosphate synthase and trehalose-6-phosphate phosphatase of *Escherichia coli. Rice Genetics Newsletter 18:* 87-89.

Goff, S. A. (1999). Rice as a model for cereal genomics. *Current Opinion in Plant Biology 2:* 86-89.

Goff, S. A., D. Ricke, T. H. Lan, G. Presting, R. Wang, M. Dunn, J. Glazebrook, A. Sessions, P. Oeller, H. Varma, et al. (2002). A draft sequence of the rice genome (*Oryza sativa* L. ssp. *japonica*). *Science 296:* 92-100.

Gong, Z., H. Koiwa, M. A. Cushman, A. Ray, D. Bufford, S. Kore-eda, T. K. Matsumota, J. Zhu, J. C. Cushman, R. A. Bressan, and P. M. Hasegawa (2001). Genes that are uniquely stress regulated in salt overly sensitive (sos) mutants. *Plant Physiology 126:* 363-375.

Grover, A. (1999). A novel approach for raising salt tolerant transgenic plants based on altering stress signaling through Ca^{2+}/calmodulin-dependent protein phosphatase calcineurin. *Current Science 76:* 136-137.

Grover, A. (2000). Water stress-responsive proteins/genes in crop plants. In *Probing photosynthesis: Mechanisms, regulation and adaptation,* M. Yunus, U. Pathre, and P. Mohanty (Eds.), pp. 397-408. New York: Taylor and Francis.

Grover, A. (2002). Molecular biology of stress responses. *Cell Stress and Chaperones 7:* 1-5.

Grover, A., M. Agarwal, S. Katiyar-Agarwal, C. Sahi, and S. Agarwal (2000). Production of high temperature tolerant transgenic plants through manipulation of membrane lipids. *Current Science 79:* 557-559.

Grover, A., P. K. Aggarwal, A. Kapoor, S. Katiyar-Agarwal, M. Agarwal, and A. Chandramouli (2003). Production of abiotic stress tolerant transgenic crops: Present accomplishments and future needs. *Current Science 84:* 355-367.

Grover, A., and A. Chandramouli (2002). Abiotic stress tolerant transgenics in the days of genomics and proteomics. *Physiology and Molecular Biology of Plants 8:* 1-19.

Grover, A., M. A. Hossain, M. E. Huq, J. D. McGee, W. J. Peacock, E. S. Dennis, and T. K. Hodges (1995). Studies on the alterations of pdc gene expression in transgenic rice. In *Proceedings of the International Rice Research Conference—fragile lives in fragile ecosystems,* pp. 911-921, Manila, Philippines.

Grover, A., A. Kapoor, S. Katiyar-Agarwal, M. Agarwal, C. Sahi, P. Jain, O. Satyalakshmi, S. Agarwal, and H. Dubey (2001b). Experimentation in biology of plant abiotic stress responses. *Proceedings of the Indian National Academy of Science B67:* 189-214.

Grover, A., A. Kapoor, D. Kumar, H. E. Shashidhar, and S. Hittalmani, (2004). Genetic improvement of abiotic stress responses: Issues, tools and concerns. In *Plant breeding—Mendelian to molecular approaches,* H. K. Jain and M. C. Kharkwal (Eds.), pp. 167-193. Narosa, India, and Dordrecht, the Netherlands: Kluwer.

Grover, A., A. Kapoor, O. Satya Lakshmi, S. Agarwal, C. Sahi, S. Katiyar-Agarwal, M. Agarwal, and H. Dubey (2001a). Understanding molecular alphabets of the plant abiotic stress responses. *Current Science 80:* 206-216.

Grover, A., S. Katiyar-Agarwal, M. Agarwal, C. Sahi, and S. Agarwal (2002). Toward the production of abiotic stress-tolerant, transgenic Pusa Basmati 1 rice plants. Proceedings of the International Rice Congress, September 16-20, 2002, Beijing, China, p. 286.

Grover, A., and D. Minhas (2000). Towards production of abiotic stress tolerant transgenic rice plants: Issues, progress and future research needs. *Proceedings of the Indian National Academy of Science B66:* 13-32.

Grover, A., A. Pareek, and S. C. Maheshwari (1993). Molecular approaches for genetically engineering plants tolerant to salt stress. *Proceedings of the Indian National Academy of Science B59:* 113-127.

Grover, A., A. Pareek, S. L. Singla, D. Minhas, S. Katiyar, S. Ghawana, H. Dubey, M. Agarwal, G. U. Rao, J. Rathee, and A. Grover (1998a). Engineering crops for tolerance against abiotic stresses through gene manipulation. *Current Science 75:* 689-696.

Grover, A., and D. Pental (1992). Interrelationship of *Oryza* species based on electrophoretic patterns of alcohol dehydrogenase. *Canadian Journal of Botany 70:* 352-358.

Grover, A., and D. Pental (2003). Breeding objectives and requirements for producing transgenics for the major field crops of India. *Current Science 84:* 310-320.

Grover, A., C. Sahi, N. Sanan, and A. Grover (1999). Taming abiotic stresses in plants through genetic engineering: Current strategies and perspective. *Plant Science 143:* 101-111.

Grover, A., N. Sanan, and C. Sahi (1998b). Genetic engineering for high-level toler-ance to abiotic stresses through overexpression of transcription factor genes: The next frontier. *Current Science 75:* 178-179.

Hahn, M., and V. Walbot (1989). Effect of cold-treatment on protein synthesis and mRNA levels in rice leaves. *Plant Physiology 91:* 930-938.

Hirano, H. Y., and Y. Sano (1998). Enhancement of *Wx* gene expression and the accumulation of amylose in response to cool temperatures during seed develop-ment in rice. *Plant Cell and Physiology 39:* 807-812.

Hoshida, H., Y. Tanaka, T. Hibino, Y. Hayashi, A. Tanaka, T. Tanaka, and T. Takabe (2000). Enhanced tolerance to salt stress in transgenic rice that overexpress chloroplast glutamine synthetase. *Plant Molecular Biology 43:* 103-111.

Hossain, M. (1996). Recent developments in the Asian rice economy: Challenges for rice research. In *Rice research in Asia: Progress and priorities,* R. E. Evenson, R. W. Werdt, and M. Hossain (Eds.), pp. 17-33. Manila, Philippines: International Rice Research Institute.

Hossain, M. A., E. Huq, A. Grover, and E. S. Dennis (1996). Characterization of pyruvate decarboxylase genes from rice. *Plant Molecular Biology 31:* 761-770.

Hossain, M. A., J. D. McGee, A. Grover, E. Dennis, W. J. Peacock, and T. K. Hodges (1994). Nucleotide sequence of a rice genomic pyruvate decarboxylase gene that lacks introns: A pseudo-gene. *Plant Physiology 106:* 1697-1698.

Huq, E., M. A. Hossain, and T. K. Hodges (1995). Cloning and sequencing of a cDNA encoding pyruvate decarboxylase2 (Accession no. U27350) from rice. *Plant Physiology 109:* 722.

International Rice Research Institute (2006). *Rice knowledge base.* http://www.knowledgebank.irri.org/ (last accessed February 26, 2007).

Jackson, M. B., and P. C. Ram (2003). Physiological and molecular basis of suscep-tibility and tolerance of rice plants to complete submergence. *Annals of Botany 91:* 227-241.

Jaglo-Ottosen, K. R., S. J. Gilmour, D. G. Zarka, O. Schabenberger, and M. F. Thomashow (1998). *Arabidopsis CBF1* over-expression induces *COR* genes and enhances freezing tolerance. *Science 280:* 104-106.

Kasuga, M., Q. Liu, S. Miura, K. Yamaguchi-Shinozaki, and K. Shinozaki (1999). Improving plant drought, salt and freezing tolerance by gene transfer of a single stress-inducible transcription factor. *Nature Biotechnology 17:* 287-291.

Katiyar-Agarwal, S., M. Agarwal, D. R. Gallie, and A. Grover (2001). Search for the cellular functions of plant *Hsp100/Clp B* genes of lima bean. *Critical Reviews in Plant Science 20:* 277-295.

Katiyar-Agarwal, S., M. Agarwal, and A. Grover (1999). Emerging trends in agri-cultural biotechnology research: Use of abiotic stress-induced promoter to drive expression of a stress resistance gene in the transgenic system leads to high level stress tolerance associated with minimal negative effects on growth. *Current Science 77:* 1577-1579.

Katiyar-Agarwal, S., M. Agarwal, and A. Grover (2003). Heat tolerant basmati rice engineered by overexpression of *hsp101* gene. *Plant Molecular Biology 51:* 677-686.

Katiyar-Agarwal, S., A. Kapoor, and A. Grover (2002). Binary cloning vectors for efficient genetic transformation of rice plants. *Current Science 82:* 873-876.

Kawasaki, S., C. Borchert, M. Deyholos, H. Wang, S. Brazille, K. Kawai, D. Galbraith, and H. J. Bohnert (2001). Gene expression profiles during the initial phase of salt stress in rice. *Plant Cell 13:* 889-905.

Khanna-Chopra, R., and S. K. Sinha (1998). Prospects of success of biotechnological approaches for improving tolerance to drought stress in crops plants. *Current Science 74:* 25-34.

Khush, G. S. (1998). Strategies for increasing crop productivity. In *Crop productivity and sustainability: Shaping the future.* V. L. Chopra, R. B. Singh, and A. Varma (Eds.), pp. 19-43. New Delhi: Oxford/IBH.

Khush, G. S. (1999). Green Revolution: Preparing for the 21st century. *Genome 42:* 646-655.

Khush, G. S., and G. H. Toenniessen (1991). *Rice biotechnology.* Wallingford, UK: CAB International, and Manila, Philippines: International Rice Research Institute.

Kim, J. C., S. H. Lee, Y. H. Cheong, C. M. Yoo, S. I. Lee, H. J. Chun, D. J. Yun, J. C. Hong, S. Y. Lee, C. O. Lim, and M. J. Cho (2001). A novel cold inducible zinc-finger protein from soybean, SCOF-1 enhances cold tolerance in transgenic plants. *The Plant Journal 25:* 247-259.

Kratsch, H. A., and R. R. Wise (2000). The ultrastructure of chilling stress. *Plant Cell and Environment 23:* 337-350.

Kumria R., B. Waie, D. Pujni, and M. V. Rajam (1999). Biotechnology of rice: Present limitations and future prospects. *Plant Cell Biotechnology and Molecular Biology 1:* 1-12.

Kyozuka, J., M. Olive, W. J. Peacock, E. S. Dennis, and K. Shimamoto (1994). Promoter elements required for developmental expression of the maize *adh1* gene in transgenic rice. *Plant Cell 6:* 799-810.

Lal, S. K., C. Lee, and M. M. Sachs (1998). Differential regulation of enolase during anaerobiosis in maize. *Plant Physiology 118:* 1285-1293.

Laszlo, A., and P. Lawrence (1983). Parallel induction and synthesis of PDC and ADH in anoxic maize roots. *Molecular and General Genetics 192:* 110-117.

Lee, B. H., S. H. Won, H. S. Lee, M. Miyao, W. I. Chung, I. J. Kim, and J. Jo (2000). Expression of the chloroplast-localized small heat shock protein by oxidative stress in rice. *Gene 245:* 283-290.

Lee, Y. L., P. L. Chang, K. W. Yeh, T. L. Jinn, C. S. Kung, W. C. Lin, Y. M. Chen, and C. Y. Lin (1995). Cloning and characterization of a cDNA encoding an 18.0-kDa class-I low molecular weight heat shock protein in rice. *Gene 165:* 223-227.

Lee, Y., and H. Kende (2001). Expression of beta-expansins is correlated with internodal elongation in deepwater rice. *Plant Physiology 127:* 645-654.

Lutts, S., J. M. Kinet, and J. Bouharmont (1996). NaCl-induced senescence in leaves of rice (*Oryza sativa* L.) cultivars differing in salinity resistance. *Annals of Botany 78:* 389-398.

Marcotte, W. R., S. H. Russell, and R. S. Quatrano (1989). Abscisic acid-responsive sequences from the *Em* gene of wheat. *Plant Cell 1:* 969-976.

Minhas, D., and A. Grover (1999a). Towards developing transgenic rice plants tolerant to flooding stress. *Proceedings of the Indian National Academy of Science B65:* 33-50.

Minhas, D., and A. Grover (1999b). Transcript levels of genes encoding various glycolytic and fermentative enzymes change in response to abiotic stresses. *Plant Science 146:* 41-51.

Mohanty, A., H. Kathuria, A. Ferjani, N. Murata, P. Mohanty, and A. K. Tyagi (2002). Transgenics of an elite *indica* variety Pusa Basmati 1 harbouring *codA* gene are highly tolerant to salt stress. *Theoretical and Applied Genetics 106:* 51-57.

Mustroph, A., and G. Albrecht (2003). Tolerance of crop plants to oxygen deficiency stress: Fermentative activity and photosynthetic capacity of entire seedlings under hypoxia and anoxia. *Physiologia Plantarum 117:* 508-520.

Mundy, J., and N. H. Chua (1988). Abscisic acid and water stress induces the expression of a novel rice gene. *EMBO Journal 7:* 2279-2286.

Mundy, J., K. Yamaguchi-Shinozaki, and N. H. Chua (1990). Nuclear proteins bind conserved elements in the abscissic acid responsive promoter of a rice RAB gene. *Proceedings of the National Academy of Sciences of the USA 87:* 1406-1410.

Murata, N., O. Ishizaki-Nishizawa, S. Higashi, H. Hayashi, Y. Tasaka, and I. Nishida (1992). Genetically engineered alteration in the chilling sensitivity of plants. *Nature 356:* 710-713.

Nozue, F., M. Umeda, Y. Nagamura, and H. Uchimiya (1996). Characterization of cDNA encoding for phosphoglucose isomerase of rice (*Oryza sativa* L.). *DNA Sequence 6:* 127-135.

Nuccio, M. L., D. Rhodes, S. D. McNeil, and A. D. Hanson (1999). Metabolic engineering of plants for osmotic stress resistance. *Current Opinion in Plant Biology 2:* 128-134.

Ono, A., T. Izawa, N. H. Chua, and K. Shimomoto (1996). The *rab16b* promoter of rice contains two distinct abscisic acid responsive elements. *Plant Physiology 112:* 483-491.

Pareek, A., S. L. Singla, and A. Grover (1995). Immunological evidence for accumulation of two novel 104 and 90 kDa HSPs in response to diverse stresses in rice and in response to high temperature stress in diverse plant genera. *Plant Molecular Biology 29:* 293-301.

Pareek, A., S. L. Singla, and A. Grover (1997). Salt responsive proteins/genes in crop plants. In *Strategies for improving salt tolerance in higher plants,* P. K. Jaiwal, R. P. Singh, and A. Gulati (Eds.), pp. 365-391. New Delhi, India: Oxford and IBH Publication Co.

Pareek, A., S. L. Singla, and A. Grover (1998a). Protein alterations associated with salinity, desiccation, high and low temperature stresses, and abscisic acid applications in seedlings of Pusa 169, a high-yielding rice *(Oryza sativa)* cultivar. *Current Science 75:* 1023-1035.

Pareek, A., S. L. Singla, and A. Grover (1998b). Protein alterations associated with salinity, desiccation, high temperature and low temperature stresses and abscisic acid application in Lal nakanda, a drought tolerant rice cultivar. *Current Science* 75: 1170-1174.

Pareek, A., S. L. Singla, and A. Grover (1999). Analysis of stress proteins at four different developmental stages in field grown rice (cultivar Pusa 169) plants. *Current Science 76:* 81-86.

Park, J. M., C. J. Park, S. B. Lee, B. K. Ham, R. Shin, and K. H. Paek (2001). Overexpression of tobacco *Tsi1* gene encoding an EREBP/AP2-type transcription factor enhances resistance against pathogen attack and osmotic stress in tobacco. *Plant Cell 13:* 1035-1046.

Perata, P., and A. Alpi (1993). Plant responses to anaerobiosis. *Plant Science 93:* 1-7.

Pla, M., A. Goday, J. Vilardell, M. J. Guiltinan, W. R. Marcotte, M. F. Niogret, R. S. Quatrano, and M. Pages (1993). The cis-regulatory element CCACGTGG is involved in ABA and water stress responses of the maize gene *rab28*. *Plant Molecular Biology 21:* 259-266.

Quimio, C. A., L. B. Torrizo, T. L. Setter, M. Ellis, A. Grover, E. M. Abrigo, N. P. Oliva, E. S. Ella, A. L. Carpena, O. Ito, et al. (2000). Enhancement of submergence tolerance in transgenic rice plants overproducing pyruvate decarboxylase. *Journal of Plant Physiology 156:* 516-521.

Rahman, M., A. Grover, W. J. Peacock, E. S. Dennis, and M. Ellis (2001). Effects of manipulation of pyruvate decarboxylase and alcohol dehydrogenase levels on the submergence tolerance of rice. *Australian Journal of Plant Physiology 28:* 1231-1241.

Reddy, A. R., W. Ramakrishna, A. Chandra Sekhar, I. Nagabhushana, P. R. Babu, M. F. Bonaldo, M. B. Soares, and J. L. Bennetzen (2002). Novel genes are enriched in normalized cDNA libraries from drought stressed seedlings of rice (*Oryza sativa* L. subsp. *indica* cv. Nagina 22). *Genome 45:* 204-211.

Ricard, B., B. Mocquot, A. Fournier, M. Delseny, and A. Pardet (1986). Expression of alcohol dehydrogenase in rice embryos under anoxia. *Plant Molecular Biology 7:* 321-329.

Ricard, B., J. Rivoal, and A. Pardet (1989). Rice cytosolic glyceraldehyde-3-phosphate dehydrogenase contains two subunits differentially regulated by anaerobiosis. *Plant Molecular Biology 12:* 131-139.

Ricard, B., J. Rivoal, A. Spiteri, and A. Pardet (1991). Anaerobic stress induces the transcription and translation of sucrose synthase in rice. *Plant Physiology 95:* 669-674.

Rice Full-length cDNA Consortium. (2003). Collection, mapping and annotation of over 28,000 cDNA clones from *japonica* rice. *Science 301:* 376-379.

Rivoal, J., B. Ricard, and A. Pardet (1990). Purification and partial characterization of pyruvate decarboxylase from *Oryza sativa* L. *European Journal of Biochemistry 194:* 791-797.

Rivoal, J., S. Thind, A. Pardet, and B. Ricard (1997). Differential induction of pyruvate decarboxylase subunits and transcripts in anoxic rice seedlings. *Plant Physiology 14:* 1021-1029.

Rohila, J. S., R. K. Jain, and R. Wu (2001). *Agrobacterium*-mediated transformation of basmati rice to express the barley *hva 1* gene for enhanced tolerance to abiotic stress. *Rice Genetics Newsletter 18:* 84-86.

Sahi, C., M. Agarwal, M. K. Reddy, S. K. Sopory, and A. Grover (2003). Isolation and expression analysis of salt stress associated expressed sequence tags from contrasting rice cultivars using PCR-based subtraction method. *Theoretical and Applied Genetics 106:* 620-628.

Saijo, Y., S. Hata, J. Kyozuka, K. Shimamoto, and K. Izui (2000). Over-expression of a single Ca^{2+}-dependent protein kinase confers both cold and salt/drought tolerance on rice plants. *The Plant Journal 23:* 319-327.

Sakamoto, A., and A. N. Murata (1998). Metabolic engineering of rice leading to biosynthesis of glycine betaine and tolerance to salt and cold. *Plant Molecular Biology 38:* 1011-1019.

Satake, T., and S. Yoshida (1978). High temperature induced sterility in *indica* rice at flowering. *Journal of Crop Science 447:* 6-17.

Sauter, M., and H. Kende (1992). Gibberellin-induced growth and regulation of the cell division cycle in deep water rice. *Planta 188:* 362-368.

Seki M., M. Narusaka, H. Abe, M. Kasuga, K. Yamaguchi-Shinozaki, P. Carninci, Y. Hayashizaki, and K. Shinozaki (2001). Monitoring the expression pattern of 1300 *Arabidopsis* genes under drought and cold stresses by using a full length cDNA microarray. *Plant Cell 13:* 61-72.

Setter, T. L., and E. S. Ella (1994). Relationship between coleoptile elongation and alcoholic fermentation in rice exposed to anoxia. I. Importance of treatment conditions and different tissues. *Annals of Botany 74:* 265-271.

Setter, T. L., T. Kupkanchankul, and H. Greenway (1989). Submergence of rice. II. Adverse effects of low CO_2 concentrations. *Australian Journal of Plant Physiology 16:* 265-278.

Setter, T. L., I. Waters, B. J. Atwell, T. Kupkanchankul, and H. Greenway (1987). Carbohydrate status of terrestrial plants during flooding. In *Plant life in aquatic and amphibious habitats,* R. M. M. Crowford (Ed.), pp. 411-433. Oxford: Blackwell.

Shimamoto, K. (Ed.). (1999). *Molecular biology of rice.* New York: Springer Verlag.

Singla, S. L., and A. Grover (1993). Antibodies raised against yeast HSP104 cross-react with a heat- and abscisic acid-regulated polypeptide in rice. *Plant Molecular Biology 22:* 1177-1180.

Singla, S. L., and A. Grover (1994). Detection and quantitation of a rapidly accumulating and predominant 104 kDa heat shock polypeptide in rice. *Plant Science 97:* 23-30.

Singla, S. L., A. Pareek, and A. Grover (1997). High temperature. In *Plant ecophysiology,* M. N. V. Prasad (Ed.), pp. 101-127. New York: John Wiley.

Singla, S. L., A. Pareek, A. K. Kush, and A. Grover (1998). Distribution patterns of the 104-kDa stress-associated protein of rice reveal its constitutive accumulation in seeds and disappearance from the just-emerged seedlings. *Plant Molecular Biology 37:* 911-919.

Slafer, G. A., and H. M. Rawson (1994). Sensitivity of wheat phasic development to major environmental factors: A re-examination of some assumptions by physiologists and modelers. *Australian Journal of Plant Physiology 21:* 393-426.

Swaminathan, M. S. (1982). Biotechnology research and third world agriculture. *Science 218:* 967-972.

Syngenta (2007). *FarmAssist pest library: Rice.* http://www.farmassist.com/ agronomic/english/library/Rice.asp (last accessed February 26, 2007).

Tashiro, T., and I. F. Wardlaw (1991). The effect of high temperature on kernel dimensions and the type and occurrence of kernel damage in rice. *Australian Journal of Agriculture and Research 42:* 485-496.

Tseng, T. S., S. S. Tzeng, K. W. Yeh, C. H. Yeh, G. C. Chang, Y. M. Chen, and C. Y. Lin (1993). The heat-shock response in rice seedlings: Isolation and expression of cDNA that encodes class I low-molecular-weight heat-shock proteins. *Plant Cell Physiology 34:* 165-168.

Tyagi, A., and J. P. Khurana (2003). Plant molecular biology and biotechnology research in the post-recombinant DNA era. *Advances in Biochemical Engineering/ Biotechnology 84:* 91-121.

Tyagi, A. K., and A. Mohanty (2000). Rice transformation for crop improvement and functional genomics. *Plant Science 158:* 1-18.

Tyagi, A. K., A. Mohanty, S. Bajaj, A. Chaudhary, and S. C. Maheswari (1999). Transgenic rice: A valuable monocot system for crop improvement and gene research. *Critical Reviews in Biotechnology 19:* 41-79.

Tzeng, S. S., K. W. Yeh, Y. M. Chen, and C. Y. Lin (1992). Two *Oryza sativa* genomic DNA clones encoding 16.9 kilodalton heat shock proteins. *Plant Physiology 99:* 1723-1725.

Umeda, M., and H. Uchimiya (1994). Differential transcript levels of genes associated with glycolysis and alcohol fermentation in rice plants (*Oryza sativa* L.) under submergence stress. *Plant Physiology 106:* 1015-1022.

Van Breusegem, F., R. Dekeyser, A. B. Garcia, B. Claes, J. Gielen, M. van Montagu, and A. B. Caplan (1994). Heat-inducible rice hsp82 and hsp70 are not always co-regulated. *Planta 193:* 57-66.

Vergara, B. S., B. Jackson, and S. K. De Datta (1976). Deep-water rice and its response to deep-water stress. In *Climate and rice,* pp. 301-319. Los Baños, Philippines: International Rice Research Institute.

Widawsky, D. A., and J. C. O'Toole (1990). *Prioritizing the rice biotechnology research agenda for eastern India.* New York: Rockefeller Foundation.

Xie, Y., and R. Wu (1989). Rice alcohol dehydrogenase genes: Anaerobic induction, organ specific expression and characterization of cDNA clones. *Plant Molecular Biology 13:* 53-68.

Xu, D., X. Duan, B. Wang, B. Hong, T. D. Ho, and R. Wu (1996). Expression of a late embryogenesis abundant protein gene, *HVA1,* from barley confers tolerance to water deficit and salt stress in transgenic rice. *Plant Physiology 110:* 249-257.

Xu, D., D. McElroy, R. W. Thornburg, and R. Wu (1993). Systemic induction of a potato pin2 promoter by wounding, methyl jasmonate and abscisic acid in transgenic rice plants. *Plant Molecular Biology 22:* 573-588.

Yamaguchi-Shinozaki, K., J. Mundy, and N. H. Chua (1990). Four tightly linked rab genes are differentially expressed in rice. *Plant Molecular Biology 14:* 29-39.

Yokoi, S., S. Hogashi, S. Kishtiani, N. Murata, and K. Toriyama (1998). Identification of the cDNA for *Arabidopsis* glycerol-3-phosphate acyltransferase (GPAT) confers unsaturation of fatty acids and chilling tolerance of photosynthesis on rice. *Molecular Breeding 4:* 269-275.

Yoshida, S., T. Satake, and D. J. Mackill (1981). High temperature stress in rice. *IRRI research paper series: No 67,* pp. 1-15. Los Baños, Philippines: International Rice Research Institute.

Young, T. E., J. Ling, C. J. Geiser-Lee, R. L. Tanguay, C. Caldwell, and D. R. Gallie (2001). Developmental and thermal regulation of the maize heat shock protein HSP101. *Plant Physiology 127:* 777-791.

Young, N. S., C. H. Yeh, Y. M. Chen, and C. Y. Lin (1999). Molecular characterization of *Oryza sativa* 16.9 kDa heat shock protein. *Biochemical Journal 344:* 31-38.

Yu, J., S. Hu, J. Wang, G. K. Wong, S. Li, B. Liu, Y. Deng, L. Dai, Y. Zhou, X. Zhang, et al. (2002). A draft sequence of the rice genome (*Oryza sativa* L. ssp. *indica*). *Science 296:* 79-92.

Zeng, L., J. A. Poss, C. Wilson, A.-S. E. Draz, G. B. Gregorio, and C. M. Grieve (2003). Evaluation of salt tolerance in rice genotypes by physiological characters. *Euphytica 129:* 281-292.

Zhu, B., J. Su, M. C. Chang, D. P. S. Verma, Y. L. Fan, and R. Wu (1998). Overexpression of a pyrroline-5-carboxylate synthetase gene and analysis of tolerance to water and salt stress in transgenic rice. *Plant Science 139:* 41-48.

Chapter 10

Rice End-Use Quality

J. S. Bao
R. Fjellstrom
C. Bergman

INTRODUCTION

Rice grain end-use quality denotes different characteristics to different people across the world and to different sectors of the food industry. Throughout the world, rice is cooked, processed, and consumed in hundreds of tremendously diverse ways, so it is little wonder that the end-use qualities required of rice vary to a large degree. These differing end-use quality characteristics fall into three categories: cooking and sensory quality, milling quality, and nutritional quality.

Most rice cultivar development programs across the world are primarily focused on producing agronomically superior cultivars that have grain characteristics traditional for their region. For example, in the southern United States, public breeding programs expend most of their efforts developing conventional long-grain types (i.e., long thin grains that cook to a firm and nonsticky texture), whereas in South Korea, conventional medium-grain types (i.e., short wide grains that cook to a soft and sticky texture) are the primary focus. Over the past few years, however, the demand for rice with specialty characteristics (e.g., aromatic) has increased across the world, whereas the world price for conventional long-grain rice has decreased (Sombilla and Hossain, 2001). Consequently, there is an increasing interest by some breeding programs to develop rice cultivars with value-added specialty characteristics suitable for use in specific industrial applications. Nearly 420,500 samples of rice and its related species are maintained in germplasm collections across the world. In these valuable genetic resources lies the opportunity for expanding the industrial applications of rice and for creating superior-quality conventional rice types using plant breeding techniques. Likewise, genetic transformation technology also offers opportunities to tailor rice cultivars to have specific end-use quality characteristics.

Creating rice cultivars with novel combinations of grain traits often requires the use of unadapted breeding parents, which consequently do not perform well under local growing conditions. Using unadapted breeding material and targeting unique grain characteristics makes the cultivar development process very challenging. As a result, some rice breeders have put together teams of food scientists, chemists, and molecular geneticists to define targeted quality characteristics and to assist in the evaluation of breeding progeny. For example, food scientists are helping to define the rice sensory attributes that today's discriminating consumers are demanding (Champagne et al., 1998). Chemists are defining the chemical composition of rice that results in specific aspects of sensory and processing quality (Bergman et al., 2000). Geneticists are studying the inheritance and genetic control of end-use quality traits, developing genetic markers that are being incorporated into breeding programs to hasten cultivar development and modifying rice characteristics using genetic transformation technology (Bergman et al., 2001; Krishnamurthy and Giroux, 2001).

The genomics era is poised to have an increasing impact on our understanding of the genetics and physiology that control rice end-use quality. Furthermore, rice is being used as a model system for other cereal grains; thus, much of what is learned about the genetics of the end-use quality of rice will also affect our knowledge of the end-use quality of other cereal grains. This chapter summarizes recent findings regarding the genetics of end-use quality traits important to conventional and specialty types of rice. The primary emphasis is placed on research studies performed at the molecular level, such as the development of genetic markers, quantitative trait locus (QTL) mapping, candidate gene analysis, and gene expression and regulation. Breeding methods and strategies using both conventional methods and molecular breeding techniques, as in molecular marker-assisted selection (MAS) and transgenic technology, are also presented.

CURRENT UNDERSTANDING

Cooking and Sensory Quality

Physicochemical criteria have been established worldwide to evaluate the cooking and sensory quality of rice—specifically, amylose content

(AC; more accurately termed *apparent* amylose content) (Juliano, 1971; Takeda et al., 1987; Williams et al., 1958), gelatinization temperature (GT) (Little et al., 1958), and gel consistency (GC) (Cagampang et al., 1973). The starch fraction known as *amylose* is the most important grain constituent that influences rice end-use quality, as it is the major determinant of cooked rice texture (Halick and Keneaster, 1956; Webb, 1975). GT is the temperature range in which starch granules swell irreversibly and burst, and is related to rice cooking time. GC measures the viscosity of pastes or gels made from milled rice flour or starch, and is an index of cooked rice texture, especially among rice varieties of high amylose content. Starch viscometric properties during heating and cooling are presently tested using a Rapid Visco Analyzer (RVA) to simulate the cooking process, and serve as an indicator of the eating and cooking characteristics of milled rice and rice flour.

AMYLOSE CONTENT

Genetic Basis and Gene Regulation

Control of the AC in rice has been the subject of numerous genetic studies. In most crosses, AC has been found to be primarily controlled by an allelic series at one locus with major effects, and by one or more modifier genes with minor effects (Bollich and Webb, 1973; Chang and Li, 1991; Kumar et al., 1987; Kumar and Khush, 1988; McKenzie and Rutger, 1983; Pooni et al., 1993). In some cases, two dominant, complementary genes have been postulated to explain the inheritance of AC in certain crosses between low- and high-AC cultivars (McKenzie and Rutger, 1983). Through QTL analysis, AC has also been reported to be controlled by a major QTL as well as by other, minor QTLs (He et al., 1999; Tan et al., 1999; Bao et al., 2000a). This major QTL is at the *Waxy (Wx)* locus, which encodes granule-bound starch synthase (GBSS). In a double haploid population derived from parents with similar AC (IR64/*Azucena*), only a minor QTL was detected for AC (Bao et al., 2002c). Therefore, the inheritance of AC appears to be complex due to pleiotropy, epistasis, cytoplasmic effects, or the triploid nature of endosperm (Pooni et al., 1993; Shi et al., 1997). For example, genes such as *dull (du)* and *amylose extender* can modify AC (Asaoka et al., 1993; Juliano, 1990; Yano et al., 1985).

The *Wx* gene encoding GBSS controls endosperm amylose synthesis in the grass family, as confirmed by the existence of *wx* mutants in which both AC and GBSS activity are low (Preiss, 1991; Smith et al., 1997).

In rice, genotypes homozygous for the Wx^a allele reportedly produce 10 times more Wx gene product than their homozygous Wx^b counterparts (Sano, 1984). The *indica* rice in this study all had the Wx^a allele and showed a higher AC, which likely explains why *indica* rice is generally higher in AC than *japonica* rice. Upon introduction of a wx^b allele into an *indica* background, AC was reduced to less than that when introduced in the wx^b *japonica* background, suggesting that the background of the Wx allele also affects AC (Mikami et al., 2000). Another report also showed a similar tendency of endosperm with the Wx^{op} allele to have a higher AC in the *japonica* background than in the *indica* background (Mikami et al., 1999). It is possible that some modifiers in the *indica* background reduce the level of expression of the Wx gene.

Wang et al. (1995) observed that AC and the level of GBSS in 31 rice cultivars was correlated with a cultivar's ability to excise the leader intron of the Wx transcript. All of the high-amylose cultivars accumulated significant amounts of completely processed Wx mRNA. All of the low-amylose cultivars accumulated both completely processed Wx mRNA and a partially processed transcript containing the leader intron, whereas waxy cultivars contained only incompletely processed, mutant wx transcripts. This observation was confirmed by reports that low-amylose cultivars have the sequence AG<u>T</u>TATA at the putative leader intron 5′ splice site, whereas all intermediate- and high-amylose cultivars have AG<u>G</u>TATA (Ayres et al., 1997; Bligh et al., 1998; Cai et al., 1998; Hirano and Sano, 1998; Isshiki et al., 1998; Wang et al., 1995). This point mutation at the 5′ junction of the first intron in the Wx gene results in an inefficient splicing of the mature transcript (Bligh et al., 1998; Cai et al., 1998; Isshiki et al., 1998).

Genetic analyses have shown that in addition to the Wx gene's role in determining AC, there are modifiers involved in the regulation of this starch fraction, as identified by mutants of such modifiers. For example, *du* or low-amylose mutant genes are known to reduce the level of Wx gene product and decrease AC in rice endosperm. As the *du* genes are independent of the Wx locus, both *cis*- and *trans*-acting factors that regulate AC through changes in the Wx gene expression could be present in rice.

To date, *du* mutants have only been reported in *japonica* rice. Among the low-amylose mutant genes, *du* controls dull endosperms, whose grain transparency is between that of waxy and nonwaxy ones. The grain appearance seen in dull-type endosperms is generally a visual indication of a lower AC. The *du2-2* gene causes a very low AC and chalky endosperms in the genetic background of *japonica* types. Near-isogenic lines with $Wx^a du2-2$ and $Wx^b du2-2$ in the background of *indica* and *japonica* have lower AC than their counterparts $Wx^a Du2-2$ and $Wx^b Du2-2$, respectively (Dung et al., 2000).

In *du-1* and *du-2* mutants, spliced Wx^b transcripts are highly reduced, whereas the processing of transcripts derived from other genes that are highly expressed in endosperm is apparently not affected. Genetic and molecular analyses of the effects of *du-1* and *du-2* on Wx^a pre-mRNA with normal splice sites indicate that these two mutations do not affect the processing of the Wx^a pre-mRNA after splicing. This suggests that *du-1* and *du-2* are mutations of genes required for the efficient splicing of mutated Wx^b pre-mRNA. Furthermore, *du-1* and *du-2* showed differential effects in endosperm versus pollen. Although both mutations caused similar effects on the splicing of Wx^b transcripts in endosperm, *du-1* caused a greater reduction of Wx^b mRNA in pollen than in endosperm, whereas *du-2* showed the reverse. Based on these results, it has been proposed that the *du-1* and *du-2* loci of rice encode tissue-specific factors that regulate the splicing of pre-mRNA (Isshiki et al., 2000).

Opaque rice genotypes do contain amylose (approximately 10%) in spite of their waxy-appearing endosperm. Genetic experiments have revealed that an allele of the *Wx* gene, Wx^{op}, controls the *opaque* trait (Mikami et al., 1999). Comparison of gene expression among near-isogenic lines (NILs) that carried four different alleles (Wx^a, Wx^b, Wx^{op}, and *wx*) found that the levels of GBSS bound to starch granules was lower in the NILs with Wx^{op} than in those with Wx^b, and showed a positive correlation between GBSS levels and endosperm AC. GBSS in anthers was also markedly reduced in the NILs with Wx^{op} and *wx*, showing an altered expression in tissue specificity in the Wx^{op} line. Analysis of the *Waxy* gene leader intron at the AGGTATA/AGTTATA polymorphism has indicated that the donor site of the Wx^{op} is the same as that of Wx^a, indicating that Wx^{op} originated from Wx^a but not from Wx^b, although the molecular mechanism for opaque endosperm is unknown (Mikami et al., 1999).

Environmental Effects

AC is affected by environmental factors—particularly by temperature—during seed development. The same cultivar grown in different environments may vary by up to 6% in AC (Juliano and Pascual, 1980). A larger variation appears to exist for the low-AC types, where cultivars that typically have 12-15% AC when grown at higher temperatures will have up to 18% at lower temperatures (Bao et al., 2000b; Larkin and Park, 1999). Genetic analyses have also shown that genotype-environment interactions for AC are significant (Shi et al., 1997). Through QTL analyses of AC, different minor QTLs were identified in different environments in addition to

Wx, which was the major QTL associated with AC identified in all environments (Bao et al., 2000a).

As with AC, the amount of GBSS increases at low temperatures compared to plants grown under warmer conditions (Hirano and Sano, 1998). *Wx* gene expression is activated reversibly in response to cool temperatures, as noted by the higher levels of *Wx* transcript at lower temperatures. The longer rice plants are exposed to cool temperatures, the higher the levels of *Wx* protein and AC. In transgenic plants with the glucuronidase (GUS) gene under the control of the *Wx* gene promoter, enhancement of GUS activity was also detected at lower temperatures, suggesting that the *Wx* promoter is temperature sensitive.

The single nucleotide polymorphism (SNP) of AGGTATA and AGTTATA at the leader intron 5' splice site displays differential temperature sensitivity (Larkin and Park, 1999). Cultivars with the sequence AGTTATA were reported to show a substantial increase in the accumulation of mature *Wx* transcripts at 18°C compared to 25°C or 32°C. The selection of leader intron 5' splice sites was also affected by temperature. A 5' splice site –93 upstream from that used in high-amylose varieties predominates at 18°C. At higher temperatures, there is increased use of a 5' splice site at –1 and a nonconsensus site at +1.

Genetic Markers Associated with the Wx Gene

There are currently three types of molecular markers that have been found to be closely linked to the *Wx* gene of rice:

1. *RFLP:* Two alleles of the *Wx* locus have been identified using RFLP analysis (Sano et al., 1986). The allele wx^a predominates in the *indica* type and its relatives, whereas wx^b is restricted to the *japonica* type (Sano et al., 1986). This marker could partially explain the difference in AC between *indica* and *japonica* subspecies. However, it is obvious that these two RFLP alleles are not adequate to explain all the observed variation in AC among commercial rice cultivars, because some *indica* rice have lower AC than *japonica* rice.
2. *Microsatellite:* A $(CT)_n$ microsatellite was identified in the 5'-untranslated region of the *wx* gene (Bligh et al., 1995). Ayres et al. (1997) identified eight *Wx* microsatellite alleles, which together explained more than 82% of the variation in AC of the nonglutinous U.S. cultivars grown in one location. Shu et al. (1999) identified seven $(CT)_n$ alleles in a total of 74 nonwaxy rice cultivars, which explained

91% of the total variation for AC. Bergman et al. (2001) reported that this microsatellite explained 88% of the variation in the AC of 198 nonwaxy U.S. cultivars and breeding lines of diverse parentage grown in four locations (see Figure 10.1 and Table 10.1). Another microsatellite, $(AATT)_n$, was reported with two alleles in the first intron of the *Wx* gene, 182 bp downstream of the $(CT)_n$ repeat (Tan and Zhang, 2001; Xiong et al., 1998). The $(AATT)_5$ allele shows lower AC than $(AATT)_6$ (Tan et al., 2001).

3. *SNP:* As mentioned before, there is a single nucleotide polymorphism (AGGTATA/AGTTATA) at the leader intron 5′ splice site of the *Wx* gene (Ayres et al., 1997; Wang et al., 1995), which explained 79.7% of the variation in the AC of nonwaxy U.S. cultivars (Ayres et al., 1997).

Genetic Marker-Assisted Breeding

The $(CT)_n$ microsatellite in the 5′-untranslated region of *Wx* gene has shown great utility for selecting breeding progeny with the desired amylose content and other AC-related traits. A relatively simple, rapid, and inexpensive alkali-based DNA extraction method has been developed

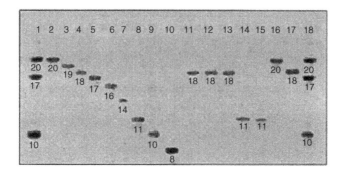

FIGURE 10.1. Polymerase chain reaction (PCR) products from a microsatellite $(CT)_n$ (cytosine and thymine) associated with the *wx* gene separated using polyacrylamide plus Spreadex gel electrophoresis. Numbers above the bands are lane numbers, whereas numbers below designate the band's microsatellite class. Lanes 1 and 18 are controls used for scoring the gel, $(CT)_{20}$, $(CT)_{17}$, and $(CT)_{10}$; lanes 2-10, one band for each known allele for the waxy microsatellite; lanes 11-13, *Bengal* (leaf material, hulled kernels, and milled kernels, respectively); lane 14, *Dixiebelle,* milled parboiled kernels; lane 15, L205, stabilized bran; lane 16, *Cypress,* instant rice kernels; lane 17, *Bengal,* crisp rice. *Source:* Bergman et al. (2001).

TABLE 10.1. Apparent amylose content of six (CT) microsatellite classes in 198 U.S. cultivars and breeding progeny.

Microsatellite class	Apparent amylose (%)
$(CT)_{10}$	25.00_a
$(CT)_{11}$	24.08_a
$(CT)_{14}$	21.59_b
$(CT)_{20}$	21.33_b
$(CT)_{17}$	14.72_c
$(CT)_{18}$	13.82_c

Notes: Apparent amylose content for each entry is the mean of samples obtained from plots grown in Texas, Mississippi, Louisiana, and Arkansas.

Values followed by the same letter are not significantly different ($p < .05$).

to facilitate the use of the *Wx* microsatellite in rice breeding populations (Bergman et al., 2001). This method is in routine use in U.S. and Australian rice breeding programs. Enhanced resolution of allele classes and separation speed is achieved by electrophoresis of polymerase chain reaction (PCR) amplification products in a polyacrylamide and Spreadex gene matrix using a triple-wide mini-electrophoresis unit for increased throughput (Figure 10.1).

Bao (2002) studied the relationship between the microsatellites and AC and other starch quality parameters with 36 F_2 plants derived from the cross between *Longtefu* A $(Wx_{11}Wx_{11})$ and 371 $(Wx_{18}Wx_{18})$. The F_2 plants could be divided into three groups according to the microsatellite markers: the $Wx_{11}Wx_{11}$, $Wx_{11}Wx_{18}$, and $Wx_{18}Wx_{18}$ groups. The F_3 seeds with the same *Wx* alleles were bulked to test the starch quality parameters, which implies that only the *Wx* allele is different among the bulked samples whereas the genetic background is identical. The results indicated that much of the variation in AC, GT, GC, and starch paste viscosity parameters could be explained by the *Wx* microsatellite.

Zhou et al. (2003) reported an example of using the *Wx* microsatellite to assist in hybrid rice quality improvement. The female parent of a number of widely used Chinese hybrids, *Zhenshan 97,* does not have the locally desired quality due to its high AC, hard GC, low GT, and chalky endosperm. The three traits considered of greatest importance for rice cooking and eating quality—AC, GC, and GT—are primarily controlled by the *Wx* locus and its linked genomic region (Tan et al., 1999). The eating and cooking quality of *Zhenshan 97* was improved by introgressing the *Wx* gene region from *Minghui 63 (wx-MH)*, a restorer line that has medium AC, soft GC, and high GT. The *wx-MH* fragment was transferred to *Zhenshan 97B* (maintainer line) by three backcrosses and one selfing, then from *Zhenshan 97B* to *Zhenshan 97A* (cytoplasmic male sterile line) by a cross and a backcross. Genetic marker-assisted selection was applied in the series to select for individuals carrying *wx-MH* to identify recombination between the *Wx* and flanking markers and also to recover the genetic background of the recurrent parent. According to the marker genotypes, the improved versions of *Zhenshan 97B* and *Zhenshan 97A*, or *Zhenshan 97B (wx-MH)* and *Zhenshan 97A (wx-MH)*, are the same as the originals except for the *Waxy* gene region of less than 6.1 cM in length. The selected lines and their hybrids with *Minghui 63*, or *Shanyou 63 (wx-MH)*, had the desired lower AC and an increased GC and GT, coupled with a reduced grain opacity. Field examinations of agronomic performance revealed that *Zhenshan 97B (wx-MH)* and *Shanyou 63 (wx-MH)* were essentially the same as the originals except for a significant decrease in grain weight.

Transgenic Engineering of the Wx Gene

Regulation of the expression of the *Wx* gene using transgenic technology could improve AC and other related starch physicochemical properties and widen the functionality of rice starch. Shimada et al. (1993) isolated the *Wx* gene between exon 4 and exon 9 and inserted it in antisense orientation between the 35 S promoter and the GUS gene of pBZ221. The resulting plasmid was introduced into rice protoplasts by electroporation. Some of the rice seeds from these transformants showed a significant reduction in AC, indicating that even when intron sequences are included, antisense constructs can reduce the level of expression of a targeted gene. Terada et al. (2000) transformed rice plants with a 2.3 kbp *Wx* cDNA having 450 bp of the *Wx* first intron in reverse orientation to rice *Wx* using the maize alcohol dehydrogenase *(adh1)* promoter. Of ten independent transgenic lines analyzed, four showed various degrees of reduction in amylose and GBSS levels in the endosperm. In two transgenic lines, a complete absence of amylose was found and an opaque endosperm observed. In one of the transgenic lines, the presence of the antisense *Wx* gene cosegregated with the reduction in AC. In the same line, a reduction in the level of *Wx* mRNA was observed in immature endosperm. Interesting enough, this reduction was observed only with mature spliced transcripts, but not with unspliced transcripts. Reduced amylose synthesis was also observed in the pollen grains of four transgenic lines. These results suggest that the integrated antisense *Wx* gene caused a reduction in amylose synthesis in endosperms and pollen grains of trans-genic rice carrying the antisense *Wx* cDNA. Another example of the transfor-mation of rice with the *Wx* gene was a study of transgene silencing (Itoh et al., 1997). Changes in the tissue-specific expression of the *Wx* gene resul-ted in the cosuppression of the transgene through gene silencing. These re-sults indicate that the manipulation of starch and other carbohydrates in rice grain is possible using antisense methodology.

GELATINIZATION TEMPERATURE

Milled rice GT is typically predicted using the alkali spreading test (Little et al., 1958). The inheritance of GT is not fully understood. One to three genes with several modifiers reportedly control this trait (Chang and Li, 1991; Hue and Choi, 1973). Quantitative inheritance of GT has also been reported (Ghosh and Govindaswamy, 1972). McKenzie and Rutger (1983) found that GT in one cross was probably controlled by a single gene, whereas in five other crosses, the segregation patterns did not conform to any identifi-able genetic model. Puri and Siddiq (1983) reported that additive gene action played a major role in the expression of this trait, although some crosses did show predominantly nonadditive gene action. GT has also been reported to be under complex genetic control by seed effects controlled by nuclear genes of endosperm cells and cytoplasmic and maternal effects controlled by genes of the maternal plant (Bao et al., 2002b; Shi et al., 1997). The genetic heri-tability of GT is high, thus allowing for early selection in breeding efforts (Bao et al., 2002b; Shi et al., 1997).

Through QTL analysis, He et al. (1999) and Bao et al. (2000a) reported that GT was controlled mainly by the *alk* gene, loosely linked (ca. 28 cM) with *Wx* on chromosome 6. Tan et al. (1999, 2001) indicated that GT, tested by the alkali spreading test or in a differential scanning calorimeter, was controlled by the *Wx* gene or by a closely linked genomic region. Zhou et al. (2003) introduced the *Wx* locus from a low-AC variety, *Minghui 63,* into a high-AC variety, *Zhenshan 97B.* The progeny were found to have a similar GT as *Minghui 63.* More recently, Umemoto et al. (2002) indicated that the starch synthase IIa (*SSIIa*) gene is located at the *alk* locus on chromosome 6. In a doubled haploid (DH) progeny (i.e., plants derived from anther cul-ture of an F_1 hybrid) population derived from parents with similar amylose content (IR64 and *Azucena*), Bao et al. (2002c) found that the starch bran-ching enzyme I (*SBEI*) gene locus on chromosome 6 controlled GT.

Environmental conditions—especially temperature—have a large effect on GT. The same cultivar grown in different regions will vary in starch GT.

In general, GT is elevated when the temperature during grain development is relatively high, and the GT is lower when this relative temperature is lower. Analyzing genotype-environment interactions, Shi et al. (1997) found GT to be primarily controlled by maternal interaction effects and cytoplasmic interaction effects. Bao et al. (2000a) reported that the *alk* locus had a major effect on GT across several environments, whereas other loci with smaller effects were identified in one or more environments.

The aforementioned studies and several other studies that examined starch mutant gene effects have indicated that the structure of amylopectin is involved in starch GT. Jane et al. (1999) analyzed amylopectin structure in relation to other physicochemical properties of starches of different botanical origins. The results showed that starches with relatively short average amylopectin branch chain lengths (e.g., degree of polymerization 11-16 in waxy rice) displayed low gelatinization temperatures compared to starches with relatively long amylopectin branches (e.g., degree of polymerization 18-21 in wheat and barley).

Nakamura et al. (2002) characterized the structure of amylopectin in the endosperm of Asian rice to determine the relationship between amylopectin structure and starch physicochemical properties. The results indicated that almost all rice amylopectin could be classified into L- or S-types. The L-type amylopectin was different from the S-type amylopectin in that the numbers of short α-1,4-glucan chains of degree of polymerization (DP) ≤ 10 were less than 20% of the total α-1,4-glucan chains of DP ≤ 24. No significant difference between both types was found in the proportion of long chains of DP ≥ 25. Among the 129 rice varieties examined, only a single variety *(Khauk Yoe)*, belonging to the tropical *japonica* rice group, had an intermediate (M-type) amylopectin structure. The proportion of amylopectin chains with DP ≤ 10 was negatively correlated with the onset temperature of starch gelatinization, whereas no correlation was observed between amylose content and starch thermal properties.

Several enzymes reportedly participate in amylopectin biosynthesis and thus may be associated with GT and the structure of amylopectin. Nishi et al. (2001) found that the *amylose extender (ae)* mutant of rice specifically altered the structure of amylopectin by reducing short chains with DP ≤ 17, with the greatest decrease in chains with DP of 8 to 12. This amylopectin alteration was associated with starch gelatinization, as determined by solubility in a urea solution. The gene encoding starch branching enzyme IIb *(SBEIIb)* is at the site of the *ae* mutation, and the activity of the SBEIIb enzyme was reduced. The activity of soluble starch synthase I (SSSI) in the *ae* mutant was also significantly lower than in the wild type, thus suggesting that the mutation had a pleiotropic effect on SSSI activity.

Other genes, such as those encoding additional soluble starch synthases (e.g., *SSSIIb*), starch branching enzymes (*SBEI* and *SBEIII*), and starch debranching enzymes (isoamylase and pullulanase) have also been reported to be involved in determining the structure of rice amylopectin (Kubo et al., 1999; Smith et al., 1997). Mutations in these genes, therefore, may also affect rice amylopectin structure and, as a consequence, alter GT.

Because there are so many genes controlling amylopectin synthesis, it is not surprising that the genetics controlling GT are complex. Incongruous results are obtained when analyzing inheritance arising from different rice genotypes carrying different alleles of starch synthesis genes. It is also understandable that the relationship between AC and GT can become confused in separate studies using different rice genotypes, as AC is controlled by the *Wx* gene, which displays linkage to the *SSSI* and *SSSIIa* genes. The correlation has been reported to be significantly negative (Tetens et al., 1997) or null (Bhattacharya et al., 1999; Nakamura et al., 2002) in different studies.

In a breeding program, screening for desirable GT can be done effectively in the early generations. However, complex phenotypic performance in the offspring of some crosses has been described (Chang and Li, 1991). Therefore, selection for GT in the later generations should be done at least once to ascertain that the desired GT has been achieved. Availability of MAS methodology for GT would be advantageous. Genetic markers closely linked to genes controlling amylopectin synthesis are hypothesized to be associated with GT. Research in support of this hypothesis was reported by Bao et al. (2002a). Waxy rice genotypes ($n = 56$) were studied to determine the relationship between microsatellites in the *SBEI* and *SSSI* genes and GT and other physicochemical properties. Three $(CT)_n$ microsatellite alleles were found at the *SBEI* locus: $(CT)_8$ and $(CT)_{10}$ together with an insertion sequence of CTCTCGGGCGA, and $(CT)_8$ alone without the insertion (Figure 10.2). Multiple microsatellites were clustered at the *SSS* locus; however, only three alleles were detected: allele $SSS\text{-}A = (AC)_2 \ldots TCC(TC)_{11} \ldots$ $(TC)_5C(ACC)_{11}$; allele $SSS\text{-}B = (AC)_3 \ldots TCT(TC)_6 \ldots (TC)_4C(ACC)_9$; and allele $SSS\text{-}C = (AC)_3 \ldots TCT(TC)_6 \ldots (TC)_4C(ACC)_8$ (Figure 10.3). Analysis of the starch physicochemical properties among different microsatellite genotypes indicated that the combination of the *SBE*-A allele combined with the *SSS*-B allele was quite different from other groups (Tables 10.2 and 10.3). A total of 15 accessions with high GT belonged to the *SBE*-A group, and 13 of them belonged to the *SSS*-B group (Bao et al., 2002a). Certainly, the co-segregation behavior of the SSR alleles and GT or other physicochemical properties should be analyzed to clarify the relationship.

Chen et al. (2003) identified two separate mutations in the *SSSIIa/alk* gene associated with low GT in a comprehensive genotype analysis of

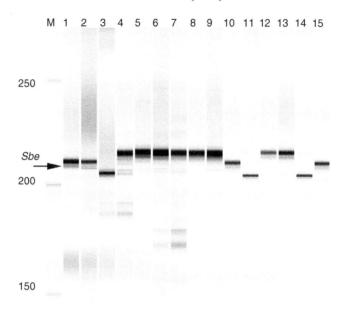

FIGURE 10.2. Polymorphism of the microsatellites in the *SBE* genes. Lane 1: *Lemont;* lane 2: *Azucena;* lane 3: *Jingxi 17;* lane 4: 371; lane 5: IR64; lane 6: *Zhaiyeqing 8;* lane 7: *Longtefu B;* lane 8: *Dian 4;* lane 9: *Shao 9915;* lane 10: *Xiangnuo no. 4;* lane 11: *Zaoshenghunuo;* lane 12: *Biyunzaonuo;* lane 13: *Zaoxiangnuo;* lane 14: *Xinguangnuo;* lane 15: PII 121. *Source:* Bao et al. (2002a).

196 international rice accessions. One mutation was common in Japanese accessions, whereas another mutation was common in low GT accessions from the United States. Both these mutations presumably encode an SSSIIa enzyme with lower activity, which results in the production of shorter chains of branched amylopectin in these accessions, causing them to have a lower GT.

Transgenic engineering with the amylopectin synthesizing genes can modify amylopectin structure and GT. The *SSSIIa* gene at the *alk* locus likely plays a critical role in the elongation of short chains within amylopectin clusters (Nakamura et al., 2002). Although GBSS, encoded by the *Wx* locus, can play a significant role in affecting GT, the *alk* locus plays the largest role in determining GT when both factors are variable in the parents of a genetic cross (He et al., 1999). Thus, the *alk* gene appears to be the best candidate to genetically engineer amylopectin structure and GT, although no efforts to date have been completed to modify this trait via transgenics.

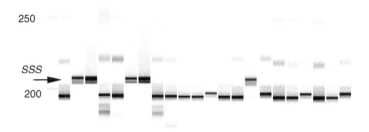

FIGURE 10.3. Polymorphism of the microsatellites in the SSS genes. Lane 1: *Xinguangnuo;* lane 2: *Cungunuo;* lane 3: *Guinuo no. 1;* lane 4: *Haocunuo;* lane 5: *Huangjinnuo;* lane 6: *Jiainuo;* lane 7: *Aiganyaxuenuo;* lane 8: *Suyuyuo;* lane 9: *Shenuo;* lane 10: *Zaoshenghunuo;* lane 11: *Chunjiangnuo no. 2;* lane 12: *Guixiangsinuo;* lane 13: *Xiangnuo no. 4;* lane 14: *Longqingzixiangnuo;* lane 15: *148nuo;* lane 16: *Zhonghuazixiangnuo;* lane 17: *Shao 9610;* lane 18: *Zaoxiangnuo;* lane 19: *Biyunzaonuo;* lane 20: *Zhenongda 454;* lane 21: *Shaonuo 9617;* lane 22: T1046. *Source:* Bao et al. (2002a).

TABLE 10.2. Comparison of the mean starch physicochemical properties of the three classes of SBE microsatellites in 56 waxy rice accessions.

Allele	Number	T_p (°C)	ΔH_g (J/g)	$\Delta T_{1/2}$ (°C)	ΔH_r (J/g)	FSV (ml/g)	PV (RVU)	HPV (RVU)	CPV (RVU)
SBE-A	34	72.6$_a$	8.5	7.5$_b$	3.6$_a$	18.3$_a$	171	95$_a$	121$_a$
SBE-B	7	69.1$_b$	8.3	8.3$_{ab}$	1.5$_b$	15.6$_b$	154	80$_b$	101$_b$
SBE-C	15	69.0$_b$	8.3	8.9$_a$	1.5$_b$	17.2$_a$	168	89$_{ab}$	112$_{ab}$

Note: Means with a different letter are significantly different ($p < .05$).

T_p: peak gelatinization temperature; ΔH_g: gelatinization enthalpy; $\Delta T_{1/2}$: width at half peak; ΔH_r: enthalpy of retrograded starch; FSV: flour swelling volume; PV: peak viscosity; HPV: hot paste viscosity; CPV: cool paste viscosity. RVU: Rapid Visco Analyzer Unit.

STARCH PASTING VISCOSITY

Amylograph studies are undertaken to simulate starch cooking and processing and to study changes occurring during cooking and cooling of aqueous starch systems. Obtaining amylograph data was made significantly

TABLE 10.3. Comparison of the mean starch physicochemical properties of the three classes of SSS microsatellites in 56 waxy rice accessions.

Allele	Number	T_p (°C)	ΔH_g (J/g)	$\Delta T_{1/2}$ (°C)	ΔH_r (J/g)	FSV (mL/g)	PV (RVU)	HPV (RVU)	CPV (RVU)
SSS-A	14	69.3$_b$	8.1$_b$	8.4$_a$	2.0$_b$	17.4$_b$	170	86$_b$	109$_b$
SSS-B	15	75.8$_a$	8.9$_a$	6.6$_b$	5.0$_a$	19.0$_a$	166	102$_a$	128$_a$
SSS-C	27	69.7$_b$	8.3$_{ab}$	8.6$_a$	1.8$_b$	17.0$_b$	168	89$_b$	113$_b$

Note: Means having a different letter are significantly different ($p < .05$).

T_p: peak gelatinization temperature; ΔH_g: gelatinization enthalpy; $\Delta T_{1/2}$: width at half peak; ΔH_r: enthalpy of retrograded starch; FSV: flour swelling volume; PV: peak viscosity; HPV: hot paste viscosity; CPV: cool paste viscosity. RVU: Rapid Visco Analyzer Unit.

easier with the introduction of the Rapid Visco Analyzer (RVA; Blankeney et al., 1991). This instrument has the advantage of requiring only a 2.5 g to 3.5 g sample and taking only 12.5 minutes per test.

To date, only a few genetic studies have reported on starch pasting viscosity parameters. Gravois and Webb (1997) studied the inheritance of RVA profiles and found that peak viscosity (PV), hot paste viscosity (HPV), and cool paste viscosity (CPV) were controlled by a single locus with additive effects. This finding was supported by both $F_{2:3}$ segregation analysis and diallele analysis. By using a more complex quantitative genetic model, the viscosity parameters of *indica* rice were found to be controlled by the effects of seed (endosperm), cytoplasm, and maternal plant. In QTL analysis of a DH mapping population derived from an *indica* (ZYQ8) and *japonica* (JX17) cross, only the *Wx* locus was significantly associated with pasting properties across two environments (Bao and Xia, 1999). Other minor QTLs were also detected, but only in one environment, suggesting that RVA pasting profiles are mainly controlled by the *Wx* gene (Bao et al., 2000c). With another DH population derived from cultivars with similar AC, no QTLs for pasting viscosity were detected at the *Wx* locus despite the fact that the parents had different *Wx* microsatellite alleles (Bao et al., 2002c).

Selection of specific *Wx* alleles is the most efficient means of breeding rice with the desired pasting viscosity characteristics. For example, RVA pasting profiles of F_3 seeds bulked with the same *Wx* alleles could match closely to parents and to bulked F_2 profiles in a cross with different *Wx* alleles and parents with different AC (Figure 10.4) (Bao, 2002). However, it should be

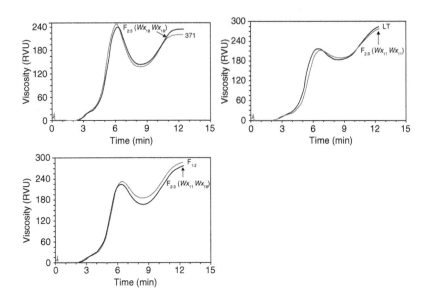

FIGURE 10.4. The RVA pasting viscosity profiles of bulked rice samples $F_{2:3}$ lines of Longtefu × 371 with the same *Wx* microsatellite markers. *Source:* Bao, 2002.

mentioned that rice accessions having similar AC can possess different pasting viscosity parameters. For example, the U.S. cultivars Dixiebelle and Jodon have similar AC but different RVA profiles. The RVA breakdown (peak value-trough) was found to correlate significantly with amylopectin fine structure in a set of rice accessions with a fairly narrow range of amylose content, but a wide variation in RVA pasting curves (Han and Hamaker, 2001). This result implies that rice starch with similar AC may vary in amylopectin structure. Therefore, actual testing of the pasting viscosity parameters may be necessary in some breeding populations even though the desired *Wx* allele has been selected.

COOKING PROPERTIES

When a sufficient sample is available, water absorption and volume expansion are commonly tested cooking characteristics of rice. These properties likely reflect variation in GT types. The test methods for these characteristics vary among laboratories, and no studies have reported on the genetic control of these parameters. On the other hand, a starch (flour) swelling volume test has been established as an indicator of wheat noodle quality. It is sometimes used for testing rice starch and starch-based foods (such as rice noodles) to predict their end-use quality (Bhattacharya et al., 1999). In a diallele analysis of *indica* rice, flour swelling volume was mainly controlled by maternal effects, and the narrow-sense maternal heritability was 74.5% (Bao et al., 2002b). QTL analysis based on a ZYQ8/JX17 DH population indicated a major QTL at the *Wx* gene on chromosome 6, explaining 36.6% of phenotypic variance in flour swelling volume, with another QTL on chromosome 7 explaining 21.4% of variance (J. Bao, unpublished data).

Aroma

The aroma of cooked rice contributes to consumer sensory acceptance of rice. Several hundred compounds can be observed in the volatiles of cooked rice, and more than 200 of these have been identified (Jezussek et al., 2002; Mahatheeranont et al., 2001). The aromatic compound 2-acetyl-1-pyrroline (2-AP) reportedly is the primary component of the popcorn-like smell of aromatic rice. Several methods have been developed to quantify 2-AP in rice (Bergman et al., 2000; Grimm et al. 2001; Mahatheeranont et al., 2001). After separating rice samples into fractions ranging from brown rice to overmilled rice, 2-AP was found to exist in an evenly distributed fashion throughout the grain (Bergman et al., 2000). Exposure to several solvents indicated that 2-AP is associated with the bound lipid fraction of rice (C. Bergman, unpublished data).

The genetics controlling aroma in rice have been well studied. All F_1 progeny from reciprocal crosses between aromatic and non-aromatic cultivars are non-aromatic (Dong et al., 2001; Pinson, 1994; Tsuzuki and Shimokawa, 1990), indicating the recessive nature of this trait. The segregation ratios in an F_2 population was 3/1 of nonaromatic to aromatic plants (Dong et al., 2001), or 13/3 of nonaromatic to aromatic (Tsuzuki and Shimokawa, 1990), fitting either a single recessive gene model or a digenic model. It has been suggested that one major gene and one inhibitor gene are involved in the inheritance of aroma (Tsuzuki and Shimokawa, 1990). Pinson's (1994) findings with four aromatic rice lines (*Jasmine 85, Amber,* PI457917, and *Dragon Eyeball 100*) revealed that *Jasmine 85* and PI457917 each contained a single gene for aroma. *Amber* and *Dragon Eyeball 100* reportedly each contained two aroma genes—a novel gene plus one allelic to the gene found in the other cultivars. Subsequent genetic marker work with this population revealed that instead of the presence of two aroma genes, there was non-Mendelian segregation of a single gene controlling the presence of aroma (Pinson et al., 2002).

Ahn et al. (1992) reported a marker on chromosome 8, identified by RFLP probe RG28, closely linked to a gene controlling aroma, designated *fgr*. This RFLP was identified using nearly isogenic lines (NILs) with or without the aroma gene. Chromosomal segments introgressed from the donor genome were distinguished by RFLPs between the NILs. Segregation analysis of the scented phenotype and the donor-derived allele indicated that RG28 mapped 4.5 cM from the *fgr* gene. Lorieux et al. (1996) carried out QTL mapping for the aromatic trait using a DH population and 2-AP quantification of the cooking water. One major and two minor QTLs controlling grain aroma in rice were found. The major gene was identified on chromosome 8, located 5.8 cM from RG28. Through trisomic analysis, a single recessive aroma gene was also located on chromosome 8 (Dong et al., 2001). For marker-aided selection of the *fgr* locus, PCR-based markers were developed from sequence analyses of the RG28 genomic clone (Garland et al., 2000) and of a highly polymorphic microsatellite near RG28 (Cordeiro et al., 2002). Bergman et al. (2002), combining information from three populations segregating for aroma, determined that the *fgr* locus was flanked by RG28 and RM223 by 3.6 cM and 1.9 cM, respectively. As the *fgr* gene on chromosome 8 is closely linked to these genetic markers, their application may facilitate early selection for the aroma gene in rice breeding programs.

Although the aforementioned results indicate that aroma is primarily controlled by genetic factors, the level of 2-AP in aromatic rice can also be affected by factors such as growing conditions, ripening and drying temperatures, and storage conditions and duration (Jezussek et al., 2002; Laksanalamai and Ilangantileke, 1994; Widjaja et al., 1996). Bergman et al. (2000) studied the levels of 2-AP in two aromatic rice genotypes grown in several environments. The levels of 2-AP were influenced by genotype, environment, and genotype-environment interactions. The strongest influence on 2-AP levels was from genotypic effects. The second largest amount of variation in 2-AP was from genotype-year-state interaction effects.

Elongation

Cooked kernel elongation (CKE) is one of the unique features of basmati-type rices. Cooked rice normally elongates a great deal lengthwise, but relatively little widthwise. Chauhan et al. (1992) estimated the genetic variance for quality traits among 45 rices and indicated that cooked grain length and grain elongation had a strong association with nonadditive gene action. Vivekanandan and Giridharan (1998) indicated via diallele analysis that linear elongation ratio, widthwise expansion ratio, and elongation index had high broad-sense heritability, but the genetic advance and genetic gain were very low. Diallele analysis with 12 *indica* rices indicated that the narrow-sense heritability for CKE (43.9%) was smaller than that of milled rice length (65.9%) and cooked rice length (61.7%), so early selection may be relatively ineffective for CKE (Bao et al., 2001).

Ahn et al. (1993) identified QTLs associated with CKE in an F_3 population derived from a cross between B8462T3-710 and the conventional recurrent parent, *Dellmont*. Analysis of variance for CKE revealed that two markers (RZ323 and RZ562) on chromosome 8 were significantly associated with this trait. Interval mapping suggested a single QTL in close proximity to RZ323. This QTL was tested in F_6 lines derived from the same cross, and the presence of the B8462T3-710 segment detected by RZ323 caused a highly significant increase in the kernel elongation ratio. Furthermore, the QTL for CKE was loosely linked with the major gene for aroma (Ahn et al., 1993). A microsatellite marker, RM44, on chromosome 8 was identified and reported to be tightly linked with a CKE QTL (Bergman et al., 2002), explaining 74% of the genetic variance for CKE. Because early phenotypic selection for CKE is not very effective, the availability of linked markers may facilitate early genotypic (marker) selection for CKE in rice breeding programs.

Kernel Hardness

Rice kernel hardness can be affected by storage changes and aging, drying and handling, kernel appearance and translucency, and grain processing and breakage during the milling process (Pomeranz and Webb, 1985; Webb et al., 1986). The biochemical and genetic bases of grain hardness and its effects on grain processing and cooking quality in sorghum, wheat, and maize have been well reviewed (Chandrashekar and Mazhar, 1999). However, studies on the genetic and biochemical basis of rice grain hardness have been limited. Puroindoline genes play a critical role in wheat *(Triticum aestivum)* kernel hardness, with mutations in them being associated with kernel texture variation (Giroux and Morris, 1997, 1998), wherein hard texture is dominantly expressed over soft texture. However, puroindoline genes appear to be absent in cereal species outside the tribe Triticeae.

Recently, a study on the potential to modify rice grain hardness by introducing puroindoline genes *(pinA and pinB)* from wheat was reported (Krishnamurthy and Giroux, 2001). Textural analysis of transgenic rice seeds from this study indicated that expression of *pinA* and *pinB* reduced rice grain hardness. After milling, flour prepared from these softer seeds

had reduced starch damage and an increased percentage of fine flour particles. These data support the hypothesis that puroindolines from wheat may
be useful in modifying the grain texture of other cereals (Krishnamurthy
and Giroux, 2001).

Appearance

The appearance of milled rice grain affects consumers' acceptance of
this product. Consumers have definite preferences in milled rice size and
shape, because they associate these kernel properties with textural properties: most medium-grain rice cooks soft and sticky, and long-grain rice
cooks firm and not sticky. Except for waxy rice, with its opaque endosperm,
consumers prefer rice that has a high degree of translucency and whiteness
and displays little chalkiness.

Grain size and shape is one of the most stable properties of a rice cultivar.
Some QTL studies have been conducted for grain length (GL), grain width
(GW), and grain shape (GS; length/width). Even though the number and
effect of QTLs are different for different populations used, a QTL on chromosome 3 has been repeatedly identified for GL, GW, and GS (Huang et al.,
1997; Lin et al., 1995; Redoña and Mackill, 1998; Tan et al., 2000). One
QTL on chromosome 5 has been identified for GW in two mapping populations from *indica-indica* crosses (Lin et al., 1995; Tan et al., 2000), but not
in two *indica-japonica* cross populations (Huang et al., 1997; Redoña and
Mackill, 1998).

Relatively high endosperm and maternal additive effects on transparency
and chalkiness indicate that these traits could be improved by selection in
early generations (Shi et al., 2002). He et al. (1999) reported that the percentage of grain area with white core (PGWC) displayed a continuous distribution in a ZYQ8/JX17 DH population with partially transgressive
segregation. For PGWC, two major QTLs were located on chromosomes 8
and 12 (He et al., 1999). White belly was reported to be primarily controlled
by a major locus on chromosome 5, located in the same genomic region as
a major QTL for grain width (Tan et al., 2000). Accurate measurement of
the chalkiness size is difficult, which was proposed to be a major source of
error for genetic analysis. A more accurate technique, using video micros-
copy and computer image analysis to measure chalkiness on grain transverse, flank, and belly sections was developed and used to re-analyze the
ZYQ8/JX17 DH population. Each of the kernel shape traits were found to
be controlled by three QTLs (Zeng et al., 2002), of which two had been previously reported (He et al., 1999). However, a mapping population derived

from parents with similar kernel shape and size variation may be needed to find loci controlling chalkiness independent of kernel shape and size.

MILLING QUALITY

The value of rough rice, in part, reflects the percentage of head rice versus total milled rice produced after milling. Milling yield is affected by both genetic and environmental factors, appearing to be a complex trait affected by hull, bran, and endosperm physicochemical characteristics. General combining ability (GCA), specific combining ability (SCA) and high-parent heterosis for head rice percentage, total milled rice percentage, and rough rice yield were evaluated in an eight-parent diallele (F_1 crosses plus parents) grown at two locations (Gravois, 1994). Genetic variation among the hybrids for total milled rice percentage was nonsignificant. GCA effects were more important than SCA effects for head rice percentage, indicating the importance of additive genetic effects in the inheritance of head rice percentage. On the other hand, Sharma and Mani (1998) indicated that non-additive gene action appeared to be predominantly controlling milling and head rice recovery for a set of diallele crosses. Another genetic analysis concluded that head rice recovery was mainly controlled by seed and cyto-plasmic effects, and milled rice recovery was controlled by seed and maternal effects (Shi and Zhu, 1998).

Many environmental factors are thought to influence milling quality, such as harvest moisture, field drainage, fluctuations in relative humidity and temperature, nitrogen fertility, disease and other stresses, and grain storage and processing conditions. Gravois et al. (1991) studied the genotype and environment effects on head rice yields and showed that heritability on a single-plot basis was low and genotype by year variance was the most important source of variation for head rice yield.

QTL analysis of milling quality traits was carried out using recombinant inbred lines (RILs) derived from an elite hybrid cross, *Shanyou 63* (Tan et al., 2001). The results showed that a major QTL in the interval between markers RM42 and C734b on chromosome 5 controls brown rice yield, and a major QTL located in the C1087-RZ403 interval on chromosome 3 controls head rice yield (Tan et al., 2001). These two loci are the major QTLs for grain width and length, respectively (see earlier). Thus, no loci controlling milling yield independent of kernel morphology traits have been reported. Well-designed studies that control for genotype variation in heading time, harvest moisture, degree of milling, and kernel morphology will be required to further dissect the genetics of milling quality.

NUTRITIONAL QUALITY

Protein Content

Proteins are the second most abundant constituent of milled rice, following starch. Proteins have two roles in rice quality, being related to the nutritive value of rice and being correlated with cooking and sensory quality. High protein content contributes to the nutritive value of rice as a food. Several reports have indicated that the protein content of rice is a quantitative trait (Sampath et al., 1968; Webb et al., 1968). Shi et al. (1996) found that protein content and protein index were controlled by genetic effects of triploid endosperm, cytoplasm, and diploid maternal plant, with estimates of narrow-sense heritabilities being 83 and 84% for protein content and protein index, respectively. Even with more complex genetic models, the narrow-sense heritabilities have been estimated to be 85% for protein content and 78% for protein index (Shi et al., 1999). Using QTL analysis, Tan et al. (2001) reported that the Wx gene was associated with protein content and was modified by several QTLs with minor effect. It is not clear whether the Wx gene actually controls protein content. The cosegregation behavior of the Wx alleles and protein content will need to be analyzed to clarify this association.

The storage proteins in rice can be classified into albumins (water-soluble proteins), globulins (salt-soluble proteins), prolamins (alcohol-soluble proteins), and glutelins (alkali-soluble proteins) by the Osborne scheme. Glutelins are the major storage proteins, accounting for 70-80% of the total grain protein content, and prolamins the least abundant, accounting only for 5%. Glutelins have a more balanced essential amino acid profile than prolamins, which are deficient in lysine and tryptophan. Few genetic studies have been conducted on each protein component. Iida et al. (1993) identified a rice mutant with relatively low glutelin content and high prolamin content. Genetic analysis indicated that the low-glutelin/high-prolamin trait was controlled by a single dominant gene. Several studies have focused on the molecular characterization of regulatory mechanisms responsible for the endosperm-specific expression of storage protein genes. Specifically, the cis-regulatory elements in the 5′ flanking region of these genes have been well characterized (Nakase et al., 1996; Takaiwa et al., 1996; Wu et al., 2000). Glutelins are encoded by a small multigene family, which has been subdivided into A and B subfamilies (Takaiwa and Oono, 1991). Maruta et al. (2002) developed rice transformed with a glutelin A antisense gene construct. The T_4 progeny plants of two transgenic lines showed a 20-40% reduction in the glutelin content of seeds and an increase in prolamin content. The total seed protein content of the two transgenic lines was not significantly different from that of the original variety. The transgenic rice with less glutelin content is targeted for use in the wine, sake, and brewing industries.

Legumes have a high seed protein content, but are deficient in methionine, cysteine, and tryptophan. Thus, it has been proposed that legume protein genes could be used to improve rice protein content and essential amino acid profile. A bean β-phaseolin gene placed under the transcriptional control of a rice glutelin promoter was transformed into rice (Zheng et al., 1995). The transgenic rice expressed the β-phaseolin seed storage protein, with levels up to 4% of the total endosperm protein. Phaseolin expression segregated as a single dominant trait with a positive gene dosage effect and was stable through three successive generations (Zheng et al., 1995). Using a similar strategy, Sindhu et al. (1997) obtained transgenic rice expressing the legumin seed storage protein from *Pisum sativum* in rice endosperm. Segregation analysis of legumin in transgenic plants suggested monogenic Mendelian inheritance, indicating a single locus or closely linked loci of gene insertion. Western blot analysis indicated that the legumin precursor was cleaved endoproteolytically into acidic and basic subunits in the rice endosperm, as in pea cotyledons (Sindhu et al., 1997).

The wild-type and methionine-modified glycinin coding sequences of soybeans *(Glycine max)* have been expressed in transgenic rice plants under the control of the rice glutelin promoter (Katsube et al., 1999). Momma et al. (1999) analyzed the chemical composition of several nutritionally and physiologically important molecules in transgenic rice expressing a soybean glycinin gene (Katsube et al., 1999) compared to a nontransgenic control. No marked differences between the two kinds of rice were found, except for protein, amino acids, and moisture levels. The protein content of the transgenic rice was about 20% higher than that of the control, with a concomitantly lower moisture content. This increased protein content mainly resulted from the increased glycinin expressed in the transgenic rice, which also resulted in increased essential amino acids (Momma et al., 1999).

Conventional plant breeding methods have been used to identify mutants with improved seed protein quality (Iida et al., 1993; Nishio and Iida, 1993). Nishio and Iida (1993) characterized rice mutants having a low content of a 16-kDa allergenic protein (globulin), which is the main allergen in rice grain for patients with atopic dermatitis. Genetic analysis showed that the low 16-kDa polypeptide content was controlled by a single recessive gene.

Lysine Content

Lysine is the most deficient essential amino acid in rice grain; consequently, it is used as an index of rice nutritional value. Shi et al. (1996)

showed that maternal effects for lysine content, lysine index, and the ratio of lysine content to protein content were more important than seed effects, and that additive genetic effects were more important than dominant effects for all of the nutrient related traits studied.

Kumamaru et al. (1997) selected 10 high-lysine mutants from N-methyl-N-nitrosourea induced mutant lines, using high-performance liquid chromatography for amino acid analysis. The lysine content of these mutants ranged from 5.10-6.38% of total amino acids, whereas that of the original cultivars had been approximately 4%. All high-lysine mutants had a floury endosperm. The segregation ratio of normal to floury endosperm in the F_2 seeds fit the expected 3:1 ratio, indicating that the high level of lysine and the floury endosperm may be controlled by the same gene. Schaeffer and Sharpe (1997) also characterized a rice mutant with enhanced grain lysine, and it was confirmed that high lysine content is genetically linked with grain chalkiness.

Lee et al. (2001) obtained transgenic rice plants with a lysine-feedback-insensitive maize $dhps$ gene under the control of a CaMV 35S or a rice glutelin promoter for overexpression and seed-specific expression, respectively. The transgenic lines (TC lines) containing mutated $dhps$ controlled by the CaMV 35S promoter possessed higher mutated DHPS transcript levels and in vitro DHPS activities in seeds than those lines containing the mutated $dhps$ gene driven by the glutelin promoter (TS lines). The content of free lysine in immature seeds of both lines was higher than that of wild-type plants. The content of free lysine in mature seeds of TC lines was higher than in wild-type plants, whereas TS lines had similar levels to the wild-type plants. It is possible that the presence of the foreign $dhps$ gene leads to an increase of lysine ketoglutarate reductase activity, resulting in enhanced lysine catabolism. However, overexpression of the mutant $dhps$ gene in a constitutive manner overcomes lysine catabolism and sustains a high lysine level in mature rice seeds.

β-Carotene Content

Vitamin A deficiency causes symptoms ranging from night blindness to xerophthalmia and keratomalacia, leading to total blindness. Immature rice endosperm is capable of synthesizing the early intermediate geranylgeranyl diphosphate, which can be used to produce the colorless carotene phytoene by the action of the enzyme phytoene synthase in rice endosperm. The synthesis of β-carotene requires the action of three additional plant enzymes: phytoene desaturase, ζ-carotene desaturase, and lycopene β-cyclase. A bacterial carotene desaturase has been reported to have a function similar to both phytoene desaturase and ζ-carotene desaturase. By introducing plant phytoene synthase, bacterial carotene desaturase, and lycopene β-cyclase genes via *Agrobacterium*-mediated transformation in a single transformation effort with three vectors, Ye et al. (2000) introduced the entire β-carotene biosynthetic pathway into rice endosperm. The transgenic rice can accumulate β-carotene in the endosperm as well as two other compounds, lutein and zeaxanthin. Recently, several lines of *indica* "golden rice" have been developed containing *psy* and *crt1* genes with or without marker via the POSITECH *(pmi)* selection system (Datta et al., 2003, 2006). Work is yet to be done to document the bioavailability of vitamin A in these transgenics and its stability under various storage and cooking conditions. Also, the yellow color of the endosperm may result in some consumers rejecting these transgenic cultivars.

Iron Content

Iron deficiency is the most widespread micronutrient deficiency worldwide. It afflicts an estimated 30% of the world population, especially where non-meat-based diets are common. Iron supplementation in the form of tablets and food fortification has not been successful in eradicating iron deficiency anemia in developing countries. Goto et al. (1999) transferred the entire coding sequence of the soybean *ferritin* (an iron-storage protein) gene into rice plants by *Agrobacterium*-mediated transformation. The iron content of T_1 seeds was as much as threefold higher than that of their untransformed counterparts. A similar approach of incorporating the *ferritin* gene into *indica* rice has been successful, with at least twofold more iron in polished seeds (Vasconcelos et al., 2003). Lucca et al. (2001) have worked out several ways to improve iron absorption. The first strategy was similar to that of Goto et al. (1999): introduce a *ferritin* gene from *Phaseolus vulgaris* into rice grains; this method increased iron content up to twofold. The second strategy was to overexpress the endogenous, cysteine-rich, metallothionein-like protein, because cysteine peptides are considered a major enhancer of iron absorption. The third strategy was to introduce thermotolerant phytase from *Aspergillus fumigatus* into the rice endosperm to increase iron bioavailability. High levels of phytic acid, which can be broken down by phytase, in cereal- and legume-based diets are the major cause of the poor absorption of iron in the human intestine. In phytase-engineered rice, the content of cysteine residues increased about sevenfold and the phytase level in the grains about 130-fold, giving a phytase activity sufficient to completely degrade phytic acid in a simulated digestion experiment (Bryant et al., 2002; Lucca et al., 2001). Rice lines expressing a chemically induced *low phytic acid* mutation of the myo-inositol-1-phosphate synthase gene also have the potential to improve iron nutrition, albeit to a small degree, as very little phytic acid remains in rice after milling (Larson et al., 2000).

FUTURE PROSPECTS

The genetic and biochemical bases for rice starch and protein quality are better understood than other end-use quality traits. Many genes involved in amylose, amylopectin, and protein synthesis in rice grains have been cloned. This research has laid the foundation for rice quality to be modified using genetic transformation technology. Also, genetic markers closely linked to starch synthesis genes have been developed, and their utility in rice quality improvement has been demonstrated. Clearly, the field of rice end-use quality has benefited from the recent developments in molecular biology.

A great deal is left to be learned about the genetic and biochemical bases of various aspects of rice end-use quality, such as milling quality and cooked grain texture. For example, fine mapping of genetic markers for most end-use quality traits has yet to be completed, and the homozygous, heterozygous, epistatic, and pleiotropic effects of the genes controlling these traits have not yet been determined. As the rice genome is becoming completely sequenced, rice genes of functional importance are being isolated, and a functional genetic map represented by expressed sequence tags of encoded genes has been constructed (Wu et al., 2002). These tools will facilitate the location of genetic factors in the vicinity of functional genes and enable the isolation and characterization of genes underlying complex end-use quality traits. The tools of functional genomics offer opportunities to identify genes related to rice grain quality traits and to characterize their expression. End-use quality mutants can be developed using techniques such as physical treatments (e.g., gamma irradiation, fast neutron), chemical treatments (e.g., EMS), and transposon, retrotransposon, or T-DNA insertion. Such mutants in combination with DNA microarray technology could lead to the rapid determination of polymorphisms associated with end-use quality traits and facilitate the cloning of the underlying mutated genes.

A challenging area of research that will likely receive attention in the future is the uncovering of the genetic basis for negatively correlated traits. For example, rice genotypes with higher yield potential often have a low head rice yield. Mutants with high protein or high lysine content characteristically appear opaque—an indicator of very low appearance quality. Through

QTL analysis, some traits may be associated with the same genomic region, indicating that it will be difficult to improve these traits simultaneously. The basis for such genetic drag, based on linkage, might be resolved with fine mapping and the development of high-density genetic linkage maps.

The scientific theories underlying the tools of the genomic era are relatively straightforward. However, the practical use of these tools with diverse germplasms can lead to unexpected results. It is broadly suspected that QTL distribution throughout any crop plant genome is cultivar dependent. This makes the breeding value of a specific allele at any QTL rather unclear. Efforts to develop public databases that incorporate genetic marker information about specific genotypes with their associated traits (e.g., Gramene, 2006; Ware et al., 2002) will surely extend genetic marker research and could assist rice breeders in the selection of breeding parent and genetic marker use.

Wild rice relatives reportedly have many unique genes suitable for crop improvement (Xiao et al., 1996). Thus, naturally occurring variation among wild relatives of cultivated rice is an underexploited resource available for breeding superior cultivars. In the future, introgression of traits from wild germplasm can be facilitated by the use of genetic markers. A strategy that has already proven successful has been to introgress portions of a wild rice genome into cultivated rice using advanced backcross transgression of QTLs (Tanksley and McCouch, 1997; Zamir, 2001).

Altered starch and protein properties are likely to be some of the early outcomes of transformation technology applied to rice. A primary goal of starch modification is to eliminate the production of chemically modified starch. Targets for starch genetic engineering can include modified starches to improve resistance to retrogradation, acid attack, shear, and high temperatures; very high amylose content to improve the strength of starch films and gels; and improvement of functional stability across environments and cultivars. Targets for protein genetic engineering include modified proteins with high lysine content and improvement of protein functional properties. Improvement in rice endosperm and bran phytochemical composition (e.g., lipids, minerals, vitamins, and antioxidants) has also been postulated to be feasible in the near future (Goffman and Bergman, 2002; Goffman et al., 2002; Zimmermann and Hurrell, 2002). For phenotypically complex traits such as milling quality and appearance quality, improvement using transformation methodology will be difficult because very little is known about the genetics and biochemistry controlling these traits. However, MAS strategies may become available to assist breeding programs as more is learned about the genomic regions controlling these traits.

POTENTIAL BARRIERS

Genomic technology is opening doors for us to be able to modify rice functionality. However, genomics is moving at a faster rate than phenomics in throughput and accuracy. We are just beginning to understand what physical and chemical properties are important to consumers and the processing industry. Thus, to fully capitalize on today's genomic tools, we must first improve our understanding of rice functional characteristics and develop rapid analytical tools capable of measuring them.

Much of genomics research is being performed by the private sector or by public-private collaborations, and most of it is taking place in developed countries. Concerns are present that the outcomes (e.g., genes or transgenic crops) from these research efforts are being protected by local or international patents, which could result in fewer agriculture research data being made available in the public domain. As a consequence, there is concern that food and agriculture markets in developing countries will become dominated by large corporations in the developed world. Farmers are considered by some to be the most at risk, as they face the need to purchase transgenic products (e.g., seeds) and their associated agrochemicals if they want to be able to compete in the global market. An alternative hypothesis is that farmers will have to pay more money for the seeds of transgenic crops, but their superior incomes from these crops will more than compensate the additional investment. Still others think that private corporations will have different policies for the developing versus the developed world, as one can point to "golden rice" technology being made available free to developing countries (Potrykus, 2001).

It is possible that some consumers will continue to reject transgenic crops, as has been seen to a great extent in the European Union. However, others may have their concerns lessened as more research into the safety of transgenic crops is reported, as consumers are provided with a choice due to laws mandating full disclosure of transgenics on food packages going into effect, and as more effective public education efforts regarding transgenics are undertaken.

REFERENCES

Ahn, S. N., C. N. Bollich, A. M. McClung, and S. D. Tanksley (1993). RFLP analysis of genomic regions associated with cooked-kernel elongation in rice. *Theoretical and Applied Genetics 87:* 27-32.
Ahn, S. N., C. N. Bollich, and S. D. Tanksley (1992). RFLP tagging of a gene for aroma in rice. *Theoretical and Applied Genetics 84:* 825-828.

Asaoka, M., K. Okunok, M. Yano, and H. Fuwa (1993). Effects of shrunken and other mutations on the properties of rice endosperm starch. *Starch 45:* 383-387.

Ayres, N. M., A. M. McClung, P. D. Larkin, H. F. J. Bligh, C. A. Jones, and W. D. Park (1997). Microsatellites and a single-nucleotide polymorphism differentiate apparent amylose classes in an extended pedigree of U.S. rice germplasm. *Theoretical and Applied Genetics 94:* 773-781.

Bao, J. S. (2002). Analysis of the relationship between *Wx* alleles and some starch quality parameters of rice (*Oryza sativa* L.). *Cereal Research Communications 30:* 397-402.

Bao, J. S., H. Corke, and M. Sun (2002a). Microsatellites in starch-synthesizing genes in relation to starch physicochemical properties in waxy rice (*Oryza sativa* L.). *Theoretical and Applied Genetics 105:* 898-905.

Bao, J. S., P. He, S. G. Li, Y. W. Xia, Y. Chen, and L. H. Zhu (2000a). Comparative mapping quantitative trait loci controlling the cooking and eating quality of rice (*Oryza sativa* L.). *Scientia Agricultura Sinica 33:* 8-13.

Bao, J. S., M. Sun, and H. Corke (2002b). Analysis of genetic behavior of some starch properties in *indica* rice (*Oryza sativa* L.): Thermal properties, gel texture, swelling volume. *Theoretical and Applied Genetics 104:* 408-413.

Bao, J. S., Y. R. Wu, B. Hu, P. Wu, H. R. Cui, and Q. Y. Shu (2002c). QTL for rice grain quality based on a DH population derived from parents with similar apparent amylose content. *Euphytica 128:* 317-324.

Bao, J. S., and Y. W. Xia (1999). Genetic control of paste viscosity characteristics in *indica* rice (*Oryza sativa* L.). *Theoretical and Applied Genetics 98:* 1120-1124.

Bao, J. S., J. K. Xie, and Y. W. Xia (2001). Genetic analysis of cooked rice elongation in *indica* rice (*Oryza sativa* L.). *Acta Agronomica Sinica 27:* 489-492.

Bao, J. S., J. J. Xu, W. Q. Wu, Q. Y. Shu, and Y. W. Xia (2000b). The changes of eating and cooking quality of *indica* early rice in different cropping seasons. *Journal of Zhejiang University (Agriculture and Life Sciences) 26:* 103-106.

Bao, J. S., X. W. Zheng, Y. W. Xia, P. He, Q. Y. Shu, X. Lu, Y. Chen, and L. H. Zhu (2000c). QTL mapping for the paste viscosity characteristics in rice (*Oryza sativa* L.). *Theoretical and Applied Genetics 100:* 280-284.

Bergman, C., J. Delgado, R. Bryant, C. Grimm, K. Cadwallader, and B. Webb. (2000). A rapid gas chromatographic technique for quantifying 2-acetyl-1-pyrroline and hexanal in rice (*Oryza sativa* L.). *Cereal Chemistry 77:* 454-458.

Bergman, C. J., J. T. Delgado, A. M. McClung, and R. G. Fjellstrom (2001). An improved method for using a microsatellite in the rice waxy gene to determine amylose class. *Cereal Chemistry 78:* 257-260.

Bergman, C. J., A. McClung, S. Pinson, and R. Fjellstrom (2002). The development of PCR based markers that are associated with cooked kernel elongation and aroma. In *Proceedings of the 29th Rice Technical Working Group Meeting,* Little Rock, AR, February 24-27, 2002, Louisiana State Agricultural Center, Crowley, LA, USA.

Bhattacharya, M., S. Y. Zee, and H. Corke (1999). Physiochemical properties related to quality of rice noodles. *Cereal Chemistry 76:* 861-867.

Bligh, H. F. J., P. D. Larkin, P. S. Roach, C. A. Jones, H. Y. Fu, and W. D. Park (1998). Use of alternate splice sites in granule-bound starch synthase mRNA from low-amylose rice varieties. *Plant Molecular Biology 38:* 407-415.

Bligh, H. F. J., R. I. Till, and C. A. Jones (1995). A microsatellite sequence closely linked to the *waxy* gene of *Oryza sativa. Euphytica 86:* 83-85.

Bollich, C. N., and B. D. Webb (1973). Inheritance of amylose in two hybrid populations of rice. *Cereal Chemistry 50:* 631-634.

Bryant, R., N. Rutger, and V. Raboy (2002). Distribution of phytic acid in grain components of low phytic acid rice mutant Lpa1. *Cereal Chemistry* p. 111, In *Proceedings of the 29th Rice Technical Working Group*, Little Rock, AR, February 24-27, 2002, Louisiana State Agricultural Center, Crowley, LA, USA.

Cagampang, G. B., C. M. Perez, and B. O. Juliano (1973). A gel consistency test for eating quality of rice. *Journal of the Science of Food and Agriculture 24:* 1589-1594.

Cai, X. L., Z. Y. Wang, Y. Xing, J. L. Zhang, and M. M. Hong (1998). Aberrant splicing of intron 1 leads to the heterogeneous 5' UTR and decreased expression of waxy gene in rice cultivars of intermediate amylose content. *The Plant Journal 14:* 459-465.

Champagne, E. T., B. G. Lyon, B. Min, B. K. Vinyard, K. L. Bett, F. E. Barton, B. D. Webb, A. M. McClung, K. A. Moldenhaur, S. Linscombe, et al. (1998). Effects of postharvest processing on texture profile analysis of cooked rice. *Cereal Chemistry 75:* 181-186.

Chandrashekar, A., and H. Mazhar (1999). The biochemical basis and implications of grain strength in sorghum and maize. *Journal of Cereal Science 30:* 193-207.

Chang, T. T., and C. C. Li (1991). Genetics and breeding. In *Rice production* (2nd ed.), B. S. Luh (Ed.), pp. 32-101. New York: Van Nostrand Reinhold.

Chauhan, J. S., V. S. Chauhan, S. B. Lodh, and A. B. Dash (1992). Environmental influence on genetic parameters of quality components in rainfed upland rice *(Oryza sativa). The Indian Journal of Agricultural Sciences 62:* 773-775.

Chen, M. H., C. J. Bergman, and R. Fjellstrom (2003). *SSSIIA locus genetic variation associated with alkali spreading value in international rice germplasm*, p. 314. XI Plant and Animal Genomes Conference, January 11-15, San Diego, CA.

Cordeiro, G. M., M. J. Christopher, R. J. Henry, and R. F. Reinke (2002). Identification of microsatellite markers for fragrance in rice by analysis of the rice genome sequence. *Molecular Breeding 9:* 245-250.

Datta K., N. Baisakh, N. Oliva, L. Torrizo, E. Abrigo, J. Tan, M. Rai, S. Rehana, S. Al-Babili, P. Beyer, et al. (2003). Bioengineered 'golden' *indica* rice cultivars with beta-carotene metabolism in the endosperm with hygromycin and mannose selection systems. *Plant Biotechnology Journal 1:* 81-90.

Datta K., M. Rai, V. Parkhi, N. Oliva, J. Tan, and S. K. Datta (2006). Improved 'golden' *indica* rice and transgeneration enhancement of metabolic target products of carotenoids (β-carotene) in transgenic elite (IR64 and BR29) *indica* 'golden' rice. *Current Science 91:* 935-939.

Dong, Y., E. Tsuzuki, and H. Terao (2001). Trisomic genetic analysis of aroma in three Japanese native rice varieties *(Oryza sativa* L.). *Euphytica 117:* 191-196.

Dung, L. V., I. Mikami, E. Amano, and Y. Sano (2000). Study on the response of *dull endosperm 2-2, du2-2*, to two *wx* alleles in rice. *Breeding Science 50:* 215-219.

Garland, S., L. Lewin, A. Blakeney, R. Reinke, and R. Henry (2000). PCR-based molecular markers for the fragrance gene in rice (*Oryza sativa* L.). *Theoretical and Applied Genetics 101:* 364-371.

Ghosh, A. K., and S. Govindaswamy (1972). Inheritance of starch iodine blue value and alkali digestion value in rice and their genetic. *Riso 21:* 123-132.

Giroux, M. J., and C. F. Morris (1997). A glycine to serine change in puroindoline b is associated with wheat grain hardness and low levels of starch-surface friabilin. *Theoretical and Applied Genetics 95:* 857-864.

Giroux, M. J., and C. F. Morris (1998). Wheat grain hardness results from highly conserved mutations in the friabilin components puroindoline a and b. *Proceedings of the National Academy of Sciences of the USA 95:* 6262-6266.

Goffman, F. D., and C. J. Bergman (2002). Total phenolics and antiradical effect of rice bran extracts. *Proceedings of the 29th Rice Technical Working Group Meeting,* Little Rock, AR, February 24-27, 2002.

Goffman, F. D., S. Pinson, and C. J. Bergman (2002). Lipid content and fatty acid profile in the bran of a germplasm collection of rice. In *Proceedings of the 29th Rice Technical Working Group Meeting,* Little Rock, AR, February 24-27, 2002.

Goto, F., T. Yoshihara, N. Shigemoto, S. Toki, and F. Takaiwa (1999). Iron fortification of rice seed by the soybean ferritin gene. *Nature Biotechnology 17:* 282-286.

Gramene (2006). *Gramene: A resource for comparative grass genomics* (version 23). http://www.gramene.org/ (last accessed February 26, 2007).

Gravois, K. A. (1994). Diallel analysis of head rice percentage, total milled rice percentage, and rough rice yield. *Crop Science 34:* 42-45.

Gravois, K. A., K. A. K. Moldenhauer, and P. C. Rohman (1991). Genetic and genotype × environment effects for rough and head rice yields. *Crop Science 31:* 907-911.

Gravois, K. A., and B. D. Webb (1997). Inheritance of long grain rice amylograph viscosity characteristics. *Euphytica 97:* 25-29.

Grimm, C. C., C. J. Bergman, J. T. Delgado, and R. Bryant (2001). Screening for 2-acetyl-1-pyrroline in the headspace of rice using SPME/ GC-MS. *Journal of Agricultural and Food Chemistry 49:* 245-249.

Halick, J. V., and K. K. Keneaster (1956). The use of a starch-iodine-blue test as a quality indicator of white milled rice. *Cereal Chemistry 33:* 315-319.

Han, X. Z., and B. R. Hamaker (2001). Amylopectin fine structure and rice starch paste breakdown. *Journal of Cereal Science 34:* 279-284.

He, P., S. G. Li, Q. Qian, Y. Q. Ma, J. Z. Li, W. M. Wang, Y. Chen, and L. H. Zhu (1999). Genetic analysis of rice grain quality. *Theoretical and Applied Genetics 98:* 502-508.

Hirano, H. Y., and Y. Sano (1998). Enhancement of *wx* gene expression and the accumulation of amylose in response to cool temperatures during seed development in rice. *Plant and Cell Physiology 39:* 807-812.

Huang, N., A. Parco, T. Mew, G. Magpantay, S. McCouch, E. Guiderdoni, J. C. Xu, P. Subudhi, E. R. Angeles, and G. S. Khush (1997). RFLP mapping of isozymes,

RAPD and QTLs for grain shape, brown planthopper resistance in a double haploid rice population. *Molecular Breeding 3:* 105-113.

Hue, M. H., and J. R. Choi (1973). The genetics of alkali digestibility in grains of - *indica japonica* hybrids. *Korean Journal of Breeding 5:* 32-36.

Iida, S., E. Amano, and T. Nishio (1993). A rice (*Oryza sativa* L.) mutant having a low content of glutelin and a high content of prolamine. *Theoretical and Applied Genetics 87:* 374-378.

Isshiki, M., K. Morino, M. Nakajima, R. O. Okagaki, S. R. Wessler, T. Izawa, and K. Shimamoto (1998). A naturally occurring functional allele of the waxy locus has a GT to TT mutation at the 5' splice site of the first intron. *The Plant Journal 15:* 133-138.

Isshiki, M., M. Nakajima, H. Satoh, and K. Shimamoto (2000). *dull:* Rice mutants with tissue-specific effects on the splicing of the *waxy* pre-mRNA. *The Plant Journal 23:* 451-460.

Itoh, K., M. Nakajima, and K. Shimamoto (1997). Silencing of *waxy* genes in rice containing *Wx* transgenes. *Molecular and General Genetics 255:* 351-358.

Jane, J. L., Y. Y. Chen, L. F. Lee, A. E. McPherson, K. S. Wong, M. Radosvljevic, and T. Kasemsuwan (1999). Effects of amylopectin branch chain length and amylose content on the gelatinization and pasting properties of starch. *Cereal Chemistry 76:* 629-637.

Jezussek, M., B. O. Juliano, and P. Schieberle (2002). Comparison of key aroma compounds in cooked brown rice varieties based on aroma extract dilution analysers. *Journal of Agricultural and Food Chemistry 50:* 1101-1105.

Juliano, B. O. (1971). A simple assay for milled rice amylose. *Cereal Science Today 16:* 334-338.

Juliano, B. O. (1990). Rice grain quality: Problems and challenges. *Cereal Foods World 35:* 245-253.

Juliano, B. O., and R. G. Pascual (1980). Quality characteristics of milled rice grown in different countries. *IRRI Research Paper Series No. 48.* Los Baños, Philippines: International Rice Research Institute.

Katsube, T., N. Kurisaka, M. Ogawa, N. Maruyama, R. Ohtsuka, S. Utsumi, and F. Takaiwa (1999). Accumulation of soybean glycinin and its assembly with the glutelins in rice. *Plant Physiology 120:* 1063-1073.

Krishnamurthy, K., and M. J. Giroux (2001). Expression of wheat puroindoline genes in transgenic rice enhances grain softness. *Nature Biotechnology 19:* 162-166.

Kubo, A., N. Fujita, K. Harada, T. Matsuda, H. Satoh, and Y. Nakamura (1999). The starch-debranching enzymes isoamylase and pullulanase are both involved in amylopectin biosynthesis in rice endosperm. *Plant Physiology 121:* 399-409.

Kumamaru, T., H. Sato, and H. Satoh (1997). High-lysine mutants of rice, *Oryza sativa* L. *Plant Breeding 116:* 245-249.

Kumar, I., and G. S. Khush (1988). Inheritance of amylose content in rice (*Oryza sativa* L.). *Euphytica 38:* 261-269.

Kumar, I., G. S. Khush, and B. O. Juliano (1987). Genetic analysis of *waxy* locus in rice (*Oryza sativa* L.). *Theoretical and Applied Genetics 73:* 481-488.

Laksanalamai, V., and S. Ilangantileke (1994). Comparison of aroma compound (2-acetyl-1-pyrroline) in leaves from pandan *(Pandanus amaryllifolius)* and Thai fragrant rice (Khao Dawk Mali-105). *Cereal Chemistry 70:* 381-384.

Larkin, P. D., and W. D. Park (1999). Transcript accumulation and utilization of alternate and non-consensus splice sites in rice granule-bound starch synthase are temperature-sensitive and controlled by a single-nucleotide polymorphism. *Plant Molecular Biology 40:* 719-727.

Larson, S. R., J. N. Rutger, K. A. Young, and V. Raboy (2000). Isolation and genetic mapping of a non-lethal rice *(Oryza sativa* L.) *low phytic acid 1* mutation. *Crop Science 40:* 1397-1405.

Lee, S. I., H. U. Kim, Y. H. Lee, S. C. Suh, Y. P. Lim, H. Y. Lee, and H. I. Kim (2001). Constitutive and seed-specific expression of a maize lysine-feedback-insensitive dihydrodipicolinate synthase gene leads to increased free lysine levels in rice seeds. *Molecular Breeding 8:* 75-84.

Lin, H. X., S. K. Min, Z. M. Xiong, H. R. Qian, J. Y. Zhuang, J. Lu, N. Huang, and K. L. Zheng (1995). RFLP mapping of QTLs for grain shape traits in *indica* rice *(Oryza sativa* L. subsp. *indica). Scientia Agricultura Sinica 28:* 1-7.

Little, R. R., G. B. Hilder, and E. H. Dawson (1958). Differential effect of dilute alkali on 25 varieties of milled white rice. *Cereal Chemistry 35:* 111-126.

Lorieux, M., M. Petrov, N. Huang, E. Guiderdoni, and A. Ghesquière (1996). Aroma in rice: Genetic analysis of a quantitative trait. *Theoretical and Applied Genetics 93:* 1145-1151.

Lucca, P., R. Hurrell, and I. Potrykus (2001). Genetic engineering approaches to improve the bioavailability and the level of iron in rice grains. *Theoretical and Applied Genetics 102:* 392-397.

Mahatheeranont, S., S. Keawsard, and K. Dumri (2001). Quantification of the rice aroma compound, 2-acetyl-1-pyrroline, in uncooked Khao Dawk Mali 105 brown rice. *Journal of Agricultural and Food Chemistry 49:* 773-779.

Maruta, Y., J. Ueki, H. Saito, N. Nitta, and H. Imaseki (2002). Transgenic rice with reduced glutelin content by transformation with glutelin A antisense gene. *Molecular Breeding 8:* 273-284.

McKenzie, K. S., and J. N. Rutger (1983). Genetic analysis of amylose content, alkali spreading score and grain dimensions in rice. *Crop Science 23:* 306-313.

Mikami, I., M. Aikawa, H. Y. Hirano, and Y. Sano (1999). Altered tissue-specific expression at the *Wx* gene of the opaque mutants in rice. *Euphytica 105:* 91-97.

Mikami, I., L. V. Dung, H. Y. Hirano, and Y. Sano (2000). Effects of the two most common *Wx* alleles on different genetic backgrounds in rice. *Plant Breeding 119:* 505-508.

Momma, K., W. Hashimoto, S. Ozawa, S. Kawai, T. Katsube, F. Takaiwa, M. Kito, S. Utsumi, and K. Murata (1999). Quality and safety evaluation of genetically engineered rice with soybean glycinin: Analyses of the grain composition and digestibility of glycinin in transgenic rice. *Bioscience, Biotechnology, and Biochemistry 63:* 314-318.

Nakamura, Y., A. Sakurai, Y. Inaba, K. Kimura, N. Iwasawa, and T. Nagamine (2002). The fine structure of amylopectin in endosperm from Asian cultivated rice can be largely classified into two classes. *Starch/ Stärke 54:* 117-131.

Nakase, M., T. Yamada, T. Kira, J. Yamaguchi, N. Aoki, R. Nakamura, T. Matsuda, and T. Adachi (1996). The same nuclear proteins bind to the 5′-flanking regions of genes for the rice seed storage protein: 16 kDa albumin, 13 kDa prolamin and type II glutelin. *Plant Molecular Biology 32:* 621-630.

Nishi, A., Y. Nakamura, N. Tanaka, and H. Satoh. (2001). Biochemical and genetic analysis of the effects of amylose-extender mutation in rice endosperm. *Plant Physiology 127:* 459-472.

Nishio, T., and S. Iida (1993). Mutants having a low content of 16 kDa allergenic protein in rice (*Oryza sativa* L.). *Theoretical and Applied Genetics 86:* 317-321.

Pinson, S. R. M. (1994). Inheritance of aroma in six rice cultivars. *Crop Science 34:* 1151-1157.

Pinson, S., R. Fjellstrom, C. Bergman, C. Grimm, and E. Champagne (2002). Inheritance of aroma: Markers help to distinguish fact from artifact. In *Proceedings of the 29th Rice Technical Working Group Meeting,* Little Rock, AR, February 24-27, 2002.

Pomeranz, Y., and B. D. Webb (1985). Rice hardness and functional properties. *Cereal Food World 30:* 784-786.

Pooni, H. S., I. Kumar, and G. S. Khush (1993). Genetic control of amylose content in selected crosses of *indica* rice. *Heredity 70:* 269-280.

Potrykus, I. (2001). Golden rice and beyond. *Plant Physiology 125:* 1157-1161.

Preiss, J. (1991). Biology and molecular biology of starch synthesis and its regulation. *Oxford Surveys of Plant Molecular and Cell Biology 7:* 59-114.

Puri, R. P., and E. A. Siddiq (1983). Studies on cooking and nutritive qualities of cultivated rice (*Oryza sativa* L.). IV. Quantitative genetic analysis of gelatinization temperature. *Genetica Agraria 37:* 335-344.

Redoña, E. D., and D. J. Mackill (1998). Quantitative trait locus analysis for rice panicle and grain characteristics. *Theoretical and Applied Genetics 96:* 957-963.

Sampath, S., S. Patnaik, and G. N. Mitra (1968). The breeding of high protein rice. *Current Science 9:* 248-249.

Sano, Y. (1984). Differential regulation of waxy gene expression in rice endosperm. *Theoretical and Applied Genetics 68:* 467-473.

Sano, Y., M. Katsumata, and K. Okuno (1986). Genetic studies of speciation in cultivated rice. 5. Inter- and intra-specific differentiation in the waxy gene expression in rice. *Euphytica 35:* 1-9.

Schaeffer, G. W., and F. T. Sharpe (1997). Electrophoretic profiles and amino acid composition of rice endosperm proteins of a mutant with enhanced lysine and total protein after backcrosses for germplasm improvements. *Theoretical and Applied Genetics 95:* 230-235.

Sharma, R. K., and S. C. Mani (1998). Combining ability analysis for physical grain quality in basmati rice. *Oryza 35:* 211-214.

Shi, C. H., J. G. Wu, X. B. Lou, J. Zhu, and P. Wu (2002). Genetic analysis of transparency and chalkiness area at different filling stages of rice (*Oryza sativa* L.) *Field Crops Research 76:* 1-9.

Shi, C. H., J. M. Xue, Y. G. Yu, X. E. Yang, and J. Zhu (1996). Analysis of genetic effects for nutrient quality traits in *indica* rice. *Theoretical and Applied Genetics 92:* 1099-1102.

Shi, C. H., and J. Zhu (1998). Genetic analysis of cytoplasmic and maternal effects for milling quality traits in *indica* rice. *Seed Science and Technology 26:* 481-488.

Shi, C. H., J. Zhu, X. Yang, Y. G. Yu, and J. G. Wu (1999). Genetic analysis for protein content in *indica* rice. *Euphytica 107:* 135-140.

Shi, C. H., J. Zhu, R. C. Zang, and G. L. Chen (1997). Genetic and heterosis analysis for cooking quality traits of *indica* rice in different environments. *Theoretical and Applied Genetics 95:* 294-300.

Shimada, H., Y. Tada, T. Kawasaki, and T. Fujimura (1993). Antisense regulation of the rice waxy gene expression using a PCR-amplified fragment of the rice genome reduces the amylose content in grain starch. *Theoretical and Applied Genetics 86:* 665-672.

Shu, Q. Y., D. X. Wu, Y. W. Xia, M. W. Gao, N. M. Ayres, P. D. Larkin, and W. D. Park (1999). Microsatellites polymorphism on the *waxy* gene locus and their relationship to amylose content in *indica* and *japonica* rice, *Oryza sativa* L. *Acta Genetica Sinica 26:* 350-358.

Sindhu, A. S., Z. W. Zheng, and N. Murai (1997). The pea seed storage protein legumin was synthesized, processed, and accumulated stably in transgenic rice endosperm. *Plant Science 130:* 189-196.

Smith, A. M., K. Denyer, and K. Martin (1997). The synthesis of the starch granule. *Annual Review of Plant Physiology and Plant Molecular Biology 48:* 67-87.

Sombilla, M. A., and M. Hossain (2001). Economics of the production and marketing of specialty rices: Recent trends and implications for technology development. In *Specialty rices of the world,* R. C. Chaudhary, D. V. Tran, and R. Duffy (Eds.), pp. 347-358. Enfield, NH: FAO Science Publishers.

Takaiwa, F., and K. Oono (1991). Genomic DNA sequences of two new genes for new storage protein glutelin in rice. *Japanese Journal of Genetics 66:* 161-171.

Takaiwa, F., U. Yamanouchi, T. Yoshihara, H. Washida, F. Tanabe, A. Kato, and K. Yamada (1996). Characterization of common cis-regulatory elements responsible for the endosperm-specific expression of members of the rice glutelin multigene family. *Plant Molecular Biology 30:* 1207-1221.

Takeda, Y., S. Hizukuri, and B. O. Juliano (1987). Structures of rice amylopectins with low and high affinities for iodine. *Carbohydrate Research 168:* 79-88.

Tan, Y. F., J. X. Li, S. B. Yu, Y. Z. Xing, C. G. Xu, and Q. Zhang (1999). The three important traits for cooking and eating quality of rice grains are controlled by a single locus in an elite rice hybrid, Shanyou 63. *Theoretical and Applied Genetics 99:* 642-648.

Tan, Y. F., M. Sun, Y. Z. Xing, J. P. Hua, X. L. Sun, Q. F. Zhang, and H. Corke (2001). Mapping quantitative trait loci for milling quality, protein content and color characteristics of rice using a recombinant inbred line population derived from an elite rice hybrid. *Theoretical and Applied Genetics 103:* 1037-1045.

Tan, Y. F., Y. Z. Xing, J. X. Li, S. B. Yu, C. G. Xu, and Q. F. Zhang (2000). Genetic bases of appearance quality of rice grains in Shanyou 63, an elite rice hybrid. *Theoretical and Applied Genetics 101:* 823-829.

Tan, Y. F., and Q. F. Zhang (2001). Correlation of simple sequence repeat (SSR) variants in the leader sequence of the *waxy* gene with amylose content of the grain in rice. *Acta Botanica Sinica 43:* 146-150.

Tanksley, S. D., and S. R. McCouch (1997). Seed banks and molecular maps: Unlocking genetic potential from the wild. *Science 277:* 1063-1066.

Terada, R., M. Nakajima, M. Isshiki, R. J. Okagaki, S. R. Wessler, and K. Shimamoto (2000). Antisense *waxy* genes with highly active promoters effectively suppress *waxy* gene expression in transgenic rice. *Plant and Cell Physiology 41:* 881-888.

Tetens, I., S. K. Biswas, L. V. Glito, K. A. Kabir, S. H. Thilsted, and N. H. Choudhury (1997). Physico-chemical characteristics as indicators of starch availability from milled rice. *Journal of Cereal Science 26:* 355-361.

Tsuzuki, E., and E. Shimokawa (1990). Inheritance of aroma in rice. *Euphytica 46:* 157-159

Umemoto, U., M. Yano, H. Satoh, A. Shomura, and Y. Nakamura (2002). Mapping of a gene responsible for the difference in amylopectin structure between *japonica*-type and *indica*-type rice varieties. *Theoretical and Applied Genetics 104:* 1-8.

Vasconcelos, M., K. Datta, N. Oliva, M. Khalekuzzaman, L. Torrizo, S. Krishnan, M. Oliveira, F. Goto, and S. K. Datta (2003). Enhanced iron and zinc accumulation in transgenic rice with the *ferritin* gene. *Plant Science 64:* 371-378.

Vivekanandan, P., and S. Giridharan (1998). Genetic variability and character association for kernel and cooking quality traits in rice. *Oryza 35:* 242-245.

Wang, Z. Y., F. Q. Zheng, G. Z. Shen, J. P. Gao, D. P. Snustad, M. G. Li, J. L. Zhang, and M. M. Hong (1995). The amylose content in rice endosperm is related to the post-transcriptional regulation of the *waxy* gene. *The Plant Journal 7:* 613-622.

Ware, D., P. Jaiswal, J. Ni, X. Pan, K. Chang, K. Clark, L. Teytelman, S. Schmidt, W. Zhao, S. Cartinhour, et al. (2002). Gramene: A resource for comparative grass genomics. *Nucleic Acids Research 30:* 103-105.

Webb, B. D. (1975). Cooking, processing and milling qualities of rice. In *Six decades of rice research in Texas,* pp. 97-106. College Station, TX: Texas Agricultural Experiment Station.

Webb, B. D., C. N. Bollich, C. R. Adair, and T. H. Johnston (1968). Characteristics of rice varieties in the U.S. Department of Agriculture collection. *Crop Science 18:* 361.

Webb, B. D., Y. Pomeranz, S. Afework, F. S. Lai, and C. N. Bollich (1986). Rice grain hardness and its relationship to some milling, cooking, and processing characteristics. *Cereal Chemistry 63:* 27-30.

Widjaja, R., J. D. Craske, and M. Wootton (1996). Changes in volatile components of paddy, brown and white fragrant rice during storage. *Journal of the Science of Food and Agriculture 71:* 214-218.

Williams, V. R., W. T. Wu, H. Y. Tsai, and H. G. Gates (1958). Varietal differences in amylose content of rice starch. *Journal of Agricultural and Food Chemistry 6:* 47-48.

Wu, C. Y., H. Washida, Y. Onodera, K. Harada, and F. Takaiwa (2000). Quantitative nature of the prolamin-box, ACGT and AACA motifs in a rice glutelin gene promoter: Minimal cis-element requirements for endosperm-specific gene expression. *The Plant Journal 23:* 415-421.

Wu, J. Z., T. Maehara, T. Shimokawa, S. Yamamoto, C. Harada, Y. Takazaki, N. Ono, Y. Mukai, K. Koike, J. Yazaki, et al. (2002). A comprehensive rice transcript map containing 6591 expressed sequence tag sites. *Plant Cell 14:* 525-535.

Xiao, J. H., S. Grandillo, S. N. Ahn, S. R. McCouch, S. D. Tanksley, J. M. Li, and L. P. Yuan (1996). Genes from wild rice improve yield. *Nature 384:* 223-224.

Xiong, L. Z., S. P. Wang, K. D. Liu, X. K. Dai, M. A. Saghai-Maroof, J. G. Hu, and Q. F. Zhang (1998). Distribution of simple sequence repeat and AFLP markers in molecular linkage map of rice. *Acta Botanica Sinica 40:* 605-614.

Yano, M., K. Okuno, J. Kawakami, H. Satoh, and T. Omura (1985). High amylose mutants of rice, *Oryza sativa* L. *Theoretical and Applied Genetics 69:* 253-257.

Ye, X., S. Al-Babili, A. Klöti, J. Zhang, P. Lucca, P. Beyer, and I. Potrykus (2000). Engineering provitamin A (β-carotene) biosynthetic pathway into (carotenoid-free) rice endosperm. *Science 287:* 303-305.

Zamir, D. (2001). Improving plant breeding with exotic genetic libraries. *Nature Reviews: Genetics 2:* 983-989.

Zeng, D. L., Q. Qian, L. Q. Ruan, S. Teng, Y. Kunihiro, H. Fujimoto, and L. H. Zhu (2002). QTL analysis of chalkiness size in three dimensions. *Chinese Journal of Rice Science 16:* 11-14.

Zheng, Z., K. Sumi, K. Tanaka, and N. Murai (1995). The bean seed storage protein β-phaseolin is synthesized, processed, and accumulated in the vacuolar type-II protein bodies of transgenic rice endosperm. *Plant Physiology 109:* 777-786.

Zhou, P. H., Y. F. Tan, Y. Q. He, C. G. Xu, and Q. Zhang (2003). Simultaneous improvement for four quality traits of Zhenshan 97, an elite parent of hybrid rice, by molecular marker-assisted selection. *Theoretical and Applied Genetics 106:* 326-331.

Zimmermann, M. B., and R. F. Hurrell (2002). Improving iron, zinc and vitamin A nutrition through plant biotechnology. *Current Opinion in Biotechnology 13:* 142-145.

Chapter 11

Genetically Improved
Nutrition-Dense Rice

S. K. Datta

INTRODUCTION

"Rice is life"—a staple crop for about three billion people throughout the world—and considered a life-saving crop for Southeast Asia. Rice provides 40-70% of the total food calories consumed in developing countries. It is also the most important source of income for rural people. In 2003, total paddy rice production was 579.5 million tons. About 95.5% of the world's rice was produced in developing countries, and 90.8% in Asia (Food and Agriculture Organization, 2003). Asian countries (except Japan and Korea) are much more dependent on rice for food energy and protein intake than other countries. However, rice is not a good source of micronutrients or proteins, especially after milling (Datta and Bouis, 2000).

Malnutrition is a severe problem in developing countries whose population depends on rice for their mineral and micronutrient requirement. In particular, vitamin A deficiency (VAD) and iron deficiency anemia (IDA) are recognized as important nutritional problems affecting millions of people, particularly women and children. In Southeast Asia, an estimated 250,000 people go blind each year due to VAD. Diet is the only source of vitamin A, as mammals cannot manufacture it on their own. Most of the

The author would like to personally thank his co-workers and colleagues. Special thanks are due to Dr. Ingo Potrykus, Dr. Peter Beyer, and Dr. Adrian Dubbock. Thanks are also due to DuPont and Dr. F. Goto for supporting the lysine and iron projects, respectively, and to Dr. Takaiwa for tissue-specific promoter work.

dietary vitamin A is of plant origin, in the form of provitamin A, which is converted to vitamin A in the body (Sivakumar, 1998). So far, none of the screened rice germplasms has shown the presence of β-carotene in polished seeds (Tan et al., 2005).

Similarly, iron deficiency affects an estimated 30% of the world population. Although brown rice contains a considerable amount of iron, the removal of the outer layers by commercial milling dramatically reduces the iron level, as most of the minerals are accumulated in the aleurone layer (Vasconcelos et al., 2003).

Conventional interventions like diet diversification and fortification have been used with limited success (Bouis et al., 2003). The unavailability of germplasm with β-carotene, the dearth of genotypes with high iron in the polished seeds, and the low effectiveness of conventional interventions suggested the improvement of mineral and micronutrient nutrition through plant biotechnology (e.g., genetic engineering) as a more sustainable strategy to combat nutritional deficiencies in human populations (Datta et al., 2003; Zimmerman and Hurrel, 2002). "In the end, the fruits of biotechnology are encapsulated in the simplest technology of all—the seed" (Phillips, 2000, p. 459).

This chapter provides an overview of the nutrition profile of the rice genome, particularly that of carotenoids and iron levels, and of conventional breeding and transgenic breeding approaches to increase such levels.

Is rice a preferred source of iron, zinc, or carotenoids? Vegetables are rich in minerals and vitamins, but they vary considerably in iron content and bioavailability. Surprisingly, most Asian countries lack the quantity of vegetables required to meet minimum vitamin A and C requirements (Figure 11.1) (Ali and Tsou, 2000). Rice is a staple food that provides 40 to 80% of the calorie requirement of Southeast Asians (Figure 11.2). Rice is cheap, can easily be grown, and can be stored for a long time. Rice with high β-carotene, iron, and zinc can make a significant impact in reducing malnutrition of the people who need it most.

RICE GENOME AND NUTRITION PROFILE

Rice belongs to the grain family, which includes other important food crops such as wheat, maize, barley, sorghum, and millet. It has a relatively small genome size (440 Mbp, compared to maize which is ~2500 Mbp). The rice genome has been sequenced (see Chapter 1). Identification of genes and characterization of gene variation by genomic sequencing will provide a comprehensive process of cloning "candidate genes" and will eventually

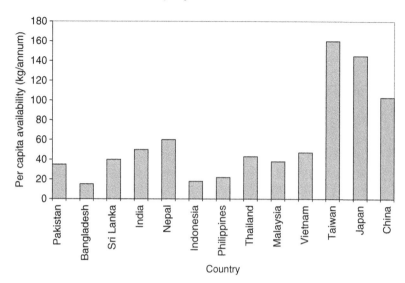

FIGURE 11.1. Availability of vegetables in selected Asian countries during 1993. *Source:* Adapted from Ali and Tsou (2000).

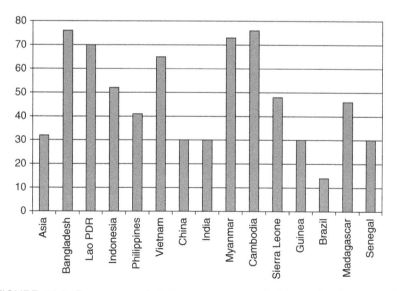

FIGURE 11.2. Percentage of dietary energy supplied from rice in selected countries. *Sources:* Adapted from www.fao.organize2004/en/f-sheetfactsheet3 .pdf; Rice Almanac 2002.

help in the biological understanding of the function of many traits, including nutritional genomics and validation of transgenes for the production of more nutritious rice.

Nutrition Profile in Rice

Brown rice (i.e., unpolished rice) has a very high nutritional value compared to other cereals, containing carotenoids and iron (see Tables 11.1-11.5). The rice grain, commonly called seed, consists of the brown rice (caryopsis) and the hull, which encloses the brown rice. The caryopsis is enveloped by the caryopsis coat, consisting of the pericarp, seed coat, and

TABLE 11.1. Nutritional composition of brown rice and other cereals.

Nutrient	Brown rice	Wheat	Barley	Sorghum	Maize	Millet
Usable protein (%)	6.3	3.5	7.9	4.8	6.6	7.5
Digestible energy (%)	96.3	86.4	81.0	79.9	87.2	87.2
Energy (kcal/100 g)	447	436	454	447	461	459
Protein % (N × 6.25)	8.5	12.3	12.8	9.6	11.4	13.4
Fat (%)	2.6	2.2	3.9	4.5	5.7	5.5
Available carbohydrate (%)	74.8	81.1	64.9	67.4	74.0	73.7
Crude fiber (%)	0.9	1.2	4.3	4.8	2.3	1.8
Ash (%)	1.6	1.6	2.2	3.0	1.6	1.8
Tannin (%)	0.1	0.5	0.8	1.9	0.5	0.7
Thiamin (mg/100 g)	0.34	0.52	0.12	0.38	0.37	0.73
Riboflavin (mg/100 g)	0.05	0.12	0.05	0.15	0.12	0.38
Niacin (mg/100 g)	4.7	4.3	3.1	3.9	2.2	2.3
Iron (mg/100 g)	3.0	5.0	7.0	10.0	4.0	8.0
Zinc (mg/100 g)	2.0	3.0	3.0	2.0	3.0	3.0
Lysine (mg/16 g N)	3.8	2.3	3.2	2.7	2.5	2.7
Threonine (g/16 g N)	3.6	2.8	2.9	3.3	3.2	3.2
Methionine + cystine (g/16 g N)	3.9	3.6	3.9	2.8	3.9	3.6
Tryptophan (g/16 g N)	1.1	1.0	1.1	1.0	0.6	3.6

Source: Adapted from http://www.ricecrc.org/reader/rice-crc/tg/nutritional.htm http://www.fao.org/docrep/t0567e/T0567E0d.htm.

TABLE 11.2. Amino acid profile of rice products (g AA per 16 g N).

Amino acid	Brown rice	Milled rice	Rice bran
Protein content (% N × 6.25)	8.7	8.5	16.8
Energy content (J/g dry matter)	18.5	18.3	23.1
Arginine	7.25	7.93	7.55
Histidine	2.43	2.24	2.53
Isoleucine	4.05	4.11	4.05
Leucine	7.92	7.86	7.29
Lysine	3.65	3.44	4.64
Methionine	2.16	2.24	2.18
Methionine + cysteine	3.32	4.19	4.43
Phenylalanine	4.94	5.09	4.53
Phenylalanine + tyrosine	8.53	8.27	7.46
Threonine	3.51	3.40	3.95
Tryptophan	1.21	1.11	0.77 (1.3)[a]
Valine	5.61	5.81	6.25
Alanine	5.58	5.51	6.92
Aspartic acid	8.95	8.95	9.33
Glutamic acid	19.49	18.54	13.22
Glycine	4.71	4.34	6.24
Proline	4.48	4.47	4.76
Serine	4.95	5.01	4.72
Amino acid score[b] (%)	62.9	59.3	80.0
Digestible energy[c] (% of total)	94.3_b	96.6_a	67.4_c
True digestibility (TD)[c] (% of diet N)	96.9_b	98.4_a	78.8_c
Biological value[c] (% of digested N)	68.9_b	67.5_b	86.6_a
Net protein use[c] (% of diet N)	66.7_b	66.4_b	68.3_a
Amino acid score × TD (%)	61.0	58.4	63.0

Source: Adapted from Juliano (2003).

[a]Based on 25-58 brown rices, 53-173 milled rices, and 20-28 rice brans.

[b]Based on 5.8 g lysine/16 g N as 100.

[c]By rows, mean values for five growing rats followed by the same subscript letter are not significantly different at the 5% level by multiple range test.

TABLE 11.3. HPLC peak retention times and spectral characteristics of lutein and β-carotene in selected and transformed rice cultivars.

Sample	HPLC time (min)	Peak R_t^a (min)	A_{max}^b (nm) (peak II, III)	Ratio^c (II/III)	Peak identification	Peak R_t^a (min)	A_{max}^b (nm) (peak II, III)	Ratio^c (II/III)	Peak identification
Lutein standard	24	11.10	450.1, 477.9	1.11	Lutein				
β-carotene standard	24					20.83	457.4, 485.2	1.12	β-carotene
Amarillo Cuba (unpolished)	24	10.41	450.1, 477.9	1.10	Lutein	19.53	457.4, 484.0	1.11	β-carotene
Dudemasino (unpolished)	24	10.34	450.1, 479.2	1.10	Lutein	19.40	458.6, 484.0	1.13	β-carotene
IR68899B (unpolished)	24	10.52	450.1, 477.9	1.12	Lutein	19.80	456.2, 488.8	1.18	β-carotene
Sirendah Kuning (unpolished)	24	10.47	450.1, 479.2	1.11	Lutein	19.74	457.4, 485.2	1.13	β-carotene
Swarna (unpolished)	24	10.50	450.1, 477.9	1.10	Lutein				
Leuang Meuang Gahn (unpolished)	24	10.27	456.2, 482.8	1.11		19.35	457.4, 488.8	1.10	β-carotene
Golden *indica* rice Nang Huong Cho Dao 3 (polished)	24	10.83	450.1, 477.9	1.11	Lutein	19.33	457.4, 484.0	1.14	β-carotene

	Retention time[a]	Absorbance maximum[b]	Ratio[c]	Compound	Retention time[a]	Absorbance maximum[b]	Ratio[c]	Compound
Golden *indica* rice IR64E26 (polished)	24	11.15	**451.4**, 479.2	1.11	Lutein	19.84	**457.4**, 482.8	1.12	β-carotene
Lutein standard	85	17.62	**444.1**, 473.1	1.09	Lutein				
β-carotene standard	85					61.52	**452.6**, 479.2	1.13	β-carotene
Bongkitan (unpolished)	85	17.57	**445.3**, 473.1	1.08	Lutein	61.77	**451.4**, 481.6	1.14	β-carotene
Calibo (unpolished)	85					62.03	**453.8**, 479.2	1.12	β-carotene
Khao Dawk Mali 105 (unpolished)	85	18.89	**444.1**, 471.9	1.09	Lutein	62.02	**452.6**, 479.2	1.12	β-carotene
Klemas (unpolished)	85	18.37	**444.1**, 473.1	1.08	Lutein	62.36	**452.6**, 479.2	1.14	β-carotene
Limpopo (unpolished)	85								
Golden rice Taipei 309 (polished)	85					61.22	**452.6**, 479.2	1.13	β-carotene
Golden *indica* rice KDGR29-104 (polished)	85	17.66	**445.3**, 473.1	1.10	Lutein	61.62	**450.1**, 477.9	1.17	β-carotene

Note: HPLC = high-performance liquid chromatography.

[a] Retention time.

[b] Absorbance maximum. Main absorbance maximum is shown in bold.

[c] Ratio of the height of the main absorbance maximum to the longest wave maximum.

313

TABLE 11.4. Iron and zinc content of some selected varieties of brown rice.

	Iron		Zinc	
Variety	Mean ± SE (mg/kg)	No. of samples	Mean ± SE (mg/kg)	No. of samples
Jalmagna	22.4 ± 1.4	5	31.8 ± 7.7	4
Zuchem	20.2 ± 1.8	4	34.2 ± 5.0	3
Xua Bue Nuo	18.8 ± 0.8	2	24.3 ± 0.7	2
Madhukar	14.4 ± 0.5	3	34.7 ± 2.8	3
IR64	11.8 ± 0.5	3	23.2 ± 1.4	3
IR36	11.8 ± 0.9	5	20.9 ± 1.4	4

Source: Adapted from Gregorio et al. (2000).

TABLE 11.5. Iron and zinc content of varieties of brown rice grown under similar conditions in eight different sets of varieties.

		Iron	Zinc
Variety set	No. of samples	Mean ± SE mg/kg (range)	
Traditional and improved varieties	140	13.2 ± 2.9 (7.8-24.4)	24.2 ± 4.6 (13.5-41.6)
IR breeding lines	350	10.7 ± 1.6 (7.5-16.8)	25.0 ± 7.6 (15.9-58.4)
Tropical *japonicas*	250	12.9 ± 1.5 (8.7-23.9)	26.3 ± 3.8 (15.0-40.1)
Popular varieties and donors	199	13.0 ± 2.5 (7.7-19.2)	25.7 ± 4.6 (15.3-37.3)
Promising lines (NCT)	83	8.8 ± 1.3 (6.3-14.5)	25.4 ± 4.2 (17.0-38.0)
New plant types	44	16.7 ± 2.1 (11.5-24.0)	29.6 ± 3.2 (23.0-36.0)
Wild rice and derivatives	21	15.6 ± 2.3 (11.8-21.0)	37.9 ± 8.6 (23.0-52.0)
Aromatic rices	51	14.6 ± 3.2 (10.8-23.2)	31.9 ± 6.0 (23.0-50.0)

Source: Adapted from Gregorio et al. (2000).

nucellar layers. Inside the nucellus layer is the aleurone, which completely encloses the embryo and endosperm. It is composed of a few layers of parenchymatous cells (Figure 11.3). The aleurone cells are filled with protein-rich bodies in a protein-carbohydrate matrix with lipid droplets (Juliano, 1984). The typical nutrient composition of rice is given in Table 11.1. Unfortunately, brown rice is rarely consumed (Doesthale et al., 1979), and polishing the rice grain by removing the outer layers entails considerable loss of carotenoids (Tan et al., 2005) and iron (Gregorio, 2002; Vasconcelos et al., 2003). A rice seedling and genetically improved and control seeds are shown in Figures 11.4a and 11.4b.

Rice Protein

Protein content of the rice grain is relatively high, but still the lowest among cereals. Milling causes considerable changes of total protein content and amino acid composition profile in the rice grain (Tables 11.1 and 11.2). Lysine is considered the major limiting essential amino acid based on the amino acid pattern and on the nutritional requirements of preschool children (FAO/WHO/UNU, 1985). The protein composition varies among rice varieties and depending on the nature of soil conditions (Juliano, 2003).

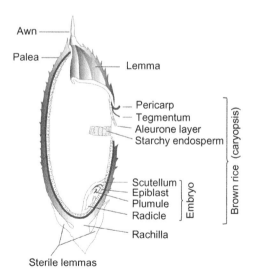

FIGURE 11.3. Cross-section of rice grain showing structural arrangement of different tissues. *Source:* International Rice Research Institute, 2002.

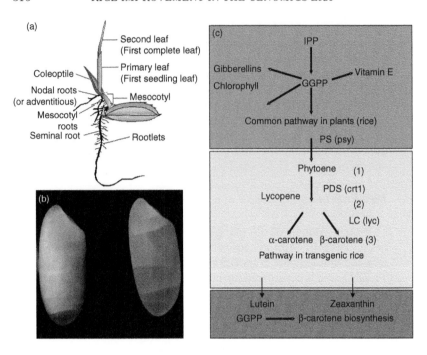

FIGURE 11.4. (a) A rice seedling; (b) β-carotene enriched seed, left, control seed, right; and (c) schematic pathway of β-carotene biosynthesis in plant. *Note:* Green tissues contain carotenoids including β-carotene but absent in endosperm tissue. See corresponding Plate 11.4 in the central gallery.

With an increasing percentage of nitrogen in the grain, there is a tendency of lysine to decrease relative to the total protein. Also, milling causes a decrease of total protein content (Juliano, 2003). (See Table 11.2.)

Carotenoid Profile in Rice

Carotenoids are a large group of lipid-soluble pigments synthesized in plants, bacteria, fungi, and algae. There are two classes of carotenoids in plants: *carotenes* (hydrocarbons) and *xanthophylls* (oxygenated derivatives of carotenes). Carotenoids function as accessory pigments in photosynthesis, serve as photoprotective compounds, and are precursors for the phytohormone abscisic acid. These also serve as antenna pigments for light harvesting and also in the dissipation of excess light energy (Demming-Adams and

Adams, 1993). Carotenoids are synthesized from geranylgeranyl diphosphate (GGPP), a C_{20} isoprenoid precursor that is also used for other important compounds in plants (Figure 11.4c). Major provitamin A carotenoids in mammals, including humans, are β-carotene, β-cryptoxanthein, and α-carotene (Graham and Rosser, 2000). Non-provitamin A carotenoids, such as lutein, zeaxanthin, and lycopene, have other health benefits (Sommerburg et al., 1998). Many traditional rice varieties worldwide are popularly known as nutrition rice, and they are often designated yellow, black, or red rice. Screening of a large number of exotic rice germplasms using high-performance liquid chromatography (HPLC) analysis showed the presence of lutein and β-carotene in 40 rice cultivars in unpolished seeds, whereas no trace of β-carotene was found in polished seeds. There was also a gradual loss of carotenoids with polishing duration (Tan et al., 2005) (Figures 11.5 and 11.6, Table 11.3). This screening work suggests that some exotic germplasms containing high lutein and other carotenoids in small amounts could be used as parental lines for rice breeding or bioengineering to improve rice carotenoid content.

TRANSGENIC APPROACH
TO IMPROVE CAROTENOIDS IN RICE

Research on carotenoid biosynthesis has been in steady progress considering its importance to human health and the availability of natural sources (Sivakumar, 1998; Sandman, 2001) (Table 11.6). Bioengineered rice (a model *japonica* rice, *Taipei 309*) enriched with genes for β-carotene biosynthesis in seeds was first reported by Ye et al. (2000). Since then, research has emphasized the following areas:

1. development of *indica* golden rice expressing β-carotene;
2. development of non-antibiotic-selected transgenic *indica* golden rice;
3. development of marker-free golden rice; and
4. phenotyping and agronomic evaluation of golden rice (Datta et al., 2003; Hoa et al., 2003).

The International Rice Research Institute (IRRI) in collaboration with scientists from National Agricultural Research Systems (NARES), the University of Freiburg, and Syngenta aims to further improve the status of golden rice. Selected *indica* lines with high β-carotene content are now being field-evaluated in Asia and the United States.

Both biolistic and *Agrobacterium*-mediated transformations were used to generate *indica* golden rice. The plasmid vectors used included pBaa3, carrying the daffodil phytoene synthase *(psy)* gene under the control of an endo-

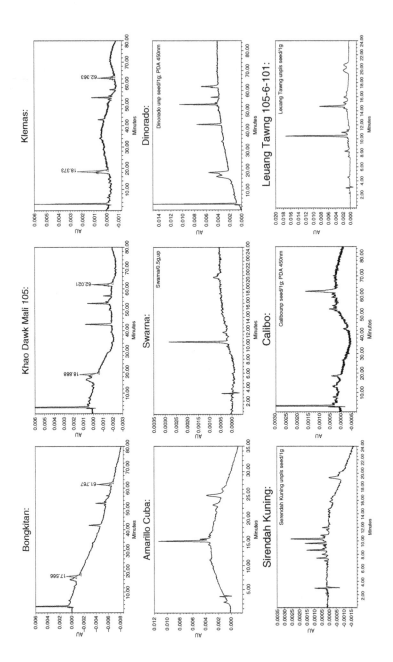

318

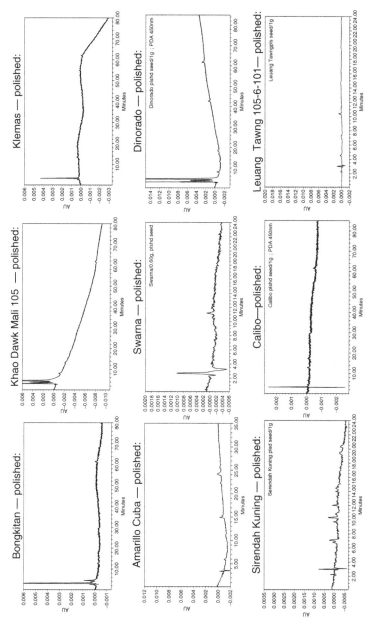

FIGURE 11.5. HPLC chromatograms and spectra of identified peaks for selected rice varieties in polished and unpolished conditions.

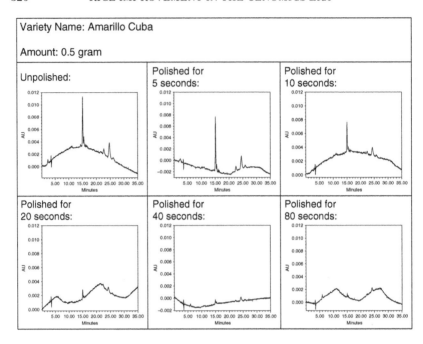

FIGURE 11.6. Gradual loss of carotenoids over polishing duration.

sperm-specific Gt1 promoter, and a bacterial phytoene desaturase *(crt1)* gene fused to a transit peptide sequence of pea-Rubisco small subunit to direct the expression of this bacterial gene into the plastids driven by the constitutive 35S promoter. The plasmid pTCL6 carried lycopene β-cyclase *(lcy)* cDNA of daffodil under the control of the 35S promoter nopaline synthase terminator. For the selectable marker gene, either of two plasmids were used: pNOV2820, which carried the phosphomannose isomerase gene driven by a constitutive cestrum promoter of yellow leaf curling virus (Syngenta international patent application #WO 01/73087 A1); or plasmid pGL2, with a *hph* gene coding for hygromycin phosphotransferase under the CaMV 35S promoter (Datta et al., 1990, 2003).

Transformation, regeneration of putative transformants, transfer to containment facilities, and molecular (PCR, Southern, RT-PCR) and biochemical analyses were performed using the methods described earlier (Datta et al., 1990, 1992, 2003). Starch and proteins are the two major storage reserves in rice seeds. Improvement of rice quality and nutrition depends primarily on the expression of individual genes, seed-specific expression, or promoters at the transcriptional and posttranscriptional levels (Qu and Takaiwa, 2004). Recently, *Agrobacterium*-mediated transformation showed an efficient development of *indica* golden rice using a Pcacar vector and POSITECH™ selection system. The PMI (phosphomannose isomerase)/ POSITECH™ selection system works well in *indica* rice, with some variation in response among genotypes (K. Datta et al., unpublished data).

HPLC analysis showed a variation of carotenoid profiles in different transgenic lines (see Figure 11.7 and Table 11.3). Several lines showed 0.8 to 3.14 μg carotenoids/g sample, including β-carotene up to 1.6 μg/g in golden rice (cv. BR29, a popular Bangladeshi variety).

EFFECT OF COOKING ON CAROTENOID PROFILE

Polished seeds of golden rice were cooked for 10 to 15 min, and HPLC chromatograms showed a reduction in the total carotenoid content by ~10% in the cooked grain compared to uncooked rice. However, the β-carotene (provitamin A) level was not much affected by the cooking process. Rather, the reduction of carotenoids in cooked rice mostly reflected the loss of lutein, zeaxanthin (carotenoids not possessing provitamin A activity), and some cryptoxanthin C. Some lines did not show any loss of β-carotene during cooking (Rai et al., 2004). More detailed studies are underway to determine the effect of storage, drying, duration of milling, and field conditions on total carotenoid profile and β-carotene levels.

EVALUATION OF IRON IN RICE GERMPLASM

It has been demonstrated that rice genotypes differ significantly in their ability to accumulate iron and zinc in seeds (Tables 11.4 and 11.5) (Graham et al., 1999; Senadhira et al., 1999; Yang et al., 1998). Experiments with different milling regimes of high-iron cultivars have shown that more than 50% of the minerals are lost during the milling process (see Figure 11.6) (Gregorio et al., 2000; Vasconcelos et al., 2003). Iron is stored in a vesicle of ferritin, consisting of 24 polypeptide subunits that can sequester some 4,000 iron atoms in a safe form as hydrous ferric oxide phosphate (Beard, 2003). It has been shown that ferritin is used by both plants and animals as the storage form of iron (Theil et al., 1997). Ferritin serves the dual purpose of providing iron for the synthesis of chlorophyll and iron proteins (ferredoxin and cytochromes) and preventing damage from free radicals produced by

TABLE 11.6. Summary of selected pioneering work in carotenoid improvement in plants.

Gene involved in carotenoid biosynthesis	Cloned/transformed	Crop species	Remarks	References
γ[a]	Cloned	Maize	Importance of such a regulatory gene in rice is conceptualized	Buckner et al., 1991
crt1 (Phytoene desaturase)	Cloned/transformed	Erwinia uredovora/ tobacco	Herbicide resistance increased	Misawa et al., 1990, 1993
crtE	Cloned	Erwinia herbicola	Coding for GGPP synthase	Math et al., 1992
A gene cluster	Cloned	Erwinia herbicola	For complete carotenoid pathway	To et al., 1994
psy	Transformed	Tomato	Resulted in dwarfism by redirecting metabolites from gibberellin pathway	Fray et al., 1995
lcy	Cloned	Daffodil	Lycopene to β-carotene	Al-Babili et al., 1996
psy	Cloned/transformed	Daffodil/ rice	Accumulation of phytoene in rice endosperm	Scheldz et al., 1996; Burkardt et al., 1997
CrtB (phytoene synthase)	Transformed	Brassica	Overexpression led to increase in carotenoids and other metabolites	Shewmaker et al., 1999
psy, crt1, lcy	Transformed	Rice	β-carotene accumulation in seeds	Ye et al., 2001
psy, crt1, lcy	Backcrossing and AC	Rice	DH lines and introgression of GR developed[a]	Baisakh et al., 2001, 2006
psy, crt1	Transformed	Rice	β-carotene (marker-free) expression in seeds	Datta et al., 2003; Parkhi et al., 2005

TABLE 11.6 (continued)

Gene involved in carotenoid biosynthesis	Cloned/transformed	Crop species	Remarks	References
psy, crt1, ferritin	Transformed	Rice	Tissue localization of β-carotene and iron in seeds	Krishnan et al., 2003
psy, crt1	Transformed	Rice	Accumulation of β-carotene in seeds	Hoa et al., 2003
psy, crt1	Backcrossing	Rice	Transfer of metabolic pathway[b]	Baisakh et al., 2006
psy, crt1	Backcrossing	Rice	Agronomic evaluation	Rai et al., 2003
psy, crt1, POSITECH[TM]	Transformed	Rice	Enhanced accumulation of β-carotene in seeds	Datta K. et al., 2006

[a]DH = Anther culture-derived doubled haploid lines.

[b]Marker-free *indica* golden rice developed.

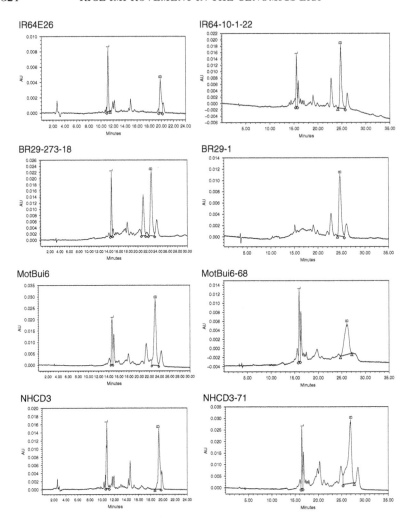

FIGURE 11.7. HPLC chromatograms showing carotenoid profile of selected golden *indica* rice. L = Lutein; B = β-carotene.

iron-dioxygen interactions. Iron in particular is partially mobilized from the leaves rather than in the seeds of rice. Therefore, it is still an open question whether plant ferritin can be mobilized from the source to the sink. Transgenic approaches using the *ferritin* gene may provide some significant improvement of high-iron rice.

DEVELOPMENT OF TRANSGENIC HIGH-IRON RICE

Brown rice contains a considerable amount of iron (12-24 µg/g). However, polishing, which removes the outer layers of the rice grain, causes considerable loss of micronutrients (Figures 11.8 and 11.9). Transgenic rice having the soybean *ferritin* gene driven by an endosperm-specific promoter *(GluB-1)* showed enhanced levels of iron after polishing (Goto et al., 1999; Lucca et al., 2001; Vasconcelos et al., 2003). Transgenic *indica* rice with the *ferritin* gene had as much as twofold higher iron content in polished rice grains (Vasconcelos et al., 2003) (Figure 11.10) Molecular (PCR and Southern) analyses confirmed the stable integration of the transgenes into two elite cultivars, BR29 and IR68144.

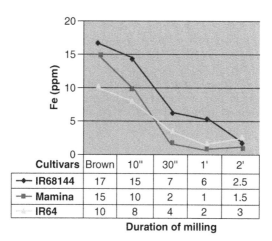

Cultivars	Brown	10"	30"	1'	2'
IR68144	17	15	7	6	2.5
Mamina	15	10	2	1	1.5
IR64	10	8	4	2	3

Duration of milling

FIGURE 11.8. Effect of milling duration on the Fe content of the rice grain.

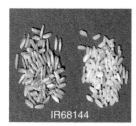

FIGURE 11.9. Brown and polished rice of different cultivars. See corresponding Plate 11.9 in the central gallery.

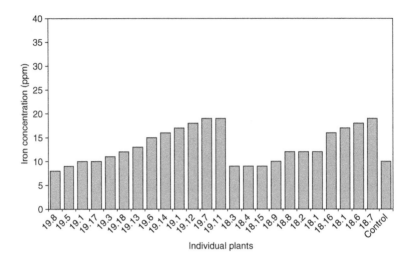

Individual plants

FIGURE 11.10. Iron concentration in T_2 seeds of lines Fr19 and Fr18, obtained by transformation of IR68144, and of control seeds, after commercial milling. Values represent an average of two independent extractions of 0.6 g of seed. *Source:* Adapted from Vasconcelos et al., 2003.

HISTOCHEMICAL LOCALIZATION OF FERRITIN PROTEIN AND β-CAROTENE IN RICE SEEDS

The distribution of the ferritin protein in rice seeds was analyzed using the immunological tissue imprinting technique; the endosperm tissue print of transformed seeds showed dark brown coloration, compared to very faint or no color in the control material. The loading of iron in the grain was also evident from histochemical techniques using a Prussian blue staining reaction. The endosperm and all other tissues of the transgenic grain showed a blue color, whereas the untransformed grain did not have any detectable blue color in the endosperm (Figure 11.11). However, iron was localized in the embryo and aleurone layers in both grain types (Krishnan et al., 2003).

For the localization of β-carotene in rice endosperm, the Carr-Price reaction was carried out with iodine derivatives under acidic conditions to form a blue color, as shown in transgenic rice grains (Figure 11.12) (Krishnan et al., 2003). Since the first publication on the synthesis of β-carotene in *japonica* rice seeds by genetic engineering using three genes—*psy* (phytoene synthase), *crt1* (phytoene desaturase), and *lcy* (lycopene cyclase), driven by constitutive and endosperm-specific promoters (Ye et al., 2000)—

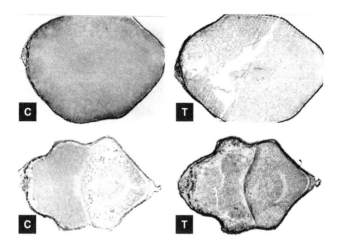

FIGURE 11.11. Iron detection in nontransgenic control (C) and transgenic rice grains (T). The accumulation of iron in control material is restricted to the aleurone layer, while iron is present in the entire grain including the endosperm in transgenic seeds. *Source:* Adapted from Krishnan et al., 2003. See corresponding Plate 11.11 in the central gallery.

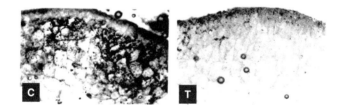

FIGURE 11.12. Transverse section of mature polished transgenic rice grain (T) showed blue color immediately after the reaction with Carr-Price reagent indicating the accumulation of β-carotene in endosperm whereas nontransgenic control (C) did not show any color reaction. See corresponding Plate 11.12 in the central gallery.

considerable progress has been made to transfer the traits of value-added nutrition to *indica* rice by genetic engineering (Datta et al., 2003; Hoa et al., 2003) and backcrossing using selected transgenic lines (Baisakh et al., 2006). Further characterization of transgenic events, particularly the transfer of vector DNA into the recipient rice genome and beyond-border transfer, was monitored by PCR (Hoa et al., 2003). The creation of marker-free *indica* golden rice IR64 has been reported (Baisakh et al., 2006), and a nonantibiotic marker, the *pmi* (phosphomannose isomerase) gene, was also used to develop golden rice (Datta et al., 2003; Hoa et al., 2003).

ESTIMATION OF IRON IN RICE GRAIN

The iron content of both polished and brown rice was estimated using either atomic absorbance mass spectrometry (AAMS) or inductively coupled plasma-optical emission spectrophotometry (ICP-OES). Pooled mean values indicated that transgenic lines had significantly higher iron in their polished seeds than the control lines (see Figure 11.10). The iron level varied from 9.0 to 19.0 µg/g dry weight in transgenic seeds, compared to 4.5 to 10 µg/g in the control seeds after polishing (Vasconcelos, unpublished data). Notably, the zinc levels in the transgenic grains were also higher than in the nontransgenic seeds.

SYNERGISM OF IRON, ZINC, AND β-CAROTENE INTERACTION AND ABSORPTION IN RICE GRAIN

There is a potential linkage and synergism of interaction and absorption of iron, zinc, and vitamin A in humans, as has been demonstrated in the literature (Christian and West, 1998; Solomons and Russell, 1980). It has been shown that synergistic interactions of zinc and vitamin A or vitamin A and iron effectively increase the absorption of the second nutrient (Garcia-Casal et al., 1998). It has been found in many rice varieties that the levels of iron and zinc are positively correlated (Table 11.7). It has also been shown that phytosiderophores mobilize not only iron but also zinc (Zhang et al., 1991).

EVALUATION OF GOLDEN RICE AND HIGH-IRON RICE UNDER GREENHOUSE CONDITIONS

Selected transgenic lines of golden rice (IR64 and BR29) and high-iron rice (IR68114) grown in greenhouses on a larger scale showed no variation of agronomic performance and phenotype as compared to the controls. Likewise, a similar level of β-carotene expression in seeds from the plants grown in pots in containment facilities and in the greenhouse was observed (Rai

TABLE 11.7. Iron and zinc concentrations in brown seeds of transgenic T_0 lines (Fr) of IR68144 and control (μg/g dry weight).

T_0 plant	Micronutrient concentration (μg/g DW)	
	Iron	Zinc
Fr 19	27.9 $\pm$ 1.0 [a]	55.5 $\pm$ 2.7[a]
Fr 25	25.4 $\pm$ 0.7 [a]	51.3 $\pm$ 1.6[a]
Fr 68	22.8 $\pm$ 0.7 [a]	45.1 $\pm$ 0.3[a]
Fr 77	21.1 $\pm$ 0.3 [a]	38.9 $\pm$ 0.3[a]
Fr 76	20.6 $\pm$ 0.7 [a]	36.1 $\pm$ 1.1[a]
Fr 15	20.5 $\pm$ 0.1 [a]	42.9 $\pm$ 0.9[a]
Fr 20	19.7 $\pm$ 3.2 [a]	40.8 $\pm$ 0.6[a]
Fr 78	18.6 $\pm$ 0.1 [a]	34.9 $\pm$ 0.1[a]
Fr 22	18.0 $\pm$ 0.6 [a]	37.4 $\pm$ 0.3[a]
Fr 28	16.2 $\pm$ 0.2 [b]	36.7 $\pm$ 0.5[b]
Control	15.7 $\pm$ 0.1	33.6 $\pm$ 0.2

Source: Modified from Vasconcelos et al. (2003).

Note: Results were obtained from two independent replications of 0.6 g seeds.

[a]Significant or [b]nonsignificant differences between the mean values were calculated by RCBD and Duncan's multiple range test at $p \leq$ .05; 0.6 g seeds were used for each sample analyzed.

et al., 2003) (Figure 11.13). Molecular (PCR, RT-PCR + Southern) analyses confirmed the stable integration of the transgenes into tropical *indica* rice (IR64, BR29, NHCD, *Mot Bui*, etc.) and expressed β-carotene in the seeds. However, introgression lines showed some variation in their carotenoid levels. More attention to selected lines having stable carotenoid levels with acceptable agronomic characters is needed (Rai et al., 2003).

BIOAVAILABILITY OF PLANT-BASED CAROTENOIDS AND IRON

Provitamin A carotenoids in fruits and vegetables are the major sources of vitamin A for a large proportion of the world's population. However, the carotenoids' substantial contribution depends on the efficient bioefficacy of provitamin A. *Bioefficacy* may be defined as the product of the fraction

FIGURE 11.13. Transgenic Golden Rice developed by biolistic (a-d) and *Agrobacterium* transformation (e-f) (β-carotene enriched IR64 and BR29 and high iron grown in greenhouse (pre-field condition) showing good agronomic performance with transgene expression: (a) transgenic BR29 plant (with gene for β-carotene; (b) expression of β-carotene in BR29 seeds (right sample); (c) transgenic IR64 plants in the greenhouse; (d) expression of β-carotene in IR64 seeds; (e) BR29 control seeds and segregating BR29 with simple genetic integration and high β-carotene expression in the seeds of the transformed line.

of the ingested amount of β-carotene that is absorbed *(bioavailability)* and the fraction that is converted to retinol in the body *(bioconversion)* (Lieshout et al., 2003).

Isotopic tracer techniques can be used for the precise estimation of the bioavailability, bioconversion, and bioefficacy of dietary carotenoids in humans. According to FAO/WHO, 6 μg β-carotene in food has the same vitamin A activity as 1 μg retinol (Food and Agriculture Organization, 1988; International Vitamin A Consultative Group, 1999). The Institute of Medicine of the National Academy of Sciences of the United States proved that 12 μg β-carotene in food has the same vitamin A activity as 1 μg retinol (Institute of Medicine, Food and Nutrition Board, 2001). Several other studies reported varied data on the conversion of β-carotene to retinol. It is obvious that more studies and data are required to draw conclusions on the conversion of provitamin A to retinol (Lieshout et al., 2003).

CAROTENOIDS AND β-CAROTENE BIOAVAILABILITY

Nutritionists have suggested that the bioavailability of β-carotene is comparatively low compared to that of vitamin A, and also that it depends on various dietary factors, such as the level of protein and fat in the diet and the cooking and processing methods used. Protein malnutrition and intestinal infections and infestations affect the absorption of vitamin A among children. The National Nutrition Monitoring Bureau of India has indicated that the average cereal intake of 106 gm per day among one- to three-year-old children may allow them to gain 400 μg retinol equivalents per day, whereas 1,600 μg of β-carotene per day is the recommended daily allowance (RDA). Golden rice alone may provide 10 to 40% of the RDA.

IRON AND ZINC BIOAVAILABILITY AND THE ROLE OF PHYTIC ACID

Plant-based foods provide less bioavailable iron and zinc, although they are very rich in trace elements such as copper, manganese, and so on. The bioavailability of iron and zinc is further reduced in the presence of the iron inhibitor phytic acid (Hunt, 2003). Improved assessment methods are required to determine the precise amount of iron and zinc bioavailability in rice- and plant-based diets in general for the benefit of vegetarians.

Nonheme iron absorption is usually inhibited by phytic acid, found in whole grains, legumes, lentils, and nuts; and polyphenols (tannic acid and chlorogenic acid) found in tea, coffee, red wine, and other cereals and other plant-based foods, including soy protein (Hallberg and Hulthén, 2000). A breeding strategy to lower phytic acid content may enhance iron bioavailability but could cause disruption of the nutrient composition of cereal grains, including phosphorus level (Graham and Welch, 1996). Increased levels of phytase, which breaks down phytic acid into phytates, may enhance the bioavailability of minerals. Phytase is present in rice grains but destroyed during boiling. Heat-stable phytase may be introduced into rice by genetic engineering, or the RNAi system may interrupt the pathway of phytic acid to enhance levels of phytase (Holm et al., 2002; Dr. Soren Rasmussen, RISO National Laboratory, personal communication). Thus, a more balanced nutrient composition achieved by altering the genetic profile of rice would be required for maximizing the nutrition benefit. Furthermore, it has been shown that the addition of ascorbic acid (Vitamin C) in the diet, which reduces the anti-nutrients, results in a higher bioavailability of iron (Tuntawifoon et al., 1990).

Certain amino acids, such as cysteine, enhance iron and zinc bioavailiut the concentration of cysteine is low in most staple foods (Hallberg, 1981). Protein quality is very important, not only for the supply of nutrients, but also because of its role in the absorption of iron and zinc. A modest increase of certain sulphur-containing amino acids (lysine and cysteine) may have a significant effect on iron and zinc bioavailability (Bouis et al., 2003; Ruel and Bouis, 1998).

IMPROVEMENT OF LYSINE AND OTHER NUTRIENTS

Lysine is synthesized from aspartate through a branch in the aspartate family pathway. Another branch of this pathway leads to the synthesis of two other essential amino acids, threonine and methionine. The entire aspartate family pathway, except for the last step of methionine production by methionine synthase, occurs in the plastid and is regulated by several feedback inhibition loops (Galili, 2002). Several isozymes of aspartate kinase (AK) are feedback-inhibited by either lysine or threonine. Similarly, dihydrodipicolinate synthase (DHPS), the first enzyme of the branch committed to lysine biosynthesis, is also feedback-inhibited by lysine. In vitro studies have revealed DHPS to be more sensitive to lysine than AK; hence, the expression of feedback-insensitive DHPS results in an overproduction of lysine, whereas overexpression of AK results in threonine overproduction. Several forms of bacterial genes—*lysC* (from *Escherichia coli*) and particularly *dapA* (from *Corynebacterium*), encoding AK and DHPS, respectively—are highly insensitive to lysine concentration, which has been reported before in soybean and canola (Falco et al., 1995). Transgenic work incorporating these genes into rice is in progress.

Transgenic rice has been produced with a lysine-rich protein gene *(lys)* cloned from a winged bean *(Psophocarpus tetragonoloba)* under the control of the maize ubiquitin promoter. The lysine content in the transgenic seeds was higher than in the wild-type seeds. One of the transgenic lines was reported to show a 16.04% increase in the lysine content of the seeds (Gao et al., 2001).

CONCLUSION

This review has highlighted the value of genetic variation in realizing breeding objectives. The genetic background of a cultivar and its adaptation in the farmer's field is very important for designing further improvements. Highly nutritious brown rice provides an important message about nutrition profiles, which challenges breeders to improve micronutrient loading in rice grains that is unaffected by polishing. Transgenic rice with the *ferritin* gene and with *psy* and *crt1* genes driven by an endosperm-specific promoter has shown promise in producing the desirable expression of transgenes in polished seeds, with selected lines showing no phenotypic difference compared to untransformed controls under greenhouse conditions. The combination of a good genetic background with multiple transgenes and allele mining has shown promising performance in field conditions and will foster the successful development of nutritious rice. A further challenge will be to evaluate the levels of bioefficacy of genetically improved, nutrition-dense rice. In the end, nutritious rice seeds enhanced with value-added characters are the simplest way of delivering essential nutrients.

REFERENCES

Ali, M., and S. Tsou (2000). The integrated research approach of the Asian vegetable research development center (AVRDC) to enhance micronutrient availability. *Food and Nutrition Bulletin 21:* 472-481.

Al-Babili, S., E. Hobeika, and P. Beyer (1996). A cDNA encoding *lycopene cyclase* (Accession No. X98796) from *Narcissus pseudonarcissus* L. (PGR 96-107). *Plant Physiology 112:* 1398.

Baisakh, N., K. Datta, M. Rai, S. Rehana, P. Beyer, I. Potrykus, and S. K. Datta (2001). Development of dihaploid transgenic "golden rice" homozygous for genes involved in the metabolic pathway for β-carotene biosynthesis. *Rice Genetics Newsletter 18:* 91-94.

Baisakh, N., S. Rehana, M. Rai, N. Oliva, J. Tan, D. Mackill, G.S. Khush, K. Datta, and S.K. Datta (2006). Marker-free transgenic (MFT) near-isogenic introgression lines (NILs) of 'golden' *indica* rice (cv IR64) with accumulation of provitamin A in the endosperm tissue. *Plant Biotechnology Journal 4:* 467-475.

Beard, J. (2003). Iron deficiency alters brain development and functioning. *Journal of Nutrition 133:* 1468-1472.

Bouis, H. E., B. M. Chassy, and J. O. Ochanda (2003). Genetically modified food crops and their contribution to human nutrition and food quality. *Trends in Food Science and Technology 14:* 191-209.

Buckner, B., T. L. Kelson, and D. S. Robertson (1991). Cloning of the γ locus of maize, a gene involved in the biosynthesis of carotenoids. *Plant Cell 2:* 867-876.

Burkhardt, P., P. Beyer, J. Wünn, A. Klöti, G. A. Armstrong, M. Schledz, J. von Lintig, and I. Potrykus (1997). Transgenic rice *(Oryza sativa)* endosperm expressing daffodil *(Narcissus pseudonarcissus)* phytoene synthase accumulates phytoene, a key intermediate of provitamin A biosynthesis. *The Plant Journal 11:* 1071-1078.

Christian, P., and K. P. West (1998). Interactions between zinc and Vitamin A: An update. *American Journal of Clinical Nutrition 68:* 435-441.

Datta, K., N. Baisakh, N. Oliva, L. Torrizo, E. Abrigo, J. Tan, M. Rai, S. Rehana, S. Al-Babili, P. Beyer, et al. (2003). Bioengineered 'golden' *indica* rice cultivars with β-carotene metabolism in the endosperm with hygromycin and mannose selection systems. *Plant Biotechnology Journal 1:* 81-90.

Datta, K., I. Potrykus, and S. K. Datta (1992). Efficient fertile plant regeneration from protoplasts of the *indica* rice breeding line IR72 *(Oryza sativa). Plant Cell Reports 11:* 229-233.

Datta, K., M. Rai, V. Parkhi, N. Oliva, J. Tan, and S. K. Datta (2006). Improved 'golden' *indica* rice and transgeneration enhancement of metabolic target products of carotenoids (B-carotene) in transgenic elite (IR64 and BR29) *indica* 'golden' rice. *Current Science 91:* 935-939.

Datta, S. K., and H. E. Bouis (2000). Application of biotechnology to improving the nutritional quality of rice. *Food and Nutrition Bulletin 21:* 451-456.

Datta, S. K., A. Peterhans, K. Datta, and I. Potrykus (1990). Genetically engineered fertile *indica*-rice plants recovered from protoplasts. *Bio/Technology 8:* 736-740.

Demming-Adams, B., and W. W. Adams (1993). Photoprotection and other responses of plants to high light stress. *Annual Review of Plant Physiology and Plant Molecular Biology 43:* 599-626.

Doesthale, Y., S. Devara, S. Rao, and B. Belavady (1979). Effect of milling on mineral and trace element composition of raw and parboiled rice. *Journal of the Science of Food and Agriculture 30:* 40-46.

Falco, S. C., T. Guida, M. Lock, J. Mauvais, C. Sanders, R. T. Ward, and P. Webber (1995). Transgenic canola and soybean seeds with increased lysine. *Bio/Technology 13:* 577-582.

FAO/WHO/UNU (1985). Energy and protein requirements: Report of a joint expert consultation (WHO, Geneva), Technical Report Series No. 724.

Food and Agriculture Organization (1988). Requirement of vitamin A, iron, folate, and vitamin B_{12}. Report of a Joint FAO/WHO Expert Consultation. *FAO Food and Nutrition Series 23.* Rome: FAO.

Food and Agriculture Organization (2003). *FAOSTAT agriculture data.* http://www.fao.org (last accessed February 26, 2007).

Fray, R. G., A. Wallace, P. D. Fraser, D. Valero, P. Hedden, P. M. Bramley, and D. Grierson (1995). Constitutive expression of a fruit *phytoene synthase* gene in transgenic tomatoes causes dwarfism by redirecting metabolites from the gibberellin pathway. *The Plant Journal 8:* 693-701.

Galili, G. (2002). New insights into the regulation and functional significance of lysine metabolism in plants. *Annual Review of Plant Physiology and Plant Molecular Biology 53:* 27-43.

Gao, F. Y., Y. X. Jing, S. H. Shen, S. P. Tian, T. Y. Kuang, and S. S. M. Sun (2001). Transfer of lysine-rice protein gene into rice and production of fertile transgenic plants. *Acta Botanica Sinica 43:* 506-511.

Garcia-Casal, M. N., M. Layrisse, L. Solano, M. A. Baron, F. Arguello, D. Lloera, J. Ramirez, I. Leets, and E. Tropper (1998). Vitamin A and β-carotene can improve nonheme iron absorption from rice, wheat and corn by humans. *Journal of Nutrition 128:* 646-650.

Goto, F., T. Yoshihara, N. Shigemoto, S. Toki, and F. Takaiwa (1999). Iron fortification of rice seed by the soybean *ferritin* gene. *Nature Biotechnology 17:* 282-286.

Graham, R. D., and J. M. Rosser (2000). Carotenoids in staple foods: Their potential to improve human nutrition. *Food and Nutrition Bulletin 21:* 404-409.

Graham, R., D. Senadhira, S. Beebe, C. Iglesias, and I. Monasterio (1999). Breeding for micronutrient density in edible potions of staple food crops: Conventional approaches. *Field Crops Research 60:* 57-80.

Graham, R. D., and R. M. Welch (1996). Breeding for staple-food crops with high micronutrient density: Long-term sustainable agricultural solutions to hidden hunger in developing countries. *Agricultural Strategies for Micronutrients Working Paper 3.* Washington DC: International Food Policy Research Institute.

Gregorio, G. B. (2002). Progress in breeding for trace minerals in staple crops. Symposium: Plant breeding: A new tool for fighting micronutrient malnutrition. *Journal of Nutrition 132*(Special issue): 500S-502S.

Gregorio, G. B., D. Senadhira, H. Htut, and R. D. Graham (2000). Breeding for trace mineral density in rice. *Food and Nutrition Bulletin 21:* 382-386.

Hallberg, L. (1981). Bioavailability of dietary iron in man. *Annual Review of Nutrition 1:* 123-127.

Hallberg, L., and L. Hulthén (2000). Prediction of dietary iron absorption: An algorithm for calculating absorption and bioavailability of dietary iron. *American Journal of Clinical Nutrition 71:* 1147-1160.

Hoa, T. T. C., S. Al-Babili, P. Schaub, I. Potrykus, and P. Beyer (2003). Golden *indica* and *japonica* rice lines amenable to deregulation. *Plant Physiology 133:* 161-169.

Holm, P. B., K. N. Kristiansen, and H. B. Pedersen (2002). Transgenic approaches in commonly consumed cereals to improve iron and zinc content and bioavailability. *Journal of Nutrition 132:* 514-516.

Hunt, J. M. (2003). Reversing productivity losses from iron deficiency: The economic case. *Journal of Nutrition 132:* 794-801.

Institute of Medicine, Food and Nutrition Board (2001). *Dietary reference intakes for vitamin A, vitamin K, arsenic, chromium, copper, iodine, iron, manganese, molybdenum, nickel, silicon, vanadium and zinc.* Washington DC: National Academy Press.

International Rice Research Institute (2002). *Rice almanac: Sourcebook for the most important economic activity on earth,* J. L. Maclean, D. C. Dawe, B. Hardy, and G. P. Hettel (Eds.). Wallingford, UK: CAB International.

International Vitamin A Consultative Group (1999). *The bioavailability of dietary carotenoids: Current concepts.* IVACG /ILSI, Washington, DC.

Juliano, B. O. (1984). Rice starch: Production, properties and uses. In *Starch: Chemistry and technology,* R. L. Whistler, J. N. BeMiller, and E. F. Paschall (Eds.), pp. 507-528. Orlando, FL: Academic Press.

Juliano, B. O. (2003). *Rice chemistry and quality.* Muñoz, Nueva Ecija, Philippines: Philippine Rice Research Institute.

Krishnan, S., K. Datta, N. Baisakh, M. de Vasconcelos, and S. K. Datta (2003). Tissue-specific localization of β-carotene and iron in transgenic *indica* rice (*Oryza sativa* L.). *Current Science 84:* 1232-1234.

Lieshout, M. V., C. E. West, and R. B. van Breemen (2003). Isotopic tracer techniques for studying the bioavailability and bioefficacy of dietary carotenoids, particularly β-carotene, in humans: A review. *American Journal of Clinical Nutrition 77:* 12-28.

Lucca, P., R. Hurrel, and I. Potrykus (2001). Genetic engineering approaches to improve the bioavailability and the level of iron in the rice grains. *Theoretical and Applied Genetics 102:* 392-397.

Math, S. K., J. E. Hearst, and C. D. Poulter (1992). The *crtE* gene in *Erwinia herbicola* encodes geranylgeranyl diphosphate synthase. *Proceedings of the National Academy of Sciences of the USA 89:* 6761-6764.

Misawa, N., M. Nakagawa, K. Kobayashi, S. Yamano, Y. Izawa, K. Nakamura, and K. Harashima (1990). Elucidation of the *Erwinia uredovora* carotenoid biosynthetic pathway by functional analysis of gene products expressed in *Escherichia coli. Journal of Bacteriology 172:* 6704-6712.

Misawa, N., H. Yamona, H. Linden, M. R. de Felipe, M. Lucas, H. Ikenga, and G. Sandmann (1993). Functional expression of the *Erwinia uredovora* biosynthesis gene *crtI* in transgenic plants showing an increase of β-carotene biosynthesis activity and resistance to the bleaching herbicide *Norflurazon. The Plant Journal 4:* 833-840.

Qu, L. Q., and F. Takaiwa (2004). Evaluation of tissue specificity and expression strength of rice seed component gene promoters in transgenic rice. *Plant Biotechnology Journal 2:* 1-13.

Parkhi V., M. Rai, J. Tan, N. Oliva, S. Rehana, A. Bandyopadhyay, L. Torrizo, V. Ghole, K. Datta, and S. K. Datta (2005). Molecular characterization of marker free transgenic *indica* rice lines that accumulate carotenoids in seed endosperm. *Molecular Genetics and Genomics 274:* 325-336.

Phillips, R. L. (2000). Biotechnology and agriculture in today's world. *Food and Nutrition Bulletin 21:* 457-459.

Rai, M., K. Datta, N. Baisakh, E. Abrigo, N. Oliva, and S. K. Datta (2003). Agronomic performance of golden *indica* rice (cv. IR64). *Rice Genetics Newsletter 20:* 30-33.

Rai, M., V. Parkhi, N. Oliva, L. Torrizo, E. Abrigo, S. Rehana, M. Khalekuzzaman, J. Tan, N. Baisakh, D. Mackill, et al. (2004). *Progress of golden rice research at IRRI.* Poster presented at the National Rice Biotechnology Network Meeting, April 17-19, 2004, New Delhi, India.

Ruel, M. T., and H. E. Bouis (1998). Plant breeding: A long-term strategy for the control of zinc deficiency in vulnerable populations. *American Journal of Clinical Nutrition 68:* 488-494.

Sandmann, G. (2001). Carotenoid biosynthesis and biotechnological application. *Archives of Biochemistry and Biophysics 385:* 4-12.

Schledz, M., S. Al-Babili, H. Lintig, S. Haubruck, S. Rabbani, H. Kleinig, and P. Beyer (1996). Phytoene synthase from *Narcissus pseudonarcissus*: Functional expression, galactolipid requirement, topological distribution in chromoplasts and induction during flowering. *The Plant Journal 10:* 781-792.

Senadhira D., G. B. Gregorio, and R. D. Graham (1999). Breeding iron and zinc-dense rice. In *Proceedings of the international workshop on micronutrient enhancement of rice for developing countries,* pp. 1-23, September 3, 1998. Stuttgart, AR: Rice Research and Extension Center.

Shewmaker, C. K., J. A. Sheehy, M. Daley, S. Colburn, and D. J. Ke (1999). Seed-specific over-expression of phytoene synthase: Increase in carotenoids and other metabolic effects. *The Plant Journal 20:* 401-412.

Sivakumar, B. (1998). Current controversies in carotene nutrition. *Indian Journal of Medical Research 108:* 157-166.

Solomons, N. W., and R. M. Russell (1980). The interaction of vitamin A and zinc: Implications for human nutrition. *American Journal of Clinical Nutrition 33:* 2031-2040.

Sommerburg, O., J. E. Keunen, A. C. Bird, and F. J. van Kuijk (1998). Fruits and vegetables that are sources for lutein and zeaxanthin: The macular pigment in human eyes. *British Journal of Ophthalmology 82:* 907-910.

Tan, J, N. Baisakh, N. Oliva, V. Parkhi, M. Rai, L. Torrizo, K. Datta, and S. K. Datta (2005). The screening of rice germplasm, including those transgenic rice lines which accumulate β-carotene in polished seeds for their carotenoids profile. *International Journal of Food Science and Technology 40:* 563-569.

Theil, E. C., J. W. Burton, and J. L. Beard (1997). A sustainable solution for dietary iron deficiency through plant biotechnology and breeding to increase seed ferritin control. *European Journal of Clinical Nutrition 21:* S28-S31.

To, K. Y., E. M. Lai, L. Y. Lee, T. P. Lin, C. H. Hung, C. L. Chen, Y. S. Chang, and S. T. Liu (1994). Analysis of the gene cluster encoding carotenoid biosynthesis in *Erwinia herbicola* EHO13. *Microbiology 140:* 331-339.

Tuntawiroon, M., N. Sritongkul, L. Rossander-Hulten, R. Pleehachenda, R. Suwanik, M. Brune, and L. Hallberg (1990). Rice and iron absorption in man. *European Journal of Clinical Nutrition 44:* 489-497.

Vasconcelos, M., K. Datta, N. Oliva, M. Khalekuzzaman, L. Torrizo, S. Krishnan, M. Oliveira, F. Goto, and S. K. Datta (2003). Enhanced iron and zinc accumulation in transgenic rice with the *ferritin* gene. *Plant Science 164:* 371-378.

Yang, X., Z. Q. Ye, C. H. Shi, M. L. Zhu, and R. D. Graham (1998). Genotypic differences in concentrations of iron, manganese, copper, and zinc in polished rice grains. *Journal of Plant Nutrition 21:* 1453-1462.

Ye, G., J. Tu, C. Hu, K. Datta, and S. K. Datta (2001). Transgenic IR72 with fused *Bt* gene *cry1AB/cry1Ac* from *Bacillus thuringiensis* is resistant against four lepidopteran species under field conditions. *Plant Biotechnology Journal 18:* 125-133.

Ye, X., S. Al-Babili, A. Kloti, J. Zhang, P. Lucca, P. Beyer, and I. Potrykus (2000). Engineering the provitamin A (β-carotene) biosynthetic pathway into (carotenoid-free) rice endosperm. *Science 287:* 303-305.

Zhang, F., M. Treeby, V. Romheld, and H. Marchner (1991). Mobilization of iron by phytosiderophores as effected by other micronutrients. *Plant and Soil 130:* 173-178.

Zimmerman, M., and R. Hurrel (2002). Improving iron, zinc, and vitamin A nutrition through plant biotechnology. *Current Opinion in Biotechnology 13:* 142-145.

Chapter 12

Expression of Human Milk Proteins in Mature Rice Grains

S. Nandi
N. Huang

INTRODUCTION

Lysozyme (LYS) and lactoferrin (LF) are the two most abundant proteins present in human milk. Human milk contains 0.05-0.4 mg/ml LYS. In addition to human milk, LYS is also found in other mammalian secretions, such as tears and saliva. Human LYS (hLYS) consists of an unglycosylated polypeptide chain with 130 amino acid residues, giving it a molecular weight of about 14 kDa. Lysozyme (EC 3.2.1.17) kills bacterial cells by cleaving the linkage between N-acetylglucosamine and N-acetylmuramic acid in the peptidoglycan layer of the bacterial cell wall. Unlike antibiotics that combat a specific type or group of bacteria, lysozyme destroys many bacteria and is therefore considered a natural antibiotic (see Jolles, 1996, for a review).

Human LF (hLF) is an 80-kDa (692 amino acid residues) iron-binding glycoprotein belonging to the transferrin family. Lactoferrin is found in an average of 1-2 mg/ml concentration in human milk and other mammalian secretions. Human LF is involved in several biological activities, including protection from pathogens, regulation of iron absorption, immune system modulation, and cellular growth promoter activity (see Lönnerdal and Iyer, 1995, and Shimazaki et al., 2000, for reviews). Of particular interest is that hLF has been shown to have antimicrobial activity against several human pathogens. Recently, LF has also been shown to have a protective effect against septic shock in germ-free piglets exposed to a lethal bacterial strain (Lee et al., 1998). Bactericidal activity of LF has also been shown against

Helicobacter pylori, a major pathogen of gastritis and peptic ulcer disease (Miehlke et al., 1996) and *Haemophilus influenzae,* a major cause of respiratory tract disease in children (Qiu et al., 1998).

Pediatricians and nutritionists generally agree that breastfeeding is the optimum way to feed infants. Breast milk not only provides the infant with a well-balanced supply of nutrients but also supplies a multitude of unique components that facilitate nutrient digestion and absorption, offer protection against microorganisms, and promote growth and development. Many of the advantages that breast-fed infants are provided by their mothers' milk are believed to be mediated by proteins in breast milk (Lönnerdal, 1985). Human milk proteins are unique, and, even if other protein sources used in infant formulas, such as skim milk, whey protein, and soy isolate, can provide amino acids at concentrations and ratios similar to those in breast milk protein, the biological activities conferred by human milk proteins cannot readily be copied. However, various physiological malfunctions of mother and/or child as well as cultural habits can prevent breastfeeding. In such cases, infant formulas have to be used, and their supplementation with biologically active human milk proteins is desirable. Expression of recombinant milk proteins in raw materials used for manufacturing infant formula may be a possible way to provide some of the beneficial factors present in breast milk.

With advances in genetic engineering, it is possible to produce human milk proteins in various expression systems including plant and mammalian cells (Chong and Langridge, 2000; Lönnerdal, 1996). Human LYS is presently purified from human milk and is therefore in limited supply. Recombinant hLYS (rhLYS) has been expressed in *Saccharomyces cerevisiae, Aspergillus oryzae,* mouse, *Nicotiana tabacum,* and *Daucus carota* (Maga et al., 1998; Nakajima et al., 1997; Takaichi and Oeda, 2000; Tsuchiya et al., 1992; Yoshimura et al., 1988), but expression levels were too low to meet the high concentration requirement of hLYS in human milk. Recombinant hLF (rhLF) has also been obtained as a fusion protein in *Aspergillus oryzae* (Ward et al., 1992) and in the baculovirus expression system (Salmon et al., 1997). An organism such as *Aspergillus*-produced protein will require a high degree of purification as well as safety and toxicity testing prior to using it as a food additive (Lönnerdal, 1996). In the plant system, rhLF has been expressed in tobacco (*Nicotiana tabacum* L. cv Bright Yellow) cells (Mitra and Zhang, 1994), tobacco plants (Salmon et al., 1998), and potato *(Solanum tuberosum)* plants (Chong and Langridge, 2000). In tobacco cell culture, the protein was truncated, whereas in tobacco and potato plants, the rhLF was processed correctly, but its expression level was very low (0.1% of total soluble protein) (Chong and Langridge, 2000).

The production of recombinant human proteins in rice grains has several advantages: (1) Rice grain proteins can accumulate an average of 9% of grain weight (Lásztity, 1996); (2) the endosperm proteins are synthesized during grain maturation and stored in protein bodies for use in the germination and seedling growth of the next plant generation; (3) grains can be stored for years without loss of functionality, and therefore downstream processing can be conducted independently of the growing season; (4) rice is generally recognized as safe (GRAS) for consumption, and rice-based foods are considered hypoallergenic (National Institutes of Health, 1984); and (5) in many countries, rice is the first solid food for infants, and rice-based infant formulas are commercially available (Bhan et al., 1988; Gastañaduy et al., 1990). Thus, rice grains expressing human milk proteins may be processed for use in infant formula or baby food without extensive purification steps.

We made an effort to express the human milk proteins in rice grains (Huang et al., 2002; Nandi et al., 2002). We obtained a very high expression level of functional LYS and LF. The biochemical, biophysical, and functional characteristics of these two proteins were identical to those of hLYS and hLF. Because rice grain ingredients are tolerated well by infants, novel infant formulas or baby foods containing rice flour or extract with recombinant functional protein are a realistic goal.

SELECTION OF PROMOTERS AND PLANT TRANSFORMATION

In our laboratory, several promoters were cloned and characterized to get high expression levels of recombinant protein and to confirm tissue-specific expression in rice (Huang, Wu et al., 2001; Hwang et al., 2001a,b, 2002; Wu et al., 2002; Yang et al., 2001). Two barley aleurone-specific promoter genes, chitinase *(Chi26)* and lipid transfer protein *(Ltp1)*, were effectively expressed in rice aleurone and embryo tissue, respectively (Hwang et al., 2001b), whereas glutelin 1 *(Gt1)* promoter was shown to be active in the endosperm only (Datta et al., 2003; Huang et al., 2002; Nandi et al., 2002; Qu and Takaiwa, 2004). A rice callus-specific expression promoter was also identified and successfully exploited (Huang, Sutliff et al., 2001; Huang, Wu et al., 2001). Our objective was to produce recombinant protein-specific tissue, and an emphasis was made on tissue-specific promoters rather than ubiquitous ones. Among the endosperm-specific promoters, we found that rice *Gt1* is the strongest. Thus, most of our constructs contained a *Gt1* promoter and a *Gt1* signal peptide, which was linked to a codon-optimized gene and ended with a nopaline synthase (NOS) terminator. We substituted the rare codons of the gene of interest with the preferred rice gene codons. It has been shown that codon-optimized genes can be expressed at higher levels in comparison to noncodon-optimized genes (Horvath et al., 2000). The expression vectors, carrying *hLYS* and *hLF* genes (Figure 12.1) were cobombarded with a selectable marker vector to generate separate lines expressing two human milk proteins.

EXPRESSION OF hLYS AND hLF IN RICE GRAIN

Protein was extracted from roots, shoots, or leaves in addition to the mature grain of transgenic plants and subjected to western blot analysis (Figure 12.2). Results indicated that there was no detectable expression of either protein in any other tissues except the grain. Because the *Gt1* promoter is expected to be active only in the immature endosperm (Okita et al., 1989), the presence of recombinant proteins in 5-day-old germinated grains showed the stability of the rhLYS and rhLF during germination.

INHERITANCE AND STABILITY OF EXPRESSION

Southern blot analysis was performed to estimate gene copy numbers, integration, and the inheritance pattern in seven consecutive generations of two selected highly expressing transgenic lines (Figure 12.3), one from each

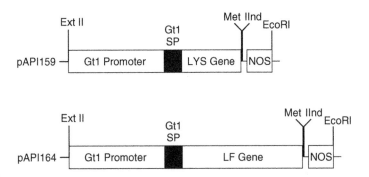

FIGURE 12.1. Diagrammatic picture and restriction map of the expression cassettes of plasmids pAPI159 and pAPI164 for expression of rhLYS and rhLF, respectively, in transgenic rice grain. *Gt1:* Rice glutelin 1 promoter; SP: Signal peptide; LYS Gene: Synthetic human lysozyme gene; LF Gene: Synthetic human lactoferrin gene; NOS: Nopaline synthase terminator/polyadenylation site.

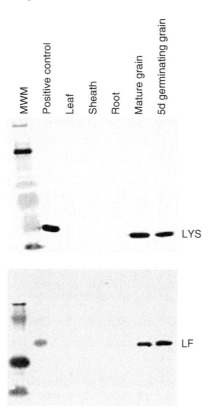

FIGURE 12.2. Analysis of rhLYS and rhLF in different rice tissues of the high expressing lines. Crude extract was separated on 18% precast SDS-PAGE. Lysozyme and lactoferrin were identified by specific antibody. Positive control: commercially available human LYS and LF used as standard.

construct. There was no homology between the rice genome (lane TP309, Figure 12.3) and the synthetic hLYS or hLF genes. The gene copy numbers in both transgenic rice lines ranged from 5 to 10 (Figure 12.3). Multiple integration of both chimeric genes into the rice genome was observed, which is a common phenomenon in transformation through particle bombardment (Cao et al., 1992; Christou, 1997; Datta et al., 1998; Tada et al., 1991). The presence of larger or smaller hybridizing fragments was probably due to the loss of one or both of the restriction enzyme recognition site(s) and/or rearrangement due to the integration of foreign DNA into the rice genome.

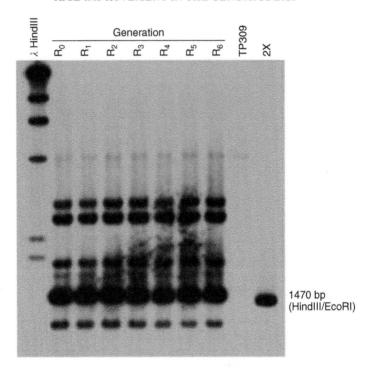

FIGURE 12.3. Southern analysis of transgenic rice plants carrying the synthetic gene of interest. DNA from seven consecutive generations of transgenic rice plants was digested with *Hind*III and *Eco*RI (see Figure 12.1). The membrane was probed with gel purified coding region of human lysozyme and lactoferrin genes. The molecular weight marker was lambda/*Hind*III (left lane). The non-transgenic plant TP309 was used as non-transformed negative control. 2× are two copies of genome equivalent.

If the transgenes were inserted in a single locus, then all the bands would co-segregate with generations. The Southern blot analysis showed that all the bands were inherited from R_0 to R_6 generation as a single linkage unit. We digested the same genomic DNA with other restriction enzymes to validate the observation.

At least 20 R_1 (first generation) grains from each line were analyzed for rhLYS and rhLF expression. Individual R_1 grains were cut in half. The endospermic half was subjected to protein expression analysis, and the corresponding positive embryonic half was germinated to generate R_1 seedlings. The seedlings were transplanted to generate R_2 grains. A transgenic plant is considered to be homozygous and expressing recombinant proteins if all grains from that plant are expressed. Homozygous lines were further confirmed by polymerase chain reaction (PCR) analysis. Based on the expression analysis and agronomic characters, selected homozygous R_2 lines were advanced to further generations. Expressions of rhLYS and rhLF of the selected homozygous lines over six generations are displayed in Figure 12.4. The amount of rhLYS and rhLF synthesized in the individual lines was constant throughout the generations tested. The highest stable expression level of rhLYS was achieved in line 159-53, which produced 0.6% rhLYS per brown rice weight, amounting to 45% of the total soluble protein extracted

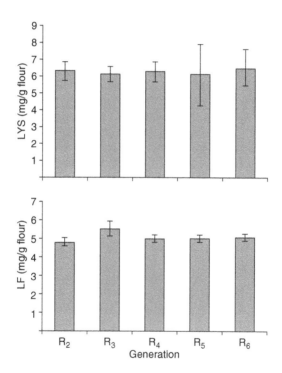

FIGURE 12.4. Stable expression of lysozyme and lactoferrin in transgenic rice grains R_2 through R_6 generation. Proteins from 1 g of brown rice flour were extracted with 40 ml of extraction buffer containing 0.35 M NaCl in PBS. Extraction was conducted at room temperature for 1 h with shaking. Homogenate was centrifuged at 14,000 rpm for 15 min at 4°C. Protein supernatant was removed and diluted as needed for activity assay or ELISA. Extraction was repeated three times and standard deviation was shown as an error bar.

from the rice grain. Similarly, quantitative analysis indicated results of up to 0.5% rhLF per brown rice weight amounting to 38% of the total soluble protein in the selected line (164-12).

MOLECULAR WEIGHT AND GLYCOSYLATION

Recombinant hLYS has a molecular mass (14 kDa) identical to that of native hLYS as confirmed by sodium dodecyl sulfate polyacrylamide gel electrophoresis (SDS-PAGE) (Figure 12.5). Furthermore, lysozyme does not have glycosylation sites. But recombinant hLF derived from transgenic rice grains was found to be glycosylated by a glycoprotein detection kit, and its mobility was close to that of hLF (Figure 12.5). To further confirm that rhLF is glycosylated, matrix-assisted laser desorption/ionization mass spectrometry (MALDI-MS) analysis showed that the rhLF has a molecular weight of 78.5 kDa, whereas the molecular weight of hLF is 80.6 kDa (Table 12.1). The higher molecular weight of rhLF compared to that of hLF without glycosylation (Salmon et al., 1998) is due to the carbohydrate moiety present in rhLF. Analysis showed that the purified rhLF contained xylose but lacked sialic acid, which is consistent with plant posttranslational modification patterns (Matsumoto et al., 1995). The influence of carbohydrate on the bioactivities of hLF or rhLF is not clear. However, it is clear that it does not affect the ability of the molecule to bind with the intestinal receptor (Kawakami and Lönnerdal, 1991), its ability to bind iron (Ward et al., 1995), or its affinity for hLYS and lipopolysaccharide (Vanberkel et al., 1995). The only reported difference between glycosylated and nonglycosylated hLF is its susceptibility to digestion by proteolytic enzymes such as trypsin (Vanberkel et al., 1995).

N-TERMINAL AMINO ACID SEQUENCE AND SURFACE CHARGES

The rice glutelin one signal sequence was engineered into the expression vector with the hope that recombinant human milk protein would be transferred into the endoplasmic reticulum and targeted into protein bodies. N-terminal sequence analysis using the Edman procedure of purified rhLYS and rhLF gave the same sequence of mature hLYS and hLF, respectively (Table 12.1). Thus, the signal peptide was removed correctly by the rice signal endopeptidase. Furthermore, reverse isoelectric focusing (IEF) gel electrophoresis revealed identical pI for recombinant and native human milk protein, indicating the same overall surface charge (Table 12.1).

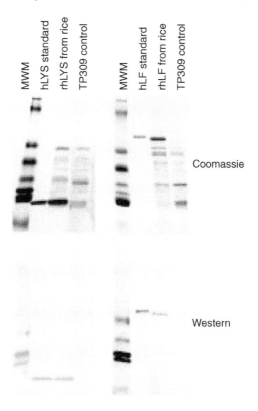

FIGURE 12.5. Coomassie and western analysis of grains from selected trans-genic lines expressing the highest amount of rhLYS (159-53) and rhLF (164-12) among all the lines tested. An equal amount of total soluble protein extracts were run on to SDS-PAGE. The same materials as in Coomassie gel were analyzed by western blot using respective antibody. Lane 1, molecular weight marker (MWM); lane 2, commercially available human LYS or LF as standard; lane 3, extract from transgenic rice grain; lane 4, extract from TP309 grain as control.

ANTIMICROBIAL ACTIVITY

The bactericidal activity of rhLYS was compared to that of native hLYS using an *Escherichia coli* strain, JM109 (Figure 12.6a). Inoculum of mid-log phase JM109 at 2×10^5 CFU (colony forming unit)/ml was incubated with filter-sterilized samples at $37°C$ for different lengths of time. One-fifth of the mixture volume was plated onto a lysogeny broth (LB) agar plate and

TABLE 12.1. Physical characterization of hLYS and hLF and rice grain-derived recombinant LYS and LF.

Source	Size (kDa)	N-terminal sequence	pI	Glycosylation	Sugar content (%)
hLYS (Sigma)	14	LysValPheGluArgCysG luLeuAlaArgThr	10.2	No	No
rhLYS (seeds)	14	LysValPheGluArg( )[a] GluLeuAlaArgThr	10.2	No	No
hLF (Sigma)	80.6	GlyArgArgArgArgSerV alGlnTrpCysAla	8.2	Yes	5.5
rhLF (seeds)	78.5	GlyArgArgArgArgSerV alGlnTrp( )[a] Ala	8.2	Yes	2.9

[a]This cycle was not defined, possibly due to an unmodified cysteine residue, which cannot form a stable derivative in Edman degradation analysis.

grown at 37°C overnight to determine the colony forming units. At a concentration of 30 μg/ml, rhLYS displayed a bactericidal time course activity similar to that of hLYS. In fact, after 1 h of incubation, almost all the bacteria had been killed, whereas untransformed rice grain extract had no killing effect (Figure 12.6a).

Human LF proteins dissolved (1.0 mg/ml) in buffer, which was untreated, pepsin treated (0.08 mg/ml at 37°C for 30 min), or pepsin (0.04 %)/pancreatin (0.016 mg/ml at 37°C for 30 min) treated, were filter-sterilized samples (Rudloff and Lönnerdal, 1992). The antimicrobial effect of rHLF (Figure 12.6b, untreated) was tested against enteropathogenic *E. coli* (EPEC) at a concentration of 10^4 CFU. Control EPEC reached 5×10^6 CFU after 12 h of incubation at 37°C, whereas EPEC with rhLF reached only 5×10^5 CFU, and as a comparison the EPEC cells with hLF produced 1×10^6 CFU. Both rhLF and hLF slowed the growth of the pathogenic EPEC cells.

Furthermore, to investigate the resistance of rhLF to protease digestion, an in vitro digestion model was developed to imitate the transit of protein through the infant gut, simulating conditions in the stomach (pepsin, low pH) and the small intestine (pancreatic enzymes, neutral pH) with moderate shaking (Rudloff and Lönnerdal, 1992). The rhLF was treated with pepsin and pancreatic enzymes, and 10^4 CFU of EPEC were exposed to the treated solution for 12 h at 37°C (Figure 12.6b). The pepsin-treated rhLF showed a higher activity than that of commercially available hLF and, as expected, lost some activity when compared to untreated rhLF. This may be due to

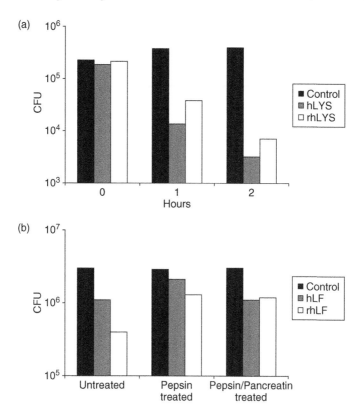

FIGURE 12.6. (a) Bactericidal activity of hLYS and rhLYS on *E. coli*. Strain JM109. The *E. coli* untransformed (TP309) grain extract, commercially available hLYS and transgenic rice grain extract containing rhLYS at 30 μg/ml, respectively, for the time indicated. At the end of the incubation, an aliquot of the mixture was plated on LB plates and colony forming units (CFU) were calculated. The data are plotted on a log scale. (b) Antimicrobial activity of rhLF against enteropathogenic *E. coli* (EPEC). The rhLF and native hLF, either treated with protease (pepsin and/or pancreatin) or untreated, were incubated with EPEC bacteria. Colony forming units (CFU) were determined after plating the EPEC cells on LB medium. The data are plotted on a log scale.

partial inactivation of hLF by pepsin. Similar antimicrobial activity was recorded when rhLF was treated with a combination of pepsin and pancreatin. All these experiments demonstrated that rice grain derived rhLF was as biologically active as commercially available hLF.

pH TREATMENT

Lysozyme activity is generally stable across a broad pH spectrum (pH 2 to 10) for the time indicated in Figure 12.7a. Both hLYS and rhLYS displayed similar pH stability. This result shows that rice grain-derived rhLYS is equally as active (approximately 200,000 units/mg) as commercially available hLYS.

The iron-binding capacity of lactoferrin is pH dependent. The stability of iron binding by rhLF toward low pH was analyzed and compared to that

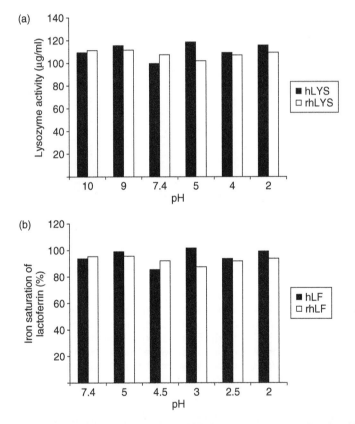

FIGURE 12.7. pH stability of hLYS and rhLYS. (a) Lysozyme was dissolved in different buffers (pH) at a concentration of 100 μg/ml. The mixture was incubated at 37°C for 30 min. The lysozyme activity was determined by activity assay. (b) The pH-dependent iron release from hLF and rice grain derived rhLF.

of hLF (Figure 12.7b). Both native hLF and rhLF were completely satu-rated with iron. Holo-hLF was incubated in buffers with pH ranging from 2 to 7.4 at room temperature for 24 h. Iron released from hLF was removed, and the iron saturation level was determined. Iron release began around pH 4 and was completed around pH 2 and was similar for both native and re-combinant hLF. The iron binding was reversible because iron-free rhLF was resaturated by raising the pH to 7.

NUTRIENT VALUE

The nutrition values of transgenic and nontransgenic grains were analyzed by standard procedures (Table 12.2). Human LF is an iron-binding protein, and each LF molecule binds two iron molecules (Fe^{3+}) (Haridas et al., 1995). The iron content of transgenic rice grains was more than twice that of non-transformed TP309 grains, whereas there were no significant differences in other tested nutrition factors between transformed and nontransformed grains (Table 12.2). The transgenic grains with increased iron content were an opaque-pinkish color. The opaque-pinkish color was observed inside as well as outside the rice endosperm. This opaque-pinkish color, segregated in Mendelian fashion, was linked with expression of rhLF and was inherited through the R_7 generation. The coloration of transgenic rice grains did not change any phenotype of transgenic rice plants. The transformed lysozyme seed did not have any visible difference in comparison to TP309. The plants of R_3 through R_7 generations were vigorous, and field performance did not show any significant difference with nontransgenic plants.

The gross nutrient composition of transgenic rice grains was similar to that of nontransformed TP309, except for increased target proteins and a twofold increase in iron content in the case of the rhLF-producing rice grain. Iron deficiency is a serious nutritional problem that affects about 30% of the world's population (World Health Organization, 1992). To address this prob-lem, researchers used the soybean ferritin gene to transform rice and achieved a two- to threefold increase in iron content in the vegetative tissue and grain endosperm of *japonica* rice (Drakakaki et al., 2000; Goto et al., 1999) and *indica* rice (Vasconcelos et al., 2003). The present study shows that a high level of expression of rhLF can also increase the level of iron in rice grains more than twofold in comparison to the wild type, probably due to each molecule of hLF binding two Fe^{3+} ions (Haridas et al., 1995; Kanyshkova et al., 1999). Our results suggest that rice with rhLF will increase the iron intake of groups ingesting this rice at a relatively modest level.

TABLE 12.2. Comparison of nutrition value per 100 grams of dehusked transgenic rice grains and TP309 as control.

Source	Carbo-hydrate (g)	Protein (g)	Fat (g)	Ca (mg)	K (mg)	Na (mg)	Fe (mg)	Water (g)	Calories	Concentration of LYS (mg)	Concentration of LF (mg)
TP309 (Control)	76.0	8.7	2.4	9	370	<10	0.87	11.3	369	0	0
Transgenic rhLYS grains	77.1	8.7	2.3	9	350	<10	0.70	12.1	365	600	0
Transgenic rhLF grains	75.7	8.7	2.2	8	330	<10	1.93	11.8	367	0	500

FACTORS THAT CONTRIBUTE
TO HIGH EXPRESSION LEVEL

We were able to obtain a very high level of expression of recombinant proteins, up to 6 and 5 g of rhLYS and rhLF per kilogram of dehusked rice grains, respectively. This level is stable for several generations (Figure 12.4). In our experience, the high level of expression with this system was likely due to a combination of strategies used, including (1) a synthetic gene with preferred rice codons, (2) a strong tissue-specific promoter, (3) a signal sequence to target protein molecules to protein bodies for storage, (4) a "position effect" determined by generating a large number of transformants, and (5) careful selection and advancement of expressed lines.

Enhancing and stabilizing the expression of foreign genes in plants by substituting rare codons with preferred codons was suggested early (Huang et al., 1990; Toenniessen, 1991) and demonstrated by the production of heat-stable β-glucanase in barley grains (Horvath et al., 2000).

Using a strong, tissue-specific promoter is another key to increasing the expression of foreign genes. Several genes for rice storage proteins have been cloned (Okita et al., 1989; Takaiwa and Oono, 1991), and, among them, the rice *Gt1* promoter is one of the strongest (Qu and Takaiwa, 2004). The *Gt1* promoter (714 bp) was cloned and engineered to control the expression of genes in transgenic rice seeds, and our results indicate this promoter was sufficient to control grain-specific expression of human milk proteins.

Studies have shown that recombinant protein expression can be increased by the use of a signal peptide (Horvath et al., 2000). The first 24 amino acids of *Gt1* were included as a signal sequence in the expression cassette. Though the sequence was predicted to be the signal peptide, it was not experimentally determined (Okita et al., 1989). The nucleotide sequence of the signal peptide was included in the expression cassette with the aim of directing the synthesized protein into the lumen of the endoplasmic reticulum and finally accumulating in the protein bodies. N-terminal sequencing of both the expressed proteins showed that the junction between *Gt1* signal sequence and mature rhLYS and rhLF was recognized and cleaved properly, exposing the correct N-terminal amino acid residues. It was hypothesized that the increased expression by the use of a signal peptide may be due to the deposition of recombinant protein in protein bodies rather than in cytosol. Our in vivo experiment showed that the foreign proteins were targeted to the protein bodies (within the endosperm) when we used the *Gt1* signal sequence as a signal peptide (unpublished results). The recombinant proteins were then protected from protease digestion by the protein body membrane and accumulated in the endosperm of the rice grain.

A position effect of foreign gene expression has long been observed and exploited in many studies (see Kusnadi et al., 1997, for a review). To exploit this effect, we generated over 100 transgenic rice plants of each construct, from which high expression lines were carefully selected and advanced.

In our selection procedure, we first emphasized the expression level, and then those lines that showed high expression and followed the Mendelian segregation ratio for a single dominant gene (3:1) in their first generation were advanced. The classical Mendelian ratio permits us assume that the target gene is in one locus, which was later confirmed by Southern blot analysis. Finally, with a combination of high expression and stable inheritance, we selected the agronomically elite lines, which performed close to the wild type (TP309), and advanced them by the single-seed descent method to produce seeds for the next generation. The agronomic selection was done in our breeding field (Figure 12.8).

UTILIZATION OF RECOMBINANT LYS AND LF

To use recombinant milk proteins for the production of infant formula and baby food, the recombinant proteins must be bioactive. The biochemical properties of rhLYS and rhLF purified from rice were found to be similar to those of native hLYS and hLF. Rice-derived rhLF is glycosylated and has

Lysozyme (Line 159-53)   Lactoferrin (Line 164-12)

FIGURE 12.8. Field performance of advanced generation (R_5) maturing transgenic plants producing either rhLYS or rhLF. See complementary Plate 12.8 in the photo gallery.

the same N-terminus as hLF. It binds the same amount of iron under various pH conditions and inhibits the growth of a human pathogen. Furthermore, rhLF showed similar resistance to in vitro protease digestion as native hLF, indicating that rhLF would possibly survive digestion in the infant gut. Indeed, it has been shown that many factors that aid in the digestion of dietary proteins, such as pepsin and pancreatic enzymes, may not function at maximum efficiency throughout early childhood. Due to this lower protease activity, many milk proteins, including hLF, are able to pass through the infant gut intact or only partially hydrolyzed.

The expression levels of recombinant milk proteins must be high enough for large-scale production, while keeping both the production and the downstream processing costs low. As discussed previously, we obtained up to 6 g of rhLYS and 5 g of rhLF per kilogram of dehusked rice grains, and the expression levels were stable for all generations tested (Figure 12.4). Assuming that dehusked rice yields are 5 tons/ha, we can harvest 30 kg and 25 kg of rhHLYS and rHLF, respectively, from each hectare of transgenic rice.

The expression of human milk proteins in rice grains provides a means to directly incorporate these proteins into infant formula without purification. This is practically and economically desirable and can be done because rice is generally recognized as safe and is currently used in infant formula. Rice is among the very first baby foods recommended for infants because it has good nutritional value and no known allergenicity (Gastañaduy et al., 1990). Thus, the proteins from rice introduce no risk and may be viewed as nutritionally sound.

FUTURE PROSPECTS

We have discussed the potential application of rhLYS and rhLF as nutraceuticals in the form of infant formula and baby food. There are also applications for these two proteins in other areas, such as food preservatives (Hughey and Johnson, 1987; Hughey et al., 1989; Proctor and Cunningham, 1988), pharmaceuticals (Proctor and Cunningham, 1988), and cosmetics (Sim et al., 2000). More significantly, rice expressing rhLYS and rhLF has also been shown to have a potential use as an animal growth promotant in an effort to eliminate or reduce the antibiotics used in animal husbandry (Humphrey et al., 2002).

The high expression level of foreign proteins in rice has implications regarding the production of valuable human proteins in an economic and safe way. Rice is consumed by a majority of the world's population and is generally recognized as safe (GRAS). This makes rice attractive as a "protein factory" to produce biomedicals and nutraceuticals for human consumption. The cloning and expression of human LYS and LF in rice grains has opened a new avenue for the bioproduction of other milk proteins. Our goal is to generate production-scale quantities of human milk proteins and other valuable proteins for human health and well-being by using rice grains as an efficient, economic, and safe "bioreactor" or, in other words, to produce value-added rice grains.

REFERENCES

Bhan, M. K., N. K. Arora, V. Khoshoo, P. Raj, S. Bhatnager, S. Sazawal, and K. Sharma (1988). Comparison of a lactose-free cereal-based formula and cow's milk in infants and children with acute gastroenteritis. *Journal of Pediatric Gastroenterology and Nutrition 7:* 208-213.

Cao, J., X. L. Duan, D. McElroy, and R. Wu (1992). Regeneration of herbicide resistant transgenic rice plants following microprojectile-mediated transformation of suspension culture cells. *Plant Cell Reports 11:* 586-591.

Chong, D. K., and W. H. Langridge (2000). Expression of full-length bioactive antimicrobial human lactoferrin in potato plants. *Transgenic Research 9:* 71-78.

Christou, P. (1997). Rice transformation: Bombardment. *Plant Molecular Biology 35:* 197-203.

Datta, K., N. Baisakh, N. Oliva, L. Torrizo, E. Abrigo, J. Tan, M. Rai, S. Rehana, S. Al-Babili, P. Beyer, et al. (2003). Bioengineered 'golden' *indica* rice cultivars with beta-carotene metabolism in the endosperm with hygromycin and mannose selection systems. *Plant Biotechnology Journal 1:* 81-90.

Datta, K., A. Vasquez, J. Tu, L. Torrizo, M. F. Alam, N. Oliva, E. Abrigo, G. S. Khush, and S. K. Datta (1998). Constitutive and tissue-specific differential expression of *cryIA(b)* gene in transgenic rice plants conferring resistance to rice insect pest. *Theoretical and Applied Genetics 97:* 20-30.

Drakakaki, G., P. Christou, and E. Stoger (2000). Constitutive expression of soybean ferritin cDNA in transgenic wheat and rice results in increased iron levels in vegetative tissues but not in seeds. *Transgenic Research 9:* 445-452.

Gastañaduy, A., A. Cordano, and G. G. Graham (1990). Acceptability, tolerance, and nutritional value of a rice-based infant formula. *Journal of Pediatric Gastroenterology and Nutrition 11:* 240-246.

Goto, F., T. Yoshihara, N. Shigemoto, S. Toki, and F. Takaiwa (1999). Iron fortification of rice seed by the soybean *ferritin* gene. *Nature Biotechnology 17:* 282-286.

Haridas, M., B. F. Anderson, and E. N. Baker (1995). Structure of human diferric lactoferrin refined at 2.2 Angstrom resolution. *Acta Crystallographica Section D—Biological Crystallography 51:* 629-646.

Horvath, H., J. Huang, O. Wong, E. Kohl, T. Okita, C. G. Kannangara, and D. von Wettstein (2000). The production of recombinant proteins in transgenic barley grains. *Proceedings of the National Academy of Sciences of the USA 97:* 1914-1919.

Huang, J., S. Nandi, Y. Wu, D. Yalda, G. Bartley, R. L. Rodriguez, B. Lönnerdal, and N. Huang (2002). Expression of natural antimicrobial human lysozyme in rice grains. *Molecular Breeding 10:* 83-94.

Huang, N., C. R. Simmons, and R. L. Rodriguez (1990). Codon usage patterns in plant genes. *Journal of China Association of Agricultural Science Societies 1:* 73-86.

Huang, J. M., T. D. Sutliff, L. Y. Wu, S. Nandi, K. Benge, M. Terashima, A. H. Ralston, W. Drohan, N. Huang, and R. L. Rodriguez (2001). Expression and purification of functional human alpha-1-antitrypsin from cultured plant cells. *Biotechnology Progress 17:* 126-133.

Huang, N., L. Y. Wu, S. Nandi, E. Bowman, J. M. Huang, T. Sutliff, and R. L. Rodriguez (2001). The tissue-specific activity of a rice beta-glucanase promoter (Gns9) is used to select rice transformants. *Plant Science 161:* 589-595.

Hughey, V. L., and E. A. Johnson (1987). Antimicrobial activity of lysozyme against bacteria involved in food spoilage and food-borne disease. *Applied and Environmental Microbiology 53:* 2165-2170.

Hughey, V. L., P. A. Wilger, and E. A. Johnson (1989). Antibacterial activity of hen egg white lysozyme against *Listeria monocytogenes* Scott A in foods. *Applied and Environmental Microbiology 55:* 631-638.

Humphrey, B. D., N. Huang, and K. C. Klasing (2002). Rice expressing lactoferrin and lysozyme has antibiotic-like properties when fed to chicks. *Journal of Nutrition 132:* 1214-1218.

Hwang, Y. S., C. McCullar, and N. Huang (2001a). Evaluation of expression cassettes in developing rice endosperm using a transient expression assay. *Plant Science 161:* 1107-1116.

Hwang, Y. S., S. Nichol, S. Nandi, J. A. Jernstedt, and N. Huang (2001b). Aleurone- and embryo-specific expression of the *beta-glucuronidase* gene controlled by the barley *Chi26* and *Ltp1* promoters in transgenic rice. *Plant Cell Reports 20:* 647-654.

Hwang, Y. S., D. Yang, C. McCullar, L. Wu, L. Chen, P. Pham, S. Nandi, and N. Huang (2002). Analysis of the rice endosperm-specific globulin promoter in transformed rice cells. *Plant Cell Reports 20:* 842-847.

Jolles, P. (1996). *Lysozymes: Model enzymes in biochemistry and biology.* Basel, Switzerland: Birkhauser Verlag.

Kanyshkova, T. G., D. V. Semenov, V. N. Buneva, and G. A. Nevinsky (1999). Human milk lactoferrin binds two DNA molecules with different affinities. *FEBS Letters 451:* 235-237.

Kawakami, H., and B. Lönnerdal (1991). Isolation and function of a receptor for human lactoferrin in human fetal intestinal brush-border membranes. *American Journal of Physiology 261:* G841-G846.

Kusnadi, A. R., Z. L. Nikolov, and J. A. Howard (1997). Production of recombinant proteins in transgenic plants: Practical considerations. *Biotechnology and Bioengineering 56:* 473-484.

Lásztity, R. (1996). *The chemistry of cereal proteins.* Boca Raton, FL: CRC Press.

Lee, W. J., J. L. Farmer, M. Hilty, and Y. B. Kim (1998). The protective effects of lactoferrin feeding against endotoxin lethal shock in germfree piglets. *Infection and Immunity 66:* 1421-1426.

Lönnerdal, B. (1985). Biochemistry and physiological function of human milk proteins. *American Journal of Clinical Nutrition 42:* 1299-1317.

Lönnerdal, B. (1996). Recombinant human milk proteins—an opportunity and a challenge. *American Journal of Clinical Nutrition 63:* 622S-626S.

Lönnerdal, B., and S. Iyer (1995). Lactoferrin: Molecular structure and biological function. *Annual Review of Nutrition 15:* 93-110.

Maga, E., G. Anderson, J. Cullor, W. Smith, and J. Murray (1998). Antimicrobial properties of human lysozyme: Transgenic mouse milk. *Journal of Food Protection 61:* 52-56.

Matsumoto, S., K. Ikura, M. Ueda, and R. Sasaki (1995). Characterization of a human glycoprotein (erythropoietin) produced in cultured tobacco cells. *Plant Molecular Biology 27:* 1163-1172.

Miehlke, S., R. Reddy, M. S. Osato, P. P. Ward, O. M. Conneely, and D. Y. Graham (1996). Direct activity of recombinant human lactoferrin against *Helicobacter pylori. Journal of Clinical Microbiology 34:* 2593-2594.

Mitra, A., and Z. Zhang (1994). Expression of a human lactoferrin cDNA in tobacco cells produces antibacterial protein(s). *Plant Physiology 106:* 977-981.

Nakajima, H., T. Muranaka, F. Ishige, K. Akutsu, and K. Oeda (1997). Fungal and bacterial disease resistance in transgenic plants expressing human lysozyme. *Plant Cell Reports 16:* 674-679.

Nandi, S., Y. Suzuki, J. Huang, D. Yalda, Y. Wu, G. B. Bartley, N. Huang, and B. Lönnerdal (2002). Expression of human lactoferrin in transgenic rice grains for application in infant formula. *Plant Science 163:* 713-722.

National Institutes of Health (1984). American Academy of Allergy and Immunology Committee on Adverse Reaction to Food and National Institute of Allergy and Infectious Diseases, pp. 130-131. *NIH Publication No. 84-2442.*

Okita, T. W., Y. S. Hwang, J. Hnilo, W. T. Kim, A. P. Aryan, R. Larson, and H. B. Krishnan (1989). Structure and expression of the rice glutelin multigene family. *Journal of Biological Chemistry 264:* 12573-12581.

Proctor, V. A., and F. E. Cunningham (1988). The chemistry of lysozyme and its use as a food preservative and a pharmaceutical. *Critical Review in Food Science and Nutrition 26:* 359-395.

Qiu, J., D. R. Hendrixson, E. N. Baker, T. F. Murphy, J. W. St. Geme, III, and A. G. Plaut (1998). Human milk lactoferrin inactivates two putative colonization factors expressed by *Haemophilus influenzae. Proceedings of the National Academy of Sciences of the USA 95:* 12641-12646.

Qu, L. F., and F. Takaiwa (2004). Evaluation of tissue specificity and expression strength of rice seed component gene promoters in transgenic rice. *Plant Biotechnology Journal 2:* 113-125.

Rudloff, S., and B. Lönnerdal (1992). Solubility and digestibility of milk proteins in infant formulas exposed to different heat treatments. *Journal of Pediatrics Gastroenterology and Nutrition 15:* 25-33.

Salmon, V., D. Legrand, B. Georges, M. C. Slomianny, B. Coddeville, and G. Spik (1997). Characterization of human lactoferrin produced in the baculovirus expression system. *Protein Expression and Purification 9:* 203-210.

Salmon, V., D. Legrand, M. C. Slomianny, I. El Yazidi, G. Spik, V. Gruber, P. Bournat, B. Olagnier, D. Mison, M. Theisen, and B. Merot (1998). Production of human lactoferrin in transgenic tobacco plants. *Protein Expression and Purification 13:* 127-135.

Shimazaki, K.-I., H. Tsuda, M. Tomita, T. Kuwata, and J.-P. Perraudin (Eds.) (2000). 4th International Conference on Lactoferrin: Structure, Function, and Applications. *Excerpta Medica International Congress series 1195.* Sapporo, Japan, May 18-22, 1999. Amsterdam: Elsevier Science.

Sim, Y. C., S. G. Lee, D. C. Lee, B. Y. Kang, K. M. Park, J. Y. Lee, M. S. Kim, I. S. Chang, and J. S. Rhee (2000). Stabilization of papain and lysozyme for application to cosmetic products. *Biotechnology Letters 22:* 137-140.

Tada, Y., M. Sakamoto, M. Matsuoka, and T. Fujimura (1991). Expression of a monocot LHCP promoter in transgenic rice. *EMBO Journal 10:* 1803-1808.

Takaichi, M., and K. Oeda (2000). Transgenic carrots with enhanced resistance against two major pathogens, *Erysiphe heraclei* and *Alternaria dauci. Plant Science 153:* 135-144.

Takaiwa, F., and K. Oono (1991). Genomic DNA sequences of 2 new genes for new storage protein glutelin in rice. *Japanese Journal of Genetics 66:* 161-171.

Toenniessen, G. H. (1991). Potentially useful genes for rice genetic engineering. In *Rice biotechnology,* G. S. Khush and G. H. Toenniessen (Eds.), pp. 253-280. Wallingford, UK: CAB International.

Tsuchiya, K., S. Tada, K. Gomi, K. Kitamoto, C. Kumagai, Y. Jigami, and G. Tamura (1992). High level expression of the synthetic human lysozyme gene in *Aspergillus oryzae. Applied Microbiology and Biotechnology 38:* 109-114.

Vanberkel, P. H. C., M. E. J. Geerts, H. A. Vanveen, P. M. Kooiman, F. R. Pieper, H. A. Deboer, and J. H. Nuijens (1995). Glycosylated and unglycosylated human lactoferrins both bind iron and show identical affinities towards human lysozyme and bacterial lipopolysaccharide, but differ in their susceptibilities towards tryptic proteolysis. *Biochemical Journal 312:* 107-114.

Vasconcelos, M., K. Datta, N. Oliva, M. Khalekuzzaman, L. Torrizo, S. Krishnan, M. Oliveira, F. Goto, and S. K. Datta (2003). Enhanced iron and zinc accumulation in transgenic rice with the *ferritin* gene. *Plant Science 64:* 371-378.

Ward, P. P., J. Y. Lo, M. Duke, G. S. May, D. R. Headon, and O. M. Conneely (1992). Production of biologically active recombinant human lactoferrin in *Aspergillus oryzae. Bio/Technology 10:* 784-789.

Ward, P. P., C. S. Piddington, G. A. Cunningham, X. D. Zhou, R. D. Wyatt, and O. M. Conneely (1995). A system for production of commercial quantities of human lactoferrin—a broad spectrum natural antibiotic. *Bio/Technology 13:* 498-503.

World Health Organization (1992). *National strategies for overcoming micronutrient malnutrition.* WHO Document no. 45/3. Geneva, Switzerland: WHO.

Wu, L., S. Nandi, L. Chen, R. L. Rodriguez, and N. Huang (2002). Expression and inheritance of nine transgenes in rice. *Transgenic Research 11:* 533-541.

Yang, D. C., L. Y. Wu, Y. S. Hwang, L. F. Chen, and N. Huang (2001). Expression of the REB transcriptional activator in rice grains improves the yield of recombinant proteins whose genes are controlled by a Reb-responsive promoter. *Proceedings of the National Academy of Sciences of the USA 98:* 11438-11443.

Yoshimura, K., A. Toibana, and K. Nakahama (1988). Human lysozyme: Sequencing of a cDNA, and expression and secretion by *Saccharomyces cerevisiae. Biochemical and Biophysical Research Communications 150:* 794-801.

Chapter 13

Nitrogen Fixation in Rice: Reality or Dream?

M. Dey
S. K. Datta

RELEVANCE OF THE GENOMIC ERA: AN INTRODUCTION

The year 2003 commemorated the fiftieth anniversary of the discovery of the DNA double helix, a discovery that dramatically changed every field of life sciences. Robust hybridization and polymerase chain reaction (PCR)-based diagnostics, recombinant DNA technology, gene technology and anti-sense techniques, and deciphering the full genomes of higher organisms are only few of the many fruits of bioinformatic and biotechnological revolutions that followed the discovery of the DNA double helix by Watson and Crick.

About two-thirds of the people in the world eat rice as the main cereal in their diet every day. To meet the demands of the bursting world population in the coming years, a substantial increase in rice production is required. Rice, being a nitrogen-intensive crop, requires a huge demand in the use of chemical fertilizers. This subsequently leads to a higher cost of rice production and serious environmental pollution (Dey and Datta, 2002). Alternatively, there exists in nature a very efficient and beneficial process of channeling atmospheric nitrogen into plants in a usable form through an endosymbiotic association between Gram-negative soil bacteria (rhizobia) and leguminous plants. For years, scientists have been keen to explore the feasibility of extending this endosymbiotic nitrogen fixation capacity to rice plants. Considering the multifaceted complexity involved in such a long-term endeavor, only a limited number of attempts in that direction have been undertaken so far with very little being accomplished.

In the genomic era when genetic manipulation across any barrier in the living kingdom is just another tool, the possibilities of genome engineering and directed evolution seem close to being achieved. At the disposal of modern plant biologists are the completely sequenced genomes of several plants, which started with *Arabidopsis,* and substantial collections of expressed sequence tags (ESTs) and genetic marker systems. The genomics revolution has led to new approaches in gene discovery and additional new technologies are being developed every day. Draft sequences of both *indica* and *japonica* rice genomes obtained by a whole-genome shotgun approach are available (Goff et al., 2002; Yu et al., 2002). Partial high-quality rice genome information obtained by the clone-by-clone strategy of the International Rice Genome Sequencing Project (IRGSP) (Sasaki and Sederoff, 2003) is also available in the public domain. Similarly, the genome-sequencing work of two model legumes, *Lotus japonicus* and *Medicago truncatula,* is in full swing. Therefore, it is now possible to perform genome-wide comparative analyses of model monocot rice with model legumes.

More and more, once-held impossible dreams are becoming reality in the realm of biology. In rice, by deploying the transgenic approach, the complete β-carotene biosynthesis pathway was engineered into the endosperm (Datta et al., 2003; Ye et al., 2000). Biotic (Tu et al., 2000) and abiotic (Garg et al., 2002) stress-tolerant rice plants were created using single or fusion gene constructs. A successful attempt to raise marker-free transgenic rice of agronomic importance beyond the laboratory scale was also made (Tu et al., 2003). On the other hand, deciphering the function of unknown genes in the sequenced genome is underway, using various forward and reverse genetic strategies—among which are insertional mutagenesis and RNAi (RNA interference)-based techniques. Therefore, it appears to be the right time to revisit the longstanding question, "How much longer will nitrogen fixation in rice continue to be a dream?"

NODULATION AND LEGUMES

An ecological and agronomically important symbiosis occurs between leguminous plants and rhizobia involving the de novo development of a specialized plant organ, the nodule. In the nodules, rhizobia fix dinitrogen into ammonia, which is assimilated by the host plant, and, in turn, rhizobia are supplied with carbon compounds (Figure 13.1). Legumes are a major

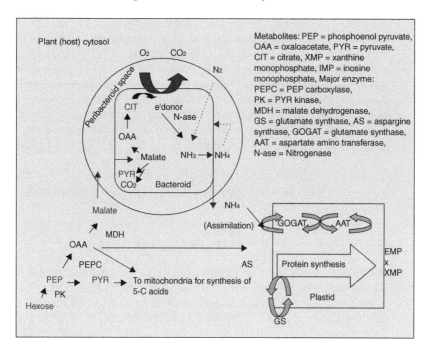

FIGURE 13.1. Simple scheme for carbon/nitrogen exchange across the peri-bacteroid membrance of legume nodules. (. . .) lines represent diffusion. *Source:* (Dey and Datta, 2002).

source of organic fertilizer, production of which on a global scale is equivalent to that produced by the entire chemical fertilizer industry (Laboratoire: Biologie moléculaire des relations plantes-microorganismes, 2002).

The nodulation process in rhizobia-legume symbiosis (also referred to as root nodule symbiosis, RNS) requires a sequence of highly regulated, complex, and coordinated events, involving the role of several genes from both partners and the exchange of signaling compounds (Day et al., 1995; Dey and Datta, 2002; Long and Staskawicz, 1993; Schultze and Kondorosi; 1998). Subsequently, rhizobia invade the host by means of an infection thread formed from curled root hairs that grow toward an emerging meristematic nodule zone in the root cortex. Enclosed by the host-derived peribacteroid membrane, bacteria are released into the nodule cells and eventually transform into dinitrogen-fixing bacteroids (Brewin, 1991). It is assumed so far that the formation of a peribacteroid membrane is not only essential to maintain a low-oxygen environment for nitrogenase to act, but also to avoid a hypersensitive response of the host. The mechanism underlying the discrimination between friend and foe in plants is not clearly understood. Nevertheless, it is believed that the whole succession to develop and maintain a functional symbiosis probably does not depend on just one mechanism responsible for defense suppression (Mithöfer, 2002). Until now, the complete signal transduction mechanism of RNS is far from being understood. However, ongoing genome analyses of the model legumes *(Medicago* and *Lotus)* have started to reveal many interesting facts (Kistner and Parniske, 2002).

AN EVOLUTIONARY PERSPECTIVE OF SYMBIOSIS

Arbuscular mycorrhiza (AM), a symbiosis formed between the majority of land plant species, including rice (Ammani and Rao, 1996; Isobe and Tsuboki, 1998) and legumes, and zygomycete fungi of the order Glomales (Brundrett, 2002) leads to an improved uptake of phosphate from the soil (Harrison, 1999; Rausch et al., 2001), and its contribution to the acquisition of nitrogen and other macro- and micronutrients might be equally significant (Hodge et al., 2001). During AM formation, fungal hyphae penetrate the epidermis of the root and grow toward the inner cortex, where they form arbuscules, highly branched structures that are thought to be the site of nutrient exchange between the two symbiotic partners. AM-like interactions were detected in early land plants; root nodule symbioses (RNS) with nitrogen-fixing bacteria evolved later. Legumes engage in RNS with a phylogenetically diverse group of Gram-negative bacteria referred to as rhizobia. Successful infection with rhizobia leading to the development of nitrogen-fixing nodules is generally host-strain specific (Perret et al., 2000), in contrast to the high degree of promiscuity exhibited by AM fungi toward plants. Both AM and RNS are endosymbioses; that is, the respective microsymbiont is hosted intracellularly. Rhizobia are recognized by their hosts via specific Nod factors (NFs; lipochitooligosaccharides; reviewed by Schultze and Kondorosi, 1996) and additional components, such as extracellular polysaccharides, lipopolysaccharides, and secreted proteins (Perret et al., 2000). Rhizobia enter rice at low frequency at points of emergence of the lateral roots (Cocking et al., 1994; Webster et al., 1997). Equivalent signal molecules of the microsymbiont have not yet been described for AM. Investigations of both AM and RNS in legume species have revealed a genetic overlap between the two types of symbioses, supporting the idea that during the evolution of RNS, parts of an evolutionarily older program were recruited from AM (Albrecht et al., 1999; Gualtieri and Bisseling, 2000; Harrison,

1999; Hirsch et al., 2001; Marsh and Schultze, 2001; Oldroyd, 2001; Provorov et al., 2002; Stougaard, 2001). A novel receptor-like kinase has been discovered that is required for the transduction of both bacterial and fungal symbiotic signals. This kinase defines an ancient signaling pathway that probably evolved in the context of arbuscular mycorrhiza and has been recruited subsequently for endosymbiosis with bacteria. An ancestral symbiotic interaction of roots with intracellular bacteria might have emerged from such recruitment, in the progenitor of the nodulating clade of plants. Analysis of symbiotic mutants of host plants and bacterial microsymbionts in RNS has revealed that present-day endosymbioses require the coordinated induction of more than one signaling pathway for development, at least some of which, such as nod-factor perception, might be very unique.

Genetics has been successful in the identification of several genes from a variety of legumes that are essential for root symbioses with both bacteria and fungi (Hirsch et al., 2001; Marsh and Schultze, 2001). These genes are referred to hereafter as the common SYM genes. Mutants in these genes are unable to support infection by symbiotic microorganisms. They comprise at least six genes from *Lotus japonicus* (Lj), *LjSYM2, LjSYM3, LjSYM4, LjSYM15, LjSYM23,* and *LjSYM30;* at least four genes from pea (Ps), *PsSYM8, PsSYM9, PsSYM19,* and *PsSYM30;* three genes from *Medicago truncatula* (Mt), *MtDMI1, MtDMI2,* and *MtDMI3;* three genes from *Melilotus alba* (Ma), *MaSYM1, MaSYM3,* and *MaSYM5;* and one gene each from alfalfa *(Medicago sativa),* bean *(Phaseolus vulgaris),* and fava bean *(Vicia faba)* (Hirsch et al., 2001; Marsh and Schultze, 2001). The cloning of orthologous genes from multiple species enabled the interconnection of genetic and physiological data obtained in different backgrounds, resulting in genetic and biochemical models that conceptualize how the common SYM genes might act together in mediating symbiotic signaling (Figure 13.2).

A growing array of sequence-based tools is helping to reveal the organization, evolution, and symbiotic relationships within and outside of legume genomes that range from tiny herbs to giant trees. Molecular phylogenies provide a context for posing hypotheses that can be tested using genomic and proteomic techniques. For example, what are the mechanisms underlying the origin of nodulation, which is suggested to have arisen three separate times (Doyle et al., 1997; Kajita et al., 2001)? If losses and gains occurred independently and several times in legume evolution, were the same proteins involved each time? Such knowledge of legume evolution can help in further molecular and cellular studies that compare rhizobial and AM symbioses.

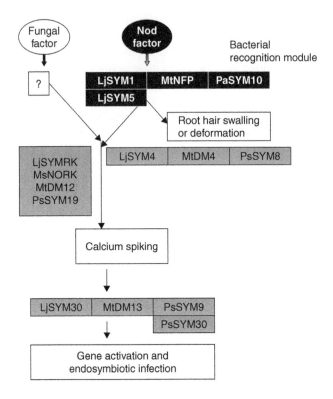

FIGURE 13.2. Pathway defined by plant genes required for both bacterial and fungal symbioses. Genes required for both root nodule symbiosis and arbuscular mycorrhiza (AM) define the common SYM pathway (depicted in gray). It is likely that the common SYM pathway is as old as AM and might be present in all AM-forming plants, whereas the genes specific for NF recognition (bacterial recognition module; in black) are possible key inventions for the process of bacterial infection in the legumes. *Source:* (Kistner and Parniske, 2002).

DISCUSSION: THE CONTEXT OF RICE

With the rapid ongoing progress of the genome sequencing projects of rice, *M. truncatula* and *L. japonicus,* it is expected that the majority of plant genes involved in the perception and transduction of bacterial and fungal signals will be cloned over the next few years. A more complete understanding of early signaling pathways should make it apparent which symbiosis genes are missing from crop plants such as rice that only form endosymbiosis with AM fungi. Once this is achieved, an already well-optimized (in rice) transgenic approach may provide the possibility of engineering into rice a complete pathway for nitrogen fixation.

The genetic overlap, as revealed already to some extent, between AM and RNS in legume species supports the idea that during the evolution of RNS, parts of an evolutionarily older program were recruited from AM. If this is true, then the fact that a rice plant can naturally enter into AM symbiosis (Isobe and Tsuboki, 1998) and that enod gene homologues are present in rice and related grasses (Kouchi et al., 1999; Reddy et al., 1999) may have significant implications on future research directions. Enod genes are early nodulin genes originally found in leguminous plants that are able to form functional nodules. These are known to encode products that are expressed before the onset of nitrogen fixation and are likely to be involved in bacterial infection and nodule development (Van Kammen, 1984). Rice cDNAs homologous to legume *enod40* and *enod93* (Reddy et al., 1998) have been isolated, although their actual function in rice has yet to be determined. Interestingly, *enod40* and a couple of other nodulin (Gianinazzi-Pearson, 1996a, 1996b) and leghemoglobin (Fruhling et al., 1997) genes are found to be induced during AM symbiosis of legumes, although their definite role has not yet been proven outside rhizobial symbiosis. Therefore, in rice, we are not sure if gene homologs like *enod40* or *enod93* might be extending their role to AM symbiosis or if *SYMRK/NORK* (which integrate presumably different fungal and bacterial signals) equivalents still operate.

The *enod40* is so far one of the best-characterized genes for which evidence of its role in nodule initiation has been provided (Charon et al., 1999; Crespi et al., 1994). It seems to induce dedifferentiation and root cortical cell division in response to Nod factors (Charon et al., 1997; Mylona et al., 1995). It is expressed in nodules at a very early stage and also in nonsymbiotic tissue and uninoculated roots. Sequences from several plant species have been determined and all of them were found to have two highly conserved regions. The first region has been suggested to code for a short peptide with biological activity (van de Sande et al., 1996), but these data could not be reproduced (Schell et al., 1999). The second region also has biological activity, but an open reading frame is missing and it was thought to act at the RNA level as a regulating sequence (Crespi et al., 1994). Expression of *enod40* in nonsymbiotic tissues hints at a probable dual role of this gene in symbiotic and nonsymbiotic organogenetic processes in plants. Considering the occurrence of *enod40* and consistent expression of its homologs in the vascular bundles of different species, a probable role of it in vascular bundle was suggested (Kouchi et al., 2000). Another very recent finding using peptide mass fingerprinting indicates the involvement of a 93-kDa binding protein (nodulin 100) in *enod40* function, which is a subunit of sucrose synthase. Based on these data, it has been suggested that *enod40* peptides are involved in the control of sucrose use in nitrogen-fixing nodules (Röhrig et al., 2002).

The *enod40* gene was transformed into rice driven by a constitutive promoter (Dey et al., 1999) that was earlier shown to express well in rice (Datta et al., 1990, 1992). Phytohormonal responses of *Mtenod40*-overexpressing and control plants in a homologous legume background *(M. truncatula)* and in the nonlegume rice were compared (Dey et al., 2004). An *enod40*-mediated root growth response, induced by ethylene inhibition, was observed in both plants. This suggests that ethylene inhibits *enod40* action in both legumes and nonlegumes. The *enod40* gene has been proposed to be involved in the transport of specific compounds into nodule primordium and in vascular bundles (Charon et al., 1999; Kouchi et al., 1999). This activity on root tissues might be inhibited by ethylene so that use of an ethylene antagonist allows a more efficient transport of critical compounds promoting root growth. Also in legumes, the effect of nitrate and the positioning of the nodules was proposed to be mediated by ethylene (Heidstra et al., 1997a, 1997b; Lee and La Rue, 1992; Ligero et al., 1991). For rice plants, however, few controlled studies have been done to clearly implicate ethylene in root growth. Like other crop plants, rice is adapted to very low internal ethylene levels (5 ppb), and its main function is related to seed setting and germination although experiments show that plant height may be affected (slowed down) at a level greater than 20 ppm (Klassen et al., 1999).

Auxin seems to interact with *enod40* function at the initiation of nodule organogenesis (Mathesius et al., 2000). But any differential root response to auxin treatments in either of the transgenic hosts (rice and *Medicago*) during phytohormonal assay (Dey et al., 2004) was not detected. However, as some physiological targets involving the *enod40* gene in these two distantly related hosts could be different, no significant difference in response was observed among the legume plants to benzylaminopurine (BAP) in either the presence or absence of nitrogen, whereas a statistically significant difference was found between control and *enod40* rice plants under nitrogen-limiting growth conditions.

Nevertheless, the fact that a limited supply of nitrogen was a prerequisite for *enod40* functioning in the *enod40*-overexpressing *Medicago* transgenics (Charon et al., 1999) and that the similar limiting condition is also a criterion for the *enod40* rice transgenics to respond to cytokinins reinforces the notion that the physiological context of *enod40* action, and eventually other early signal transduction steps of the nodule developmental program in legumes, may be similar in rice. When it comes to energy costs of the host plant, bioengineered C_4 rice (Ku et al., 1999) may have an edge over C_3 rice. However, in general, the photosynthetic capacity exceeds the storage capacity in rice, meaning that, given a sink (e.g., the nitrogen fixation site such as a nodule), rice might be in a position to spend some extra energy without cutting down its yield (Dey and Datta, 2002).

Nitrogen fixation in rice by rhizobia may still be a far-off dream. As outlined previously, however, recent findings that rice does possess many abilities that only a few years ago we would have strictly attributed to legumes, coupled with the data on evolutionary perspective of RNS and AM symbiosis, affirm that it might be worthwhile to pursue such a dream.

REFERENCES

Albrecht, C., R. Geurts, and T. Bisseling (1999). Legume nodulation and mycorrhizae formation: Two extremes in host specificity meet. *EMBO Journal 18:* 281-288.

Ammani, K., and A. S. Rao (1996). Effect of two arbuscular mycorrhizal fungi *Aculospora spinosa* and *A. scrobiculata* on upland rice varieties. *Microbiological Research 151:* 235-237.

Brewin, N. (1991). Development of the legume root nodule. *Annual Review of Cell Biology 7:* 191-226.

Brundrett, M. (2002). Coevolution of roots and mycorrhizas of land plants. *The New Phytologist 154:* 275-304.

Charon, C., C. Johansson, E. Kondorosi, A. Kondorosi, and M. Crespi (1997). *enod40* induces dedifferentiation and division of root cortical cells in legumes. *Proceedings of the National Academy of Sciences of the USA 94:* 8901-8906.

Charon, C., C. Sousa, M. Crespi, and A. Kondorosi (1999). Alteration of *enod40* expression modifies *Medicago truncatula* root nodule development induced by *Sinorhizobium meliloti. Plant Cell 11:* 1953-1965.

Cocking, E. C., G. Webster, C. A. Batchelor, and M. R. Davey (1994). Nodulation of non-legume crops: A new look. *Agro-Food-Industry Hi-Tech 1:* 21-24.

Crespi, M. D., E. Jurkevitch, M. Poiret, Y. Aubenton-Carafa, G. Petrovics, E. Kondorosi, and A. Kondorosi (1994). *ENOD40,* a gene expressed during nodule organogenesis, codes for a non-translatable RNA involved in plant growth. *EMBO Journal 13:* 5099-5112.

Datta, K., N. Baisakh, N. Oliva, L. Torrizo, E. Abrigo, J. Tan, M. Rai, S. Rehana, S. Al-Babili, P. Beyer, et al. (2003). Bioengineered 'golden' *indica* rice cultivars with β-carotene metabolism in the endosperm with hygromycin and mannose selection systems. *Plant Biotechnology Journal 1:* 81-90.

Datta, S. K., K. Datta, N. Soltanifar, G. Donn, and I. Potrykus (1992). Herbicide-resistant *indica* rice plants from IRRI breeding line IR72 after PEG mediated transformation of protoplasts. *Plant Molecular Biology 20:* 619-629.

Datta, S. K., A. Peterhans, K. Datta, and I. Potrykus (1990). Genetically engineered fertile *indica* rice recovered from protoplasts. *Bio/Technology 8:* 736-740.

Day, D. A., L. Whitehead, J. H. M. Hendriks, and S. D. Tyerman (1995). Nitrogen and carbon exchange across symbiotic membranes from soybean nodules In *Nitrogen fixation: Fundamentals and applications,* I. A. Tikhonovich, N. A. Provorov, V. I. Romanov, and W. E. Newton (Eds.), pp. 557-564. Netherlands: Kluver Academic Publishers.

Dey, M., A. Complainville, C. Charon, L. Torrizo, A. Kondorosi, M. Crespi, and S. K. Datta (2004). Phytohormonal responses of *enod40*-overexpressing plants in *Medicago truncatula* and rice. *Physiologia Plantarum 120:* 132-139.

Dey, M., and S. K. Datta (2002). Promiscuity of hosting nitrogen fixation in rice— an overview from legume perspective. *Critical Reviews in Biotechnology 22:* 281-314.

Dey, M., L. B. Torrizo, R. K. Chaudhuri, P. M. Reddy, J. K. Ladha, K. Datta, and S. K. Datta (1999). Transgenic rice harbouring legume *ENOD40* gene. *Rice Genetics Newsletter 16:* 147-149.

Doyle, J. J., J. L. Doyle, J. A. Ballenger, E. E. Dickson, T. Kajita, and H. Ohashi. (1997). A phylogeny of the chloroplast gene *rbcL* in the Leguminosae—taxonomic correlations and insights into the evolution of nodulation. *American Journal of Botany 84:* 541-554.

Fruhling, M., H. Roussel, V. Gianinazzi-Pearson, A. Puhler, and A. M. Perlick (1997). The *Vicia faba* leghaemoglobin gene *VfLb29* is induced in root nodules and in roots colonized by the arbuscular mycorrhizal fungus *Glomus fasciculatum. Molecular Plant-Microbe Interactions 10:* 124-131.

Garg, A. K., J. K. Kim, T. G. Owens, A. P. Ranwala, Y. D. Choi, L. V. Kochian, and R. J. Wu (2002). Trehalose accumulation in rice plants confers high tolerance levels to different abiotic stresses. *Proceedings of the National Academy of Sciences of the USA 99:* 15898-15903.

Gianinazzi-Pearson, V. (1996a). Arbuscular mycorrhizae: Getting to the roots of the symbiosis. *Plant Cell 8:* 1871-1883.

Gianinazzi-Pearson, V. (1996b). Plant cell responses to arbuscular mycorrhizal fungi: Getting to the roots of symbiosis. *Plant Cell 8:* 1899-1913.

Goff, S. A., D. Ricke, T. H. Lan, G. Presting, R. Wang, M. Dunn, J. Glazebrook, A. Sessions, P. Oeller, H. Varma, et al. (2002). A draft sequence of the rice genome (*Oryza sativa* L. ssp. *japonica*). *Science 296:* 92-100.

Gualtieri, G., and T. Bisseling (2000). The evolution of nodulation. *Plant Molecular Biology 42:* 181-194.

Harrison, M. J. (1999). Molecular and cellular aspects of the arbuscular mycorrhizal symbiosis. *Annual Review of Plant Physiology and Plant Molecular Biology 50:* 361-389.

Heidstra, R., G. Nilsen, F. Martinez-Abarca, A. Van Kammen, and T. Bisseling (1997a). Nod factor-induced expression of leghemoglobin to study the mechanism of NH_4NO_3 inhibition on root hair deformation. *Molecular Plant-Microbe Interactions 10:* 215-220.

Heidstra, R., W. C. Yang, Y. Yalcin, S. Pech, and A. M. Emons (1997b). Ethylene provides positional information on cortical cell division but is not involved in Nod factor induced root hair tip growth in *Rhizobium*-legume interactions. *Development 124:* 1781-1787.

Hirsch, A. M., M. R. Lum, and J. A. Downie (2001). What makes the rhizobia-legume symbiosis so special? *Plant Physiology 127:* 1484-1492.

Hodge, A., C. Campbell, and A. H. Fitter (2001). An arbuscular mycorrhizal fungus accelerates decomposition and acquires nitrogen directly from organic material. *Nature 413:* 297-299.

Isobe, K., and Y. Tsuboki (1998). The relationship between growth promotion by arbuscular mycorrhizal fungi and root morphology and phosphorus absorption in gramineous and leguminous crops. *Japanese Journal of Crop Science 67:* 347-352.

Kajita, T., H. Ohashi, Y. Tateishi, C. D. Bailey, and J. J. Doyle (2001). rbcL and legume phylogeny, with particular reference to Phaseoleae, Millettieae, and allies. *Systematic Botany 26:* 515-536.

Kistner, C., and M. Parniske (2002). Evolution of signal transduction in intracellular symbiosis. *Trends in Plant Science 7:* 511-518.

Klassen, S. P., B. Bugbee, and W. F. Campbell (1999). Ethylene sensitivity of crop plants: Implications. In *ASGSB annual meeting abstract #53,* p. 53. Logan: Utah State University, Crop Physiology Laboratory.

Kouchi, H., K. Takane, R. B. So, J. K. Ladha, and P. M. Reddy (1999). Rice *ENOD40*: Isolation and expression analysis in rice and transgenic soybean root nodules. *The Plant Journal 18:* 121-129.

Kouchi, H., K. Takane, R. So, P. M. Reddy, and J. K. Ladha (2000). Characterization of rice *ENOD40*: Do *ENOD40s* accomplish analogous functions in legumes and nonlegumes? In *The quest for nitrogen fixation in rice,* J. K. Ladha and P. M Reddy (Eds.), pp. 263-272. Los Baños, Philippines: International Rice Research Institute.

Ku, M. S. B., S. Agarie, M. Nomura, H. Fukayama, H. Tsuchida, K. Ono, S. Hirose, S. Toki, M. Miyao, and M. Matsuoka (1999). High level expression of maize phosphoenolpyruvate carboxylase in transgenic rice plants. *Nature Biotechnology 17:* 76-80.

Laboratoire: Biologie moléculaire des relations plantes-microorganismes. (2002). http://www.toulouse.inra.fr/lbmrpm/eng/symbiose.htm (last accessed April 6, 2002).

Lee, K. H., and T. A. La Rue (1992). Exogenous ethylene inhibits nodulation of *Pisum sativum* L. cv. Sparkle. *Plant Physiology 100:* 1759-1763.

Ligero, F. J., M. Caba, C. Lluch, and J. Olivares (1991). Nitrate inhibition of nodulation can be overcome by the ethylene inhibitor aminoethoxyvinylglycine. *Plant Physiology 97:* 1221-1225.

Long, S. R., and B. J. Staskawicz (1993). Prokaryotic plant parasites. *Cell 73:* 921-935.

Marsh, J. F., and M. Schultze. (2001). Analysis of arbuscular mycorrhizas using symbiosis-defective plant mutants. *New Phytology 150:* 525-532.

Mathesius, U., C. Charon, B. G. Rolfe, A. Kondorosi, and M. Crespi (2000). Temporal and spatial order of events during the induction of cortical cell division in white clover by *Rhizobium leguminosarum* cv. trifolii inoculation or cytokinin addition. *Molecular Plant-Microbe Interactions 13:* 617-628.

Mithöfer, A. (2002). Suppression of plant defense in rhizobia-legume symbiosis. *Trends in Plant Science 7:* 440-444.

Mylona, P., K. Pawlowski, and T. Bisseling (1995). Symbiotic nitrogen fixation. *Plant Cell 7:* 869-885.

Oldroyd, G. E. D. (2001). Dissecting symbiosis: Developments in Nod factor signal transduction. *Annals of Botany 87:* 709-718.

Perret, X., C. Staehelin, and W. J. Broughton (2000). Molecular basis of symbiotic promiscuity. *Microbiology and Molecular Biology Reviews 64:* 180-201.

Provorov, N. A., A. Y. Borisov, and I. A. Tikhonovich (2002). Developmental genetics and evolution of symbiotic structures in nitrogen-fixing nodules and arbuscular mycorrhiza. *Journal of Theoretical Biology 214:* 215-232.

Rausch, C., P. Daram, S. Brunner, J. Jansa, M. Laloi, G. Leggewie, N. Amrhein, and M. Bucher (2001). A phosphate transporter expressed in arbuscule-containing cells in potato. *Nature 414:* 462-470.

Reddy, P. M., R. K. Aggarwal, M. C. Ramos, J. K. Ladha, D. S. Brar, and H. Kouchi (1999). Widespread occurrence of the homologs of the early nodulin *(ENOD)* genes in *Oryza* species and related grasses. *Biochemical and Biophysical Research Communications 258:* 148-154.

Reddy, P. M., H. Kouchi, and J. K. Ladha (1998). Isolation, analysis and expression of homologs of the soybean early nodulin gene *Gmenod93 (GmN93)* from rice. *Biochimica et Biophysica Acta 1443:* 386-392.

Röhrig, H., J. Schmidt, E. Miklashevichs, J. Schell, and M. John (2002). Soybean *ENOD40* encodes two peptides that bind to sucrose synthase. *Proceedings of the National Academy of Sciences of the USA 99:* 1915-1920.

Sasaki, T., and R. R. Sederoff (2003). Genome studies and molecular genetics: The rice genome and comparative genomics of higher plants. *Current Opinion in Plant Biology 6:* 97-100.

Schell, J., T. Bisseling, M. Dulz, H. Franssen, K. Fritze, M. John, T. Leinow, A. Lebnick, E. Miklashevichs, K. Pawlowski, et al. (1999). Reevaluation of phytohormone-independent division of tobacco protoplast-derived cells. *The Plant Journal 17:* 461-466.

Schultze, M., and A. Kondorosi (1996). The role of lipochitooligosaccharides in root nodule organogenesis and plant cell growth. *Current Opinion in Genetics and Development 6:* 631-638.

Schultze, M., and A. Kondorosi (1998). Regulation of symbiotic root nodule development. *Annual Review of Genetics 32:* 33-57.

Stougaard, J. (2001). Genetics and genomics of root symbiosis. *Current Opinion in Plant Biology 4:* 328-335.

Tu, J., K. Datta, N. Oliva, G. Zhang, C. Xu, G. S. Khush, Q. Zhang, and S. K. Datta (2003). Site-independently integrated transgenes in the elite restorer rice line

Minghui 63 allow removal of a selectable marker from the gene of interest by self-segregation. *Plant Biotechnology Journal 1:* 155-165.

Tu, J., K. Datta, Q. Zhang, and S. K. Datta (2000). Field performance of transgenic elite commercial hybrid rice expressing *Bacillus thuringiensis* δ-endotoxin. *Nature Biotechnology 18:* 1101-1104.

Van de Sande, K., K. Pawlowski, L. Czaja, U. Weineke, J. Schell, J. Schmidt, R. Walden, M. Matvienko, J. Wellink, A. B. Van Kammen, et al. (1996). Modifications of phytohormone response by a peptide encoded by *ENOD40* of legumes and a non-legume. *Science 273:* 370-373.

Van Kammen, A. (1984). Suggested nomenclature for plant genes involved in nodulation and symbiosis. *Plant Molecular Biology Reporter 2:* 43-45.

Webster, G., C. Gough, J. Vasse, C. A. Batchelor, K. J. O'Callaghan, S. L. Kothari, M. R. Davey, J. Denarie, and E. C. Cocking (1997). Interactions of rhizobia with rice and wheat. *Plant and Soil 194:* 115-122.

Ye, X., S. Al-Babili, A. Klöti, J. Zhang, P. Lucca, P. Beyer, and I. Potrykus (2000). Engineering the provitamin A (β-carotene) biosynthetic pathway into (carotenoid-free) rice endosperm. *Science 287:* 303-305.

Yu, J., S. Hu, J. Wang, G. Wong, S. Li, B. Liu, Y. Deng, L. Dai, Y. Zhou, X. Zhang, et al. (2002). A draft sequence of the rice genome (*Oryza sativa* L. ssp. *indica*). *Science 296:* 79-92.

Chapter 14

Rice for Sustainable Food and Nutritions Security

M. S. Swaminathan
S. Appa Rao

INTRODUCTION: CHALLENGES AHEAD

Rice occupies and will continue to occupy a pivotal place in global food and livelihood security systems. Of the annual world production of 596.485 million tons from 155.128 million hectares (ha), Asia produces 540.621 million tons from 138.563 million ha. The average rice yield in Asia is 3.9 tons/ha, compared to 3.8 tons/ha worldwide, 6.4 tons in Japan, 6.3 tons in China, 2.1 tons in India, and 10.1 tons in Australia (Riceweb.org). Per capita consumption of rice in Asia ranges from 132 to 449 g/day (Table 14.1). The world's population is predicted to reach close to 7 billion by the year 2010 and around 8 billion by 2020, at which time 82% of the people will live in developing countries (Borlaug, 2002; Evans, 1998). The UN forecasts that the world's population will reach 9.4 billion by 2050; on the medium projection, there will be only 1.2 billion people in the more developed regions, compared to 8.2 billion in the currently less developed areas, 5.5 billion of them in Asia and 2 billion in Africa (United Nations Development Programme, 2001). The world must develop the capacity to feed 10 billion people within the next 40 to 50 years, predominantly within Asia and Africa as the first priority. The second one is eliminating chronic undernutrition, which still afflicts many people in the world, by producing enough food for all.

More than 90% of rice is produced and consumed in Asia. However, it is a staple food crop of 40% of the world's population. Under conditions of

TABLE 14.1. Per capita availability of milled rice and contribution of rice to dietary energy and protein in selected rice-eating countries (1997-1999).

Country	Availability of milled rice g/person/day	% Contribution of rice	
		Energy	Protein
Myanmar	187.0	76.0	68.0
Cambodia	448.6	76.7	69.6
Vietnam	147.0	68.0	62.0
Bangladesh	441.2	75.6	66.0
Indonesia	413.6	51.4	42.9
Thailand	132.0	58.0	48.0
Korea, PDR	259.0	33.5	21.0
China	251.0	30.4	19.5
Korea, Republic of	259.0	33.5	21.0
Nepal	26.23	38.5	29.4
Sri Lanka	255.3	38.4	37.0
Philippines	267.4	40.9	30.1
Malaysia	254.2	29.8	20.4
India	207.9	30.9	24.1

Source: FAOSTAT as of December 19, 2002 (Food and Agriculture Organization of the United Nations, 2002).

expanding urbanization and per capita income, a minimum annual growth rate of 2.5% will be needed in food grain production in India to balance the food budget. This will involve producing 240 million tons by the year 2007, as compared to the existing production of about 200 million tons. Recent estimates of rice demand indicate that a compound annual growth rate of 1.25% is needed to meet the expected world rice consumption in 2020. In most developing countries, agriculture occupies an important place in national livelihood security systems (Table 14.2). In spite of a steady decline in agriculture as a percentage of the gross domestic product (GDP) (Table 14.3), the share of agriculture in the workforce remains high. The situation is different in industrialized countries (Gardner et al., 2000). In many countries where agriculture is a major source of employment, low income (Table 14.4) due to the lack of jobs, resulting in inadequate purchasing power, is now the primary cause of food insecurity at the household level.

TABLE 14.2. Importance of agriculture in the domestic economy of selected countries.

| Country | Share of agriculture in GDP | | Share of agriculture in workforce (%) |
	1990	1999	
India	31	28	65
China	27	19	71
Bangladesh	28	21	62
Pakistan	26	26	48
Sri Lanka	20	17	47
Philippines	22	17	42
Egypt	19	17	33
USA	2	2	3
France	3	2	4
Italy	3	3	7
Germany	1	1	3

Source: World Bank (2002).

TABLE 14.3. Annual real GDP growth rates in selected developing countries of Asia.

Country/region	1996 (%)	2002 (%)
Bangladesh	5.0	3.2
China	9.6	6.8
India	7.3	5.2
Indonesia	8.0	3.5
Malaysia	10.0	2.5
Pakistan	2.9	4.4
Philippines	5.7	3.2
Thailand	5.9	2.0
Vietnam	9.3	4.8
Developing Asia	8.3	5.6

Source: International Monetary Fund (2001).

TABLE 14.4. Regional comparison of income poverty in developing countries.

Region	1987	1990	1993	1996	1998
East Asia and Pacific	417.5	452.4	431.9	265.1	278.3
South Asia	474.4	495.1	505.1	531.7	522.0
Asia and Pacific	891.9	947.5	937.0	796.8	800.3
Europe and Central Asia	1.1	7.1	18.3	23.8	24.0
Latin America and the Caribbean	63.7	73.8	70.8	76.0	78.2
Middle East and North Africa	9.3	5.7	5.0	5.0	5.5
Sub-Saharan Africa	217.2	242.3	273.3	289.0	290.9
Asia and Pacific (% of world total)	75.4	74.2	71.8	66.9	66.8
Total	1183.2	1276.4	1304.3	1190.6	1198.9

Source: International Fund for Agricultural Development (2001).

Therefore, agricultural intensification, diversification, and value addition are essential for providing opportunities for skilled jobs in the farm sector. Importing food rather than improving productivity will only aggravate the job famine in predominantly agricultural countries.

A quantum leap in rice yield was possible using semidwarf varieties that responded to increased use of fertilizers, pesticides, herbicides, and a host of other chemicals, along with water. Rice production increased by 2.3% per year from 1968 to 2001 after the release of IR8 (Hossain and Narciso, 2002). In India, rice production grew 1.11% from 1994 to 2001 (Venkataramani, 2002). The slowdown in productivity growth of irrigated rice is due to a decline in the nitrogen-supplying capacity of intensively cultivated wetland soils, which is around 30%, over a 20-year period, at all nitrogen levels. Two other macronutrients demanded by the rice plant are phosphorus and potassium; their deficiencies are becoming widespread across Asia in areas not considered deficient. Balanced use of macronutrients, such as nitrogen, phosphorus, and potassium, along with slow-release fertilizers, should be encouraged because they improve fertilizer use efficiency over time.

The biological pathways for raising the yield ceiling include both an increase in total biomass and a higher harvest index. It is essential that the capacity of the plant to produce higher biomass per day be enhanced, because the scope for yield improvement through the harvest index pathway has been practically exhausted. Such a trend in yield improvement is continuing, with the commercial exploitation of hybrid vigor providing added opportunities for raising the yield ceiling further in *indica* rice (Peng et al., 1999; Swaminathan, 1996). In addition to the morphological and physiological attributes (Table 14.5), tolerance or resistance to a wide range of biotic and abiotic stresses will also be necessary. Several donors from tropical germplasm are available in varieties with good genetic background (Table 14.6). This will call for pyramiding of genes from diverse genetic material. Several useful traits from wild rice were identified and transferred to cultivated rice (Table 14.7) using various biotechnological approaches (Khush, 2003).

Pyramiding major genes is possible due to the identification of quantitative trait loci (QTLs) for several traits. The pyramided lines that were developed with two or more genes showed a wider spectrum and a higher level of resistance than the lines with only a single gene (Khush, 2003). Sanchez et al. (2000) used sequence-tagged site (STS) markers to pyramid three genes for bacterial blight (BB) resistance in an elite breeding line of rice. The pyramided lines having three or four genes in combination showed an increased and wider spectrum of resistance to BB than those having a single resistance gene. Singh et al. (2001) used marker-assisted selection (MAS) to pyramid genes for BB resistance into a high-yielding *indica* rice cultivar, PR 106, that is susceptible to BB. Marker-assisted selection was also employed to pyramid genes for resistance to blast and gall midge. The greatest strength of rice improvement research is the availability of a wide range of germplasm in the International Rice Gene Bank at the International Rice Research Institute (IRRI) (Jackson et al., 1997) as well as in the ex situ gene banks of several countries such as India, China, and Japan. Some of the donors of traits essential for elevating and stabilizing rice yields are indicated in

TABLE 14.5. Attributes necessary for increasing the yield potential.

Agronomic characteristic	Trait desired
Habit	Semidwarf (105 cm)
Stem	Very strong
Tillers and foliage	Moderate (10); all productive; stay green; high-tissue nitrogen
Root system	Semideep and well developed
Flowering	Synchronous; prolonged grain filling (35 days)
Panicle	Heavy (175 grains/panicle); high spikelet fertility
Grain	Medium fine; 25 g/1,000 grains
Milling recovery	More than 75%

TABLE 14.6. Donors for various traits being used for developing new plant types in rice.

Trait	*Indica*/tropical *japonica*
Short stature	IR5, MD2, Shend Nung 89-366
Sturdy stem	Ghorbaran, Pusa 44-37, Sengkeu, Sipapak, Sirah Barch
Heavy panicle	G1291, G1298, Daringan, Djawa serang, Ketan Gebat
Prolonged grain filling	IR65740-AC1-3, IR65601-10-1-2, BG 380, BG90-2, Darinagan
Photosynthetic efficiency	AC4491, Mahsuri, Ptb10, T141, NS1281, T90, Bam3
Translocation efficiency	Ptb10
Low photorespiration	B76, Patna 23
Slow leaf senescence	G110, Pusa 1021, Pusa 1266 and some *indica/japonica* derivatives
High-dry-matter production	Swarnaprabha, Intan, Pankaj, Manasarovar
High-density grains	Badshabhog, Roupsali
High harvest index	Vijaya, T 141

Source: Directorate of Rice Research, Hyderabad, India.

Table 14.6. With the advent of molecular breeding technologies, it is also becoming possible to transfer genes from wild *Oryza* species (Table 14.7). Potential useful traits in *Oryza* species were identified at IRRI (Table 14.8). The yields recorded in India with some of the new hybrids are given in Table 14.9.

Pingali et al. (1997) described in detail the steps needed to increase rice production in Asia to meet future needs. If global warming and the associated changes in temperature, precipitation, and sea level do occur, the position of rice in national and global food security systems will increase, because rice has the ability to grow under very diverse environmental conditions. Rice is by far the best-adapted crop to lowland soils that are prone to flooding during the rainy season. Pingali et al. drew attention to the following challenges faced by rice researchers and development agencies:

- Productivity gains from the exploitation of green revolution technologies are close to exhaustion.
- In the absence of further technological change, Asian farmers face increasing costs per ton of rice produced.

- Adverse agricultural externalities are increasing due to the lack of a holistic perspective on the part of the farm resource base management.
- Despite an anticipated decline in per capita rice consumption, aggregate Asian demand for rice is expected to increase by 50 to 60% between 1990 and 2025 due to both an increase in population and a reduction in poverty.
- Economic growth and the commercialization of agricultural systems could reduce the competitiveness of rice relative to other crops and other farm enterprises.
- An upward shift in the rice yield frontier is necessary to meet future rice requirements and to sustain farm-level profits.

TABLE 14.7. Useful traits from wild species used for rice improvement.

Species	Useful traits
Oryza alta	High biomass
O. australiensis	BPH resistance; drought tolerance
O. baarthii	BB BB, Bacterial blight resistance and sheath blight resistance; drought avoidance
O. brachyantha	Yellow stemborer, BB, BPH, and leaf folder resistance
O. eichengeri	BPH, GLH, and WBPH resistance
O. grandiglumis	High biomass
O. granulata	Yellow stemborer, BLB, and BPH resistance; shade tolerance; adaptation to aerobic soils
O. latifolia	High biomass; sheath blight resistance
O. longistaminata	Drought tolerance
O. meridionalis	Elongation ability
O. meyeriana	Shade tolerance; adaptation to aerobic soils
O. minuta	BPH, GLH, WBPH, BB, (BB) sheath blight, and blast resistance
O. nivara	Grassy stunt virus and sheath blight resistance
O. officinalis	BPH, GLH, and WBPH resistance
O. punctata	BPH resistance
O. ridleyi	Shade tolerance; stemborer, blast, and BB resistance
O. rufipogon	Source of CMS; sheath blight resistance

Source: Directorate of Rice Research, Hyderabad, India.

Notes: BB = bacterial blight; BPH = brown planthopper; GLH = green leafhopper; CMS = cytoplasmic male sterility; WBPH = white-backed planthopper.

TABLE 14.8. Chromosome number, genomic composition, and potential useful traits of *Oryza* species.[a]

Species	2n	Genome	Number of accessions	Distribution	Useful or potentially useful traits
O. sativa complex					
O. sativa L.	24	AA	84,186	Worldwide	Cultigen
O. glaberrima Steud.	24	A^gA^g	1,299	West Africa	Cultigen; tolerance to drought, acidity, and iron toxicity; resistance to RYMV, African gall midge, nematodes; weed competitiveness
O. nivara Sharma et Shastry	24	AA	1,130	Tropical and subtropical Asia	Resistance to grassy stunt virus
O. rufipogon Griff.	24	AA	858	Tropical and subtropical Asia, tropical Australia	Resistance to BB, tungro virus; tolerance to aluminum and soil acidity; source of CMS
O. breviligulata A. Chev. et Roehr. (O. barthii)	24	A^gA^g	214	Africa	Resistance to GLH, BB; drought avoidance
O. longistaminata A. Chev. et Roehr	24	A^lA^l	203	Africa	Resistance to BB, nematodes; drought avoidance
O. meridionalis Ng	24	A^mA^m	46	Tropical Australia	Elongation ability; drought avoidance
O glumaepatula Steud.	24	$A^{gp}A^{gp}$	54	South and Central America	Elongation ability; source of CMS
O. officinalis complex					
O. punctata Kotschy ex Steud.	24, 48	BB, BBCC	59	Africa	Resistance to BPH, zigzag leafhopper

Species	2n	Genome	No.	Distribution	Useful traits
O. minuta J.S. Pesl. ex C.B. Presl.	48	BBCC	63	Philippines and Papua New Guinea	Resistance to BB, blast, BPH, GLH; tolerance to Shb
O. officinalis Wall ex Watt	24	CC	265	Tropical and subtropical Asia, tropical Australia	Resistance to thrips, BPH, GLH, WBPH, BB, stem rot
O. rhizomatis Vaughan	24	CC	19	Sri Lanka	Drought avoidance; rhizomatous
O. eichingeri A. Peter	24	CC	29	South Asia and East Africa	Resistance to BPH, WBPH, GLH
O. latifolia Desv.	48	CCDD	40	South and Central America	Resistance to BPH; high biomass production
O. alta Swallen	48	CCDD	6	South and Central America	Resistance to striped stemborer; high biomass production
O. grandiglumis (Doell) Prod.	48	CCDD	10	South and Central America	High biomass production
O. australiensis Domin.	24	EE	36	Tropical Australia	Resistance to BPH, BB; drought avoidance
O. meyeriana complex					
O. granulata Nees et Arn. ex Watt	24	GG	24	South and Southeast Asia	Shade tolerance; adaptation to aerobic soil
O. meyeriana (Zoll. Et Mor. ex Steud.) Baill.	24	GG	11	Southeast Asia	Shade tolerance; adaptation to aerobic soil
O. ridleyi complex					
O. longiglumis Jansen	48	HHJJ	6	Irian Jaya, Indonesia, and Papua New Guinea	Resistance to blast, BB
O. ridleyi Hook. F.	48	HHJJ	15	South Asia	Resistance to BB, blast, stemborer, whorl maggot

TABLE 14.8 *(continued)*

Species	2n	Genome	Number of accessions	Distribution	Useful or potentially useful traits
Unclassified					
O. brachyantha A. Chev. et Roehr	24	FF	19	Africa	Resistance to BB, yellow stemborer, leaf folder, whorl maggot; tolerance to laterite soil
O. schlechteri Pilger	48	HHKK	1	Papua New Guinea	Stoloniferous
Related genera			15		

Source: Dr. D. S. Brar, IRRI (personal communication).

Notes: BB = bacterial blight; BPH = brown planthopper; CMS = cytoplasmic male sterility; GLH = green leafhopper; RYMV = rice yellow mottle virus; Shb = sheath blight; WBPH = white-backed planthopper.

[a]Accessions maintained at IRRI, Philippines.

TABLE 14.9. Promising hybrids identified in India.

Hybrid	Mean yield (tons/ha)	Yield advantage (tons/ha)
KRH-2	5.52	0.92
PHB-71	5.35	0.73
Sahyadri	5.29	0.67
PA-6201	5.20	0.58
NSD-2	5.17	0.56

Source: Mishra (2003).

Compounding these problems are the potential dangers arising from the diminishing investment in research in institutions devoted entirely to national and international public good and the expanding IPR regime. The question now is how much more improvement can we bring about in productivity without ecological harm? In other words, can we launch an evergreen revolution in rice in the coming years, marked by sustained advances in productivity, profitability, stability, and sustainability of rice farming systems (Swaminathan, 1996, 2000, 2002a)? How can we also increase the role of rice in the nutritional security of families dependent on it for their dietary energy supply? How can rice production be insulated from the adverse impact of potential changes in precipitation, temperature, and in sea level? Above all, how can we maintain and strengthen international cooperation in rice improvement?

RICE FOR FOOD AND NUTRITION SECURITY

As an immediate measure for strengthening food security at the level of individuals and households, there is no better option than initiating a systematic effort in each agroclimatic zone to identify and remove the constraints responsible for the prevailing yield gaps. Community-based institutions and local organizations should be fully involved both in identifying constraints that limit production and in removing them.

Food-Based Approach to Nutrition Security

Under this approach, food security was considered essentially in terms of production. It was assumed that adequate food production would ensure adequate availability of food in the market as well as in the household. In the 1970s, it became clear that availability alone does not lead to food security. It is becoming evident that even if availability and access are satisfactory, the biological absorption of food in the body is related to the consumption of clean drinking water as well as to environmental hygiene. Finally, even if physical and economic access to food is assured, ecological factors will determine the long-term sustainability of food security systems. Thus, today we have to look at food security from the viewpoint of physical, social, economic, and environmental access (Paroda, 2001; Swaminathan, 2001). The whole-cycle approach described by the M. S. Swaminathan Research Foundation (MSSRF, 2002) emphasizes the needs and ways to address all age groups. Such an approach will involve the following steps (Swaminathan, 2002a, 2002b).

Food Availability

This is a function of both home production and imports. There is no time to relax on the food production front. The present global surplus of food grains is the result of inadequate consumption on the part of the poor and should not be mistaken as a sign of overproduction. Mainstreaming the nutritional dimension in the design of cropping and farming systems is essential. Developing nations should aim to achieve revolutions in the following five areas to sustain and expand the gains already achieved—productivity, quality, income and employment, small farm management, and enlarging the food basket.

Food Access

Lack of purchasing power reduces the availability of food even though food is available. Inadequate livelihood opportunities in rural areas are responsible for household nutrition insecurity (Latham, 1997). For example, India today has over 65 million tons of wheat and rice in government godowns (warehouses); yet poverty-induced hunger affects over 200 million people (Table 14.2). It is endemic in south Asia and sub-Saharan Africa (International Fund for Agricultural Development, 2001; Ramalingaswami et al., 1997; Swaminathan, 1999a). Macroeconomic policies, at the national and global levels, should be conducive to fostering job-led economic growth based on microenterprises supported by microcredit (Swaminathan, 2002d). Where poverty is pervasive, suitable measures to provide a needed food entitlement should be introduced.

Food Absorption

The lack of access to clean drinking water, poor environmental hygiene, and an inadequate health care infrastructure lead to poor assimilation of food consumed. Nutrition security cannot be achieved without environmental hygiene, primary health care, and clean drinking water. Culinary habits also need careful evaluation because some methods of cooking may lead to the loss of vital nutrients.

Transient Hunger

Because considerable seasonal variation in body weight is possible due to changes in weather (Huang and Bouis, 1996), any strategy for nutrition security should provide for steps to meet such transient hunger. The Indian state of Maharashtra introduced an employment guarantee plan nearly 25 years ago to assist the poor in earning their daily bread during seasons when opportunities for wage employment are low. Similarly, there is a need for mainstreaming considerations of gender, age, and occupation into the national nutrition strategy.

We should accelerate our efforts to improve agricultural production through yield improvement, higher factor productivity, and better postharvest management. Advances in science and technology, such as, for example, in biotechnology, information, communication, space technologies, nuclear and renewable energy technologies, and management science, have opened up opportunities for achieving an evergreen revolution, that is, sustainable advances in crop productivity per units of land, water, and time without associated ecological harm.

Thus, sustainable food security was defined as "physical, economic, social, and ecological access to balanced diets and safe drinking water, so as to enable every individual to lead a productive and healthy life in perpetuity" (Swaminathan, 2001). Achieving such a form of food security will require synergy between technology and public policy. The most important among the internal threats is the damage to the ecological foundations essential for sustained agricultural advances, such as land, water, forests, and biodiversity. The external threats include the unequal trade bargain inherent in the World Trade Organization (WTO) agreement of 1994, the rapid expansion of proprietary science, and potential adverse changes in temperature, precipitation, sea level, and ultraviolet B.

Increasing Production and Productivity

Bridging the Yield Gap

Maximum rice yields reported by India, China, and the IRRI range from 11 to 13 tons/ha. However, the average climate-adjusted yield potential is about 8 tons/ha for inbred varieties and 8.8 tons/ha for hybrids in intensive double-cropping areas (Hill et al., 2001). The gap between potential and actual yields is higher in most rice farming systems; the present average yield is just 40% of what can be achieved, even with the currently available technologies (Hill et al., 2001). This is because of imperfect adaptation of cultivars to local environments, insufficient provision of nutrients and water, and incomplete control of pests, diseases, and weeds. There is considerable scope for further investment in land improvement through drainage, terracing, control of acidification, and so on, where these have not yet been introduced. Although irrigated areas are making good progress, there is a need for more intensive research and development in rain-fed lowland and upland areas. Therefore, a massive effort should be made to launch a productivity revolution in farming. An integrated approach is necessary to remove the technological, infrastructure, and social and policy constraints responsible for the productivity gap and, in some cases, productivity decline. Reducing the cost of production through ecotechnologies and improving income through efficient production and postharvest technologies will help to enhance opportunities for both skilled employment and farm income. Public policies should not only pay attention to agrarian reform and input and output pricing, but also to reaching the unreached in technology dissemination through training, technoinfrastructure, and trade. Public policy research should receive as much attention as agronomic research. Hence, mutually reinforcing packages of technology, services, and public policies will be needed. Future agricultural production programs will have to be based on a three-pronged strategy designed to foster an evergreen revolution, which will lead to increased production without associated ecological and social harm. These strategies include defending the gains already achieved, extending the gains, and making new gains.

Defending the Gains Already Achieved

There is a need for stepping up maintenance research to ensure that new strains of pests and pathogens do not cause crop losses and to prevent the introduction of invasive alien species. Water harvesting, watershed development, and economic and efficient water use can help to enhance productivity and income considerably. Where water is scarce, crops with high value and low watering requirements should be promoted. As pulses and oilseeds are important income-earning and soil-enriching crops, they should be included in rice farming systems. In areas such as the Red River delta in Vietnam, where seven crops of rice are grown in two years, pulses should be promoted in crop rotation to enrich soil properties.

Extending the Gains

This is done to rain-fed and semiarid hilly and island areas, which have so far been bypassed by modern yield enhancement technologies. Regional imbalances in agricultural development are growing, based largely on the availability of assured irrigation on the one hand, and assured and remunerative marketing opportunities on the other. The introduction of ecoregional technology missions, aimed at providing appropriate packages of technology, technoinfrastructure, services and input, and output pricing and marketing policies, will help to include the excluded in agricultural progress. Technologies for elevating and stabilizing yields are available for semiarid and dry farming areas (Ryan and Spencer, 2001). Therefore, the emphasis should be on farming systems that can optimize the benefits of natural resources in a sustainable manner and not merely on cropping systems. Because upland rice farming areas are also ideal for the cultivation of high-value pulses and oilseeds with low watering requirements, they should be promoted.

Making New Gains

Farming systems intensification, diversification, and value addition should be promoted. Watershed and wasteland atlases should be used for developing improved farming systems, which can provide more income and jobs. Adding value to primary products should be done at the village level. This will call for appropriate institutional structures that can offer key centralized services to small and marginal farm families and provide them with economies of scale in ecofarming (i.e., integrated pest management, integrated nutrient supply, scientific water management, precision farming, as well as marketing). A quantum leap in sophistication of management of all production factors will be required to sustain yield gains (Rockwood, 2001; Swaminathan, 2001) from the present levels to the commercially feasible threshold of about 80% yield potential.

Hybrids. Because commercial rice hybrids result in 10-15% more yield than pure-line varieties, they should be encouraged. In India, the yield advantage of hybrids ranged from 1.0 to 1.5 tons/ha, leading to a profit of 2,781-6,291 rupees (Rs; Mishra, 2003). In addition, hybrid rice seed production also enhances employment potential. There is a need to identify new sources of male sterility and restorers because about 95% of the male sterile lines used in commercial production in China and other countries have a wild abortive (WA) type of cytoplasm. The newly developed cytoplasmic male sterility (CMS) line from *Oryza perennis* is different from that of WA. The search for new stable sources of sterile cytoplasm as in *O. glumaepatula* and for restorers of these CMS sources should continue. The search for linkage between the genes for resistance to biotic and abiotic stresses and molecular markers should continue to facilitate selection for resistance during the transfer of resistance to other elite breeding lines of rice.

Photosynthetic efficiency improvement. Stimulation of photosynthesis, reduction in photorespiration, and enhancement of source-sink relationships need to be explored, because they are the major yield-determining factors. Increased sink size is the most important trait for increasing rice yield potential. Promising avenues under investigation include incorporating genes that confer C_4 photosynthesis into the rice plant and improving lodging resistance so that higher leaf nitrogen concentration can be maintained to improve radiation use efficiency during the most rapid crop growth periods. Modifying plant architecture for better light interception, assimilating partitioning, and using solar energy, nutrients, and water more efficiently are necessary to improve rice yield.

To improve the photosynthetic efficiency of rice plants, the transfer of C_4 traits into C_3 rice is being explored. Ku et al. (1999) used *Agrobacterium*-mediated transformation and introduced into rice a gene for phosphoenolpyruvate carboxylase (PEPC) from maize, which catalyzes the initial fixation of atmospheric CO_2 in C_4 plants. The level of expression of the maize PEPC in transgenic rice plants correlated with the amount of transcript and the copy number of the inserted maize gene. Transgenic rice plants exhibited reduced O_2 inhibition of photosynthesis and higher photosynthetic rates compared to those of untransformed plants. These findings demonstrate a successful strategy for introducing the key biochemical component of the C_4 pathway of photosynthesis into rice.

PRODUCING MORE SUSTAINABLY: SHIFTING TO AN ERA OF PRECISION FARMING

Precision Farming

Rice researchers and farmers must move rapidly to an era of precision farming, which will help to reduce the cost of production and improve productivity on an ecologically sustainable basis. They should launch a movement for achieving an evergreen revolution in rice farming systems based on eco-logically sustainable and location-specific precision farming technologies. Precision farming methods, which can help to enhance income and yield per drop of water and per units of land and time, need to be standardized, demonstrated, and popularized rapidly, if a reduction in the cost of produc-tion is to be achieved without a reduction in yield. A responsive, field-specific management approach will require farmers to monitor crop growth stage, nitrogen status, and pest pressure to precisely identify when nitrogen topdressing, insecticide, or fungicide applications are required. Farmers need to monitor crop growth and nitrogen status and have access to predictions of growth stage, crop stage, and yield potential from crop simulation models that use real-time weather data and weather projections. This information is crucial for estimating the nitrogen fertilizer requirement and the proper tim-ing of nitrogen topdressings and prophylactic treatment against endemic diseases when weather conditions are conducive to disease progression. The revolution in information technology should make it feasible for small rice farmers in Asia to access needed information. Without access to this infor-mation, it will not be possible to sustain the rate of yield gain needed to meet rice demand. A precise match of genotype to environment is needed while utilizing field-specific tactics to ensure that input requirements are met without deficiency or excess time and space.

Small-Farm Management

Institutional structures, which will confer upon farm families with small holdings the advantages of economies of scale during both the production and the postharvest phases of agriculture, are urgently needed. For exam-ple, thanks to the cooperative method of organization of milk processing and marketing, India is now a world leader in milk production. Strategic partnerships with the private sector will help farmers' organizations to have access to assured and remunerative marketing opportunities.

There are greater opportunities for achieving higher yields per units of land, water, and time, provided rice farmers are able to shift to precision farming methods. The five vital areas of research, development, and exten-sion that need attention from the point of view of achieving environmen-tally sustainable advances in rice productivity are

- Soil health and fertility management
- Water management
- Integrated plant health management
- Energy management
- Postharvest management

Soil Health and Fertility Management

Several studies have shown that the recovery of applied urea in lowland rice can be as low as 20% during the main growing season. Also, about one-third of applied nitrogen is immobilized in the soil. All over south Asia, about one-third to one-half of fertilizer nitrogen applied to rice crops is lost by leaching, ammonia volatilization, denitrification, and surface runoff. In the United States, global positioning satellite (GPS) technology is being used to measure soil health properties, such as soil salinity. The use of chlorophyll meters in the management of nitrogen is becoming more widespread. The Silsoe Research Institute in the UK has developed "plant-scale husbandry," which involves the use of a high-tech tractor to operate nozzles that release precise doses of herbicides, pesticides, and fertilizers. Researchers at Silsoe feel that this method could help to cut down the use of chemicals by 90%. Nutrient use efficiency in India could be achieved (Mishra, 2003) with the following methods:

1. Significant amounts of nitrogen could be saved by leaf color chart-based applications.
2. One-time application of slow-release nitrogen fertilizer is better than four splits of urea.
3. Combined use of inorganic fertilizer and organic manures such as farmyard manure (FYM) and compost (up to 25%) is better than using only mineral fertilizers.
4. Pre-Kharif legume crops could provide significant quantities of nitrogen to Kharif rice.
5. Crop residue incorporation under rice cropping systems improves yield and soil properties.

Pulses and Oilseeds in Rice Farming Systems

A minimum of 20 kg of nitrogen is needed to enable rice plants to produce 1 ton of rice. At this rate, it will be environmentally disastrous, particularly with reference to groundwater quality, if farmers supply all the needed nutrients for high yields through mineral fertilizers. Pulses and oilseeds are very important for their contribution to human and animal nutrition (Johansen et al., 2000), as components of indigenous cropping systems, and as restorers of soil fertility. Promotion of these crops in rice-based production systems of south and Southeast Asia and introduction of interventions that would lead to increased rice and legume production would be an advantage. Inclusion of legumes in the system helps in conserving the natural resource base, particularly soil fertility and groundwater. They can play a significant role in enhancing the factor productivity of a production system. Substantial increases in the production of rice and pulses can be achieved by promoting high-yield, short-duration varieties of legumes and fine-tuning the management aspects (Greenland, 1997; Johansen et al., 2000). There is a need to explore the feasibility of endowing rice plants with endosymbiotic nitrogen fixation capacity because it has a significant impact on the global economy and will help to improve the environment. Many authors (Dey and Datta, 2002; Quispel, 1991; Swaminathan, 1986) have reviewed the present status and suggested various options to achieve this.

Evergreen Revolution

Rice scientists should foster an evergreen revolution in rice through partnerships for the development and dissemination of precision farming technologies. The major goals that were proposed by the Food and Agriculture Organization of the United Nations (FAO)-sponsored International Network for an Evergreen Revolution in Rice by Swaminathan (2002b) are as follows:

- Initiate an integrated gene management program.
- Improve productivity per unit of input, particularly of nutrients and water, and thereby reduce the cost of production.
- Substitute, to the extent possible, knowledge and farm-produced inputs for capital and market-purchased chemicals.
- Enhance the ecological and social sustainability of high-yield technologies.
- Increase farmers' income and opportunities for skilled employment.
- Establish an information grid and information villages for empowering women and men engaged in rice farming with new knowledge and skills, thereby conferring on rice farmers the strengths of knowledge societies.

Water Management

Due to scarcity and competition for water, more research that ties together management of scarce water resources, agronomic practices, and development and selection of suitable rice varieties is needed. New technology and management practices are needed to increase rice productivity in many of the water-stressed areas of the world bypassed by the green revolution. Water management practices for rain-fed and drought-prone areas include a combination of breeding for drought tolerance and managing limited water supplies to ensure that adequate water is available at critical stages of growth such as flowering and grain filling. Through better management practices at the farm level, there appears to be ample scope for increasing the productivity of water. There may be a need for systems that produce rice in aerated soil that is saturated with water only when heavy rainfall causes ponding or after intermittent flood irrigation. Flush irrigation to saturate soil and then allow soil moisture depletion until a subsequent irrigation would allow even further increase in water use efficiency. We have no option except to produce more food, feed grain, fiber, fuel wood, and other agricultural commodities per unit of arable land and irrigation water, because both arable land and irrigation water will be shrinking resources in per capita terms. Therefore, we should not lose further time in initiating integrated efforts to develop and master technologies that can help to usher in an evergreen revolution or sustainable advances in productivity. The year 2003 marked the fiftieth anniversary of the discovery of the DNA double helix by Watson and Crick, which revolutionized the concept of rice production in the genomics era. Hence, this seems like an appropriate time to move for achieving an evergreen revolution in rice farming systems based on ecologically sustainable and location specific precision farming technologies.

Integrated Plant Health Management

Five diseases (blast, bacterial blight, sheath blight, tungro, and grassy stunt) and four insects (brown planthopper, green leafhopper, stem borer, and gall midge) are of major importance for rice in tropical and subtropical Asia. Most of the modern varieties of rice contain moderate resistance to one or more of these major disease and insect pests. Resistance to bacterial blight (BB) has been achieved by marker-assisted breeding (Huang et al., 1997; Singh et al., 2001). Resistance to BB, sheath blight, and stem borer has been achieved by transgene pyramiding (Datta et al., 2002).

The development of rice varieties with durable resistance to bacterial blight caused by *Xanthomonas oryzae* pv. *oryzae* and for blast caused by *Pyricularia oryzae* was possible because resistance genes have been tagged with molecular markers. Several examples of transgenic rice plants with agronomically important genes are available (Table 14.10). In several cases, durable resistance to blast is believed to be associated with quantitative or polygenic inheritance. Under these conditions, there is little or no gain in fitness for a pathogen variant to overcome only a fraction of the polygenes. Breeders should aim at incorporating quantitative or polygenic resistance into rice varieties. Rice plants transformed with *Bacillus thuringiensis (Bt)* were found to be highly toxic to striped stemborer and yellow stemborer (Datta et al., 1998; Ye et al., 2001). Sources of resistance to some diseases

TABLE 14.10. Some examples of transgenic rice plants with agronomically important genes.

Transgene	Transfer method	Useful trait
Bar	Microprojectile bombardment	Tolerance to herbicide
Bar	PEG mediated	Tolerance to herbicide
Coat protein	Electroporation	Resistance to stripe virus
Coat protein	Particle bombardment	Resistance to rice tungro spherical virus
Chitinase	PEG mediated	Sheath blight resistance
crylA(b)	Electroporation	Resistance to striped stemborer
crylA(b)	Particle bombardment	Resistance to yellow stemborer and striped stemborer
crylA(c)	Particle bombardment	Resistance to yellow stemborer
cry1A(b), crylA(c)	*Agrobacterium* mediated	Resistance to striped stemborer and yellow stemborer
cry1A(b), crylA(c)	Particle bombardment	Resistance to yellow stemborer
cry1A(c), cry2A, gna	Particle bombardment	Resistance to stemborer, leaf folder, and brown planthopper
CpTi	PEG mediated	Resistance to striped stemborer and pink stemborer
Na	Particle bombardment	Insecticidal activity for brown planthopper
Corn cystatin (CC)	Electroporation	Insecticidal activity for *Sitophilus zeamais*
Xa21	Particle bombardment	Resistance to bacterial blight
Coda	Electroporation	Increased tolerance to salt
Soybean ferritin	*Agrobacterium* mediated	Increased iron content in seed
Psy	Particle bombardment	Phytoene accumulation in rice endosperm
psy, crt1, lcy	*Agrobacterium* mediated	Provitamin A

Note: PEG = polyethylene glycol.

have been identified within cultivated rice germplasm. However, sources of resistance to sheath blight are not available, and there are only a few donors for resistance to tungro disease. A coat protein gene for rice stripe virus was introduced into two *japonica* varieties, and the transformants exhibited a significant level of resistance to virus infection that was inherited by the progenies. Modern biotechnological tools are available (Khush, 2003) to overcome some of the constraints of conventional tools (Table 14.11). Several useful elite lines and improved cultivars were developed using different biotechnological tools (Table 14.12).

Energy Management

Because fossil fuels are not unlimited, alternate sources of energy should be explored. Renewable sources of energy such as solar, wind, and organically produced biogases should be promoted. Increasing hydroelectric power from large dams greatly influences the environment and affects biodiversity. Efficient and environmental friendly means of providing energy need to be explored such as blending methanol from sugarcane and other plants.

TABLE 14.11. Overcoming some of the constraints of conventional rice breeding using biotechnology tools.

Constraint	Biotechnology applications
Limited genetic variability for resistance to stemborers	Transgenic rice carrying *Bt* genes show enhanced resistance to stemborers
Limited genetic variability for resistance to sheath blight	Transgenic rice carrying chitinase genes show resistance to sheath blight
Lack of genetic variability for β-carotene	Transgenic 'golden' rice carrying phytoene synthase *(psy)*, phytoene desaturase *(crt1)*, and lycopene cyclase *(lcy)* show provitamin activity in rice seeds
Low selection efficiency to pyramid genes for durable resistance to pests	Marker-assisted selection practiced to pyramid genes for bacterial blight, blast, and gall midge resistance
Long breeding cycle of rice varieties	Varieties developed in shorter period through anther culture
Narrow gene pool for resistance to pests	Genes from wild species with broad spectrum of resistance to bacterial blight, brown planthopper, and tungro incorporated into elite breeding lines of rice
Characterization of pathogen population difficult and laborious	DNA fingerprinting practiced to characterize pathogen populations and for gene deployment for durable resistance

TABLE 14.12. Some examples of biotechnology products in rice.

Biotechnology tools	Products
Anther culture	Several improved cultivars and elite breeding lines developed and released for commercial cultivation in China, Korea, and the Philippines.
Molecular marker-assisted selection	Marker-assisted selection practiced. Pyramided lines with durable resistance-carrying genes for resistance to bacterial blight, blast, and gall midge developed in India, China, the Philippines, and Indonesia. Some of the pyramided lines have been field tested and are in the process of release for commercial cultivation.
Alien introgression	Elite breeding lines of rice carrying genes from wild species for resistance to bacterial blight, blast, brown planthopper, and tungro and tolerance to acidic conditions have been developed. Varieties resistant to brown planthopper and tolerant to acidity have been released in Vietnam.
Somaclonal variation	Early-maturing, high-yield, and blast-resistant varieties released in Hungary and Japan.
Transformation	Novel genes inserted into rice cultivars for tolerance to herbicide and resistance to stemborer, sheath blight, and bacterial blight, along with provitamin A for improved nutritional quality. Field tests of transgenic rice carrying *Bt* gene for stemborer resistance have been made in China. However, no commercial release of transgenic rice has been made so far.

Postharvest Management

The scope for reducing postharvest losses remains unclear. Pimentel (1989) estimated such losses to be about 20% on a world scale, ranging from 9% in the United States to 40-50% in some developing countries. For solving the food supply problems of developing countries, there is a need for reducing postharvest losses, which will also lead to price stability.

RESEARCH STRATEGIES AND PRIORITIES

These strategies include integrated gene management (IGM), integrated efforts in feeding and breeding rice for high productivity, information empowerment, overcoming hidden hunger caused by micronutrient deficiencies, and promoting rice as a substrate for oral vaccines.

Integrated Gene Management

The IGM program in rice should be based on the following three goals of the Convention on Biological Diversity (CBD):

- Conservation
- Sustainable use
- Equitable sharing of benefits

Conservation

In this area, rice research organizations should strive to strengthen the continuum in the following three major methods of conservation of agrobiodiversity:

- in situ
- in situ on farm
- ex situ

National rice research systems should have well-defined plans and programs in the areas of ex situ preservation in gene banks and in situ on-farm conservation by farmers through participatory breeding and market linkages. The more than 100,000 strains of rice available today are the result of the conservation ethics of farm and tribal families, most of whom are from Asian countries. India is the largest contributor to this collection, followed by Laos (Appa Rao et al., 2002a). Swaminathan (2002c,d) is a strong proponent of recognizing and rewarding farmers for their invaluable contributions to conservation of rice varieties that are essential for rice improvement.

Sustainable Use

The vast ex situ collections in gene banks are used in a limited way. Adding through participatory breeding and varietal selection, characterization, and evaluation could promote their use. Farmers' knowledge about the samples collected and conserved would also promote their use. Lao farmers assign names to varieties that describe the most important traits—maturity, types of endosperm (glutinous or nonglutinous), adaptation (wet lands, dry lands, garden lands), and others (Appa Rao et al., 2002b). Some of the names such as "forgets husband" (good taste), "dog stares at it" (bad taste), and "aroma" (aromatic flowers) help to identify appropriate accessions from the vast collections. Integrated Mendelian and molecular breeding and deriving of novel genetic combinations by rice breeders for developing location-specific varieties are designed to promote ecologically desirable agricultural practices. Molecular linkage maps have made it possible to identify and study the effects of the individual loci that control a quantitatively inherited trait. Such QTLs can help to improve polygenically controlled traits (Tanksley and McCouch, 1997).

Efforts should be made to prevent nutritious crops from becoming "lost crops" through participatory breeding, by creating an economic stake in their conservation, and by including them in crop rotations in rice-based cropping systems. A distinctive contribution should be in enlarging the composition of the food basket by including minor and underutilized crops, which are often rich in micronutrients, in rice farming systems. Such crops are often ecologically well adapted and can be of considerable significance to household nutrition security.

Equity in Sharing Benefits

Equity in sharing benefits should be promoted through the following:

1. Strengthen steps to prevent misappropriation of germplasm held in trust under agreement with the FAO.
2. Assist the FAO in finalizing the revised international undertaking on genetic resources and assist in getting it included as a protocol under the CBD.
3. Work with the FAO in promoting a multilateral system of exchange of genetic resources in crops of importance to food and nutrition security.
4. Intensify efforts in gathering information relevant to the equitable sharing of benefits with the conservers of genetic resources and holders of traditional knowledge, so that the concept of farmers' rights becomes a reality.
5. Promote the integration of the principles of equity and ethics in the use of genetic resources and information at the international level.

Integrated Efforts in Feeding and Breeding Rice for High Productivity

Recent estimates of rice demand indicate that a compound annual growth rate of 1.25 is needed to meet expected rice consumption in 2020. The projected increase in rice demand must be met entirely by greater output per unit area on existing rice land. Meeting projected rice demand will depend on sustaining an adequate rate of gain in average rice yields on existing arable land. With increasing population, global rice production needs to be increased from the 1995 level of 460 million tons to 980 million tons by 2020 AD.

At current levels of nitrogen use efficiency, this will involve a doubling of the 10 million tons of nitrogenous fertilizers that are currently being used each year for rice production worldwide. Rice suffers from a mismatch between nitrogen used and nitrogen supplied as fertilizers, resulting in a heavy loss of applied nitrogenous fertilizers. Improved water management and agronomic practices will help to reduce losses. In addition, two basic approaches have been proposed to solve this problem. One is to regulate the timing of nitrogen application based on the need of the plants, thus partly increasing the efficiency of the use of applied nitrogen. The second is the introduction of integrated nutrient supply systems involving green manures, biofertilizers, compost, and other forms of organic manures, along with the minimum essential quantities of mineral fertilizers.

Increase the Ability of the Rice System to Fix Its Own Nitrogen

New frontiers of science offer exciting opportunities for investigating the possibilities of incorporating nitrogen fixation capacity in rice (Dey and Datta, 2002; Dey et al., 2004). It is now well over 100 years since the existence of microorganisms capable of biological fixation of atmospheric nitrogen was experimentally proved and the nitrogen-fixing capacity of legume-rhizobia symbiosis established. Since then, this symbiotic system has been well understood and exploited as an effective means of raising the nitrogen status of soil and for providing nitrogen for crops and pastures. Optimism that nonleguminous crops could similarly be benefited was fueled in the 1970s and 1980s with the discovery of several nitrogen-fixing organisms forming specific associations with nonlegumes. Several approaches, including the use of molecular mapping and breeding methods, have raised fresh hopes that success in this field could be achieved in the foreseeable future. Several discoveries in this area suggest that symbiotic nitrogen fixation can be extended to nonlegumes. The transfer of *nif* genes, together with other genes necessary for functional nitrogen fixation, into chloroplasts was believed to be the best strategy to achieve this end. But the complexity of gene regulation remains a great hindrance to achieving a functional nitrogen-fixing transgenic system (Quispel, 1991; Swaminathan, 1982). Significant advances have been made in recent years in the induction of - nodule-like structures in rice roots by rhizobia and establishing an endophytic system in rice (Ladha et al., 1997). These results, along with recent technical advances involving the induction of nodular structures on the roots of cereal crops such as wheat and rice (Bruijn et al., 1995), offer the prospect that dependable symbiosis with free-living diazotrophs, such as azospirillum, or with rhizobia can eventually be achieved. Ultimately, we should package all such opportunities into an integrated soil health care and fertility system. This is a vital component of an evergreen revolution. Yuan (1998) described the opportunities for the spread of hybrid rice. Research on feeding for high yield should proceed concurrently with breeding for high yield. Otherwise, to realize the yield potential of "super-rices," large doses of mineral fertilizers will have to be applied, with harmful long-term ecological consequences.

Knowledge and Information Empowerment

A global knowledge system for rice would offer tremendous opportunities to improve the efficiency and effectiveness of networks. Information empowerment holds the key to successful ecological rice farming. The following steps will be useful in this context:

1. Develop an interactive and two-way learning system.
2. Help to train value adders who can convert generic information into location-specific information.
3. Develop a system that can reach the unreached and include the excluded, particularly farm and rural women, in terms of information and skill empowerment.
4. Harness modern information technology to spread awareness and understanding of dying wisdom and traditional knowledge systems, particularly with reference to water harvesting and coping mechanisms for unfavorable weather conditions.
5. Establish information villages in principal rice-growing areas for explaining the benefits of modern information technology to rice farming families, particularly with reference to new materials, management practices, and marketing opportunities. This will be an essential prerequisite for initiating an era of precision farming in rice.

Overcoming Hidden Hunger Caused by Micronutrient Deficiencies

The challenge of micronutrient deficiencies in the diet is becoming great. Iodine, vitamin A, and iron deficiencies are serious in many parts of the developing world (Table 14.13). Worldwide, iron deficiency affects over 1 billion children and adults. Recent analyses from the U.S. Institute of

TABLE 14.13. Estimated percentage of iron deficiency-induced anemia in various regions.

| Region | Children | | Women |
	0-4 Years	5-12 Years	15-49 Years
South Asia	56	50	58
Africa	56	49	44
Latin America	26	26	17
Industrialized countries	12	7	1
World	43	37	35

Source: http://www.idpas.org/regions/html

Medicine (Burkhardt et al., 1997; Earl and Woteki, 1994; Swaminathan, 1999b, 2002a) highlight the effect of severe anemia in accounting for up to one in five maternal deaths. Maternal anemia is pandemic and is associated with high maternal mortality risk (MMR); anemia during infancy, compounded by maternal undernutrition, leads to poor brain development. Iron deficiency is also a major cause of permanent brain damage and death in children and limits the work capacity of adults (Smith and Haddad, 2000; Swaminathan, 2002b). There is not enough appreciation of the serious adverse implications to future generations arising from the high incidence of low birth weight (LBW) among newborn babies. Low birth weight is a major contributor to stunting and affects brain development in children. The present century will be a knowledge century, with agriculture and industry becoming more knowledge intensive. Denial of opportunities for the full expression of one's innate genetic potential for mental development even at birth is the cruelest form of inequity that can prevail in any society (Smith and Haddad, 2000). We must take steps to eliminate such inequity at birth, which leads to a denial of opportunities to nearly one out of every three children born in south Asia for performing their legitimate roles in the emerging knowledge era.

    Wherever rice is the staple, a multipronged strategy for the elimination of hidden hunger should be developed by rice scientists. The IRRI has undertaken research on enriching rice biofortified with provitamin A, iron, and other micronutrients (for more details, read the chapter by S. K. Datta ongenetically improved nutrition-dense rice). Fortification, promotion of balanced diets, new semiprocessed foods involving an appropriate blend of rice and micro-nutrient rich millets (Swaminathan, 1999b), and genetic improvement could all form part of an integrated strategy to combat the following major nutritional problems in predominantly rice-eating families.

* Protein-energy malnutrition
* Nutritional anemia (iron deficiency)
* Vitamin A deficiency
* Iodine deficiency
* Dietary deficiencies of thiamin, riboflavin, fat, calcium, vitamin C, and zinc

Swaminathan (2002a) suggested that the International Rice Commission could include the nutrition security aspect as an integral part of the international network. We must fight the serious threat to the intellectual capital of developing countries caused by low-birth-weight children and hidden hunger. Some of the research areas worthy of attention in this context are described next.

Breeding for Nutritional Quality

Nutritive quality is as important as cooking quality for countries in tropical Asia, where rice is the principal source of dietary protein, vitamins (B1), and minerals (Fe, Ca) (Juliano and Villareal, 1993). Rice provides about 40% of the protein in the Asian diet. Among the cereal proteins, rice protein is considered to be biologically the richest by virtue of its high digestibility (88%), high lysine content ($\pm4\%$), and relatively better net protein utilization. Yet, it is nutritionally handicapped on account of two factors: (1) its inherently low protein content (6-8%) and (2) an inevitable milling loss of as much as 15-20%. Unlike in other cereals, increased protein content in rice does not result in decreased protein quality because all of its fractions (glutelin, 65%; globulin and albumin, 15%; and lysine-cysteine-rich prolamin, 14%) are rich in lysine and other essential amino acids (Pehu and Siddiq, 1986; Sood and Siddiq, 1986; Vilawan and Siddiq, 1973). Even a marginal increase of 2 percentage points of protein, therefore, would mean a 10-15% increase in the nutritionally rich protein intake in our diet.

Evaluation of germplasm (Juliano and Villareal, 1993) reveals a wide variability in protein content with several valuable donor sources being available for high protein content (14-16%). As early as the 1970s screening of breeding lines/improved varieties led to the identification of a few genotypes combining high yields with moderately high protein content. Among the high-yield varieties, "Improved Sabarmati," developed at the Indian Agricultural Research Institute in New Delhi, has been found to contain 10.5-11.0% protein (Sood and Siddiq, 1986). Breeding for high protein content undertaken by IRRI and India and other countries has so far met with limited success. The complex mode of inheritance, proneness to profound environmental influence, and nonlinear relationship between yield and protein content have contributed to this slow progress. With low protein content remaining partially dominant over high protein content, the additive and nonadditive components of gene action have been found to govern this trait. Studies on the extent of genotype × environment interaction reveal considerable variation among high-protein genotypes over diverse environments (Pehu and Siddiq, 1986).

Milling, essential for improving palatability, digestibility, storage life, and cooking quality, removes considerable quantities of protein, vitamins (B_1), and minerals (Fe, Ca), which are largely located in the outer layers of the endosperm. There are two ways to minimize their loss: (1) exploiting the genetic variation in the distribution pattern of protein bodies and (2) taking into account the differential resistance of the outer layers to scraping. Microscopic screening of a transverse section of the endosperm of the rice germplasm revealed wide variation in the distribution pattern of protein bodies (Vilawan and Siddiq, 1973). In conflict with the general notion that the higher the protein content, the better and deeper the distribution pattern, genotypes with protein bodies located deep in the core of the endosperm have been identified in strains with low protein content. Furthermore, the finding that the percentage of protein loss does not depend on the distribution pattern adds another dimension to the problem. High protein content in milled samples of some of the varieties, for instance, appears to be due to higher resistance of the outer layers of the endosperm to milling. Among several factors, the thickness and texture of the aleurone layer seems to determine the level of milling loss (Pehu and Siddiq, 1986). Once their genetics and response to selection are established, breeding for low protein loss may become feasible.

Breeding for nutritional improvement was recommended at the 19th Session of the International Rice Commission, which called for an increase in focus on strategies to combat malnutrition (Gopalan, 2001; Philip et al., 2000; Swaminathan, 1999b). There are four categories of direct interventions believed to be successful in reducing micronutrient malnutrition: supplementation, fortification, dietary diversification, and disease reduction (Huang and Bouis, 1996). The nutritional status of populations will focus on the potential for improving malnutrition, primarily micronutrient malnutrition through genetic improvement.

Golden Rice

Refer to Chapter 11 on genetically improved nutrition-dense rice by S. K. Datta.

Iron Deficiency

Refer to Chapter 11 on genetically improved nutrition-dense rice by S. K. Datta.

Rice As a Substrate for Oral Vaccines

Scientists at the Biotech Foundation, Thomas Jefferson University, headed by Professor Hillary Koprowski, have extensively studied the use of tobacco plants with genetically engineered plant viruses for the production of agents against human immunodeficiency virus (HIV), rabies virus, colorectal cancer, and bovine viral diarrhea. Studies conducted in animals showed that these tobacco plant-produced agents protect animals against infections by other diseases (Yusibov et al., 1997). Significant progress was made in developing oral vaccines for bacterial and viral diarrhea, cholera, and malaria using genetically engineered bananas and potatoes (Arntzen and Mason, 1997). Rice offers several advantages for being an ideal system for oral vaccine production (Yusibov et al., 1997). With universal agronomic production technology, relatively easy and established transformation and transgenic systems, available information on gene expression and specific promoters, expected higher-level recombinant proteins, and with no reported toxic substance being produced by it, rice could be a viable and acceptable crop species for production of oral vaccines. If such research is undertaken and results in the desired result, several diseases can be controlled in regions where rice is a basic food for young children.

Molecular Maps

The availability of a comprehensive molecular genetic map in rice composed of more than 2,300 DNA markers has been a major advance in rice genetics. Singh et al. (1996) mapped centromeres on the molecular genetic map of rice and determined the correct orientation of linkage groups. Rice has become a model system for genomics research due to its comparatively small genome size, the synteny of its genome with other cereals, the availability of densely populated molecular maps containing more than 2,300 DNA markers, its characterized yeast artificial chromosome (YAC) and bacterial artificial chromosome (BAC) libraries, the large-scale analysis of expressed sequence tags (ESTs), the vast amount of genetic resources, and the comparative ease of transformation (Khush and Brar, 1998; Tanksley and McCouch, 1997). We have at our disposal a number of simple and elegant tools that enable us to manipulate the rice genome to elicit desirable responses—tolerance to moisture stress, salinity-alkalinity, heat, levels of photosynthetic efficiency, dry-matter accumulation, and partitioning. Efforts for identification of unique genes such as salt tolerance in mangroves have been undertaken using large-scale genome sequencing and differential expression analysis.

Isolation and Characterization of Salt-Tolerant Genes

Coastal ecosystems that suffer from the twin problems of low productivity and uncertain yield are an important part of the natural resources base of many countries. Growing population pressure, increasing soil erosion and water pollution caused by intensive farm practices, seawater intrusion, and the attendant soil and water quality problems caused by groundwater depletion have led to various stresses on the coastal ecosystems. MSSRF's anticipatory research program aims at developing characterized prebreeding genetic material capable of offering resistance/tolerance to coastal stress for grassroots-level breeders for developing location-specific crop varieties.

Identification of novel genes from the mangrove species was undertaken by developing gene libraries enriched with stress-induced genes and screening for potential genes conferring stress tolerance (Parida, 2003). Four cDNA libraries have been constructed from the salt-treated *Avicennia marina*. By screening with heterologous probes from other organisms or through reverse transcriptase polymerase chain reaction (RT-PCR) probes, few potential stress-tolerant genes were isolated from the cDNA libraries and fully sequenced. A number of full-length genes of practical importance to abiotic stress tolerance have been identified, sequenced, and characterized.

Transformation vectors have already been constructed for dicots with 35S promoter (pGA643 series) and for monocots with ubiquitin promoter (pCAMBIA series) incorporating *BADH, SOD1, LTP1,* and *GlyI* from the mangrove species. Transformation with constructs containing *SOD1, LTP1,* and *GlyI* in rice, *Brassica,* and tobacco are at different stages of development. Putative transformed plants were identified using PCR and Southern blot techniques. Subsequent generations are being raised to study the efficacy of these transgenics in addition to the pure lines being raised.

Using the cDNA library constructed from *Avicennia marina,* MSSRF has been able to sequence over 1,600 ESTs (Parida, 2003). Sequences of these clones were deposited in the databases at the National Center for Biotechnology Information (NCBI). This is the first-ever bulk submission from any laboratory in India. Having identified full-length or novel genes, sequencing is being undertaken at the 3' extremity. To characterize the rest of the insert, subcloning of these clones onto a plasmid vector is being carried out for functional analysis and DNA sequencing.

Biotechnology—tissue culture, gene mapping, gene transfer, and other techniques, particularly anther culture—has now become an important avenue for advances in rice breeding. Owing to the advent of molecular mapping and the ability to scan the genomes of wild species for new and useful genes, we may now be in a position to unlock the genetic potential of these germplasm resources. Exotic germplasm is a likely source of new and valuable genes capable of increasing yield and other complex traits important to agriculture.

Health and Environmental Effects

Malaria

Lowland rice provides an excellent environment for the *Anopheles* mosquito, the vector that transmits malaria. The rampant occurrence of malaria in many of the rice-producing countries is mainly because rice fields provide an excellent breeding ground for the mosquitoes. The commonly used insecticide dichloro-diphenyl-trichloroethane (DDT), although banned in most countries, is still used in some developing countries. It has a cumulative effect in the human body, contaminates water sources, and destroys the aquatic fauna and human health.

Greenhouse Gases

Ominous signs of global climate change can be seen in the increasing concentration of greenhouse gases. Methane and nitrous oxide are produced by the emission of CO_2. Wet rice fields are known to emit considerable amounts of methane. High rates of nitrogen fertilizer application and low uptake efficiency by the rice plant would promote loss of nitrogen from denitrification, contributing to the greenhouse gas load in the atmosphere and nitrate losses to ground- and surfacewater resources. Less expensive technologies for minimizing the emission of methane must be explored. CO_2 concentration and temperature associated with the greenhouse effect can influence photosynthesis.

CONCLUSIONS

National food security systems should ensure that the following are realized:

1. Every individual should have physical, economic, social, and environmental access to a balanced diet, which includes the necessary macro- and micronutrients; safe drinking water; sanitation; environmental hygiene; primary health care; and education, so as to lead a healthy and productive life.
2. Food should originate from efficient and environmentally benign production technologies that conserve and enhance the natural resource base of rice.
3. A massive effort should be made to launch a productivity revolution in farming. An integrated approach is necessary to remove the technological, infrastructure, and social and policy constraints responsible for the productivity gap. A systematic effort in each agroclimatic zone to identify and remove the constraints responsible for the prevailing yield gaps should be initiated. Community-based institutions and other local organizations should be fully involved both in identifying and in removing constraints that limit production.
4. Hybrid rice would significantly contribute toward global food security and environmental protection because hybrids generate 15-20% more yield than pure-line varieties. However, grain quality needs to be improved.
5. Agricultural intensification, diversification, and value addition are essential for providing opportunities for skilled jobs in the farm sector.
6. Due to scarcity and competition for water, more research that ties together management of scarce water resources, agronomic practices, and development and selection of suitable rice varieties is needed.
7. Precise balance between crop nitrogen demand and nitrogen supply from indigenous soil resources and applied fertilizer is required to increase yield, optimize profit, and minimize environmental concerns associated with nitrogen losses. Management of nutrients, insect pests, and diseases must respond to seasonal conditions rather than follow a set of standard guidelines.

8. A precise match of genotype to environment is needed while utilizing field-specific tactics to ensure that input requirements are met without deficiency or excess in terms of time and space.

9. We should initiate a new millennium project on precision farming to achieve sustained high productivity by rice farmers and nutrition security for rice consumers.

10. Research must be focused on inter alia enhancement of heterosis and a quick approach to homozygosity, identification and tagging of quantitative trait loci, marker-assisted selection, alien gene introgression for biotic and abiotic stresses, and development of designer rice plants with all the nutrients in the required quantity and quality.

11. Environmental-related problems may occur at low rates, but their assessment over years is critically needed for sustainable productivity and biosafety.

12. New developments in transgenics or genetically modified organisms offer a choice of crop varieties with high yields and pest-resistant and drought-tolerant properties. Identification of novel resistance genes to endemic diseases such as sheath blight and their transfer to rice using molecular approaches needs attention.

13. All new rice varieties should undergo complete nutrient analysis for the assessment of the adequacy of nutrient intake and to provide a benchmark for assessing the impact of the new varieties.

14. An end-to-end approach, with scientific attention encompassing all the links in the production-processing-marketing-consumption chain, is essential if higher growth rates in productivity are to be achieved without associated ecological harm.

15. At MSSRF, recombinant DNA technology was successfully used for the identification and isolation of stress-tolerant genes from *Avicennia marina* and the wild rice *Porteresia coarctata.* Some of these isolated genes were characterized and analyzed for their expression levels in saline conditions. There is a need for a global partnership in genomics and molecular breeding in rice, bringing together advanced and applied research institutions, to ensure that the ongoing Rockefeller Foundation-supported Rice Biotechnology Program is continued.

16. A policy framework for biosafety and gene deployment must be established to help avoid risks and to promote the safe use of molecular breeding techniques.

17. There is a need to explore the feasibility of endowing the rice plant with endosymbiotic nitrogen fixation capacity because this will have

a significant impact on the global economy and may help to improve the environment.

Thus, rice researchers face significant challenges to meet the need for enhancing the productivity, profitability, and stability of rice farming systems on an ecologically sustainable basis (Greenland, 1977; Swaminathan, 2000). This chapter describes some of the scientific opportunities available to meet these challenges. Enhanced rice production in southern Asia and Southeast Asia is not only necessary for food security but also for the security of the livelihoods of millions of small and marginal farmers and landless labor families. In addition to grain, the rice crop provides an equal quantity of other biomass. We should add value to the use of straw, bran, and husk. There are numerous market-based opportunities for such value addition in rice cropping systems. Rice research institutions should develop an integrated biomass utilization strategy.

The dependence on rice in our food basket will increase if there are changes in climate such as alterations in precipitation and temperature as well as a rise in sea level. Rice, unlike wheat, has a wide range of adaptation to growing conditions, which is an asset in an era of global change. Anticipatory research for meeting the challenges of climate change such as identifying and transferring genes from salt-tolerant species such as *Porteresia coarctata* is equally important. Exotic germplasm is a likely source of new and valuable genes capable of increasing yield and other complex traits important to rice production. In another 25 years, the importance of rice in sustainable food security systems will increase, because this unique crop, unlike wheat, is capable of growing under a wide range of latitudes and altitudes and will thereby become a blessing under conditions of changes in weather and sea levels induced by global warming.

REFERENCES

Appa Rao, S., C. Bounphanousay, J. M. Schiller, and M. T. Jackson (2002a). Collection, classification, and conservation of cultivated and wild rices of the Lao PDR. *Genetic Resources and Crop Evolution 49:* 75-81.

Appa Rao, S., C. Bounphanousay, J. M. Schiller, and M. T. Jackson (2002b). Naming of traditional rice varieties by farmers in the Lao PDR. *Genetic Resources and Crop Evolution 49:* 83-88.

Arntzen, C. J., and H. S. Mason (1997). Oral vaccine production in edible tissues of transgenic plants. In *New generation vaccines,* M. M. Levine, G. C. Woodrow, J. B. Kaper, and G. S. Cobon (Eds.), pp. 263-277. New York: Marcel Dekker.

Borlaug, N. E. (2002). Feeding a world of 10 billion people: The miracle ahead. *In Vitro Cell Development Biology—Plant 38:* 221-228.

Bruijn, F. J., Y. Jing, and F. B Dazzo (1995). Potentials and pitfalls of trying to extend symbiotic interactions of nitrogen fixing organisms to presently nonnodulated plants—such as rice. *Plant and Soil 174:* 225-240.

Burkhardt, P., P. Beyer, J. Wunn, A. Kloti, G. A. Armstrong, M. Scheldz, J. von Linting, and I. Potrykus (1997). Transgenic rice *(Oryza sativa)* endosperm expressing daffodil *(Narcissus pseudonarcissus)* phytoene synthase accumulates phytoene, a key intermediate of provitamin A biosynthesis. *The Plant Journal 11:* 1071-1078.

Datta, K., N. Baisakh, K. Maung Thet, J. Tu, and S. K Datta (2002). Pyramiding transgenes for multiple resistance in rice against bacterial blight, yellow stem borer and sheath blight. *Theoretical and Applied Genetics 106:* 1-8.

Datta, K., A. Vasquez, J. Tu, L. Torrizo, M. F. Alam, N. Oliva, E. Abrigo, G. S. Khush, and S. K. Datta (1998). Constitutive and tissue-specific differential expression of *CryIA(b)* gene in transgenic rice plants conferring enhanced resistance to insect pests. *Theoretical and Applied Genetics 97:* 20-30.

Dey, M., A. Complainville, C. Charon, L. Torrizo, A. Kondorosi, M. Crespi, and S. K. Datta. (2004). Phytohormonal responses of *enod40*-overexpressing plants in *Medicago truncatula* and rice. *Physiologia Plantarum 120:* 132-139.

Dey, M., and S. K. Datta (2002). Promiscuity of hosting nitrogen fixation in rice: An overview from the legume perspective. *Critical Reviews in Biotechnology 22:* 281-314.

Earl, R., and C. E. Woteki (Eds.) (1994). *Iron deficiency anemia: Recommended guidelines for the prevention, detection and management among U.S. children and women of childbearing age.* Washington, DC: Institute of Medicine, National Academy Press.

Evans, L. T. (1998). *Feeding the ten billion: Plants and population growth.* Cambridge, UK: Cambridge University Press.

Food and Agriculture Organization of the United Nations (2002). *FAOSTAT.* http://faostat.fao.org/ (last accessed February 26, 2007).

Gardner, G., B. Halweil, and J. Peterson (2000). Overfed and underfed: The global epidemic of malnutrition. *World Watch Paper 150.* Washington, DC: World Watch Institute.

Gopalan, C. (2001). *Combating vitamin A deficiency and micronutrient malnutrition through dietary improvement,* pp. 4-16. Chennai, India: M. S. Swaminathan Research Foundation.

Greenland, D. J. (1997). *The sustainability of rice farming.* Wallingford, UK: CAB International.

Hill, J., B. Hardy, and P. Fredenburg (2001). *Rice research: The way forward.* Manila, Philippines: International Rice Research Institute.

Hossain, M., and J. Narciso (2002). *Global rice market: Trends and perspectives.* Special lecture delivered at the annual conference of the Indian Society of Agricultural Economics, December 19-21, 2002, New Delhi, India.

Huang, J., and H. Bouis (1996). *Structural changes in the demand for food in Asia*. Washington DC: IFPRI.

Huang, N, E. R. Angeles, J. Domingo, G. Magpantay, S. Singh, G. Zhang, N. Kumaradivel, J. Bennett, and G. S. Khush (1997). Pyramiding of bacterial blight resistance genes in rice: Marker-assisted selection using RFLP and PCR. *Theoretical and Applied Genetics 95:* 313-320.

International Fund for Agricultural Development (2001). *Assessment of rural poverty in Asia and the Pacific Region. August 2001*. Rome: IFAD.

International Monetary Fund (2001). *World economic outlook: The global economy after September 11, December 2001*. http://www.imf.org/external/pubs/ft/weo/2001/03/ (last accessed February 26, 2007).

Jackson, M. T., G. C. Loresto, S. Appa Rao, M. Jones, E. P. Guimares, and N. Q. Ng (1997). Rice. In *Biodiversity in trust: Conservation and use of plant genetic resources in CGIAR centers*. D. Fuccillo, L. Sears, and P. Stapleton (Eds.). Cambridge, UK: Cambridge University Press.

Johansen, C., J. M. Duxbury, S. M. Virmani, C. L. L. Gowda, S. Pande, and P. K. Joshi (Eds.) (2000). *Legumes in rice and wheat cropping systems of the Indo-Gangetic Plain—Constraints and opportunities*. ICRISAT Patancheru, India and Cornell University, Ithaca, New York, USA.

Juliano, B. O., and C. P. Villareal (1993). *Grain quality evaluation of world rices*. Los Baños, Philippines: IRRI.

Khush, G. S. (2003). *Rice biotechnology*. Paper presented at the Interdisciplinary dialogue on the legacy of Watson and Crick: 50 years later, January 9-12, 2003, Chennai, India: M. S. Swaminathan Research Foundation.

Khush, G. S., and D. S Brar (1998). The application of biotechnology to rice. In *Agricultural biotechnology in international development*, C. L. Eves and B. M. Bedford (Eds.), pp. 21-121. Wallingford, UK: CAB International.

Ku, M. S. B., S. Agarie, M. Nomura, H. Fukayama, H. Tsuchida, K. Ono, S. Hirose, S. Toki, M. Miyao, and M. Matsuoka (1999). High-level expression of maize phosphoenolpyruvate carboxylase in transgenic rice plants. *Nature Biotechnology 17:* 76-80.

Ladha, J. K., F. J. Bruijn, and K. A. Malik (1997). Assessing opportunities for nitrogen fixation in rice. *Plant and Soil 194:* 1-10.

Latham, M. C. (1997). Human nutrition in the developing world. *Food and Nutrition Series No. 29*. Rome: Food and Agriculture Organization of the United Nations.

Mishra, B. (2003). *Rice improvement in India: Application of biotechnology*. Paper presented at the Interdisciplinary dialogue on the legacy of Watson and Crick: 50 years later, January 9-12, 2003. Chennai, India: M. S. Swaminathan Research Foundation.

M. S. Swaminathan Research Foundation (2002). *The mangrove decade and beyond: Activities, lessons and challenges in mangrove conservation and management, 1990-2001*. Manual no. 5. Chennai: MSSRF.

Parida, A. (2003). *Transfer of genes for salinity tolerance from mangrove tree species to annual species*. Paper presented at the interdisciplinary dialogue on the legacy of Watson and Crick: 50 years later, January 9-12, 2003. Chennai, India: M. S. Swaminathan Research Foundation.

Paroda, R. S. (2001). *Plant genetic resources for food and nutritional security*. New Delhi: Indian Society of Plant Genetic Resources.

Pehu, E., and E. A. Siddiq (1986). Studies on variability for resistance to milling and protein loss on milling and various grain characteristics influencing them in rice. (*Oryza sativa* L.). *Cereal Research Communications 14:* 297-304.

Peng, S., K. G. Cassman, S. S. Virmani, J. Sheehy, and G. S. Khush (1999). Yield potential trends of tropical rice since the release of IR8 and the challenge of increasing rice yield potential. *Crop Science 39:* 1552-1559.

Philip, J., S. Smitasiri, M. ul Haq, J. Tagwireyi, K. R. Norum, R. Uauy, and M. S. Swaminathan (2000). Ending malnutrition by 2020: An agenda for change in the millennium. *Food and Nutrition Bulletin, United Nations University, Tokyo.* Vol. 21(13), supplement.

Pimentel, D. (1989). Ecological systems, natural resources, and food supplies. In *Food and Natural Resources,* D. Pimentel and C. W. Hall (Eds.), pp. 1-29. New York: Academic Press.

Pingali, P. L., H. Hossain, and R. V. Gerpacio (1997). *Asian rice bowls: The returning crisis?* Wallingford, UK: CAB International, in association with the International Rice Research Institute.

Quispel, A. (1991). A critical evaluation of prospects for nitrogen fixation with non-legumes. *Plant and Soil 137:* 1-11.

Ramalingaswami, V., U. Johnson, and J. Rohde (1997). Malnutrition: A South Asian enigma. In *Malnutrition in South Asia: A regional profile,* G. Stuart (Ed.), pp. 11-22, Rosa Publication No.5. Kathmandu, Nepal: UNICEF Regional Office for South Asia.

Rockwood W. G. (Ed.) (2001). *Rice research and production in the 21st century: Symposium honoring Robert R. Chandler, Jr.* Los Baños, Philippines: IRRI.

Ryan, J. G., and D. C. Spencer (2001). *Future challenges and opportunities for agricultural R&D in the semi-arid tropics*. Patancheru, India: ICRISAT.

Sanchez, A. C., D. S. Brar, N. Huang, Z. Li, and G. S. Khush (2000). Sequence tagged site marker-assisted selection for three bacterial blight resistance genes in rice. *Crop Science 40:* 792-797.

Singh, K., T. Ishii, A. Parco, N. Huang, D. S. Brar, and G. S. Khush (1996). Centromere mapping and orientation of the molecular linkage map of rice (*Oryza sativa* L.). *Proceedings of the National Academy of Sciences of the USA 93:* 6163-6168.

Singh, S., J. S. Sidhu, N. Huang, Y. Vikal, Z. Li, D. S. Brar, H. S. Dhaliwal, and G. S. Khush (2001). Pyramiding three bacterial blight resistance genes into *indica* rice cultivar PR106. *Theoretical and Applied Genetics 102:* 1011-1015.

Smith, L. C., and L. Haddad (2000). *Overcoming child malnutrition in developing countries: Past achievements and future choices*. Washington, DC: International Food Policy Research Institute.

Sood, B. C., and E. A. Siddiq (1986). Genetic analysis of crude protein content in rice. *Indian Journal of Agricultural Sciences 56:* 796-797.

Swaminathan, M. S. (1982). Biotechnology research and third world agriculture. *Science 218:* 967-972.

Swaminathan, M. S. (1986). Building national and global nutrition security systems. In *Global aspects of food production,* M. S. Swaminathan and S. K. Sinha (Eds.), pp. 417-449. Riverton, NJ: International Rice Research Institute and Tycooly International.

Swaminathan, M. S. (1996). *Sustainable agriculture: Towards an evergreen revolution.* Delhi: Konark.

Swaminathan, M. S. (Ed.) (1999a). *Enlarging the basis of food security: The role of underutilized species.* International Workshop held at the M. S. Swaminathan Research Foundation, Chennai, India, February 17-19, 1999.

Swaminathan, M. S. (1999b). Issues and challenges in sustainable increased rice production and the role of rice in human nutrition in the world. In *Proceedings of the 19th session of the International Rice Commission: Assessment and orientation towards the 21st century,* pp. 7-17. Rome: FAO.

Swaminathan, M. S. (2000). An evergreen revolution. *Biologist 47*(2): 85-89.

Swaminathan, M. S. (2001). Food security and sustainable development. *Current Science 81:* 948-954.

Swaminathan, M. S. (2002a). *Building a national nutrition security system.* Paper presented at India-ASEAN Eminent Persons Lecture Series 11, FAO, Bangkok, January 2002.

Swaminathan, M. S. (2002b). *Nutrition in the third millennium: Countries in transition.* Plenary lecture, 17th International Congress on Nutrition, Vienna, August 27-31, 2002.

Swaminathan, M. S. (2002c). The past, present and future contributions of farmers to the conservation and development of genetic diversity. In *Managing plant genetic diversity,* J. M. M. Engels, V. Ramanatha Rao, A. H. D. Brown, and M. T. Jackson (Eds.), pp. 23-31. Rome: IPGRI.

Swaminathan, M. S. (2002d). *From Rio de Janeiro to Johannesburg: Action today and not just promises for tomorrow.* Chennai, India: East West Books (Madras).

Tanksley, S. D., and S. R. McCouch (1997). Seed banks and molecular maps: Unlocking genetic potential from the wild. *Science 227:* 1063-1066.

United Nations Development Programme (2001). *Making new technologies work for human development.* New York: Oxford University Press.

Venkataramani, G. (2002). Policies need to be farmer-friendly. In *Survey of Indian agriculture 2002,* pp. 5-7, *The Hindu.*

Vilawan, S., and E. A. Siddiq (1973). Study on the mutational manipulation of protein characteristics in rice. *Theoretical and Applied Genetics 43:* 276-280.

World Bank (2002). *World development report 2001/2002.* http://www.worldbank. org/wdr/2001/ (last accessed February 26, 2007).

Ye, G. Y., J. Tu, C. Hu, K. Datta, and S. K. Datta (2001). Transgenic IR72 with fused *Bt* gene *cry1Ab/cry1Ac* from *Bacillus thuringiensis* is resistant against four Lepidopteran species under field conditions. *Plant Biotechnology 18:* 125-133.

Yuan, L. P. (1998). *Hybrid rice development and use: Innovative approach and challenges.* 19th session of the International Rice Commission, Cairo, Egypt, September 7-10, 1998.

Yusibov, V, A. Modelska, K. Steplewski, M. Agadjanyan, D. Weiner, D. Hooper, and H. Koprowski (1997). Antigens produced in plants by infection with chimeric plant viruses immunized against rabies virus and HIV-1. *Proceedings of the National Academy of Sciences of the USA 94:* 5784-5788.

Chapter 15

Food Safety of Transgenic Rice

B. M. Chassy

INTRODUCTION

Rice, maize, and wheat are the world's most important agricultural crops. Almost 3 billion people rely on rice as their major source of food. Rice is the most productive cereal grain crop, and it can be planted across a wide range of geographies and climatic conditions. In spite of this remarkable status in food and agriculture, rice cultivation faces a number of significant challenges. The world's growing population is projected to reach more than 7.5-8.5 billion by 2025 (United Nations, 2002). This population growth will add more than a billion people who depend on rice for the majority of their energy and protein. Drought, salinization of soils, erosion, and the lack of significant areas of unused arable land, coupled with pests and diseases, impose constraints on increasing rice production (Food and Agriculture Organization of the United Nations, 2004). Reductions in chemical inputs, labor, and energy, which could lower the cost of production and enhance long-term sustainability, are equally important.

Although rice is unparalleled in importance as a food source, it is not a perfect food. Rice can supply sufficient energy and protein to meet the needs of human nutrition. It is also a good source of thiamin, riboflavin, and niacin. Rice alone does not, however, provide sufficient levels of several nutrients. In particular, iron and vitamin A deficiency are common among populations who depend on rice (Food and Agriculture Organization of the United Nations, 2002b).

The need to produce more rice in a more environmentally friendly and sustainable manner as well as the need to improve the nutritional quality of rice has prompted rice breeders to attempt to introduce a number of improvements into rice. The foregoing chapters have detailed the use of modern methods of genomics, bioinformatics, and molecular biology coupled with the traditional tools of the plant breeder to explore the development of new and improved rice varieties. The list of potential enhancements is formidable; among them are insect resistance, fungal disease resistance, bacterial blight resistance, improved weed management, enhanced tolerance to abiotic stresses, nutritional enhancements, improvements in rice quality, the production of higher-value proteins such as milk proteins in rice, and eventually rice plants that can fix their own nitrogen.

Many, but certainly not all, of these projected improvements will depend on the use of a new and sometimes controversial technology for their development. Over the course of the past 30 years, molecular biologists and biotechnologists have learned how to isolate a gene that encodes a specific trait and transfer that gene into other organisms. These techniques of recombinant DNA technology have been used to produce numerous kinds of transgenic plants, including the rice varieties described in the previous chapters. Transgenic plants are also called genetically modified organisms (GMOs) and the foods derived from them, GM foods. Other terms that have been applied are plants derived through (modern) biotechnology, biotech plants, and bioengineered plants.

New technologies are customarily treated with care until the implications of their use are fully understood. Governments around the world have developed regulatory systems and tasked them to conduct a science-based risk assessment of transgenic crops. More than 60 varieties of transgenic crops have been approved for use around the globe since 1994. Approximately 102 million hectares of transgenic crops were grown by 10.3 million farmers in 22 countries in 2006 (James, 2006). One-third of the area planted and more than half the farmers were in developing countries. Soybeans (64% transgenic), cotton (38% transgenic), canola (18% transgenic), and maize (17% transgenic) comprise the principal transgenic crop plantings. It is also important to recognize that a number of countries have approved new transgenic crops but do not plant them whereas others either have no regulatory approval system in place or have rejected the technology altogether. Thus, there is significant remaining opportunity for the transgenic crops already available (James, 2006).

This spectacular rate of adoption and the positive agricultural outcomes observed in countries such as the United States, Canada, and Argentina leave no doubt that biotechnology provides the molecular plant breeder with a new and powerful tool. Genes can now be quickly and efficiently moved into a plant from literally any living organism rather than from just those with which the plant can be crossed. Advocates of this new technology point out that it is more precise and defined because the sequence of the introduced gene(s) is known and its encoded product(s) well characterized. They note that the nature of the inserted DNA, its location in the genome, and the product(s) it produces can all be characterized at the molecular level. Noting that plant breeders have been introducing new traits for millennia by inexact methods that always produce undefined recombinant DNA and that could include chemical or radiation mutagenesis, they conclude that the transgenic plants produced in this manner are as safe as, if not safer than, those produced by other methods of plant breeding. Skeptics have countered that these newly introduced genes may have uncertain functions and unanticipated consequences in an unfamiliar host and that the very act of insertion of foreign DNA into the host chromosome may have produced local or even global unintended effects.

None of the transgenic rice improvements described in the previous chapters will be useful to the food and agricultural systems—and ultimately to people—unless two important conditions are satisfied: (1) It must be shown that they are as safe as any other variety of rice that is planted and eaten and (2) rice farmers and consumers must accept them. This chapter will describe the general principles of food safety assessment of transgenic crops, discuss some general and specific safety issues associated with various improved rice varieties, and conclude with a brief discussion of factors that will influence consumer acceptance.

PRINCIPLES OF SAFETY ASSESSMENT
OF TRANSGENIC CROPS

It is important to recognize that food is never absolutely safe. One hundred percent safety is therefore an unachievable goal. Many of the foods that we eat contain toxicants, antinutrients, or potential allergens. Foods may be contaminated with pathogenic microbes or viruses. Individuals may have idiosyncratic sensitivities or reactions to certain foods, or they may eat more of a food than they can tolerate. Millions of consumers simply eat too much food and suffer from obesity, whereas others make food choices that do not provide them with all the essential nutrients and a beneficial array of healthful food components such as fiber and antioxidants. Occasionally, plant breeders may even produce a plant that is more toxic than its progenitor: Elevated levels of psoralens in celery, solanine in potatoes, and curcurbitacin in squash are three examples (Cellini et al., 2004; Kuiper et al., 2001). Even organic farmers have produced poisonous squash for their food safety-oriented consumers (DeGregori, 2004). Food safety authorities seek to minimize food risks for consumers and to ensure that new foods introduced into the marketplace have a reasonable certainty of producing no harm when used as intended. Novel foods and food ingredients, that is to say those that have not previously been consumed by humans, are subjected to a premarket safety review in most countries. The food safety assessment paradigm is applied on a case-by-case basis. Transgenic plants and the foods and feeds derived from them have generally been regarded as novel foods.

In the United States, as the first transgenic crops were being developed, regulators worked to develop a coordinated framework for biotechnology regulation (Council for Agricultural Science and Technology, 2001). The National Academy of Sciences (NAS, 1987) and the National Research Council (NRC, 1989) concluded that biotechnology poses no new or unusual risks and that any new risks should be of the same kinds as those that are associated with crops produced by conventional genetic breeding. The U.S. Department of Agriculture (USDA), the U.S. Environmental Protection Agency (EPA), and the U.S. Food and Drug Administration (FDA) were each assigned a role in the premarket safety assessment process. In practice, most of the first wave of transgenic crops were designed to be essentially identical to their conventional counterparts while offering the consumer delayed ripening (Flavr-Savr tomato) and the producer insect resistance (*Bt* maize) or herbicide tolerance (HT cotton, HT canola, HT soybean; Institute of Food Technologists, 2000).

The safety evaluation paradigm applied to these crops became known as substantial equivalence (SE) (Organization for Economic Cooperation and Development, 1993). SE seeks to identify differences between two varieties so that the safety implications of the observed differences could be studied (Food and Agriculture Organization of the United Nations, 2000). SE has not been without its critics, and as a result, the paradigm has been perfected in recent years (Food and Agriculture Organization of the United Nations, 2000; Millstone et al., 1999). It has been recently suggested that the term "substantial equivalence" be replaced with the term "comparative assessment," because this better describes the actual safety assessment process (Chassy et al., 2004; Kok and Kuiper, 2003; König et al., 2004). The SE paradigm asserts that the ingredients and components that are shared in common between two foods pose equivalent risks. Such risks are usually minimal in commonly consumed foods and ingredients. SE further asserts that only the observed differences are of material interest to a safety assessment.

The process of producing a transgenic plant is conceptually simple, although it may prove costly and challenging in the laboratory and field. In concept, a defined segment of DNA is introduced into the plant, the newly introduced piece of DNA expresses one or more proteins, and the newly introduced protein(s) may or may not evoke further changes. In some cases, such as occurs with the insertion of an insecticidal protein for example, no further products are produced. The protein itself confers resistance to insects. In other cases such as the introduction of a new enzyme, a new product or family of products would appear. Golden rice strains producing β-carotene are examples in which the safety of the newly produced enzymes and the products that they produce must be assessed (Datta et al., 2003; Potrykus, 2001; Ye et al., 2000). Some transgenic plants contain no new proteins or metabolites; gene knockouts and gene-silencing constructs often fall into this category. Finally, plants inserted with a viral coat protein may contain no new components because virus-infected plants contain many viral proteins, including the coat protein—often at very high concentrations. Because virus-protected plants no longer contain virus particles, they may be said to have fewer components than their conventional counterparts (see the following discussion). The food safety assessment process can therefore be divided into three generic questions—keeping in mind that "safe" is always relative to other foods that we consume:

1. Is the inserted DNA reasonably safe to consume?
2. Are the newly introduced products reasonably safe?
3. Are intended and unintended changes reasonably safe?

These three questions are answered by evaluating the characteristics listed in Table 15.1.

Several recent comprehensive reviews on the food safety assessment paradigm and its application to transgenic foods are available (Chassy, 2001, 2002; Chassy et al., 2004; Cockburn, 2002; König et al., 2004; Kuiper et al., 2001). A number of comprehensive expert panel reports have also appeared (Food and Agriculture Organization of the United Nations, 2000; Institute of Food Technologists, 2000). Several of these reports have analyzed the pros and cons of transgenic crops in an attempt to reconcile opposing views and establish common ground. These documents review and comment on the peer-reviewed scientific literature (GM Science Review Panel, 2003, 2004; International Council for Science, 2003). The two UK GM Science reviews contain comprehensive bibliographies and a balanced robust discussion of the issues, including a thorough discussion of food safety upon which one can reasonably conclude that transgenic crops raise no new or unusual safety issues. The Food and Agriculture Organization of the United Nations (FAO) has also recently published an assessment of the prospects for biotechnology to help agriculture and food security in the developing world (Food and Agriculture Organization of the United Nations, 2004). This report con-

TABLE 15.1. Characteristics evaluated in a novel food safety assessment.

Number	Characteristic evaluated	Evidence required
1	Safety of the source organism or food	Knowledge of source organism
2	History of safe use	Knowledge of source as food (may have no prior use)
3	Estimation of dietary intake (exposure)	Human dietary intake surveys and models projecting future consumption
4	Safety of DNA ingestion	Generally recognized as safe (GRAS; see discussion)
5	Safety of antibiotic resistance marker	Corresponding antibiotic not indispensable in medicine (see discussion); resistance common
6	Level of expression and identity of the introduced product (including DNA)	Assay and molecular characterization of newly expressed material (protein)
7	Potential for toxicity (protein product)	Short-term acute toxicity testing in animals (rodents, chicks)
8	Potential for allergenicity (protein product)	Allergenicity of source, digestive stability, similarity to known allergens
9	Safety of compositional changes	Assay by accepted analytical methods and characterization of risks/benefits
10	Impact on nutritional value and nonnutritive health-beneficial compounds	Assay by accepted analytical methods and characterization of risks/benefits
11	Changes in natural toxicants, antinutrients	Assay by accepted analytical methods and characterization of risks/benefits
12	Significance of intended and unintended effects	Weight of all evidence and characterization of risks/benefits (hazard and exposure)
13	Long-term animal studies	Not applicable (see discussion)
14	Human feeding trials	Not applicable (see discussion)

cludes that, although it is not a panacea, biotechnology can be a powerful tool for improvement if appropriately regulated on a case-by-case basis. The report also agrees with the food safety assessment that crops developed to date have been safe. An excellent literature database, which includes case studies of transgenic crops and training programs, can be found on the AgBios Web site (AgBios, 2007).

SAFETY QUESTIONS COMMON TO ALL NOVEL FOODS

Donor and History

The safety of the donor of the gene(s) transferred into the transgenic plant is an important first consideration of the evaluation. Obviously, if genes are isolated from organisms that are toxic to humans or that cause allergies, researchers need to ensure that the genes they select do not encode such traits. Even though there is no fundamental reason why genes isolated from toxic organisms cannot be safely used if the safety of the products that they encode is demonstrated, developers usually avoid such sources unless there is a compelling reason to use them. Although a history of safe use does not by itself ensure safety, it gives developers an added level of security. It is important to note that although a gene may be isolated from a source that is normally safe to consume, consumption in a transgenic may result in increased exposure.

Genes isolated from nonallergenic foods that are commonly encountered in the diet are less likely to encode toxins or allergens, but many of the traits that are most interesting to the plant breeder can be found in organisms that are not commonly eaten and therefore whose safety for consumption is unknown. History of safe use is also not a precise term. Thus, in practice, it matters little if the transgenic plant contains genes isolated from a commonly eaten food or from some rare exotic organism because the same steps of safety assessment will be followed. A history of safe use provides an additional assurance of safety but is not taken as a confirmation of safety. The primary reasons for noting the source organism and history of use is to establish whether there exist a priori special safety concerns.

Databases containing DNA and protein sequence information are growing at a nearly exponential rate. Several excellent databases are accessible at no cost to users around the world via the Internet. Genbank (National Center for Biotechnology Information, 2006) and Swissprot (BioPerl, 2006) are two examples. One of the criteria used for the selection and screening of candidate traits for introduction into plants is that the newly expressed protein should contain no extensive sequence homology with known toxins or food allergens. The presence of similarity to a known toxin or allergen does not necessarily eliminate a protein from consideration, but it may serve as a trigger for further analysis.

Levels of Consumption

It is a principle of risk characterization in toxicology that Risk = Hazard ×
Exposure. The harm that is done by exposure to any toxicant is proportional
both to its potency and the dose (exposure). Setting aside potential toxic
effects, it is also important in the case of products that are nutritionally enhanced to project intake levels in the target nutritionally at-risk population
for whom the improved food was designed to determine if it will have nutritional efficacy (Chassy et al., 2004). Human diets vary by factors such as
geography, age, gender, income, ethnicity, food availability, and individual
preference. It is thus necessary to establish the expected dietary intake (e.g.,
exposure) for the newly developed food or food ingredient for each population subgroup. Dietary survey data are beginning to become available for
most countries (Food and Agriculture Organization of the United Nations,
2002a; U.S. Department of Agriculture, 2002a; World Health Organization,
2002).

It is often the case that the poor, especially those in less well-developed
regions, eat simple diets that lack diversity. Although this lack of dietary
diversity makes estimation of dietary intake somewhat easier, it raises an interesting safety concern. Should the fact that a food constitutes a large portion of the diet change the way the safety evaluation is conducted or require
us to raise the safety standard? For example, 40% of the maize supply in the
United States is *Bt* maize, but most U.S. consumers do not eat very much
whole maize. The majority of U.S. maize is consumed as animal feed or is
used in the manufacture of oil and starch that contain no *Bt* proteins. Nonetheless, the safety evaluation of *Bt* maize was stringent enough to ensure
safety for consumers for whom maize comprises more than half of their dietary intake. The evidence, in fact, indicated that even the highest levels of
consumption would have more than a millionfold safety factor. This wide
margin of safety is necessary because a small portion of the U.S. population
derives a significant portion of their diet from whole maize (Marasas et al.,
2004).

The safety assessment process seeks to establish that consumption by
the highest intake consumer is safe (see Table 15.2). Even consumers with
exaggerated levels of consumption should be safe if a food passes the food
safety review. The criterion applies specifically to *Bt* rice and other transgenic rice varieties because rice can comprise more than 50% of the dietary
intake in many parts of the world.

TABLE 15.2. Absence of acute toxicity of transgenic proteins.

Protein crop	NOAEL	Proteins (mg/kg)	Stable in SGF
Cry1Ab	Maize	4,000	No
Cry1Ac	Cotton, tomato	4,200	No
Cry 2Aa	Cotton	3,000	No
Cry 2Ab	Cotton, maize	3,700	No
Cry3A	Potato	5,200	No
Cry3Bb1	Maize	3,850	No
CP4EPSPS	Soybean, cotton, canola, maize, sugar beet	572	No
NPTII	Cotton, potato, tomato	5,000	No

Source: Compiled from Betz et al., 2000; Astwood et al., 1996; Thomas et al., 2004.

Note: NOAEL = no observable adverse effect; SGF = simulated gastric fluid.

DNA Safety

DNA is generally recognized as safe (GRAS) for human and animal consumption. No toxicity or other adverse effects has ever been attributed to DNA ingestion. Humans typically consume a few hundred milligrams/day of DNA (Beever and Kemp, 2000). The majority of ingested DNA is very rapidly hydrolyzed by the digestive system, however, small quantities may escape digestion for some time (Hohlweg and Doerfler, 2001). Apparently, the plant matrix may protect some DNA from digestion. Evidence shows that small fragments of DNA can be absorbed into the human or animal body where they can be found in lymphocytes and may even be transmitted to fetuses in utero (Chowdhury et al., 2003; Doerfler, 2000; Schubbert et al., 1994, 1997, 1998). Insertion of whole genes and subsequent gene expression, however, has not been demonstrated to occur in vivo. It is highly unlikely that plant DNA would be incorporated and expressed in a mammalian somatic cell (Beever and Kemp, 2000; GM Science Review Panel, 2003). Horizontal gene transfer from plants to animals or bacteria has been reviewed by authors from the European Network on the Safety of Genetically Modified Crops who found transgenic DNA is no different from any other plant DNA, that it presents no special safety concerns, and that it represents very minimal risk (Van de Eede, 2004).

It has also been speculated that the CMV 35S promoter element that is used to drive gene expression in many transgenic plants could present safety problems. In particular, it has been claimed (without any supporting evidence) that strong promoters present special hazards and that the 35S sequence also carries a recombinational hot spot. It was suggested that the promoter could (1) inadvertently activate or enhance expression of growth regulator genes that may be located at the site of integration, leading to possible hyperplasia/malignancy; (2) reactivate dormant viruses; or (3) recombine, resulting in viruses with novel phenotypes (Ho et al., 1999). Recent studies show that the promoter lacks a recombinational hotspot and that it is not a strong promoter in mammalian cells (Vlasák et al., 2003). Vlasák et al. conclude that their work "indicates that the potential hazards associated with the use of [CaMV] p35S may not be so serious as it is sometimes maintained. . . . it is questionable whether [the] relatively low transcription activity of [the] CaMV 35S promoter can induce such hazardous events in mammalian cells," p. 201.

Hull et al. (2000) explain that the CaMV virus is naturally present in a wide range of traditional food crops (crucifers), including cabbage, cauliflower, oilseed rape, mustard, and other brassicas and can be routinely found in extremely high numbers (approximately 10,000 copies per cell)—compared to only one or a few copies per cell for transgenic crops. As a result, the consumption of these vegetables would result in the ingestion of vastly more copies of the CaMV 35S promoter than the consumption of transgenic crops containing the introduced promoter, because CaMV is naturally present in many food crops. CaMV is a caulimovirus that also is found in groundnuts, soybeans, and cassava as well as brassicas. Hull and colleagues (2000) conclude: ". . . there is no evidence that the CaMV 35S promoter will increase the risk over those already existing from the breeding and cultivation of conventional crops. There is no evidence that the 35S promoter, or other retroelement promoters, will have any direct effects, in spite of being consumed in much larger quantities than would be from transgenes in GM crops," p. 4.

Although there seems little reason to believe that the rDNA present in transgenic plants could harm human and animal health (Beever and Kemp, 2000; GM Science Review Panel, 2003, 2004), a great deal of concern has been directed at the antibiotic resistance marker genes that are often contained in the inserted DNA. The fear has arisen that marker genes would jump from plants to soil or gut bacteria and then into human pathogens, giving rise to the spread of antibiotic-resistant bacteria. Such transfers are highly unlikely to occur for a number of reasons: DNA degrades quickly in the soil or gastrointestinal (GI) system, bacteria would need to be competent to take up DNA, the marker DNA would need to compete with a large excess of plant DNA for uptake by bacteria, complete marker genes would need to be inserted, and the plant-adapted resistance genes would need to express in the host bacteria. This is a highly unlikely sequence of events; moreover, repeated attempts to demonstrate transfer in vitro and in soil and agricultural fields have failed to produce antibiotic-resistant bacteria, although it has been demonstrated that marker genes can be inserted by marker rescue (homologous recombination) at a very low frequency into bacteria already containing the same marker gene sequences (GM Science Review Panel, 2003).

Although hypothetical horizontal transfer of antibiotic resistance genes from plant to bacteria might conceivably occur at some very low frequency, it is important to recognize that antibiotic resistance genes are already widespread in natural ecosystems. Medical problems related to the spread of antibiotic resistance and the emergence of multiple antibiotic-resistant organisms can be attributed to the widespread misuse of antibiotics by consumers, by professionals in medical and dental practice, and possibly in agriculture (Food and Agriculture Organization of the United Nations, 2000). Overuse and the improper application of antibiotics are strong selective forces for the enrichment of microbial populations that already harbor resistance genes for resistant organisms. Furthermore, markers that confer resistance to clinically important antibiotics for which there are not readily available substitutes are not used in the development of transgenic plants. Kanamycin resistance, conferred by insertion of the gene *nptII*, and hygromycin resistance are the most commonly used resistance markers (Institute of Food Technologists, 2000). A recent report by a scientific panel reporting to the member states and EU Commission concluded that there is no rationale for inhibiting or restricting the use of genes in this category, either for field experimentation or for the purpose of placing on the market (European Food Safety Authority, 2004).

There are also marker systems that can be used for the selection of transformants that do not rely on the use of antibiotic resistance genes. For example, herbicide tolerance is commonly used as a selective marker. Plant cells are normally sensitive to mannose, but the insertion of a gene encoding phosphomannose isomerase makes them able to metabolize mannose and tolerate its presence in the medium (Schiermeier, 2000). This technology has been successfully applied to rice transformation (Datta et al., 2003; Nicholl et al., 2004). Plastid engineering is also an attractive alternative because high levels of expression can be achieved (Daniell, 2002; Maliga, 2001), and the antibiotic resistance genes can subsequently be excised using the cyclization recombination gene-locus of X (CRE-LOX) system

(Iamthan and Day, 2000). Cotransformation of a trait and a marker located on two separate vectors followed by segregation has also proven to be an effective technique to avoid introduction of antibiotic resistance marker genes (Komari et al., 1996).

Characterization of Introduced Products

The inserted DNA is typically characterized at the molecular level using DNA sequencing, polymerase chain reaction (PCR), and Southern blot analysis to establish that the intended sequence has been inserted without any additions, deletions, or rearrangements that adversely alter the coding sequence and its expression elements. It is not unusual to find that multiple copies of genes and/or multiple small fragments of DNA have been inserted into the plant genome. The presence of these additional pieces of DNA does not raise safety concerns, provided that it can be shown that they do not interrupt essential genes or result in new fusion proteins being formed. The inserted sequences and the adjacent insertion sites can be inspected for the presence of open reading frames (ORFs). If ORFs are found, reverse transcriptase PCR (RT-PCR) or northern blots can be used to assess if they are transcribed. If they are transcribed and proteins are synthesized from ORFs of unknown function, the safety of the new proteins should be evaluated.

The level of expression of inserted genes can be assessed by northern blots or RT-PCR. Expression is normally best established, however, by assaying the plant material for the functional properties of the desired trait. Both the level of expression and the specificity can be characterized as means to demonstrate that the trait has been successfully transferred and that it has retained its characteristics in the new host. An excellent case study that illustrates molecular characterization of inserted DNA, expression, and cloned product is available on the Internet (AgBios, 2007).

Inserted proteins are also subjected to detailed molecular characterization. The protein can sometimes be isolated from plant material in sufficient quantities to allow determination of molecular size on sodium dodecyl sulfate (SDS) gels. The possible presence of glycosylation or other posttranscriptional modifications is also evaluated. Western blots are an aid in this characterization if antisera are available. Additional characterization such as matrix-assisted laser desorption/ionization-time-of-flight (MALDI-TOF) mass spectrometry and N-terminal sequencing can provide assurance that changes in protein sequence have not occurred. It is often not possible to isolate sufficient quantities of protein from plant material to allow a full biochemical characterization. Moreover, the larger quantities of proteins that would be required for toxicological analysis using the acute dose animal feeding studies described later are almost always isolated from bacteria or yeast clones containing the same inserted gene. To date, no evidence suggests that evaluation of the safety of a protein produced in an alternative host does not provide a reasonable assurance of safety.

Toxicology of Inserted Proteins

Very few proteins are known to cause adverse reactions in humans when administered via oral ingestion. Reactions that have been observed (i.e., toxicity and anaphylaxis) are almost always immediate and acute (König et al., 2004; Sjoblad et al., 1992). Many proteins are degraded or partially degraded during food processing and preparation. Most ingested proteins are quickly denatured by the low pH in the stomach and rapidly degraded by proteolytic enzymes. To act as a toxicant or an allergen, a protein must persist in digestive fluids for a sufficient period of time to exert a physiological impact (Astwood et al., 1996). Two criteria are therefore used to assess the safety of inserted proteins: (1) acute toxicity in animals and (2) digestibility.

Proteins that have been inserted into the herbicide-tolerant (HT) and insect-protected *(Bt)* crops that are used in agriculture today have been subjected to extensive toxicological screening. The absence of oral toxicity of *Bt* proteins is well characterized because preparations of *Bt* spores have been in use for many years as crop protectants (Betz et al., 2000). The data in Table 15.2 demonstrate that extremely high doses of several *Bt* proteins (Cry1Ab, Cry3Bb1), the enzyme enolpyruvylshikimate-3-phosphate synthase (HT), and the antibiotic resistance marker enzyme neomycin phosphotransferase (NPTII) produce no observable adverse effect (NOAEL) in rodents (Astwood et al., 1996). These levels represent between 500- and 10,000,000-fold higher levels of exposure than would be encountered in the diets of humans or animals (Astwood and Fuchs, 2001; Betz et al., 2000; Cockburn, 2002).

The active protein components synthesized in transgenic plants are often present at 0.005-0.1% of the plant's cellular protein, and sometimes they are barely detectable in the portion of the plant that is actually consumed. The fractions derived from a transgenic plant and consumed by humans often contain negligible amounts of protein (i.e., starch, vegetable oil). The actual calculated levels of human dietary exposure are usually very low; at the most, exposures amount to no more than a few micrograms of protein a day. The acute toxicity studies support the conclusion that these proteins would pose no problem to human health when incorporated into the diet. As noted previously, one of the principal reasons for lack of biological activity of these proteins in animals and humans is their relative instability in gastric fluid. An in vitro analysis using simulated gastric fluid (SGF) can be used to determine the potential survival of a protein in vivo (Thomas et al., 2004). A recent multilaboratory study published a standard reproducible method for performing SGF analysis (Thomas et al., 2004). The proteins that show no observable toxic effects when administered to rodents at high doses (Table 15.2) are all rapidly degraded in gastric fluid (Astwood et al., 1996; Thomas et al., 2004).

It is not difficult to understand why NPTII and CP4EPSPS are not toxic. They are enzymes that catalyze biochemical reactions that are not associated with any known toxicity. They are also commonly consumed in the diet. It may be less obvious why a plant-incorporated pesticide such as a *Bt* toxin that is incorporated to kill specific insect pests is innocuous for humans and animals to consume. The answer lies in the exquisite biological specificity of the Cry or *Bt* proteins. The *Bt*s are a large family of related insecticidal molecules, each of which has a narrow and specifically defined group of organisms upon which they are active. It has also been demonstrated that *Bt* toxins do not bind to mammalian gastrointestinal cells in vitro (Betz et al., 2000). Regarding the safety of *Bt* consumption, the EPA concluded:

> Furthermore, the *Bacillus thuringiensis* delta-endotoxins affect insects via a well known mechanism in which they bind to unique receptor sites on the cell membrane of the insect gut, thereby forming pores and disrupting the osmotic balance. There are no known equivalent receptor sites in mammalian species which could be affected, regardless of the age of the individual. Thus, there is a reasonable certainty that no harm will result to infants and children from dietary exposures to residues of *Bacillus thuringiensis*. (Environmental Protection Agency, 1998)

Allergenicity Assessment

Food allergy affects 2-3% of adults and 5-6% of children in the United States. Eight foods cause more than 90% of reported food allergies: eggs, fish, milk, nuts, peanuts, shellfish, soy, and wheat (Food and Agriculture Organization of the United Nations, 2001). Allergy sufferers first develop the capacity for an immunoglobulin E (IgE)-mediated immune response through sensitization to a specific food component. Subsequent dietary exposure results in an allergic response than can be as simple as mild discomfort or as serious as anaphylactic shock and subsequent death. The sensitizing molecule is typically a major protein component of the offending food. Although subjects may eventually be desensitized, food allergy sufferers need to avoid food products containing ingredient(s) to which they are sensitive. The development of transgenic crops containing novel proteins raises several issues related to food allergy: (1) it is theoretically possible to transfer a gene encoding an allergen into a nonallergenic food, thereby making it a food allergen; (2) it is conceivable that a novel protein could be introduced to which some individuals would become sensitized de novo; and (3) although it is a remote possibility, it is also conceivable that elevating the content of an endogenous normally nonallergenic protein in a food could allow it to become an allergen by increasing the level of exposure. In 1996, an International Life Sciences Institute (ILSI) expert report proposed a decision tree that would be useful in assessing the potential allergenicity of a newly developed transgenic food (Metcalfe et al., 1996). The decision tree was modified by a Food and Agriculture Organization of the United Nations/World Health Organization (FAO/WHO) expert panel (Food and Agriculture Organization of the United Nations, 2000) and was further modified in by the FAO/WHO expert panel in 2001 (Food and Agriculture Organization of the United Nations, 2001).

The decision tree requires that the sequence of inserted protein(s) be compared to those of known food allergens (Figure 15.1). Databases of known food allergens are available online. Proteins that show significant sequence homology to known food allergens can be subjected to further bioinformatics analysis. The 2001 FAO/WHO expert report suggested that candidate proteins should be scanned for sequences of six amino acids that also occur in known allergens. Subsequent research has demonstrated that more than 70% of proteins, including many that are commonly consumed in the human diet without causing food allergy, can be matched to known allergens using this strategy (Hileman et al., 2002). Statistical analysis shows that matching 6-mers should occur frequently when comparing proteins of average size and amino acid composition, whereas screening with an 8-amino-acid-wide match window results in very infrequent occurrence of false positives while allowing the recognition of known food allergens (Hileman et al., 2002).

In an attempt to improve the essential sulfur amino acid content in soybeans, a 2S albumin-encoding gene from Brazil nuts was transferred into soybeans. Subsequently, testing showed that the developers had inserted a previously unknown allergenic protein (Nordlee et al., 1996). Fortunately, the availability of serum from Brazil nut-sensitive individuals allowed investigators to show that a potential allergen had been inserted. Although development was immediately discontinued, the experience demonstrates

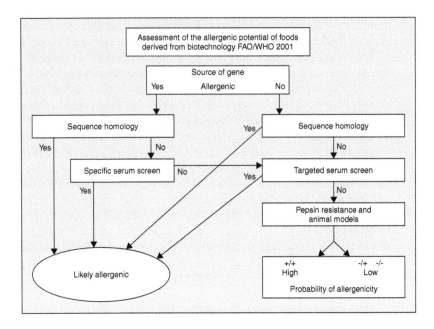

FIGURE 15.1. FAO/WHO 2001 expert report decision tree.

that known allergens can be detected in a premarket safety assessment—in this case early in the development process.

Most developers of transgenic crops do not isolate genes from plants known to cause food allergies in order to avoid any chance that an allergen will be transferred. In this case, if a bioinformatics screen demonstrates that the novel protein has little similarity to known allergens, the sensitivity to digestion is assessed in vitro using SGF (Thomas et al., 2004). It should be noted that humans consume hundreds of thousands of distinct proteins of which very few are allergens. Most food allergens are stable to digestion or contain large fragments that are stable to digestion. It cannot be concluded that because a protein is indigestible it is likely to be an allergen because many indigestible proteins are not food allergens. No validated animal model system for evaluation of allergenic potential is available, so it is currently impossible to perform in vivo testing of potential allergenicity (König et al., 2004). If a protein is isolated from a plant that is not known to cause food allergy, if the protein lacks structural similarity to known allergens, and if the protein is rapidly degraded in SGF, it can be said that the weight of evidence points to the conclusion that there is a low probability that it will become an allergen.

Rice is often substituted for other grains in the diets of allergy sufferers, because rice is generally considered a hypoallergenic food. Rice allergy is rare, although it may occur at frequencies of up to 10% in atopic individuals who often suffer from dermatitis, respiratory allergy, and/or asthma. The major rice allergens are a multigene family of 14- to 16-kDa proteins. Minor 33- and 60-kDa allergens have also been reported. Protease treatments can inactivate rice allergens and make rice acceptable in the diet of allergic atopic individuals. Transgenic rice strains with depressed levels of 14- to 16-kDa antigens have been developed using antisense RNA technology (Tada et al., 1996). Recently, a transgenic rice variety that might be able to desensitize allergy sufferers has been reported (Betterhumans, 2003).

Composition, Nutrition, and Efficacy

Determination of the composition of a novel food is the cornerstone of the food safety assessment process. As noted previously, the risk characterization process focuses on the safety implications of any compositional differences observed between a new food and its traditional comparator. A list of analytes developed by the Organization for Economic Cooperation and Development (OECD) for the detailed nutrient evaluation of maize is similar to that for most foods, although additions/deletions are necessary for specific crops on a case-by-case basis (Organization for Economic Cooperation and Development, 2002):

Proximate analysis (protein, lipid, carbohydrate, fiber, ash, moisture)
Amino acids
Fatty acids
Fat-soluble vitamins
Water-soluble vitamins
Minerals for which a need in human nutrition has been established
Known beneficial nonnutritive substances
Known antinutrients
Toxicants

The food we eat provides macro- and micronutrients, an array of nonnutritive health beneficial components, and in some instances deleterious materials such as endogenous or exogenous toxicants. Composition analysis can account for essentially 100% of the mass of a specific food, and if all of the components in all of the foods consumed by an individual are summed, the total content of the diet can be determined. Knowledge of composition and dietary intake is essential to determine (1) if changes have taken place in the nutritional value of a food during plant breeding; (2) if an unintended loss of a beneficial component has occurred; (3) if an increase in a potentially deleterious component has occurred; and (4) when the nutritional value has been intentionally changed, if the change is both safe and efficacious in the intended improvement of the dietary intake of a nutrient(s) (Chassy et al., 2004). The safety assessment paradigm therefore weighs heavily on the evidence provided by compositional analysis. A second essential factor is the level of exposure in the diet to nonnutritive substances—including any newly introduced components—as well as the overall intake of nutrients (König et al., 2004).

Food composition databases can be used to compute dietary intake (U.S. Department of Agriculture, 2002b); however, these databases are often not adequate to determine if composition has been altered during the plant-breeding process. The food composition databases present mean values observed for many accessions of a particular food tabulated by each of the analytes measured. The composition of crops varies by variety; place of planting, including soil type; year of planting; and cultural conditions. To determine if the composition of a new variety has been altered, it must be determined if the compositional score for each analyte falls within the normal range observed for that analyte in that crop. It is not possible to draw conclusions about unintended effects related to changes in composition without understanding the natural variability in natural product composition. Until recently, very little information was available about natural variability in composition. The International Life Sciences Institute has recently placed a free and easily accessible crop composition database online (International Life Sciences Institute, 2006). The utility of the database has been described in a recent publication (Ridley et al., 2004). It can be seen in Figure 15.2 that the protein concentration of hybrid maize varied between 5.5 and 12.5% in multiple plots on three continents over three years. The results of amino acid analysis reported in Figure 15.3 demonstrate that the content of individual amino acids varies widely as well; for example, glutamic acid varied from about 13 to 28 mg/g and leucine from about 8 to 20 mg/g.

In many instances, the content of secondary metabolites that are of interest for their nutritional, health-beneficial, or health-protective effects may be even more variable than macronutrients. The FDA has recently approved a health claim for soya. Many consumers are interested in increasing the soya and isoflavone contents of their diets in response to reports of various potential health benefits. The data presented in Table 15.3 show that soybeans vary greatly in isoflavone content. When the same cultivar was grown at

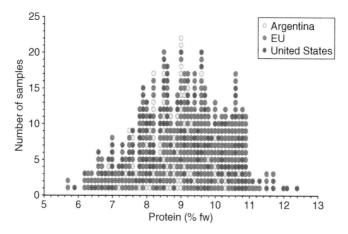

FIGURE 15.2. Distribution of protein values in the ILSI database. The numbers of samples having the measured fresh weight (fw) of protein are plotted for maize samples taken over a three-year period from multiple test fields on three continents. *Source:* Adapted from Ridley et al., 2004. See corresponding Plate 15.2 in the central gallery.

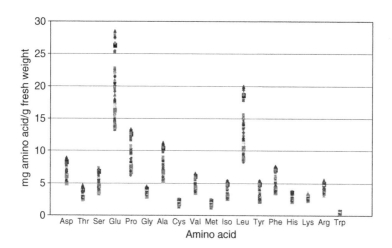

FIGURE 15.3. Natural variability in amino acid composition in reference maize hybrids. Seven varieties of maize were grown at six locations during one season. *Source:* Adapted from Ridley et al., 2004. See corresponding Plate 15.3 in the central gallery.

TABLE 15.3. Levels of isoflavones in soybeans.

Variety	Location	Total isoflavones[a]	Ratio
Hardin	Girard, IL	46.9[b]	–
Hardin	Urbana, IL	81.9	1.7
Hardin	Pontiac, IL	156.1	3.3
Hardin	Dekalb, IL	170.8	3.6
Hardin	Urbana, IL	115.9[b]	—
Amcor	Urbana, IL	149.8	1.3
Sprite	Urbana, IL	309.3	2.7
Century	Urbana, IL	250.2	2.2

Source: Eldridge and Kwolek (1983).

[a]mg isoflavones/10 g fresh weight; samples all grown in 1980.

[b]Least significant difference at 0.05 probability level = 71.7 for location and 14.7 for variety.

four different sites, a nearly fourfold difference in isoflavone content was observed. Almost threefold differences were observed when four commercial varieties of soybean were compared.

Whole Food Testing and Animal Studies

Testing in animals is an essential element of a toxicological safety assessment; animal testing methods for the testing of whole foods are not particularly useful, although they may provide some added level of assurance of safety. The underlying reasons for the dramatic difference in power of animal testing are summarized in Table 15.4. Analytical techniques have far greater power than animal studies for the detection of individual compounds. Moreover, animal studies on whole foods suffer from the lack of a specific targeted hypothesis as well as the possible presence of a variety of confounders. It is much easier to design a study to investigate an effect on a specific target organ, enzyme level, or serum metabolite concentration than it is to compare the health outcomes of two diets. Although several whole-food studies have been reported (Chassy et al., 2004), it may be more appropriate to use animal studies to probe for suspected metabolic or toxic effects on a case-by-case basis than it would be to require them in every case. It is also not clear that whole-food animal studies would extrapolate very well to

TABLE 15.4. Applicability of toxicological testing to whole foods in animals.

Chemical toxicology	Testing whole foods
Single chemical	Complex mixture (unknown structures)
Highest dose level should produce an adverse effect	Highest dose that does not cause rejection or nutritional imbalance
Low doses usually less than 1% of the diet	High doses, usually more than10% of the diet
Easy to achieve a dose high enough to assure an adequate safety factor (>100 normal human intake)	Difficult or impossible to achieve doses more than a few multiples of human intake; therefore, conventional toxicological safety factor cannot be assigned
Acute effects obvious	Acute effects, other than those caused by nutritional imbalance, nearly always absent
Possible to study specific routes of metabolism and excretion of toxic compound(s)	Complex metabolism of many ingredients; some unidentified
Cause/effect relatively clear	Effects usually absent or, if observed, confused by multiple possible causes
High sensitivity	Low sensitivity
Good reproducibility	Poor reproducibility

human populations that have great genetic and dietary diversity. Animal studies are therefore not likely to be an effective screen for unintended effects.

Nonetheless, several whole animal feeding studies have appeared (Chassy et al., 2004; König et al., 2004). For example, the rat animal model has been used to evaluate the safety of glyphosate-tolerant maize (Hammond et al., 2004), and the 42-day broiler model (Brake and Vlachos, 1998) has been used to evaluate the safety of *Bt* maize. Although animal feeding studies are of limited value in establishing safety, they are essential in assessing the nutritional value of a food. To demonstrate that the intended changes produce the desired nutritional effect, nutritionally enhanced crop plants need to be tested in animal model systems. The International Life Sciences Institute has published a guide of best practices that should be followed in the design of animal studies (International Life Sciences Institute, 2003).

Animal studies are of value in evaluating the feed performance of a newly developed transgenic crop. A number of such studies have been reported (Chassy et al., 2004; Chesson and Flachowsky, 2003; Clark and

Ipharraguerre, 2001; Flachowsky and Aulrich 2001). All transgenic crops evaluated to date have displayed feed performance characteristics comparable to those observed for their conventional counterparts. Furthermore, it has been demonstrated that neither inserted rDNA nor newly introduced transgenic proteins can be found in milk, flesh, or eggs isolated from animals or poultry fed with transgenic crops.

Understanding Unintended Effects

Perhaps the most misunderstood and misrepresented hazard of transgenic foods is the possible introduction of undetected and unintended changes that could lead to adverse effects. Critics assert that these unintended effects will go undetected because although a safety evaluation can minimize known hazards, it is impossible to anticipate and control the unknown, such as, for example, the activation of a gene that encodes a cryptic toxin (Schubert, 2002). In fact, unintended effects can and do occur in all plant breeding. Recall the previously cited examples: elevated levels of psoralens in celery, solanine in potatoes, and curcurbitacin in squash (Cellini et al., 2004; Kuiper et al., 2001). Critics also assume that any and all change is bad. In fact, changes can be beneficial, neutral, or deleterious; thus, change per se is not a matter for concern. Moreover, more significant changes in DNA occur in conventional plant breeding (Beachy et al., 2002; National Academy of Sciences, 2004b). The fact that only one or two genes are inserted into an organism containing 20-30,000 genes makes modern biotechnology a more precise science from which there are likely to arise fewer rather than greater numbers of unintended effects.

Plant breeders typically cull more than 99% of the specimens they produce in breeding experiments. A rigorous agronomic, phenotypic, and functional evaluation must be passed before an experimental plant advances to field evaluations. In addition, it may be required to pass composition screens, and it must look and taste good if it is to be accepted by consumers. The high degree of physical and chemical similarity provides strong evidence that significant undetected changes have not occurred. Successive back crosses against an elite strain of the same crop that performs well in the field further reduces the chance that unintended changes will remain undetected in newly developed plants (Cellini et al., 2004; National Academy of Sciences, 2004b). Over the years crop breeding has proven to be a reasonably safe science, perhaps in no small measure because the starting materials are often wholesome plants that have been safely consumed for centuries.

Two recent reviews on the topic of unintended effects have reached similar conclusions (Cellini et al., 2004; National Academy of Sciences, 2004b): (1) All plant breeding produces unintended effects, (2) the kinds of unintended changes that take place in the preparation of transgenic plants are the same as in conventional breeding, (3) no new or different risks are associated with the production of transgenic crops, (4) the conventional plant-breeding procedures associated with the preparation of a transgenic plant may evoke more unintended changes than the introduction of the transgene itself, and (5) every novel plant should be evaluated on a case-by-case basis, regardless of the process used to produce it. It is illuminating to read the press release issued by the National Academy of Sciences when the new report was made public (National Academy of Sciences, 2004a). It states in part that

Federal agencies should assess the safety of genetically altered foods— whether produced by genetic engineering or by other techniques, such as conventional breeding for desirable traits—on a case-by-case basis to determine whether unintended changes in their composition could adversely affect human health, says a new report from the National Academies' National Research Council and Institute of Medicine. The scope of each safety evaluation should not be based solely on the technique used to alter the food, said the committee that wrote the report, because even traditional methods such as cross-breeding can cause unexpected changes. Instead, greater scrutiny should be given to foods containing new compounds or unusual amounts of naturally occurring substances, regardless of the method used to create them. The report offers a framework to guide federal agencies in selecting the route of safety assessment. A new genetically modified food whose composition is very similar to a commonly used conventional version may warrant little or no additional safety evaluation. But if an unknown substance has been detected in a food, a more detailed analysis should be conducted to determine whether an allergen or toxin may be present. Likewise, foods with nutrient levels that fall outside the normal range should be assessed for their potential impact on consumers' diets and health.

It is also worth noting that a product should probably not be approved for introduction if significant unanswered questions remain regarding its safety. This is because postmarket monitoring is unlikely to detect adverse affects in a large heterogeneous population (Chassy et al., 2004; König et al., 2004). Postmarket surveillance may, however, be of value in confirming that anticipated levels of intake were accurately projected. Such studies will often be confounded by consumers' inexact recall of their consumption habits, the presence of large numbers of products that might contain transgenic products, and consumers' ever-changing patterns of dietary intake. Perhaps only in the case of populations that consume only a few dietary components will postmarket surveillance be likely to detect adverse effects. Populations that consume a primarily rice-based diet meet that criterion, but such populations often suffer from undernutrition or malnutrition that would render interpretation of postmarket health data difficult at best.

Safety Issues for Transgenic Rice

As described in the preceding sections, a series of questions is applied to all transgenic crops in the course of the food safety assessment (Table 15.1). These questions seek to establish that the introduced DNA does not give rise to unexpected proteins, that the proteins introduced have retained their native structure and function, that the proteins will not give rise to toxicity or allergenicity at the projected levels of consumer exposure, that the food is as nutritious as its conventional counterpart, and that no unintended changes have taken place that would lead to adverse reactions. If a marker is present that has been widely used in approved commercialized crops, no further questions regarding marker safety must be answered. The safety of the experimental transgenic rice varieties reported in the literature and described in the preceding chapters could be effectively assessed by the risk characterization paradigm described here. One important underlying assumption is that it is unnecessary to repeat evaluations of genes and constructions that have been previously published and approved by regulatory authorities. The safety issues associated with several new rice varieties are summarized in the following paragraphs.

What may be less than obvious to some plant breeders is that food safety should be built into the laboratory development process. In fact, it could be argued that it is pointless to prepare a transgenic rice plant containing a transgenic protein whose safety is not fairly well established. Safety can be established through a prior history of safe use, by bioinformatics assessment, by in vitro testing (e.g., digestibility), or by some combination of these methods. A good argument can be made for engaging food safety experts early in the plant breeding process.

Insect-Protected Rice

The food safety of insect protection conferred by the introduction of genes encoding highly specific pesticidal proteins isolated from *Bacillus thuringiensis* is well documented, and a number of these crops have been approved for commercial use in multiple countries. Any of the Cry proteins listed in Table 15.2 could be introduced in transgenic rice, and the safety established without the need for direct testing of the toxicity of the Cry protein product. The key concerns in the testing of a *Bt* rice variety that has incorporated one or two *cry* genes found in previously commercialized crops are maintaining the structural and functional identity of the inserted protein, determining the level of expression of the protein, and demonstrating that the composition of the rice falls within the range of values normally observed for rice varieties. Insect-protected rice may have additional health benefits (see the section on reducing mycotoxin exposure). The use of insect protectant strategies that have not previously been employed in approved transgenic crops would, of course, require a more complete food safety assessment as described previously in the section on safety questions common to all novel foods.

Herbicide-Tolerant Rice

Tolerance to the broad-spectrum herbicide glyphosate mediated by CP4EPSPS is found in the most widely planted commercial transgenic crops; approximately 60% of the soybeans harvested in 2003 were HT soybeans. The herbicide-tolerant form of the enzyme used in transgenic constructs is present at very low concentrations in the plant and is readily digested in the GI tract. The protein belongs to a supergene family, many of whose members are commonly encountered in the diet. HT technology can be considered reasonably safe if it is demonstrated that CP4EPSPS is the only new protein introduced and that no other unintended compositional changes have occurred (Harrison et al., 1996). The use of HT strategies that have not previously been employed in approved transgenic crops would, of course, require a more complete food safety assessment as described previously in the section on safety questions common to all novel foods.

Virus-Resistant Rice

Virus-resistant crops present few food safety concerns because the only element that has been changed is that large numbers of unwanted virus particles are no longer present in the plant. Virus-resistant papayas and squash have been approved and commercialized in the United States. If a virus-resistant rice were produced by gene knockout or gene silencing, it would probably only be necessary to demonstrate that no new proteins were introduced and that the composition was not significantly changed. Such virus-resistant rice could be considered equivalent to conventional rice.

Blight- and Fungus-Resistant Rice

Bacterial leaf blight-resistant rice that relies on the introduction of genes cloned from wild rice that encode proteins that confer bacterial resistance may not pose additional risks, although these genes have not previously been used in commercialized crops or approved by safety regulators (Datta, 2002). The major question posed by the introduction of such protectant systems is the safety of the newly introduced protein(s). In this regard, if the genes are isolated from a source that has a history of safe use in food at a similar level of exposure, a measure of assurance of safety is provided. Often, however, the genes will be derived from a source that is not a food with a history of safe use. In either case, it will be necessary to determine if the incorporated proteins have any resemblance to known toxicants, allergens, or antinutrients. If similarity to toxicants is found, additional studies (tissue culture, animal models) will need to be performed to establish that the protein is not toxic. The inserted protein(s) should be rapidly digested in SGF. If a close structural similarity to known allergens is found, the protein can be tested against serum isolated from patients who are allergic to the allergen it resembles. A negative test observed with a sufficient number of subjects would allow development to proceed with caution. As noted previously, many of these assessments can be performed before transgenic events are created in the laboratory. Once a transgenic candidate is in hand, the level of expression and structural and functional identity need to be documented. It should be noted that blight- and fungal-resistant rice may have the gratuitous side effect of resulting in harvested rice that is lower in mycotoxin content (see the section on reducing mycotoxin exposure). This possibility should be evaluated when such varieties are tested in the field.

Nutritionally Enhanced Rice

Nutritionally enhanced varieties are potentially one of the most useful applications of biotechnology to rice. The first and to date still the most notable example of a nutritionally enhanced rice is Golden rice, into which genes conferring the ability to synthesize β-carotene were inserted (Datta et al., 2003; Potrykus, 2001; Ye et al., 2000). It has been asserted in the mass media that transgenic plants that bear significant compositional changes will present complex problems for regulators. This view is based on the flawed assumption that a food is always perfect as it is and any change is therefore bad; a further underlying assumption is that humans do not know enough to alter the "perfect" design of nature and will invariably make mistakes when they "play God with nature."

A recent study analyzed the safety assessment procedure that could be applied to nutritionally enhanced crops (Chassy et al., 2004). It reached quite a different set of conclusions than noted previously: (1) The assessment paradigm applied to date is sufficiently robust to support the evaluation of crops with significantly altered composition; (2) because no components new to human nutrition are present, these crops may be safer than transgenics to which wholly novel proteins have been inserted; and (3) the major evaluative questions focus on the impact of the changes on the composition of the diet and consequent health outcomes. In the case of nutritionally enhanced rice, both intended and unintended changes could be beneficial rather than deleterious. The major focus of the safety assessment of nutritionally enhanced rice varieties will be on their composition and predicted intakes and health outcomes. These data can be supported by well-designed animal nutrition studies. To have an impact on health, the nutrient intended to be supplied by a nutritional enhancement must be bioavailable, well absorbed, and consumed in sufficient quantity to improve health. It must also be shown that health will not be affected by compensatory losses in the concentration of other important macro- or micronutrients. The introduction of a nutritionally enhanced or biofortified rice could be followed by postintroduction surveillance to establish that it is consumed in intended quantities and that it produces the desired nutritional outcome.

Prospects for Mycotoxin Reduction in Rice

The mycotoxins ochratoxin A, aflatoxins, citrinin, citreoviridin, and fumonisin have been found in harvested rice (Council for Agricultural Science and Technology, 2003). Mycotoxins are the most toxic and carcinogenic chemical compounds known. They can cause acute illness and death through exposure to high doses, and chronic exposure to extremely small quantities may cause liver and kidney disease, spontaneous abortions, hemorrhagic disease, and various forms of cancer and tumors. Fumonisin has recently been shown to interfere with folic acid uptake and to contribute to an increased incidence of neural tube and other birth defects (Marasas et al., 2004). The extent of morbidity and mortality that can be attributed to the consumption of mycotoxin-contaminated rice is not known because the necessary studies have not been conducted. It is known that moldy rice is often consumed in the less-developed regions, especially during times of food shortage. The common practice of diverting moldy rice to baking, beverage production, or animal feed does not remove the mycotoxin from the food chain because it returns in products derived from moldy rice or from animals or products from animals fed moldy rice.

Munkvold and colleagues made the interesting observation that *Bt* maize consistently contained lower levels of fumonisin over a three-year study (Munkvold et al., 1999). These observations have been confirmed in subsequent studies (Hammond et al., 2004). It appears that the reduction in insect damage found in insect-protected plants usually significantly lowers but does not eliminate fumonisin. It has been suggested that although insect damage facilitates fungal growth, it is not a prerequisite (Hammond et al., 2004; Munkvold et al., 1999). This raises the interesting possibility that insect-protected rice could have lower levels of mycotoxin—a possibility that has yet to be evaluated in the field.

As noted previously, introduction of blight and fungus resistance to rice may also lower mycotoxin contamination. The introduction of mycotoxin-detoxifying enzymes is also a possibility. A gene encoding a fumonisin ester hydrolase cloned from a saprophytic microbe present in maize ears is being expressed for fumonisin detoxification (Duvick et al., 1994). A gene isolated from mycotoxin-producing fusaria encoding trichothecene 3-*o*-acetyltransferase significantly decreases toxicity of trichothecenes in yeast and could be evaluated in rice (Kimura et al., 1998). The ribosomal protein that is the target of trichothecene has been modified in a rice gene to produce a trichothecene-tolerant rice ribosomal protein. Tobacco containing the modified rice gene provided transgenic cell cultures with tolerance to the trichothecene DON (Harris and Gleddie, 2001). It is unknown if the gene confers the same phenotype on rice. An understanding of the potential widespread public health effects of mycotoxins justifies survey research in countries with high levels of rice consumption to determine if mycotoxins are a public health risk. Reduction of insect damage, prevention of mold growth, and proper postharvest storage could prove effective at reducing mycotoxin-associated morbidity and mortality (Council for Agricultural Science and Technology, 2003).

CONSUMER ACCEPTANCE
OF TRANSGENIC CROPS AND FOODS:
THE WAY FORWARD

Researchers have developed transgenic crops using the techniques of modern molecular biology (i.e., cell culture, recombinant DNA technology, genomics, bioinformatics, and gene transfer techniques). Modern biotechnology has given science a powerful new tool for the improvement of crops with respect to increased yield and profitability, reduction of the environmental impact of agriculture, and improvements in consumer quality traits such as nutritional value. Modern biotechnology is now widely used in research because its application provides a powerful, rapid, well-defined, and precise tool for developers.

Experience has taught us that it is wise to be cautious with any new technology. Thus, systems were set in place to ensure that the safety of new transgenic varieties would be carefully scrutinized by government regulators before they were allowed onto the market. A number of scientific societies and medical societies have reviewed the safety of biotechnology and come to the conclusion that the products developed using this new research tool are as safe as any other (Food and Agriculture Organization of the United Nations, 2000, 2004; GM Science Review Panel, 2003, 2004; Institute of Food Technologists, 2000; International Council for Science, 2003; National Academy of Sciences, 2004b). A recent EU review of 15 years and $64 million devoted to biosafety research on transgenic crops concluded that not a single study has found any new risks to human health or the environment, beyond the usual uncertainties of conventional plant breeding. Indeed, the use of more precise technology and the greater regulatory scrutiny probably make them even safer than conventional plants and foods (European Commission, 2001). A recent review conducted by the National Academy of Sciences reached a similar conclusion (National Academy of Sciences, 2004b).

During the past decade, commercial planting of transgenic crops has expanded by 10-15% a year in the countries that have adopted the technology. Today, more than 10.2 million farmers are planting more than 100 million hectares of transgenic crops in 22 countries; 6 million of these are in developing countries (James, 2006). Four transgenic crops (soya, maize, cotton, and canola) are now planted on 10% of the world's farmed land. Improved yields and profitability along with reductions in the environmental impacts of agriculture have been clearly documented. This success has occurred without incurring the many negative food safety and environmental effects hypothesized by skeptics and critics. Considering the monumental challenges facing agriculture, the need to double production due to population growth while improving sustainability and reducing environmental impact (Food and Agriculture Organization of the United Nations, 2002b, 2004; United Nations, 2002), one would think that this phenomenal success would have been praised by the press, policymakers, and the public at large.

These new transgenic crops were instead greeted with a well-financed and highly organized global misinformation campaign designed to frighten consumers regarding the safety and wisdom of using this new technology. Though based on the false premise that crop varieties produced through this technology are somehow so different that they represent new species if not new life forms, the well-orchestrated antibiotechnology campaign has managed to scare consumers all over the world. Consumers were told that there was no proof that these crops were safe to eat because they had not been proven to be absolutely safe, that their long-term effects could not be anticipated, and that they had been rushed to the market by an industry-government-academy conspiracy that had greater desire for profits than it had respect for consumer or environmental safety. The media have often been willing conduits of misinformation (Braun and Moses, 2004; Marks and Kalaitzandonakes, 2001). Scientists have often been unwilling to participate in social discourse regarding biotechnology, and when they have chosen to do so, they have often been ineffective. Thus, the public discourse and debate in virtually every country has been framed by the opponents of biotechnology.

The net effect of negative messages and adverse publicity on consumer attitudes could have been foretold. Consumers are uncomfortable with uncertainty and confused with regard to whom they can trust to provide them honest information about biotechnology. Consumers admit that they do not understand the technical and scientific issues surrounding biotechnology; often they say that they have not heard much at all about it (Bonny, 2003; International Food Information Council, 2003). Not surprisingly, consumer attitudes in the EU are decidedly negative for a variety of complex reasons (Bonny, 2003), and although acceptance of agricultural biotechnology in the United States is gradually climbing, approval is still only in the 70-80% range, depending on the question and application at hand (International Food Information Council, 2003). Fewer surveys have been reported in the major rice-eating countries of the world, but there is little reason to suspect that consumer opinion will be any different than it is in the United States or the EU, given the global nature of the campaign by nongovernmental organizations (NGOs) against transgenic crops. For example, in March 2000 it was reported that the Philippines faces opposition to genetic engineering in agriculture led by a coalition of NGOs. The public debate centers on transgenic

Bt rice and its potential contribution to future food security in Asia. The results of a small survey demonstrated that NGOs and other public interest groups are generally not in favor of transgenic rice, whereas scientists generally are in favor. Most political decision makers described very high expectations with regard to the potential of genetic engineering for solving problems in the Philippine rice economy. However, their attitude with regard to the risks and benefits of *Bt* rice was one of ambivalence (Aerni et al., 2000).

There may also be reasons for adoption or nonadoption of biotechnology that are far more complex than simple NGO opposition. For example, transgenic crops are currently planted in six European countries and by more than 9 million farmers in the developing countries (James, 2006), and EU consumers believe that biotechnology will improve agriculture (Bonny, 2003). On the other hand, some EU farmers benefit from the nonadoption of transgenic crops, as do European chemical input manufacturers (Graff and Zilberman, 2004). It can be argued, in fact, that the moratorium on the introduction of new transgenic crops into the EU that has just ended, as well as the general negative consumer attitudes, has been profitable for EU food retailers and manufacturers (Kalaitzandonakes and Bijman, 2003). Although substantial numbers of EU farmers and consumers approve transgenic crops, the EU position on these new crops probably will not change until the economics no longer favor their nonadoption. It is less than clear that the nontariff trade barriers against transgenic crops erected by the EU will be brought down by the U.S. complaint to the World Trade Organization (WTO) because winning the case will not change consumer attitudes. If the United States prevails in its WTO action, EU leaders may, however, have an excuse to enact policies that are more favorable to transgenic crops.

It is difficult to know if the same market forces that have shaped the biotechnology controversy in other parts of the world will drive down adoption in the world's major rice-consuming countries. It has been suggested that this will not be the case in areas of the world where increased production of rice will be especially critical in the coming years. China appears to be on a fast track to approval and marketing of *Bt* rice because it needs to significantly improve rice production (Jia and Jayaraman, 2004). Many Asian countries have developed the capacity to harness this new technology, and they appear to be moving deliberately forward.

How can newly improved varieties of rice avoid the same rocky road that confronted the first products of this safe new technology? It seems clear that farmers, civil organizations, policymakers, the media, and the public need to be informed and consulted from the very beginning of the development process. Openness and transparency of the process will be key factors in building trust. All the stakeholders need to be at the table, or it will be perceived that something is being hidden. Because only scientists understand the complex science, they must learn to explain it to the public. But scientific judgments will be left to science only if scientists and regulators communicate to the media and the wider public, and only if scientists and regulators are trusted. The way through the controversy also involves three significant educational needs: (1) Scientists need to learn how to communicate, (2) the media need to learn more about science, and (3) policymakers and the public need to become more science literate so that they can think critically about the numerous complex technological issues with which they will be confronted in the coming decades. Europe has begun to embark on programs designed to meet these goals (Braun and Moses, 2004). Eventually, we may also learn to accept that in free societies there will be naysayers about every new technology, and although we respect their rights, society as a whole must move on—with or without them because 100% agreement will never be achieved on any issue.

For transgenic rice to have a future in world agriculture, scientists will need to become better communicators and educators. They must learn to actively seek out and confront misinformation rather than retreat from confrontation. They must also learn that science operates in a social, cultural, economic, ethical, and political context. Science is but one voice in the social discourse, but it is an essential one. If rice breeders fail to become active advocates for sound science, their contributions will remain locked in their laboratories.

REFERENCES

Aerni, P., S. Anwander Phan-Huy, and P. Rieder (2000). Public acceptance of transgenic *Bt* rice in the Philippines. *Journal of Environmental Assessment Policy and Management 2:* 99-118.

AgBios (2007). Home page. http://www.agbios.com/ main.php (last accessed February 26, 2007).

Astwood, J., and R. Fuchs (2001). Status and safety of biotech crops. In *American Chemical Society symposium series 774: Agrochemical discovery: Insect, weed, and fungal control,* D. R. Baker and N. K. Umetsu (Eds.), pp. 152-164. Washington, DC: American Chenucak Society.

Astwood, J. D., L. N. Leach, and R. L. Fuchs (1996). Stability of food allergens to digestion in vitro. *Nature Biotechnology 14:* 1269-1273.

Beachy, R., J. L. Bennetzen, B. M. Chassy, M. Chrispeels, J. Chory, J. R. Ecker, J. P. Noel, S. A. Kay, C. Dean, C. Lamb, et al. (2002). Divergent perspectives on GM food. *Nature Biotechnology 20:* 1195-1196.

Beever, D. E., and C. F. Kemp (2000). Safety issues associated with the DNA in animal feed derived from genetically modified crops. A review of the scientific and regulatory procedures. *Nutrition Abstracts and Reviews Series B: Livestock Feeds and Feeding 70:* 175-181.

Betterhumans (2003). *Allergy-fighting transgenic rice to be sold in Japan.* http:// archives.betterhumans.com/News/2884/Default.aspx (last accessed February 26, 2007).

Betz, F. S., B. G. Hammond, and R. L. Fuchs (2000). Safety and advantages of *Bacillus thuringiensis*—protected plants to control insect pests. *Regulatory Toxicology and Pharmacology 32:* 156-173.

BioPerl (2006). *SwissProt documentation.* http://doc.bioperl.org/releases/bioperl-1.0.1/Bio/DB/SwissProt.html (last accessed February 26, 2007).

Bonny, S. (2003). Why are most Europeans opposed to GMOs? Factors explaining rejection in France and Europe. *Electronic Journal of Biotechnology 6:* 51-71. http://www.ejbiotechnology.info/content/vol6/issue1/full/4/ (last accessed February 26, 2007).

Brake, J., and D. Vlachos (1998). Evaluation of transgenic event 176 *Bt* corn in broiler chickens. *Poultry Science 77:* 648-653.

Braun, R., and V. Moses (2004). A public policy on biotechnology education: What might be relevant and effective? *Current Opinion in Biotechnology 15:* 1-4.

Cellini, F., A. Cheson, I. Colquhoun, A. Constable, H. V. Davies, K. H. Engel, A. M. R. Gatehouse, S. Kärenlampi, E. J. Kok, J. J Leguay, et al. (2004). Unintended effects and their detection in genetically modified crops. *Food and Chemical Toxicology 42:* 1089-1125.

Chassy, B. (2001). Food safety assessment of current and future plant biotechnology products. In *Biotechnology and safety assessment, J.* Thomas and R. Fuchs (Eds.), pp. 87-115. San Diego: Academic Press.

Chassy, B. M. (2002). Food safety evaluation of crops produced through biotechnology. *Journal of the American College of Nutrition 21:* 166S-173S.

Chassy, B. M., J. J. Hlywka, G. A. Kleter, E. J. Kok, H. A. Kuiper, M. McGloughlin, I. C. Munro, R. H. Phipps, and J. E. Reid (2004). Nutritional and safety assessments of foods and feeds nutritionally improved through biotechnology. *Comprehensive Reviews in Food Science and Food Safety 3:* 35-104. http://www. blackwell-synergy.com/toc/crfs/3/2 (last accessed February 26, 2007).

Chesson, A., and G. Flachowsky (2003). Transgenic plants in poultry nutrition. *World's Poultry Science Journal 59:* 201-207.

Chowdhury, E. H., H. Kuribara, A. Hino, P. Sultana, O. Mikami, N. Shimada, K. S. Guruge, M. Saito, and Y. Nakajima (2003). Detection of corn intrinsic and recombinant DNA fragments and Cry1Ab protein in the gastrointestinal contents of pigs fed genetically modified corn *Bt*11. *Journal of Animal Science 81:* 2546-2551.

Clark, J. H., and I. R. Ipharraguerre (2001). Livestock performance: Feeding biotech crops. *Journal of Dairy Science 84*(Supplement E): E9-E18.

Cockburn, A. (2002). Assuring the safety of genetically modified (GM) foods: The importance of an holistic, integrative approach. *Journal of Biotechnology 98:* 79-106.

Council for Agricultural Science and Technology (2001). *Evaluation of the U.S. regulatory process for crops developed through biotechnology.* Washington, DC: CAST.

Council for Agricultural Science and Technology (2003). *Mycotoxins: Risks in plant, animal, and human systems.* Ames, IA: CAST.

Daniell, H. (2002). Molecular strategies for gene containment in transgenic crops. *Nature Biotechnology 20:* 581-586.

Datta, S. K. (2002). Bioengineered rice for plant protection. In *Biotechnology and genetic engineering reviews,* Vol. 19, S. Harding (Ed.), pp. 339-354. Andover, UK: Intercept.

Datta, K., N. Baisakh, N. Oliva, L. Torrizo, E. Abrigo, J. Tan, M. Rai, S. Rehana, S. Al-Babili, P. Beyer, et al. (2003). Bioengineered 'golden' *indica* rice cultivars with β-carotene metabolism in the endosperm with hygromycin and mannose selection systems. *Plant Biotechnology Journal 1:* 81-90.

DeGregori, T. R. (2004). Genetic engineering not significantly more dangerous than conventional breeding. *The Daily Cougar 69*(141). http://www.stp.uh.edu/vol69/141/opinion/oped1.html (last accessed February 26, 2007).

Doerfler, W. (2000). *Foreign DNA in mammalian systems.* Weinheim, Germany: Wiley-VCH.

Duvick, J., T. Rood, and S. Grant (1994). Isolation of fumonisin-metabolizing fungi from maize seed. In *Proceedings of the Fifth International Mycological Congress,* p. 56. Vancouver, Canada, August 14-21, 1994.

Eldridge, A. C., and W. F. Kwolek (1983). Soybean isoflavones: Effect of environment and variety on composition. *Journal of Agricultural and Food Chemistry 31:* 394-396.

Environmental Protection Agency (1998). *EPA registration eligibility decision (RED) Bacillus thuringiensis,* EPA 738-R-98-004, March 1998.

European Commission (2001). *EC-sponsored research into the safety of genetically modified organisms: A review of results.* http://europa.eu.int/comm/research/quality-of-life/gmo/index.html (last accessed February 26, 2007).

European Food Safety Authority (2004). Opinion of the scientific panel on genetically modified organisms on the use of antibiotic resistance genes as marker genes in genetically modified plants. *The EFSA Journal 48:* 1-18. http://www.efsa.eu.int/science/gmo/gmo_opinions/384_en.html (last accessed February 26, 2007).

Flachowsky, G., and K. Aulrich (2001). Nutritional assessment of feeds from genetically modified organisms. *Journal of Animal and Feed Sciences 10*(Supplement 1): 181-194.

Food and Agriculture Organization of the United Nations (2000). *Safety aspects of genetically modified foods of plant origin. Report of a joint FAO/WHO expert consultation on foods derived from biotechnology.* http://www.fao.org/ag/agn/food/risk_biotech_aspects_en.stm (last accessed February 26, 2007).

Food and Agriculture Organization of the United Nations (2001). *Evaluation of allergenicity of genetically modified foods. Report of a joint FAO/WHO expert consultation of allergenicity of foods derived from biotechnology.* http://www. fao.org/docrep/007/y0820e/y0820e00.htm (last accessed February 26, 2007).

Food and Agriculture Organization of the United Nations (2002a). *Nutrition country profiles.* http://www.fao.org/ag/agn/nutrition/profiles_en.stm (last accessed February 26, 2007).

Food and Agriculture Organization of the United Nations (2002b). *The state of food insecurity in the world, 2002.* http://www.fao.org/docrep/005/y7352e/y7352e00 .htm (last accessed February 26, 2007).

Food and Agriculture Organization of the United Nations (2004). *The state of food and agriculture 2003-2004.* http://www.fao.org/docrep/006/Y5160E/ Y5160E00.htm (last accessed February 26, 2007).

GM Science Review Panel (2003) *GM science review: First report.* London: UK Department of Trade and Industry. http://www.gmsciencedebate.org.uk/report/ pdf/gmsci-report1-full.pdf (last accessed February 26, 2007).

GM Science Review Panel (2004) *GM science review: Second report.* London: UK Department of Trade and Industry. http://www.gmsciencedebate.org.uk/report/ pdf/gmsci-report2-full.pdf (last accessed February 26, 2007).

Graff, G. D., and D. Zilberman (2004). Explaining Europe's resistance to agricultural biotechnology. *Agricultural and Resource Economics 7:* 1-4.

Hammond, B., K. Campbell, K. Pilcher, D. Clinton, T. Degooyer, A. Robinson, B. Mcmillen, S. Spangler, S. Riordan, L. Rice, and J. Richard (2004). Lower fumonisin mycotoxin levels in the grain of *Bt* corn grown in the United States in 2000-2002. *Journal of Agricultural Food Chemistry 52:* 1390-1397.

Harris, L. J., and S. C. Gleddie (2001). A modified *Rp13* gene from rice confers tolerance of the *Fusarium graminearum* mycotoxin deoxynivalenol to transgenic tobacco. *Physiological and Molecular Plant Pathology 58:* 173-181.

Harrison, L., M. Bailey, M. Naylor, J. Ream, B. Hammond, D. Nida, B. Burnette, T. Nickson, T. Mitsky, M. Taylor, et al. (1996). The expressed protein in glyphosate-tolerance soybean, 5-enolpryruvyl-shikimate-3-phosphate synthase from *Agrobacterium* sp. strain CP4, is rapidly digested in vitro and is not toxic to acutely gavaged mice. *Journal of Nutrition 126:* 728-740.

Hileman, R., A. Silvanovich, R. Goodman, E. Rice, G. Holleschak, J. Astwood, and S. Hefle (2002). Bioinformatic methods for allergenicity assessment using a comprehensive allergen database. *International Archives of Allergy and Immunology 128:* 280-291.

Ho, M. W., A. Ryan, and J. Cummins (1999). Cauliflower mosaic viral promoter—a recipe for disaster? *Microbial Ecology in Health and Disease 11:* 194-197.

Hohlweg, U., and W. Doerfler (2001). On the fate of plant or other foreign genes upon the uptake in food or after intramuscular injection in mice. *Molecular Genetics and Genomics 265:* 225-233.

Hull, R., S. N. Covey, and P. Dale (2000). Genetically modified plants and the 35S promoter: Assessing the risks and promoting the debate. *Microbial Ecology in Health and Disease 12:* 1-5.

Iamtham, S., and A. Day (2000). Removal of antibiotic resistance genes from transgenic tobacco plastids. *Nature Biotechnology 18:* 1172-1176.

Institute of Food Technologists (2000). *IFT expert report on biotechnology and foods.* Chicago: IFT. http://members.ift.org/IFT/Research/IFTExpertReports/biotechfoods_report.htm (last accessed February 26, 2007).

International Council for Science (2003). *New genetics, food and agriculture: Scientific discoveries—societal dilemmas.* Paris: ICSU. http://www.icsu.org/Gestion/img/ICSU_DOC_DOWNLOAD/90_DD_FILE_ICSU_GMO%20report_May%202003.pdf (last accessed February 26, 2007).

International Food Information Council (2003). *Americans' acceptance of food biotechnology matches growers' increased adoption of biotech crop.* IFIC Survey, May 2003. http://www.ific.org/research/biotechres03.cfm (last accessed February 26, 2007).

International Life Sciences Institute (2003). *Best practices for the conduct of animal studies to evaluate crops genetically modified for input traits.* Washington, DC: ILSI.

International Life Sciences Institute (2006). *Crop composition database* (Version 3.0). http://www.cropcomposition.org/ (last accessed February 26, 2007).

James, C. (2006). *Global status of commercialized Biotech/GM crops: 2006.* ISAAA Briefs No. 35-2006. Ithaca, NY: International Service for the Acquisition of Agri-Biotech Applications. http://www.isaaa.org/Briefs/35/index.htm (last accessed February 26, 2007).

Jia, H., and K. S. Jayaraman (2004). China ramps up efforts to commercialize GM rice. *Nature Biotechnology 22:* 642.

Kalaitzandonakes, N., and J. Bijman (2003). So who's driving biotech acceptance? *Nature Biotechnology 21:* 366-369.

Kimura, M., I. Kaneko, M. Komiyama, A. Takatsuki, H. Koshino, K. Yoneyama, and I. Yamaguchi (1998). Trichothecene 3-O-acetyltransferase protects both the producing organism and transformed yeast from related mycotoxins. *Journal of Biological Chemistry 273:* 1654-1661.

Kok, E., and H. Kuiper (2003). Comparative safety assessment for biotech crops. *Trends in Biotechnology 21:* 439-444.

Komari, T., Y. Hiei, Y. Saito, N. Murai, and T. Kumashiro (1996). Vectors carrying two separate T-DNAs for co-transformation of higher plants mediated by *Agrobacterium tumefaciens* and segregation of transformants free from selection markers. *The Plant Journal 10:* 165-174.

König, A., A. Cockburn, R. W. R. Crevel, E. Debruyne, R. Grafstroem, U. Hammerling, I. Kimber, I. Knudsen, H. A. Kuiper, A. A. C. M. Peijnenburg, et al. (2004). Assessment of the safety of foods derived from genetically modified (GM) crops. *Food and Chemical Toxicology 42:* 1047-1088.

Kuiper, H. A., G. A. Kleter, P. J. M. Noteboom, and E. J. Kok (2001). Assessment of the food safety issues related to genetically modified foods. *The Plant Journal 27:* 503-528.

Maliga, P. (2001). Plastid engineering bears fruit. *Nature Biotechnology 19:* 826-827.

Marasas, W. F. O., R. T. Riley, K. A. Hendricks, V. L. Stevens, T. W. Sadler, J. Gelineau-van Waes, S. A. Missmer, J. Cabrera, O. Torres, W. C. A. Gelderblom, et al. (2004). Fumonisins disrupt sphingolipid metabolism, folate transport, and neural tube development in embryo culture and in vivo: A potential risk factor for human neural tube defects among populations consuming fumonisin-contaminated maize. *Journal of Nutrition 134:* 711-716.

Marks, L. A., and N. Kalaitzandonakes (2001). Mass media communications about agrobiotechnology. *AgBioForum 3-4:* 199-208. http://www.agbioforum. org/v4n34/v4n34a08-marks.htm (last accessed February 26, 2007).

Metcalfe, D. D., J. D. Astwood, R. Townsend, H. A. Sampson, S. L. Taylor, and R. L. Fuchs (1996). Assessment of the allergenic potential of foods derived from genetically engineered crop plants. *Critical Reviews in Food Science and Nutrition 36:* S165-S186.

Millstone, E., E. Brunner, and S. Mayer (1999). Beyond substantial equivalence. *Nature 401:* 525-526.

Munkvold, G. P., R. L. Hellmich, and L. G. Rice (1999). Comparison of fumonisin concentrations in kernels of transgenic maize *Bt* hybrids and nontransgenic hybrids. *Plant Disease 83:* 130-138.

National Academy of Sciences (1987). *Introduction of recombinant DNA-engineered organisms into the environment: Key issues.* Washington, DC: National Academy Press.

National Academy of Sciences (2004a). *News Release: Composition of altered food products, not method used to create them, should be basis for federal safety assessment.* http://www8.nationalacademies.org/onpinews/newsitem.aspx?Record ID=10977 (last accessed February 26, 2007).

National Academy of Sciences (2004b). *Safety of genetically engineered foods: Approaches to assessing unintended health effects.* Washington, DC: National Academy Press. http://books.nap.edu/catalog/10977.html?onpi_newsdoc0727 2004 (last accessed February 26, 2007).

National Center for Biotechnology Information (2006). *GenBank overview.* http:// www.ncbi.nlm.nih.gov/Genbank/index.html (last accessed February 26, 2007).

National Research Council (1989). *Field testing genetically modified organisms: Framework for decisions.* Washington, DC: National Academy Press.

Nicholl, D., J. Cromer, J. Adams, T., Cherry, E. van Grinsven, F. Woldeyes, and E. Dunder (2004). Rice transformation at Syngenta. Abstract P-2099. *World Congress on in vitro biology,* San Francisco, CA, May 22-26, 2004, p. 67-A.

Nordlee, J. A., S. L. Taylor, J. A. Townsend, L. A. Thomas, and R. K. Bush (1996). Identification of a Brazil nut allergen in transgenic soybeans. *New England Journal of Medicine 334:* 688-694.

Organization for Economic Cooperation and Development (1993). *Safety evaluation of foods derived by modern biotechnology: Concepts and principles.* Paris: OECD.

Organization for Economic Cooperation and Development (2002). Consensus document on compositional considerations for new varieties of maize *(Zea mays):* Key food and feed nutrients, anti-nutrients and secondary plant metabolites.

OECD Environmental Health and Safety Publications. Series on the Safety of Novel Foods and Feeds. No. 6. Paris: OECD.

Potrykus, I. (2001). Golden rice and beyond. Plant Physiology 125: 1157-1161.

Ridley, W. P., R. D. Shillito, I. Coats, H. Y. Steiner, M. Shawgo, A. Phillips, P. Dussold, and L. Kurtyka (2004). Development of the International Life Sciences Institute Crop Composition Database. Journal of Food Composition and Analysis 17: 423-438.

Schiermeier, Q. (2000). Novartis pins hopes for GM seeds on new marker system. Nature 406: 924.

Schubbert, R., U. Hohlweg, D. Renz, and W. Doerfler (1998). On the fate of orally ingested foreign DNA in mice: Chromosomal association and placental transmission to the fetus. Molecular and General Genetics 259: 569-576.

Schubbert, R., C. Lettmann, and W. Doerfler (1994). Ingested foreign (phage M13) DNA survives transiently in the gastrointestinal tract and enters the bloodstream of mice. Molecular and General Genetics 242: 495-504.

Schubbert, R., D. Renz, B. Schmitz, and W. Doerfler (1997). Foreign (M13) DNA ingested by mice reaches peripheral leukocytes, spleen, and liver via the intestinal wall mucosa and can be covalently linked to mouse DNA. Proceedings of the National Academy of Sciences of the USA 94: 961-966.

Schubert, D. (2002). A different perspective on GM food. Nature Biotechnology 20: 969.

Sjoblad, R. D., J. T. McClintock, and R. Engler (1992). Toxicological considerations for protein components of biological pesticide products. Regulatory Toxicology and Pharmacology 15: 3-9.

Tada, Y., M. Nakase, T. Adachi, R. Nakamura, H. Shimada, M. Takahashi, T. Fujimura, and T. Matsuda (1996). Reduction of 14-16 kDa allergenic proteins in transgenic rice plants by antisense gene. FEBS Letters 391: 341-345.

Thomas, K., M. Aalbers, G. A. Bannon, M. Bartels, R. J. Dearman, D. J. Esdaile, T. J. Fu, C. M. Gesatt, N. Hadfield, C. Hatzos, et al. (2004). A multi-laboratory evaluation of a common in vitro pepsin digestion assay protocol used in assessing the safety of novel proteins. Regulatory Toxicology and Pharmacology 39: 87-98.

United Nations (2002). World population prospects the 2002 revision. New York: UN Department of Economic and Social Affairs, Population Division. http://www .un.org/esa/population/publications/wpp2002/WPP2002_Vol3.htm (last accessed February 26, 2007).

U.S. Department of Agriculture (2002a). Data tables. Beltsville, MD: Food Survey Research Group. http://www.barc.usda.gov/bhnrc/foodsurvey/home.htm (last accessed February 26, 2007).

U.S. Department of Agriculture (2002b). National nutrient database for standard reference (Release 15). http://www.ars.usda.gov/ba/bhnrc/ndl (last accessed February 26, 2007).

Van de Eede, G., H. Aarts, H. J. Buhk, G. Corthier, H. J. Flint, W. Hammes, B. Jaconson, T. Midtveldt, J. van der Vossen, A. von Wright, et al. (2004). The relevance of gene transfer to the safety of food and feed derived from genetically modified (GM) plants. Food and Chemical Toxicology 42: 1127-1156.

Vlasák, J., M. Ä, A. Pavlík, D. Pavingerová, and J. Bíza (2003). Comparison of hCMV immediate early and CaMV 35S promoters in both plant and human cells. *Journal of Biotechnology 103:* 197-202.

World Health Organization (2002). *Nutrition databases.* Geneva, Switzerland: Global Database on National Nutrition Policies and Programs. http://www.who .int/nutrition/databases/en/ (last accessed February 26, 2007).

Ye, X; S. Al-Babili, A. Kloti, J. Zhang, P. Lucca, P. Beyer, and I. Potrykus (2000). Engineering the provitamin A (beta-carotene) biosynthetic pathway into (caro-tenoid-free) rice endosperm. *Science 287:* 303-305.

Index

Printed and bound by CPI Group (UK) Ltd, Croydon, CR0 4YY

23/10/2024

01778263-0001